Lecture Notes in Economics and Mathematical Systems

687

More information about this series at http://www.springer.com/series/300

Gustav Feichtinger • Raimund M. Kovacevic •
Gernot Tragler
Editors

Control Systems and Mathematical Methods in Economics

Essays in Honor of Vladimir M. Veliov

Springer

Editors
Gustav Feichtinger
Operations Research and Control Systems
Vienna University of Technology
Vienna, Austria

Raimund M. Kovacevic
Operations Research and Control Systems
Vienna University of Technology
Vienna, Austria

Gernot Tragler
Operations Research and Control Systems
Vienna University of Technology
Vienna, Austria

ISSN 0075-8442 ISSN 2196-9957 (electronic)
Lecture Notes in Economics and Mathematical Systems
ISBN 978-3-319-75168-9 ISBN 978-3-319-75169-6 (eBook)
https://doi.org/10.1007/978-3-319-75169-6

Library of Congress Control Number: 2018941820

Printed on acid-free paper

This Springer imprint is published by the registered company Springer International Publishing AG part of Springer Nature.
The registered company address is: Gewerbestrasse 11, 6330 Cham, Switzerland

Vladimir M. Veliov

Preface

Since the days of Lev Pontryagin and his associates, optimal control is a discipline having shown a tremendous upswing—not only with regard to its mathematical foundations but also with respect to numerous fields of applications, which have given rise to highly active research areas. There are, however, not many scholars who have been able to make contributions to both mathematical developments and (socio-)economic applications.

Vladimir Veliov is one of them. During his whole scientific career, he has contributed with highly influential research on mathematical aspects of optimal control theory and applications in economics and operations research. One of his trademarks is the impressive profoundness and the breadth of his research. The present volume, which is published on the occasion of his 65th anniversary, reflects this diversity in an excellent way.

This book comprises twenty original chapters of highly distinguished researchers, most of whom have already collaborated with Vladimir Veliov. The chapters are embraced in one methodological part and two parts focusing on different highly topical areas of applications in economics and operations research.

The first part, "Mathematical Methods," is devoted to state-of-the-art mathematical foundations of optimal control theory. The mathematical aspects in this part cover stability theory for difference inclusions, metric regularity, generalized duality theory, the Bolza problem from a functional analytic view, and fractional calculus.

A.L. Dontchev opens the mathematical part by presenting several open problems concerning regularity properties of solutions of optimal control problems with constraints. In particular, Dontchev discusses singular perturbations, the two-norm discrepancy, discrete approximations and necessary optimality conditions, and finally metric regularity and radius theorems.

A.A. Davydov and Yu.A. Kasten develop the non-local normal form for the main symbol of (linear) second-order mixed type (elliptic/hyperbolic) PDEs. Here, they consider the Cibrario-Tricomi case with periodic coefficients.

R. Baier and E. Farkhi prove Filippov-type stability theorems for discrete difference inclusions. Compared to the standard case of a Lipschitz right-hand side, they analyze inclusions under weaker conditions, i.e., one-sided Lipschitz or strengthened one-sided Lipschitz. Such inclusions can be obtained by the Euler discretization of differential inclusions with perturbations in the set of initial points.

R. Cibulka and T. Roubal give an analysis of strong metric regularity for a nonmonotonic generalized equation with applications to non-regular electrical circuits. They show the existence of a Lipschitz selection of a solution mapping. Moreover, they investigate the accuracy of an inexact Euler–Newton continuation method for tracking solution trajectories.

F. Gozzi, R. Monte, and M.E. Tessitore study a class of infinite horizon optimal control problems with incentive constraints in the discrete time case. More specifically, they establish sufficient conditions under which the value function associated with such problems satisfies the dynamic programming principle.

M.I. Krastanov and N.K. Ribarska study the classical problem of the calculus of variations under the assumption that the integrand is a continuous function. A non-smooth variant of the classical du Bois-Reymond lemma is presented. Under suitable additional assumptions, a non-smooth version of the classical Euler equation is proved.

A.B. Kurzhanski gives an in-depth analysis of the interrelations between problems of optimal state-constrained control and problems of optimal state estimation. Apart from the duality in mathematical sense, he states a system duality and indicates the correct functional spaces for solving such problems in a unified framework.

S. Harizanov and S. Margenov present a study which is motivated by the recent development in the fractional calculus and its applications. In particular, they present and analyze the properties of positive approximations of the inverse of fractional powers obtained by a technique which is based on best uniform rational approximations. Sufficient conditions for positiveness are proven, complemented by sharp error estimates. The theoretical results are supported by representative numerical tests.

Next, A.V. Dmitruk and N.P. Osmolovskii consider very general cone-constrained optimization problems in Banach spaces and prove a necessary optimality condition in the form of a Lagrange multiplier rule. Such results have a broad field of applications; in particular, they can be applied to optimal control problems with, e.g., state and mixed control-state constraints or with age structure.

We close the first part with the chapter by L. Grüne, S. Pirkelmann, and M. Stieler who consider the turnpike property for infinite horizon undiscounted optimal control problems in discrete time and with time-varying data. They show that, under suitable conditions, a time-varying strict dissipativity notion implies the turnpike property and a continuity property of the optimal value function. They also discuss the relation of strict dissipativity to necessary optimality conditions and illustrate their results by an example.

The other two parts of this book deal with various applications of control theory, among which we find population dynamics, population economics, epidemiology, optimal growth theory, resource and energy economics, environmental management, and climate change. Further topics are optimal liquidity, dynamics of the firm, and wealth inequality.

In part two, "Economics and Environmental Models," S. Aseev and T. Manzoor begin with studying an optimal growth model for a single-resource economy under logistic growth with a Cobb–Douglas-type production function and exogenously driven knowledge stock. The optimal paths of the resulting infinite-horizon control problem with an unbounded set of control constraints are characterized for all possible parameter values. As a main finding, there are only two qualitatively different types of behaviour, depending on the relative size of the growth rate and of the social discount rate.

F. Wirl addresses the largely ignored puzzle that a good like natural gas is characterized by huge price differences between the USA, Europe, and Japan. The chapter introduces the risk of the US government imposing export regulations in order to protect the interest of local firms and consumers. It analyzes the corresponding stochastic and dynamic rational expectation equilibrium by applying a trick from Kamien and Schwartz, and it shows the persistence of apparent arbitrage.

Next, P. Brunovsky, M. Halicka, and M. Mitas apply the standard maximum principle to analyze a problem arising in mathematical finance. The optimal solution trajectories show some unexpected features.

H. Dawid, R.F. Hartl and P.M. Kort consider the effect of investment in solar panels on optimal dynamic firm behavior. To do so, an optimal control model is analyzed that has as state variables goodwill and green capital stock. Following current practice in companies like Tesla and Google, they take into account that the use of green energy has positive goodwill effects. As a solution, they find an optimal trajectory that overshoots before reaching a stable steady state.

In the final chapter of part two by W. Semmler, H. Maurer, and A. Bonen, the funding of climate policies through taxation and by the issuing of clima bonds is studied. Applying a new numerical solution technique for an appropriate optimal control model, valuable insights for the design of efficient climate financing policies are derived.

This book closes with part three, "Population Dynamics and Spatial Models," in which R. Boucekkine, B. Martinez, and J.R. Ruiz-Tamarit begin with presenting a demo-economic optimal growth model. Assuming both intragenerational as well as intergenerational altruism and child-rearing costs, some interesting results on the transitional dynamics and comparative statics are calculated.

To investigate the effect of demography on wealth inequality, M. Sanchez-Romero, S. Wrzaczek, A. Prskawetz, and G. Feichtinger propose an economic growth model with overlapping generations in which individuals are altruistic toward their children and differ with respect to the age of their parent (generational gap). The proposed model avoids the unrealistic assumption of a representative agent, and hence, it is more suitable for understanding the wealth accumulation process. Using realistic demographic data, the model predicts that the decline in fertility reduces wealth inequality, while increases in life expectancy have a small and non-monotonic effect on wealth inequality.

Based on classic epidemiological compartmental approaches, the next chapter by S. Bernard, T. Cesar, and A. Pietrus proposes a new model for the propagation of rumor within a social network by taking into account the different possible changes of classes of the individuals of a social network being represented by ignorants, spreaders, stiflers, and their subclasses. They analyze admissible equilibrium states and their stability and give criteria for persistence of the model.

Before turning to the final chapter, A. Xepapadeas and A.N. Yannacopoulos investigate an optimal control model for resource management with pollution externalities in a spatial context. Spatial differentiated instruments as price or quantities-zoning systems turn out to be important for transport phenomena in agglomerations.

This book closes with the chapter by L.-I. Aniţa, S. Aniţa, V. Capasso, and A.-M. Moşneagu who analyze a control problem with structured population dynamics subject to diffusion. The optimal choice of a harvesting region is considered together with the optimal harvesting strategy within the chosen sub-region. Moreover, they also give a necessary and a sufficient condition for zero stability (eradication) in this context.

We are convinced that this book will be particularly interesting for both pure and applied mathematicians working on the theory and applications of optimal control in areas including—but not restricted to—economics, operations research, environmental management, and population economics.

Finally, we wish to express our sincere gratitude to all the authors of this book for their contributions and the referees for their constructive suggestions on how to improve the individual chapters.

Vienna, Austria Gustav Feichtinger
Vienna, Austria Raimund M. Kovacevic
Vienna, Austria Gernot Tragler

Contents

Contributors

Laura-Iulia Aniţa Faculty of Physics, "Alexandru Ioan Cuza" University of Iaşi, Iaşi, Romania

Sebastian Aniţa Faculty of Mathematics, "Alexandru Ioan Cuza" University of Iaşi, Iaşi, Romania

"Octav Mayer" Institute of Mathematics of the Romanian Academy, Iaşi, Romania

Sergey Aseev Steklov Mathematical Institute of Russian Academy of Sciences, Moscow, Russia

International Institute for Applied Systems Analysis, Laxenburg, Austria

Krasovskii Institute of Mathematics and Mechanics, Ural Branch of Russian Academy of Sciences, Yekaterinburg, Russia

Robert Baier University of Bayreuth, Bayreuth, Germany

Séverine Bernard Université des Antilles, LAboratoire de Mathématiques Informatique et Applications EA4540, Pointe-à-Pitre cedex, FWI, Guadeloupe

Anthony Bonen Labour Market Information Council, Ottawa, ON, Canada

Raouf Boucekkine Aix-Marseille University, CNRS, EHESS, Centrale Marseille, AMSE and IMRA, IUF, Marseille, France

Pavol Brunovský Department of Applied Mathematics and Statistics, Comenius University in Bratislava, Bratislava, Slovakia

Vincenzo Capasso ADAMSS (Centre for Advanced Applied Mathematical and Statistical Sciences), Universitá degli Studi di Milano, Milan, Italy

Ténissia Cesar Université des Antilles, Laboratoire de Mathématiques Informatique et Applications EA4540, Pointe-à-Pitre cedex, FWI, Guadeloupe

Radek Cibulka NTIS - New Technologies for the Information Society and Department of Mathematics, Faculty of Applied Sciences, University of West Bohemia, Pilsen, Czech Republic

Alexey A. Davydov National University of Science and Technology "MISIS", Lomonosov Moscow State University, Moscow, Russia

Herbert Dawid Department of Business Administration and Economics and Center for Mathematical Economics, Bielefeld University, Bielefeld, Germany

Andrei V. Dmitruk Russian Academy of Sciences, Central Economics and Mathematics Institute, Moscow, Russia

Lomonosov Moscow State University, Moscow, Russia

Moscow Institute of Physics and Technology, Dolgoprudny, Russia

Asen L. Dontchev Mathematical Reviews, American Mathematical Society, Ann Arbor, MI, USA

Elza Farkhi School of Mathematical Sciences, Sackler Faculty of Exact Sciences, Tel Aviv University, Tel Aviv, Israel

Gustav Feichtinger Wittgenstein Centre for Demography and Global Human Capital (IIASA, VID/ÖAW, WU), Vienna Institute of Demography/Austrian Academy of Sciences, Vienna, Austria

Vienna University of Technology (TU Wien), Vienna, Austria

Fausto Gozzi Dipartimento di Economia e Finanza, Università Luiss - Roma, Rome, Italy

Lars Grüne Chair of Applied Mathematics, Mathematical Institute, University of Bayreuth, Bayreuth, Germany

Margaréta Halická Department of Applied Mathematics and Statistics, Comenius University in Bratislava, Bratislava, Slovakia

Stanislav Harizanov Institute of Information and Communication Technologies, Bulgarian Academy of Sciences, Sofia, Bulgaria

Richard F. Hartl Department of Business Administration, Production and Operations Management, University of Vienna, Vienna, Austria

Yu. A. Kasten Vladimir State University named after Alexander and Nikolay Stoletovs, Vladimir, Russia

Peter M. Kort Department of Econometrics and Operations Research & Center, Tilburg University, Tilburg, The Netherlands

Department of Economics, University of Antwerp, Antwerp, Belgium

Mikhail I. Krastanov Faculty of Mathematics and Informatics, University of Sofia, Sofia, Bulgaria

Alexander B. Kurzhanski Moscow State University, Moscow, Russia
University of California at Berkeley, Berkeley, CA, USA

Talha Manzoor Department of Electrical Engineering, Center for Water Informatics & Technology, Lahore University of Management Sciences, Lahore, Pakistan

Svetozar Margenov Institute of Information and Communication Technologies, Bulgarian Academy of Sciences, Sofia, Bulgaria

Blanca Martínez Department of Economics, Universidad Complutense de Madrid, Madrid, Spain

Instituto Complutense de Análisis Económico (ICAE), Madrid, Spain

Helmut Maurer Institute for Analysis and Numerics, University of Münster, Münster, Germany

Mario Mitas Department of Applied Mathematics and Statistics, Comenius University in Bratislava, Bratislava, Slovakia

Roberto Monte Dipartimento di Ingegneria Informatica e Ingegneria Civilei, Macroarea di Ingegneria, Università di Roma "Tor Vergata", Rome, Italy

Ana-Maria Moşneagu Faculty of Physics, "Alexandru Ioan Cuza" University of Iaşi, Iaşi, Romania

Nicolai P. Osmolovskii Department of Informatics and Mathematics, Kazimierz Pulaski University of Technology and Humanities in Radom, Radom, Poland

Department of Applied Mathematics, Moscow State University of Civil Engineering, Moscow, Russia

Alain Pietrus Université des Antilles, LAboratoire de Mathématiques Informatique et Applications EA4540, Pointe-à-Pitre cedex, FWI, Guadeloupe

Simon Pirkelmann Chair of Applied Mathematics, Mathematical Institute, University of Bayreuth, Bayreuth, Germany

Alexia Prskawetz Wittgenstein Centre for Demography and Global Human Capital (IIASA, VID/ÖAW, WU), Vienna Institute of Demography/Austrian Academy of Sciences, Vienna, Austria

Vienna University of Technology (TU Wien), Vienna, Austria

Nadezhda K. Ribarska Institute of Mathematics and Informatics, Bulgarian Academy of Sciences, Sofia, Bulgaria

Tomáš Roubal Department of Mathematics, Faculty of Applied Sciences, University of West Bohemia, Pilsen, Czech Republic

J. Ramon Ruiz-Tamarit Department of Economic Analysis, Universitat de València, València, Spain

IRES Department of Economics, Université Catholique de Louvain, Louvain-la-Neuve, Belgium

Miguel Sánchez-Romero Wittgenstein Centre for Demography and Global Human Capital (IIASA, VID/ÖAW, WU), Vienna Institute of Demography/Austrian Academy of Sciences, Vienna, Austria

Willi Semmler New School for Social Research, New York, NY, USA

International Institute for Applied Systems Analysis, Laxenburg, Austria

University of Bielefeld, Bielefeld, Germany

Marleen Stieler Chair of Applied Mathematics, Mathematical Institute, University of Bayreuth, Bayreuth, Germany

M. Elisabetta Tessitore Dipartimento di Economia e Finanza, Macroarea di Economia, Università di Roma "Tor Vergata", Rome, Italy

Franz Wirl University of Vienna, Faculty of Business, Economics and Statistics, Vienna, Austria

Stefan Wrzaczek Wittgenstein Centre for Demography and Global Human Capital (IIASA, VID/ÖAW, WU), Vienna Institute of Demography/Austrian Academy of Sciences, Vienna, Austria

Vienna University of Technology (TU Wien), Vienna, Austria

Anastasios Xepapadeas Athens University of Business and Economics, Athens, Greece

Athanasios N. Yannacopoulos Athens University of Business and Economics, Athens, Greece

Part I
Mathematical Methods

On Some Open Problems in Optimal Control

Asen L. Dontchev

To Vlado, with friendship and respect

Abstract Several open problems are presented concerning regularity properties of solutions of optimal control problems with constraints.

In the early 1980s I was very lucky to share an office with Vladimir Veliov in the Institute of Mathematics and Mechanics, Bulgarian Academy of Sciences, where he was then a PhD student, and I was a "junior research collaborator III degree", an approximate translation of the most junior research position in Bulgaria at that time. Collaboration with Vladimir came naturally and I have greatly benefited, in all those years, from his ideas, attention to details and rigorous mathematical thinking. Here are several problems that I have encountered in our joint work with Vladimir throughout the years, which I tried to solve, but failed, and which he would have solved, but was not really interested.

A. L. Dontchev (✉)
Mathematical Reviews, American Mathematical Society, Ann Arbor, MI, USA
e-mail: dontchev@umich.edu

G. Feichtinger et al. (eds.), *Control Systems and Mathematical Methods in Economics*, Lecture Notes in Economics and Mathematical Systems 687,
https://doi.org/10.1007/978-3-319-75169-6_1

1 Singular Perturbations

In our first[1] joint paper Dontchev and Veliov (1983) with Vladimir we considered the limit of the reachable set of a singularly perturbed control system. Specifically, we focused on the linear control system

$$\begin{cases} \dot{x}(t) &= A_{11}x(t) + A_{12}y(t) + B_1u(t), \quad x(0) = x_0, \\ \lambda\dot{y}(t) &= A_{21}x(t) + A_{22}y(t) + B_2u(t), \quad y(0) = y_0, \end{cases} \tag{1}$$

where $x(t) \in \mathbb{R}^n$, $y(t) \in \mathbb{R}^m$, $u(t) \in \mathbb{R}^d$, A_{ij} and B_i, $i, j \in \{1, 2\}$, are constant in time t matrices (for simplicity) with respective dimensions, and the admissible controls are measurable functions on $[0, 1]$ with values in a compact set $U \subset \mathbb{R}^d$ for a.e. $t \in [0, 1]$. Systems of the form (1) describe evolution in two different time scales: a slow part associated with the variable x, and a fast part associated with y. The parameter λ represents a perturbation called *singular*, because setting $\lambda = 0$ eliminates the dynamics of the fast states. Skipping to the important part, we were interested in the continuity properties of the reachable set at a fixed time, say $t = 1$, when $\lambda \searrow 0$. The reachable set at time t of the set of all points that are values at t of solutions of (1) corresponding to admissible controls; in our case it is a compact and convex set.

In the second half of the last century the singular perturbations in control had been a hot subject both in the former USSR, building on a fundamental result by Tikhonov (1948), and in the USA, because of the potential for applications in control engineering actively advocated by Kokotovich and his associates, see the book (Kokotovich et al. 1999). After removing the control in (1), e.g., by setting $u = 0$, Tikhonov's theorem says that when the eigenvalues of the matrix A_{22} have negative real parts, then the solution of (1) (which is unique without controls) converges point-wisely to the solution of the reduced system obtained for $\lambda = 0$, except at $t = 0$ for y. Already in its original statement, Tikhonov's theorem deals with a more general nonlinear differential equation, but I shall not go into this here.

Somewhat surprisingly at a first sight, we found that, under the assumptions of Tikhonov's theorem, the reachable set of system (1) is convergent, in the Hausdorff metric, to a set which could be considerably larger than the set obtained by setting $\lambda = 0$. This happens already for simple examples of systems with $n = 2$ and $m = 1$ and is a generic property for more general systems. We also established a formula of the limit of the reachable set. Since the control system (1) can be reformulated as a differential inclusion, this result means that Tikhonov's theorem is not valid for differential inclusions, in general. This observation has been later extended to much broader settings in a series of papers by Z. Artstein with collaborators.

[1]Technically, the paper Dontchev and Veliov (1982) appeared before Dontchev and Veliov (1983) but it was completed after the submission of Dontchev and Veliov (1983).

In the same paper Dontchev and Veliov (1983) we considered system (1) subject to both state and control constraints. In the case when state constraints are imposed only on the slow state x, e.g., $x(t) \in X$ for all $t \in [0, 1]$, for some convex and closed set $X \subset \mathbb{R}^n$, we found the limit of the reachable set for the slow states. However, the case of state constraints on the fast states y turned out to be much harder.

Vladimir found the following example, given in Dontchev and Veliov (1983):

$$\begin{cases} \dot{x} &= y_1, \quad x(0) = -1, \\ \lambda \dot{y}_1 &= -y_1 + y_2, \quad y_1(0) = 0, \\ \lambda \dot{y}_2 &= -y_2 + u, \quad y_2(0) = e, \end{cases} \tag{2}$$

$$t \in [0, 1], \quad u(t) \in [0, 1], \quad y_1(t) \in [-1, 1].$$

For $\lambda = 0$ the reduced problem becomes

$$\dot{x} = u, \quad x(0) = -1, \quad u(t) \in [0, 1],$$

and the corresponding reachable set for the slow variable x at $t = 1$ is the interval $[-1, 0]$. The solution of the fast part of the singularly perturbed system (2) has the form

$$y_2^\lambda(t) = e^{1-t/\lambda} + \frac{1}{\lambda}\int_0^t e^{-(t-s)/\lambda} u(s) ds \geq e^{1-t/\lambda},$$

$$y_1^\lambda(t) = \frac{1}{\lambda}\int_0^t e^{-(t-s)/\lambda} y_2(s) ds \geq \frac{t}{\lambda} e^{1-t/\lambda},$$

Observe that $\max_{t\in[0,1]} y_1^\lambda(t) \geq 1$, hence, for $\lambda > 0$, in order to satisfy the constraint $y_1(t) \in [0, 1]$ for each $t \in [0, 1]$ the only admissible control is $u(t) \equiv 0$. The corresponding slow state then is

$$x^\lambda(t) = -1 + \lambda e - (t + \lambda)e^{1-t/\lambda},$$

hence the limit when $\lambda \searrow 0$ of the reachable set for x is just the point -1. That is, in contrast to the case of control constraints only, the limit of the reachable set for the slow state x is a proper subset of the reachable set for x at $\lambda = 0$.

It is an open problem to describe the limiting behavior of the reachable set, or the part of it corresponding to the slow states, of a singularly perturbed control system in the presence of state and control constraints.

Observe that the initial conditions for y_1 and y_2 do not satisfy the reduced equation obtained by setting $\lambda = 0$ in (2); in this case for the second equation we have $-0 + e = 0$. The referee of this paper asked whether this inconsistency could be the cause of the discrepancy in the reachable sets. The answer however is no; indeed, if we take instead $y_1(0) = e$ the initial condition satisfies the reduces equation but we still have

$$y_1^\lambda(t) = e^{1-t/\lambda} + \frac{1}{\lambda}\int_0^t e^{-(t-s)/\lambda} y_2(s)ds \geq \frac{t}{\lambda}e^{1-t/\lambda}.$$

This is expected since the asymptotic stability of the fast subsystem eliminates the dependence of the state for times $t > 0$ on the initial conditions.

2 The Two-Norm Discrepancy

Consider the following linear-quadratic optimal control problem with inequality state constraints

$$\text{minimize} \int_0^1 (x^T Qx + u^T Ru)dt \tag{3}$$

subject to

$$\begin{aligned} &\dot{x} = Ax + Bu, \quad x(0) = p, \\ &Kx(t) + b \leq 0, \quad t \in [0, 1], \\ &u \in L^2[0, 1], \quad x \in W^{1,2}[0, 1], \end{aligned}$$

where $x(t) \in \mathbb{R}^n$, $u(t) \in \mathbb{R}^d$, the matrices A, B and the vector b are constant in time t and have respective dimensions, the matrix K has k rows, Q is symmetric and positive semidefinite while R is symmetric and positive definite. Further, L^2 is the usual Lebesque space of square integrable on $[0, 1]$ functions, and $W^{1,2}$ is the space of absolutely continuous functions on $[0, 1]$ whose derivatives are in L^2. The problem is parameterized by the initial state $x(0)$ being a parameter $p \in \mathbb{R}^n$ which varies in a neighborhood of its reference value $p = 0$. Assume that for $x(0) = p = 0$ there exists a state trajectory $\tilde{x}$ such that $K\tilde{x}(t) + b < 0$ for all $t \in [0, 1]$. Then, it is well known that for each p in a neighborhood of 0 there exists a unique solution (x_p, u_p) of this problem. Our goal is to determine the continuity properties of the function $p \mapsto (x_p, u_p)$ at $p = 0$.

Denote

$$\mathcal{A}(t) = \{i \in \{1, 2, \ldots, k\} \mid Kx_0(t) + b = 0\},$$

that is, $\mathcal{A}(t)$ is the set of indices of the active constraints for the reference trajectory x_0 at t. Assume that the set $\mathcal{A}(0)$ is empty and there exists a constant $\alpha > 0$ such that

$$\left| \sum_{i \in \mathcal{A}(t)} v_i K_i B \right| \geq \alpha |v_{\mathcal{A}(t)}| \tag{4}$$

for each $t \in [0, 1]$ for which $\mathcal{A}(t) \neq \emptyset$ and for each choice of $v \in \mathbb{R}^{\mathcal{A}(t)}$. For a more general problem with state and control constraints, Hager (1979) obtained a result which implies that under (4) that for each p the optimal control u_p is Lipschitz continuous as a function of time. Based on Hager's work, it was proved in Dontchev (1983, Theorem 2.6) that there exists a neighborhood P of 0 and a constant $c > 0$ such that for each $p, p' \in P$ one has

$$\|u_p - u_{p'}\|_{L^2} + \|x_p - x_{p'}\|_{W^{1,2}} \leq c\|p - p'\|, \tag{5}$$

that is, the solution mapping of (3) is Lipschitz continuous around $p = 0$ in the L^2 norm for the control and in the $W^{1,2}$ norm for the state. If instead of state constraint there are control constraints, even in a more general form $u(t) \in U$ for a.t. $t \in [0, 1]$ for some closed and convex set U, it was moreover proved that (5) holds in L^∞ for the control and in $W^{1,\infty}$ for the state. But there is still no proof of Lipschitz continuity in those spaces in the presence of state constraints.

Why is this problem important? First, the L^∞ space is clearly a more natural space for the control in this problem because the optimal controls are in fact Lipschitz continuous functions in time. But much more important is that if such a result is true, then it could be extended to much more general problems involving nonlinear control systems and non-quadratic integrands. The reasoning is as follows:

In nonlinear analysis, the path from linear equations to nonlinear equations is well paved by the implicit function theorem. In optimization problem with constraints, the optimality conditions are described by variational inequalities, or even more general inclusions, so one needs an implicit function theorem for variational inequalities. In 1980 Robinson (1980) obtained such a theorem, a version of which I present next. Recall that a mapping H acting from a Banach space X to a Banach space Y with $(\bar{x}, \bar{y}) \in \operatorname{gph} H$ is said to have a *single-valued localization around* $\bar{x}$ *for* $\bar{y}$ if there exist neighborhoods U of $\bar{x}$ and V of $\bar{y}$ such that the truncated inverse $U \ni x \mapsto H(x) \cap V$ is a function on U. Also recall that a mapping $F : X \rightrightarrows Y$ is said to be *strongly metrically regular* at $\bar{x}$ for $\bar{y}$ when its inverse F^{-1} has a single-valued localization around $\bar{y}$ for $\bar{x}$ which is Lipschitz continuous around $\bar{y}$. In optimization, the strong metric regularity is usually obtained under a coercivity condition, which is a strong form of the second-order sufficient optimality condition.

The following version of Robinson's theorem is a part of Dontchev and Rockafellar (2014, Theorem 5F.4):

Theorem 1 *Let X, Y be Banach spaces and let P be a metric space. For $f : P \times X \to Y$ and $F : X \rightrightarrows Y$, consider the generalized equation $f(p, x) + F(x) \ni 0$ where p is a parameter with a reference value $\bar{p}$, and let S be the associated solution mapping S with $\bar{x} \in S(\bar{p})$. Let f be Fréchet differentiable in x around $(\bar{p}, \bar{x})$ such that both f and $D_x f$ are continuous around $(\bar{p}, \bar{x})$ and f is Lipschitz continuous around $\bar{p}$ uniformly in x around $\bar{x}$. Suppose that the linearized mapping $x \mapsto f(\bar{p}, \bar{x}) + D_x(\bar{p}, \bar{x})(x - \bar{x}) + F(x)$ is strongly metrically regular at $\bar{x}$ for 0. Then the solution mapping S has a single-valued localization around $\bar{p}$ for $\bar{x}$ which is Lipschitz continuous near $\bar{p}$.*

Consider now the following optimal control problem for a nonlinear system, which is a generalization of (3):

$$\text{minimize} \int_0^1 (x^T Qx + u^T Ru)dt \tag{6}$$

subject to

$$\dot{x} = f(x, u), \quad x(0) = p,$$
$$Kx(t) + b \leq 0,$$
$$u \in L^2[0, 1], \quad x \in W^{1,2}[0, 1],$$

where f is continuously differentiable everywhere and all other data are as in problem (3). In order to apply Theorem 1, one needs to linearize the system in the space where the optimality mapping is strongly regular, that is, the space for which the solution of the linear-quadratic approximation of the problem is Lipschitz continuous. But we only know from Dontchev (1983) that the optimal control is Lipschitz continuous in L^2. In order to use Theorem 1 in L^2 one needs to linearize in L^2, that is, to have that the function $W^{1,2} \times L^2 \ni (x, u) \mapsto f(x, u) \in L^\infty$ is continuously Fréchet differentiable. This could be obtained under additional assumptions on f and its derivatives, such as Lipschitz continuity on the entire space $\mathbb{R}^{n+m}$. Such an assumption has been used in the literature, e.g. in Vasil'ev (1981), but it is too restrictive; for example it rules out functions f that are polynomial in u of order > 1. However, if we are able to prove Lipschitz continuity of the optimal controls for problem (3) in L^∞, then there would not be a problem to extend it to the problem (6) and beyond, because in that case, in order to have Lipschitz continuity of $W^{1,\infty} \times L^\infty \ni (x, u) \mapsto f(x, u) \in L^\infty$ it is sufficient to require f be continuously differentiable.

In some works from the 90s the difficulty in obtaining Lipschitz continuity in L^∞ for the control in state constrained problems was explained with the gap between the L^2 norm in which coercivity condition is stated and the L^∞ norm where the functions involved are differentiable; this gap was called the *two-norm discrepancy*, see e.g. Malanowski (1994). But it is still not known whether the desired property, Lipschitz continuity of the optimal controls in L^∞, holds, e.g., for the linear-quadratic problem (3). Note that there is no two-norm discrepancy when the state constraints are replaced by control constraints.

3 Discrete Approximations and Necessary Optimality Conditions

As mentioned in the preceding section, Hager (1979) was the first to give conditions under which the optimal control for a state and control constrained problem is Lipschitz continuous in time. This condition was exploited in Dontchev (1981) to obtain a $O(h)$ error estimate in the L^2 norm for the Euler discretization of a linear-convex optimal control problem. Based on a modified version of Robinson's theorem, this estimate was proved valid in Dontchev and Hager (2001) for a nonlinear optimal control problem. By an embedding result, the L^2 error estimate becomes $O(h^{2/3})$ in the L^∞ norm. A further extension to Runge-Kutta methods was obtained in Dontchev et al. (2000). Under some stronger conditions, the sharp $O(h)$ error estimate in L^∞ for a problem with pure state constraints was announced recently in Bonnans and Festa (2017). But the theory of discrete approximations of optimal control problems for ordinary differential equations is still far from being complete. One of the problems in this area, on which Vladimir has been working recently, see Haunschmied et al. (2014) which is based on Quincampoix and Veliov (2013), is to relax the coercivity condition used in the works cited above.

Since the time of Euler, discrete approximation have been used not only for numerical computations but also to obtain purely theoretical results. In optimal control, discrete approximations can be used to obtain necessary optimality conditions, by first stating such conditions for the discretized problem and then passing to the limit. To be specific, consider the following optimal control problem:

$$\text{Minimize} \quad J(x,u) = \int_0^1 \psi(x,u)dt$$

subject to

$$\dot{x} = f(x,u), \quad x(0) = 0,$$

$$x(t) \in X \text{ for } t \in [0,1], \quad u(t) \in U \text{ for a.e. } t \in [0,1], \quad x \in W^{1,\infty} \quad u \in L^\infty,$$

where X and U are closed and convex sets in $\mathbb{R}^n$ and $\mathbb{R}^m$, respectively, and the functions ψ and f are sufficiently smooth. Applying the Euler discretization to this problem with a step-size h over a mesh $i = ih$ and let $N = 1/h$, results in the discrete-time problem

$$\text{Minimize} \quad J_N(x,u) = \sum_{i=0}^{N-1} h\psi(x^i, u^i) \tag{7}$$

subject to

$$x^{i+1} = x^i + hf(x^i, u^i) \text{ for } i = 0, 1, \ldots, N-1, \quad x^0 = 0,$$

$$x^i \in X \text{ for } i = 1, \ldots, N, \quad u^i \in U \text{ for } i = 0, \ldots, N-1.$$

The first-order necessary condition for problem (7) is provided by the Karush-Kuhn-Tucker theorem, see e.g. (Rockafellar 2009, Theorem 6.14). In terms of the Lagrangian

$$L(x,u,q) = J_N(x,u) + \sum_{i=0}^{N-1} \langle q_i, x_{i+1} - hf(x_i,u_i)\rangle,$$

where $q_i, i = 0, 1, \cdots, N-1$ are the Lagrange multipliers (costate), the following first-order necessary optimality involving the state equation, an "adjoint inclusion", and a variational inequality for the control are well known:

$$\begin{cases} x_{i+1} &= x_i + hf(x_i,u_i), \quad i = 0,1,\ldots,N-1, \quad x_0 = p, \\ -q_{i-1} &\in -q_i + h\nabla_x H(x_i,u_i,q_i) + N_X(x_i), \quad i = 1,\ldots,N-1, \quad q_{N-1} = 0, \\ 0 &\in \nabla_u H(x_i,u_i,q_i) + N_U(u_i), \quad i = 0,\ldots,N-1, \end{cases} \tag{8}$$

where H is the Hamiltonian defined as $H(x,u,q) = \psi(x,u) - q^{\mathsf{T}} f(x,u)$ and N_C is the normal cone mapping associated with a closed and convex set C. The question now is whether one can obtain a necessary optimality condition for the continuous problem by passing to zero with h in (8). Note that the right side of the adjoint inclusion is unbounded and the usual approach, by embedding (8) in a continuous problem with piecewise linear state and piecewise constant control and then apply a compactness argument, might not work. But what if one assumes that the optimal control of the continuous problem is Lipschitz continuous in time and a coercivity condition holds? Indeed, the convergence results obtained in Dontchev and Hager (2001) suggests it might be possible to describe the limit in (8) when $h \searrow 0$.

4 Metric Regularity and Radius Theorems

Metric regularity is an older sibling of the strong metric regularity, which is present already in the Banach open mapping principle but formally introduced only in the 1980s. Consider a mapping $\mathcal{F} : X \rightrightarrows Y$, X and Y being metric spaces, and a point $(\bar{x}, \bar{y})$ in its graph gph $\mathcal{F}$ such that gph $\mathcal{F}$ is locally closed at $(\bar{x}, \bar{y})$, meaning that there exists a neighborhood W of $(\bar{x}, \bar{y})$ such that the set gph $\mathcal{F} \cap W$ is closed in W. Then $\mathcal{F}$ is said to be *metrically regular* at $\bar{x}$ for $\bar{y}$ when there is a constant $\kappa \geq 0$ together with neighborhoods U of $\bar{x}$ and V of $\bar{y}$ such that

$$d\big(x, \mathcal{F}^{-1}(y)\big) \leq \kappa d\big(y, \mathcal{F}(x)\big) \quad \text{for every} \quad (x,y) \in U \times V. \tag{9}$$

The infimum of the set of values κ for which (9) holds is called the *modulus* of metric regularity and denoted by reg $(\mathcal{F}; \bar{x}\,|\,\bar{y})$. Strong metric regularity is obtained from metric regularity by additionally assuming that the inverse $\mathcal{F}^{-1}$ has a graphical localization around $\bar{y}$ for $\bar{x}$ which is nowhere multivalued; in that case reg $(\mathcal{F}; \bar{x} \,|\, \bar{y})$ is the Lipschitz modulus of the graphical localization of $\mathcal{F}^{-1}$ at $\bar{y}$ for $\bar{x}$.

In the paper Dontchev and Rockafellar (1996) it was shown that, for variational inequalities over polyhedral convex sets, metric regularity is equivalent to strong metric regularity. The proof given in Dontchev and Rockafellar (1996) uses polyhedral combinatorics. Very recently, Ioffe (2016) gave a new proof of this result, still in finite dimensions, heavily using recent advances in variational analysis. Turning to optimal control, it is plausible to expect that for optimal control problem with state and/or control constraints over convex polyhedral sets, and possibly assuming some regularity in time of the optimal control, these two concept would become equivalent as well. This problem was conjectured in Dontchev and Veliov (2009), where also the following open problem was stated: is metric regularity of an optimality mapping stable under discrete approximation involving approximations of element of a function space by a finite-dimensional subspace. In this line, Vladimir found an example in which metric regularity of a mapping is not inherited by the restriction of this mapping on a subspace.

The next open problem is about the radius of metric regularity of mappings describing optimality conditions.

A basic result in linear algebra, known as the Eckart–Young theorem, says that for any nonsingular matrix A,

$$\inf\{\|B\| \mid A + B \text{ singular } \} = \frac{1}{\|A^{-1}\|}.$$

This result is closely connected with the concept of conditioning of linear equations. A far reaching generalization of this result was proved in Dontchev et al. (2003) for the property of metric regularity of a set-valued mappings acting between Euclidean spaces. The following theorem combines a couple of results from Dontchev and Rockafellar (2014, Section 6A).

Theorem 2 *Consider a mapping $F : \mathbb{R}^n \rightrightarrows \mathbb{R}^m$ which is [strongly] metrically regular at $\bar{x}$ for $\bar{y}$. Then*

$$\inf_{B \in \mathcal{L}(X,Y)} \Big\{ \|B\| \mid F + B \textit{ is not [strongly] metrically regular at } \bar{x} \textit{ for } \bar{y} + B\bar{x} \Big\} = \frac{1}{\operatorname{reg}(F; \bar{x} \,|\, \bar{y})}.$$

Moreover, the infimum remains unchanged when either taken with respect to linear mappings of rank 1 or enlarged to all functions f that are Lipschitz continuous around $\bar{x}$, with $\|B\|$ replaced by the Lipschitz modulus lip $(f; \bar{x})$ *of f at $\bar{x}$.*

The proof of this theorem in Dontchev et al. (2003) uses the coderivative criterion for metric regularity while the proof in Dontchev and Rockafellar (2014) employs the graphical derivative criterion. Both those criteria use in an essential way the compactness of the unit ball in $\mathbb{R}^n$. In contrast, it was shown in Dontchev and Rockafellar (2014, Corollary 6A.5) that, for the case where F is a differentiable function, Theorem 2 remains valid in Banach spaces. Also note that this result is obtained for a general set-valued mapping which does not mean that it would be valid for particular mappings that appear, e.g., in optimality systems for optimal control problems. This comes from the fact that the perturbation should preserve the structure of the mapping, which changes the entire picture. It is an open problem to obtain a radius theorem for an optimal control problem, which would provide a basis for determining a concept of conditioning of such problems. Further down this road is developing techniques for preconditioning, and so forth.

At the end of this paper I would like to cite the reviews of two papers of Vladimir that appeared in MathSciNet. The first one is by M. Valadier, for the paper Veliov (1987):

> The purpose of this paper is to give very precise information about the convexity of the attainable set D of a linear control system. Namely, under some hypotheses there exist submanifolds Γ_i of the boundary ∂D of D such that a suitably defined index of convexity has bounded values on the sets $\partial D \setminus \Gamma_i$. The paper ends with an application to the gradient method of Frank and Wolfe. There are only five references but none of them correspond to the main part of the paper. If the technical results are completely new it is a very original paper.

The second is for Veliov (2010) written by Z. Artstein who mentions an open problem. I suspect the problem is still open.

> The paper examines the error resulting when the class of all measurable controls is replaced by the family of piecewise constant controls, uniformly meshed over a finite time interval. To that end, the equation is assumed to be linear in the control, the controls belong to a compact set in $\mathbb{R}^m$, and the state trajectories are assumed to be within a prescribed compact set in $\mathbb{R}^n$. Approximation estimates are sought for both the state trajectories and the reachable set. Employing fairly delicate estimates, it is verified in this paper that under a Lipschitz differentiability condition and the commutativity of the Lie brackets of the controlled vector fields, and the exterior ball property of the reachable set, for the latter there exists a piecewise constant approximation scheme that yields an error that is only of order h^2, with h being the mesh size. In the general smoothly differentiable case, it is verified that the approximation of the bundle of trajectories is of order h, and one can do better in regard to the approximation of the reachable set, namely, getting an order $h^{1.5}$. It is noted that the problem whether the latter estimate is strict is still open.

Acknowledgements Supported by the NSF Grant 156229, by Austrian Science Foundation (FWF) Grant P26640-N25, and the Australian Research Council, project DP160100854.

References

J.F. Bonnans, A. Festa, Error estimates for the Euler discretization of an optimal control problem with first-order state constraints. SIAM J. Numer. Anal. **55**, 445–471 (2017)

A.L. Dontchev, Error estimates for a discrete approximation to constrained control problems. SIAM J. Numer. Anal. **18**, 500–514 (1981)

A.L. Dontchev, *Perturbations, Approximations and Sensitivity Analysis of Optimal Control Systems*. Lecture Notes in Control and Information Sciences, vol. 52 (Springer, Berlin, 1983)

A.L. Dontchev, W.W. Hager, The Euler approximation in state constrained optimal control. Math. Comput. **70**, 173–203 (2001)

A.L. Dontchev, R.T. Rockafellar, Characterizations of strong regularity for variational inequalities over polyhedral convex sets. SIAM J. Optim. **6**, 1087–1105 (1996)

A.L. Dontchev, R.T. Rockafellar, *Implicit Functions and Solution Mappings. A View from Variational Analysis*. Springer Monographs in Mathematics, 2nd edn. (Springer, Dordrecht, 2014)

A.L. Dontchev, V.M. Veliov, A singularly perturbed optimal control problem with fixed final state and constrained control. Control Cybern. **11**, 19–28 (1982)

A.L. Dontchev, V.M. Veliov, Singular perturbation in Mayer's problem for linear systems. SIAM J. Control Optim. **21**, 566–581 (1983)

A.L. Dontchev, V.M. Veliov, Metric regularity under approximations. Control Cybern. **38**(4B), 1283–1303 (2009)

A.L. Dontchev, W.W. Hager, V.M. Veliov, Second-order Runge-Kutta approximations in control constrained optimal control. SIAM J. Numer. Anal. **38**(1), 202–226 (2000)

A.L. Dontchev, A.S. Lewis, R.T. Rockafellar, The radius of metric regularity. Trans. Am. Math. Soc. **355**, 49–517 (2003)

W. Hager, Lipschitz continuity for constrained processes. SIAM J. Control Optim. **17**, 321–338 (1979)

J.L. Haunschmied, A. Pietrus, V.M. Veliov, *The Euler Method for Linear Control Systems Revisited. Large-Scale Scientific Computing*. Lecture Notes in Computer Science, vol. 8353 (Springer, Heidelberg, 2014), pp. 90–97

A.D. Ioffe, On variational inequalities over polyhedral sets. Math. Program. Ser. B **168**, 261–278 (2018)

P. Kokotovich, H.K. Khalil, J. O'Reilly, *Singular Perturbation Methods in Control. Analysis and Design*. Classics in Applied Mathematics, vol. 25 (SIAM, Philadelphia, PA, 1999). Corrected reprint of the 1986 original

K. Malanowski, Regularity of solutions in stability analysis of optimization and optimal control problems. Control Cybern. **23**(1–2), 61–86 (1994)

M. Quincampoix, V.M. Veliov, Metric regularity and stability of optimal control problems for linear systems. SIAM J. Control Optim. **51**, 4118–4137 (2013)

S.M. Robinson, Strongly regular generalized equations. Math. Oper. Res. **5**, 43–62 (1980)

R.T. Rockafellar, R.J.-B. Wets, *Variational Analysis* (Springer, Berlin, 2009)

A.N. Tikhonov, On the dependence of the solutions of differential equations on a small parameter. Mat. Sb. (N.S.) **22**, 193–204 (1948). In Russian

F.P. Vasil'ev, *Method for Solving Extremal Problems* (Nauka, Moscow 1981)

V.M. Veliov, On the convexity of integrals of multivalued mappings: applications in control theory. J. Optim. Theory Appl. **54**, 541–563 (1987)

V.M. Veliov, On the relationship between continuous- and discrete-time control systems. Cent. Eur. J. Oper. Res. **18**, 511–523 (2010)

On Nonlocal Normal Forms of Linear Second Order Mixed Type PDEs on the Plane

Alexy A. Davydov and Yu. A. Kasten

Abstract Here we propose the nonlocal normal form of main symbol of linear second order mixed type PDEs on the plane for Cibrario-Tricomi case with periodic coefficients. In particular that provides the normal form for equation, which describes an infinitesimal bending of typical rotation surface or sufficiently close to the one near its parabolic line.

1 Introduction

The beginning of theory of local normal forms of generic linear second order partial equations on the plane goes back to middle of eighteenth century when d'Alembert and Euler proposed the form of wave equation and Laplace equation to describe the motion of the string and the velocity potential of an incompressible fluid, respectively. By the end of nineteenth century these two normal forms, which represents the elliptic and hyperbolic type equations, were well known and used in analysis of various applied problems. But already by this time there had appeared the understanding that a generic linear second order partial equations could change type in its domain of definition. Now such equations are called of *mixed type*.

The systematic studies of mixed type equations were started by young Italian mathematician Francesco Tricomi in his treatise (Tricomi 1923), that had provided the ideology and the motivation for many studies of followers. In this treatise Tricomi introduced a local normal form $u_{yy} + yu_{xx}$ for the main symbol of linear second order mixed type partial equation on the plane and the respective model

A. A. Davydov (✉)
National University of Science and Technology "MISIS", Lomonosov Moscow State University, Moscow, Russia
e-mail: davydov@mi.ras.ru

Yu. A. Kasten
Vladimir State University named after Alexander and Nikolay Stoletovs, Vladimir, Russia
e-mail: julikasten@bk.ru

G. Feichtinger et al. (eds.), *Control Systems and Mathematical Methods in Economics*, Lecture Notes in Economics and Mathematical Systems 687,
https://doi.org/10.1007/978-3-319-75169-6_2

equation

$$u_{yy} + yu_{xx} = 0. \tag{1}$$

The last equation changes its type on the x-axis and is of elliptic and hyperbolic type in domains $y > 0$ and $y < 0$, respectively. At each point $(x_0, 0)$ of the axis this equation has two outgoing characteristics, which lie in the domain $y \leq 0$, are defined by the equation

$$dx^2 + ydy^2 = 0.$$

and have the form

$$9(x - x_0)^2 = -4y^3.$$

Together with the model equation F.Tricomi introduced new type of boundary value problems and proved the respective theorem of existence and uniqueness of solution. The considered special domain is bounded by intersecting characteristics outgoing from two points of type change line and by smooth arc, which lies in the elliptic domain and connects these points, and the boundary conditions are of Dirichlet type and are defined on this arc and on one of the two arcs of characteristics. Now this problem is called Tricomi one (Smirnov 1978; Rassias 1990).

The derivation of normal form (1) was also done in treatise (Tricomi 1923) but it was not complete. The correct ground for the form was provided a little bit later on by Cibrario (1932), who was also Italian mathematician.

The complete local classification of characteristic net of generic linear second order mixed type PDE's on the plane was finished in the end of twentieth century in Davydov and Rosales-Gonzales (1996, 1998). Besides the cases of classical Laplace, wave and Cibrario-Tricomi equations mentioned above the classification includes three new normal forms, which correspond to mixed type equations and appear by the typical tangency of the respective characteristic directions field with the type change line.

The new three normal forms have real parameters (invariants) up to smooth change of coordinates and multiplication of the equation by non-vanishing smooth function (Davydov 1985; Davydov and Rosales-Gonzales 1996, 1998), but topologically the respective normal forms of characteristic net have no any ones (Davydov 1985; Kuz'min 1992). Note that the respective normal forms of characteristic net play important role in the analysis of controllability of generic control system and dynamic inequalities on surfaces and of the stability of their controllability (Davydov 1989, 1992, 1995).

Recently some analogous classifications were obtained for generic families of such equations (Bruce and Tari 1997; Bruce et al. 2000; Davydov et al. 2008; Davydov and Trinh Thi Diep 2011; Bogaevsky 2014). However, a reasonably complete smooth classifications, even for typical one-parameter families, is not obtained yet.

The question on non-local normal forms of generic linear second order partial equations on the plane had appeared naturally after the paper (Grishina and Davydov 2007), in which the structural stability in classical Andronov-Pontryagin-Peixoto sense were proved for generic simplest dynamic inequality on the plane

$$(\dot{x} - v(x, y))^2 + (\dot{y} - u(x, y))^2 \leq f(x, y)$$

with bounded domain of its definition (=$\{f \geq 0\}$) or with small time local controllability property outside some compact. Such an inequality describes the simple motion of non-inertial object in the plane with smooth drift (v, u) and maximum proper velocity $\sqrt{f}$. For example, the domain of definition of dynamic inequality

$$(\dot{x} - 2)^2 + (\dot{y})^2 \leq 1 - x^2 - y^2 \tag{2}$$

coincides with unit disk $x^2 + y^2 \leq 1$, and in this disk the set of its orbits is structurally stable. Indeed in this disk the motion with admissible velocities is from left to right (see Fig. 1). The positive or negative orbit of a point is bounded by the respective arc of the unit circle and the trajectories of admissible velocities of limit directions, which are outgoing from the point or incoming into it, respectively. It is easy to see that for a sufficiently C^2-close dynamic inequality we have the

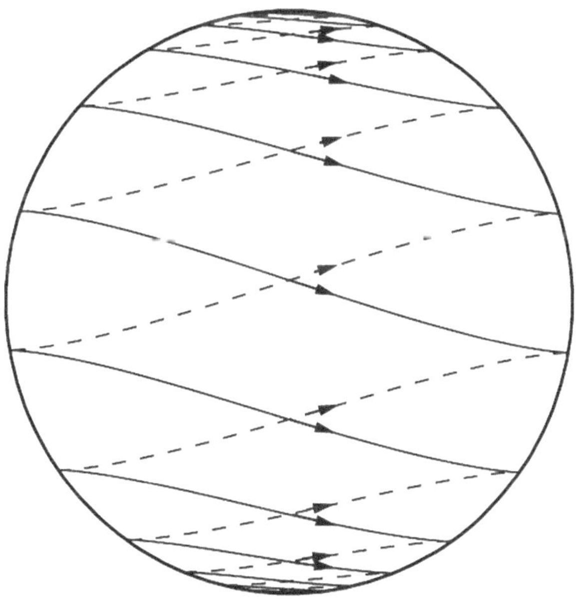

Fig. 1 Structural stability

same structure of orbits, and its family of orbits is reduced to the initial one by homeomorphism of the plane which is close to the identity.

Note that picture Fig. 1 also presents a characteristic net of linear mixed type equation on the plane

$$u_{xx} + K(x, y)(1 - x^2 - y^2)u_{yy} = 0 \tag{3}$$

with some smooth negative function K (that could be easily found inside the disk by calculating the limiting direction field of inequality (2)).

While the topological structure of characteristic net could be important for the equation solvability or for some boundary value problems, the homeomorphism, which keeps the net structure, could not preserve properties of the equation solutions. That poses the problem to search nonlocal normal forms for mixed type equations at least for the simplest configurations of characteristic net like on Fig. 1, for example, for Eq. (3).

One of the first steps in this direction was done in paper (Kasten 2013) where the model equation

$$u_{rr} + (1 - r)u_{\phi\phi} = 0$$

was proposed. This equation has near the unit circle behaviour of characteristic net, which is analogous to the one provide by Cibrario-Tricomi equation near its type change line $y = 0$. But in this paper the normal form for model equation was grounded only in the hyperbolic domain near the circle. Here we propose the extension of this normal form to a neighbourhood of the circle and do one important reduction step to provide the full justification for the form used in paper (Kasten 2013) to analyse a new boundary value problems in the theory of mixed type equations.

2 Main Result, Useful Notions and Statements

Here we introduce some notions, formulate some useful statement and the main result. We work with characteristic equation of linear second order partial equation on the plane in order do not discuss the terms including low derivatives of unknown function.

2.1 *Flat Functions*

A smooth function on a smooth manifold is called *flat at a point* if at this point its infinite jet is zero, and *flat on a subset* of a manifold if the function is flat at each point of the subset. For example, on the (x, y)-plane the function

$$f(x, y) = \begin{cases} 0, \ x \leq 0 \\ e^{-1/|x|}, \ x \neq 0 \end{cases}$$

is flat on the line $x = 0$.

The following statement is used below.

Proposition 1 *Let two smooth functions on the arithmetical space coincide on half-space, then the difference of these functions is flat on the boundary of this half-space.*

Proof Denote the difference by d. Assume that d is not flat at a point P of the boundary of half-space. In such a case some finite jet $j^k d(P)$, where k is some natural, is not zero. Consider the smaller k with such property and take linear local coordinates $x = (x_1, \ldots, x_n)$ in our space with the origin at point P such that the half-space under consideration is $x_1 \geq 0$.

Taylor expansion of the difference at the origin up to order k is homogeneous polynomial T of order k,

$$T(x) := \sum_{|\alpha|=k} a_\alpha x^\alpha$$

where $\alpha = (\alpha_1, \alpha_2, \ldots, \alpha_n)$ is multi-index with integer $\alpha_i \geq 0$, $|\alpha| = \alpha_1 + \alpha_2 + \cdots + \alpha_n$, $x^\alpha = x_1^{\alpha_1} x_2^{\alpha_2} \ldots x_n^{\alpha_n}$, and all a_α are real numbers.

This polynomial is not zero due to the selection of k. Hence there exits vector v such that $T(v) \neq 0$. It is easy to see that $v_1 \neq 0$ due to selection of local coordinates. Consider now the restriction of the difference d to the line $x = tv$ with $t \in \mathbb{R}$, which has to have a form

$$d(x = tv) = t^k [A + \ldots]$$

where $A = \sum_{|\alpha|=k} a_\alpha v^\alpha \neq 0$ and "…" stay for the terms at least of order 1 with respect to t. Hence near the origin for sufficiently small t we have

$$|d(tv)| \geq |A||t|^k/2,$$

and so the difference is not zero near the origin in the half-space $x_1 \geq 0$. That contradicts to the proposition condition about the coincidence of the functions in the half-space $x_1 \geq 0$ near the origin.

Thus our assumption is wrong and the proposition statement is true. □

Remark 1 It is clear that an analogous statement is true for smooth functions on manifold, when the functions coincide on an open set with smooth boundary.

2.2 Normal Form

We start from the characteristic equation proposed and grounded in Kasten (2013)

$$d\phi^2 + (1 - r)dr^2 = 0 \tag{4}$$

Due to Proposition 1 any other characteristic equation with smooth coefficients, which coincides with (4) near the circle $r = 1$ in the domain $r \geq 1$, has the form

$$(1 + a(r, \phi))d\phi^2 + b(r, \phi)drd\phi + (1 - r)(1 + c(r, \phi))dr^2 = 0, \tag{5}$$

where $a,\ b,\ c$ are smooth functions, which are flat on $r = 1$. Besides these functions are 2π-periodic with respect to ϕ.

Theorem 1 *Equation (5) is reduced to Eq. (4) in some neighbourhood of circle $r = 1$ by some coordinates change of the form*

$$\tilde{r} = r + R(r, \phi)),\ \tilde{\phi} = \phi + \Phi(r, \phi)) \tag{6}$$

with some smooth functions R and Φ, which are 2π-periodic in ϕ and flat on $r = 1$, if near this circle an uniformly elliptic equation

$$[D(r, \phi)(1 + u_r)]_r + [D^{-1}(r, \phi)u_\phi]_\phi = 0 \tag{7}$$

with a smooth function D being equal to 1 in the domain $r \geq 1$ has smooth solution, which is zero in this domain.

This theorem is proved in the next section. Its proof is done on the base of some combination of approaches proposed in the Cibrario paper (Cibrario 1932) and singularity theory tools.

Remark 2 Note that this theorem reduces the task of the construction of coordinates, which are smooth perturbation of initial ones in the domain $r \leq 1$ and reduces the equation to the needed normal form in this domain near the circle $r = 1$, to famous problem about the existence of smooth extension of a solution of uniformly elliptic equation trough a hypersurface. For Eq. (7) such an extension is unique (Kondratiev and Landis 1988). But we have not found in the literature any clear statement about the existence of such an extension for the equation of our type. But due to the domain, in which such the extension is needed, could be taken like a small neighborhood of the circle such an extension has to exists (Hormander 1955).

3 Proof of Theorem 1

Here we prove our main result.

3.1 Preliminary Work

The following statement is useful.

Lemma 1 *If a smooth function f on $m+1$ real variables $x \in \mathbb{R}$ and $y \in \mathbb{R}^m$ is flat on the plane $x = 0$ then for any positive real ν there exists smooth function F on $m+1$ real variables such that for $x \geq 0$ we have*

$$f(x, y) = F(x^{\nu}, y) \tag{8}$$

Indeed, for $z \neq 0$ let us define $F(z, y) := f(|z|^{1/\nu}, y)$. It is clear that outside the hyperplane $z = 0$ the function F is smooth. But f is flat on $x = 0$, and hence the function F is also flat on $z = 0$. Thus the function F is smooth. That proves the statement of Lemma 1.

Using this lemma we could rewrite Eq. (5) near the circle $r = 1$ in the domain $r \leq 1$ in the form

$$(1 + A(z, \phi))d\phi^2 + B(z, \phi)dzd\phi + \frac{4}{9}(1 + C(z, \phi))dz^2 = 0,$$

with $z := (1 - r)^{3/2}$ and some smooth functions A, B and C, which are flat on $z = 0$. After the division on the first coefficient $1 + A$ this equation takes the form

$$d\phi^2 + B(z, \phi)dzd\phi + \frac{4}{9}(1 + C(z, \phi))dz^2 = 0, \tag{9}$$

with some new functions B and C with the same properties. Equation (9) is uniformly elliptic near the circle $z = 0$. To prove the theorem we work with this equation in the domain $z \geq 0$ near the type change line and control that the coordinate changes used are identical up to the adding of smooth functions being flat on this line.

Now we remove the function B by an appropriate change of coordinates. To do that it is sufficient to take (in the domain $z \geq 0$) new coordinate $\tilde{\phi}$, which is locally near the type change line is the solution of equation

$$2\tilde{\phi}_z - B\tilde{\phi}_\phi = 0$$

with initial value ϕ on the line. Note that such solution exists near type change line and is unique because the line has no characteristic points of the last equation

(Arnold 1983). Also it is equal to ϕ up to flat function on the type change line because the function B is flat on this line. So from now on we work with Eq. (9) with $B \equiv 0$ and in the form

$$d\phi^2 + (1 + C(z, \phi))dz^2 = 0, \tag{10}$$

which could be obtained from the form (9) with $B \equiv 0$ by rescaling of coordinate z.

To proof Theorem 1 it is sufficient to find coordinates of type (6) in domain $z \geq 0$ near the circle $z = 0$ (=type change line)

$$\tilde{z} = z + Z(z, \phi), \quad \tilde{\phi} = \phi + \Phi(z, \phi) \tag{11}$$

with some smooth functions Z and Φ, which are 2π - periodic with respect to second argument and flat on the circle $z = 0$, such that in new coordinates Eq. (10) takes the form ("tilde" is omitted)

$$d\phi^2 + dz^2 = 0$$

up to multiplication on a smooth non-vanishing function. Indeed the last equation by transformation defined by

$$\hat{\phi} = \phi, \quad \frac{2}{3}(1 - \hat{r})^{3/2} = z$$

is reduced in the domain $z \geq 0$ near the circle $z = 0$ to the form

$$d\hat{\phi}^2 + (1 - \hat{r})d\hat{r}^2 = 0$$

in the domain $\hat{r} \leq 1$ near the circle $\hat{r} = 1$. Due to the flatness of function Φ and Z on the circle $\tilde{z} = 0$ the final change of coordinates

$$(r, \phi) \mapsto (\hat{r}, \hat{\phi})$$

is smooth in the domain $r \leq 1$ near the circle $r = 1$ and could be smoothly extended to some neighborhood of this circle as the identical map in the domain $r \geq 1$.

Thus, it is sufficient to proof the existence of coordinates (6) with properties requested.

3.2 Construction of Needed Coordinates

To construct the coordinate transformation requested we calculate from (11) the differentials

$$dz = \frac{(1 + \Phi_\phi)d\tilde{z} - Z_\phi d\tilde{\phi}}{\Delta}, \quad d\phi = \frac{-\Phi_z d\tilde{z} + (1 + Z_z)d\tilde{\phi}}{\Delta},$$

where

$$\Delta = (1 + Z_z)(1 + \Phi_\phi) - Z_\phi \Phi_z,$$

and substitute them to Eq. (10). After additional multiplication of the obtained equation on Δ^2 we get

$$\left[(1+Z_z)^2 + (1+C)Z_\phi^2\right] d\tilde{\phi}^2 - 2\left[(1+Z_z)\Phi_z + (1+C)(1+\Phi_\phi)Z_\phi\right] dz d\phi +$$
$$\left[\Phi_z^2 + (1+C)(1+\Phi_\phi)^2\right] = 0.$$

One needs to find smooth Z and Φ being flat on $z = 0$ and such that in the last equation the middle coefficient is zero and two others are equals identically. That leads to the system

$$\begin{cases} (1+Z_z)\Phi_z + (1+C)(1+\Phi_\phi)Z_\phi = 0 \\ (1+Z_z)^2 + (1+C)Z_\phi^2 = \Phi_z^2 + (1+C)(1+\Phi_\phi)^2 \end{cases} \tag{12}$$

Substituting from the first equation of this system

$$\Phi_z = -\frac{(1+C)(1+\Phi_\phi)Z_\phi}{1+Z_z}$$

to the second we get the following equation

$$(1+Z_z)^2 + (1+C)Z_\phi^2 = \left[\frac{(1+C)(1+\Phi_\phi)Z_\phi}{1+Z_z}\right]^2 + (1+C)(1+\Phi_\phi)^2$$

or

$$(1+Z_z)^2 + (1+C)Z_\phi^2 = \left[\frac{(1+C)(1+\Phi_\phi)}{1+Z_z}\right]^2 [(1+C)Z_\phi^2 + (1+Z_z)^2].$$

We search a function Z which is flat on $z = 0$. Hence for such a function the expression in left hand side of the last equation is not zero near the circle $z = 0$. Dividing the last equation on this expression and using the (needed) flatness of Z and Φ on the circle we get the equation

$$\sqrt{1+C}(1+\Phi_\phi) = 1 + Z_z.$$

This equation together with the first equation of (12) leads to system

$$\begin{cases} -\Phi_z = \sqrt{1+C} Z_\phi \\ 1 + \Phi_\phi = \frac{1}{\sqrt{1+C}}(1+Z_z) \end{cases} \tag{13}$$

The integrability condition for this system gives the following second order partial equation

$$(\sqrt{1+C}Z_{\phi})_{\phi} + \left(\frac{1}{\sqrt{1+C}}(1+Z_z)\right)_z = 0, \tag{14}$$

where the function D, $D(z,\phi) = \sqrt{1+C(z,\phi)}$, is smooth and equal to 1 in the domain $z \leq 0$ ($r \geq 1$). In this domain near this circle the last equation has zero solution $Z = 0$. Hence, if this solution could be smoothly extended to a neighbourhood of this circle then the extension is flat on the circle itself due to Proposition 1, and so the needed transformation of coordinates exists.

Thus the statement of Theorem 1 is true.

Acknowledgements This research is done by partial financial support by Ministry of Education and Science of the Russian Federation under the project 1.638.2016/FPM.

References

V.I. Arnold, *Geometrical Methods in the Theory of Ordinary Differential Equations* (Springer, New York, Berlin, 1983), 353 p.

I.A. Bogaevsky, Implicit ordinary differential equations: bifurcations and sharpening of equivalence. Izvestiya: Math. **78**(6), 1063–1078 (2014). http://dx.doi.org/10.1070/IM2014v078n06ABEH002720

J.W. Bruce, F. Tari, Generic 1-parameter families of binary differential equations. Discrete Contin. Dynam. Syst. **3**, 79–90 (1997)

J.W. Bruce, F. Tari, G.J. Fletcher, Bifurcations of binary differential equations. Proc. R. Soc. Edinb. Sect. A **130**, 485–506 (2000)

M. Cibrario, Sulla reduzione a forma canonica delle equazioni lineari alle derivative parzialy di secondo ordine di tipo misto. Rend. Lombardo **65**, 889–906 (1932)

A.A. Davydov, Normal form of a differential equation, not solvable for the derivative, in a neighbourhood of a singular point. Funct. Anal. Appl. **19**(2), 81–89 (1985). https://doi.org/10.1007/BF01078387

A.A. Davydov, Singularities of fields of limiting directions of two-dimensional control systems. Mat. USSR-Sb. **64**(2), 471–492 (1989). http://dx.doi.org/10.1070/SM1989v064n02ABEH003321

A.A. Davydov, Structural stability of control systems on orientable surfaces. Mat. USSR-Sb. **72**(1), 1–28 (1992)

A.A. Davydov, Local controllability of typical dynamical inequalities on surfaces. Proc. Steklov Inst. Math. **209**, 73–106 (1995)

A.A. Davydov, E. Rosales-Gonzales, Complete classification of generic linear second-order partial differential equations in the plane. Dokl. Math. **350**(2), 151–154 (1996)

A.A. Davydov, E. Rosales-Gonzales, Smooth normal forms of folded resonance saddles and nodes and complete classification of generic linear second order PDE's on the plane, in *International Conference on Differential Equation, Lisboa 1995*, ed. by L. Magalhaes, C. Rocha, L. Sanchez (World Scientific, Singapore, 1998), pp. 59–68

A.A. Davydov, L. Trinh Thi Diep, Reduction theorem and normal forms of linear second order mixed type PDE families in the plane. TWMS J. Pure Appl. Math. **2**(1), 44–53 (2011)

A.A. Davydov, G. Ishikawa, S. Izumiya, W.-Z. Sun, Generic singularities of implicit systems of first order differential equations on the plane. Jpn. J. Math. **3**(1), 93–119 (2008)
Yu.A. Grishina, A.A. Davydov, Structural stability of simplest dynamical inequalities. Proc. Steklov Inst. Math. **256**, 80–91 (2007)
L. Hormander, On the theory of general partial differential operators. Acta Mat. **94**, 161–248 (1955)
J.A. Kasten, Solvability of the boundary value problem for a Tricomi type equation in the exterior of a disk. J. Math. Sci. **188**(3), 268–272 (2013)
V.A. Kondratiev, E.M. Landis, Qualitative theory of second order linear partial differential equations, in *Partial Differential Equations – 3*. Itogi Nauki i Tekhniki. Ser. Sovrem. Probl. Mat. Fund. Napr., vol. 32 (VINITI, Moscow, 1988), pp. 99–215
A.G. Kuz'min, *Non-classical Equations of Mixed Type and Their Applications in Gas Dynamics* (Birkhauser, Basel, 1992). ISBN:0-8176-2573-9
J.M. Rassias, *Lecture Notes on Mixed Type Partial Differential Equations* (World Scientific, Singapore, 1990), pp. 1–144
M.M. Smirnov, *Equations of Mixed Type* (American Mathematical Society, Providence, RI, 1978)
F. Tricomi, Sulle equazioni lineari alle derivate partziali di 2^0 ordine di tipo misto. Memorie della R. Acc. dei Lincei, S.V. **XIV** (1923)

Discrete Filippov-Type Stability for One-Sided Lipschitzian Difference Inclusions

Robert Baier and Elza Farkhi

Abstract We state and prove Filippov-type stability theorems for discrete difference inclusions obtained by the Euler discretization of a differential inclusion with perturbations in the set of initial points, in the right-hand side and in the state variable. We study the cases in which the right-hand side of the inclusion is not necessarily Lipschitz, but satisfies a weaker one-sided Lipschitz (OSL) or strengthened one-sided Lipschitz (SOSL) condition. The obtained estimates imply stability of the discrete solutions for infinite number of fixed time steps if the OSL constant is negative and the perturbations are bounded in certain norms. We show a better order of stability for SOSL right-hand sides and apply our theorems to estimate the distance from the solutions of other difference methods, as for the implicit Euler scheme to the set of solutions of the Euler scheme. We also prove a discrete relaxation stability theorem for the considered difference inclusion, which also extends a theorem of Grammel (Set-Valued Anal. 11(1):1–8, 2003) from the class of Lipschitz maps to the wider class of OSL ones.

1 Introduction

We regard the *differential inclusion*

$$\dot{x}(t) \in F(x(t)) \subset \mathbb{R}^n \quad (\text{a.e. } t \in I := [t_0, T]), \quad x(t_0) - x^0 \in X_0 \tag{1}$$

R. Baier
University of Bayreuth, Bayreuth, Germany
e-mail: robert.baier@uni-bayreuth.de

E. Farkhi (✉)
School of Mathematical Sciences, Sackler Faculty of Exact Sciences, Tel Aviv University, Tel Aviv, Israel
e-mail: elza@post.tau.ac.il

G. Feichtinger et al. (eds.), *Control Systems and Mathematical Methods in Economics*, Lecture Notes in Economics and Mathematical Systems 687,
https://doi.org/10.1007/978-3-319-75169-6_3

and its *(set-valued) Euler discretization*

$$\eta^{j+1} \in \eta^j + hF(\eta^j), \quad j = 0, 1, \ldots, N-1, \quad \eta^0 = x^0 \in X_0, \tag{2}$$

where the initial set $X_0 \subset \mathbb{R}^n$ is compact and nonempty, the *step size* is given by $h := \frac{T-t_0}{N}$ for some $N \in \mathbb{N}$ and the *grid points* $t_j := t_0 + jh$, $j = 0, 1, \ldots, N$, form a *grid* $\mathcal{G}_h$ and a partition of I in N *subintervals* $I_j := [t_j, t_{j+1}]$, $j = 0, \ldots, N-1$. For the sake of simplicity we consider here the autonomous case, although the results may be reformulated also for maps F depending additionally on the time t.

We denote by $\mathcal{S}$ the *set of solutions* of (1) restricted to the grid $\mathcal{G}_h$ and by $\mathcal{S}_h$ the set of the solutions of (2). These sets are considered in the space of *grid functions* $\eta_h := \{\eta^j\}_{j=0}^N$ with the usual Euclidean norm (see below).

In the classical Filippov Theorem (Filippov 1967) it is supposed that the map F is Lipschitz in the state variable and existence and exponential Lipschitz stability of the set of solutions of (1) with respect to perturbations in the initial condition and the right-hand side is derived. The perturbed inclusion studied by Filippov in 1967 is

$$\dot{y}(t) \in F(y(t)) + \overline{\varepsilon}(t) \quad (\text{a.e. } t \in I), \quad y(t_0) = y^0 \in X_0 \tag{3}$$

with $\overline{\varepsilon}(t) \in \mathbb{R}^n$. For Lipschitz continuous multifunction F, the same stability rate as for the *perturbations* $\overline{\varepsilon}(t)$ in (3), called here *'outer' (set) perturbations,* holds also for the inclusion with *'inner' (state) perturbations*

$$\dot{y}(t) \in F\big(y(t) + \overline{\delta}(t)\big), \quad (\text{a.e. } t \in I), \quad y(t_0) = y^0 \in X_0, \tag{4}$$

where $\overline{\delta}(t) \in \mathbb{R}^n$. Removing the Lipschitz continuity usually leads to the loss of stability with respect to these perturbations. Fortunately, if the map F is *one-sided Lipschitz (OSL)*, the stability in the problem with inner and outer perturbation

$$\dot{y}(t) \in F\big(y(t) + \overline{\delta}(t)\big) + \overline{\varepsilon}(t) \quad (\text{a.e. } t \in I), \quad y(t_0) = y^0 \in X_0, \tag{5}$$

is preserved, possibly in a weaker form (Donchev and Farkhi 1998, 2000).

The *OSL condition* for single-valued functions $f : \mathbb{R}^n \to \mathbb{R}^n$ with *constant* $\mu \in \mathbb{R}$,

$$\langle x - y, f(x) - f(y) \rangle \le \mu |x - y|^2 \quad (x, y \in \mathbb{R}^n), \tag{6}$$

is known in numerical analysis (see e.g., Dekker and Verwer 1984, Auzinger et al. 1990 and in Hairer and Wanner (1996, Sec. IV.12)), where $|\cdot|$ denotes the usual Euclidean norm in $\mathbb{R}^n$. In Hilbert and Banach spaces this concept was already known under the name *dissipative* respectively *monotonic/accretive* operators (see e.g., Lumer and Phillips 1961, Browder 1967a, Browder 1967b, Martin 1970).

Here are the definitions of the two one-sided Lipschitz properties for set-valued maps investigated here.

Definition 1 (Donchev and Farkhi 1998) A set-valued map $F : \mathbb{R}^n \rightrightarrows \mathbb{R}^n$ is called *one-sided Lipschitz (OSL)* with *(OSL) constant* $\mu \in \mathbb{R}$, i.e., for all $x, y \in \mathbb{R}^n$ and all $\xi \in F(x)$ there exists $\zeta \in F(y)$ with

$$\langle x - y, \xi - \zeta \rangle \le \mu |x - y|^2 . \tag{7}$$

For set-valued maps the OSL condition was first introduced in a stronger (uniform) form by Kastner-Maresch and Lempio in Kastner-Maresch (1990b), Lempio (1992), and in a weaker (relaxed) abstract form in Banach spaces by Donchev and Ivanov (1991, 1992). The condition of Kastner-Maresch (1990b), Lempio (1992), called here *uniform one-sided Lipschitz (UOSL)*, requires that (7) is satisfied for *all* $x, y, \xi \in F(x), \zeta \in F(y)$. This condition implies uniqueness of the solution of (1) and allows convergence order 1 for 1d problems (Lempio 1992) or, provided that the solution is piecewise smooth, for implicit Runge-Kutta methods with special stability properties (Kastner-Maresch 1990b, 1992).

In Donchev and Farkhi (1998) the most used explicit form of the OSL condition for set-valued maps in $\mathbb{R}^n$ was coined and the Filippov theorem (Filippov 1967) (with outer perturbations) was extended to the case of OSL right-hand side of the inclusion. In Donchev and Farkhi (2000) a more general Filippov theorem is proved for the inclusion (5) with OSL right-hand side and with both outer and inner perturbations. Then, Hölder one half rate of stability with respect to the inner perturbations is obtained. This result is applied there to obtain order of convergence $\mathcal{O}(\sqrt{h})$ for the Hausdorff distance between the sets of solutions of (1) and (2). The same order appears first in Lempio and Veliov (1998) for an OSL map $F(\cdot)$. Various generalizations of the OSL condition and of this important theorem may be found in Donchev and Farkhi (2009). We also refer the reader to the overview papers on OSL (Lempio and Veliov 1998; Donchev 2002, 2004; Baier and Farkhi 2013).

Definition 2 (Lempio and Veliov 1998) A set-valued map $F : \mathbb{R}^n \rightrightarrows \mathbb{R}^n$ satisfies the *strengthened one-sided Lipschitz (SOSL) condition* with a *(SOSL) constant* $\mu \in \mathbb{R}$, if for all $x, y \in \mathbb{R}^n$ and all $\xi \in F(x)$ there exists $\zeta \in F(y)$ such that whenever $x_i > y_i$ for some $i \in \{1, \ldots, n\}$ we have the inequality

$$\xi_i - \zeta_i \le \mu |x - y|_\infty \tag{8}$$

and whenever $x_i < y_i$ for some $i = 1, \ldots, n$ we have the inequality

$$\zeta_i - \xi_i \le \mu |x - y|_\infty . \tag{9}$$

Here, $| \cdot |_\infty$ is the *maximum norm* and z_i denotes the i-th *coordinate of a vector* $z := (z_1, \ldots, z_n)^\top \in \mathbb{R}^n$.

The maximum norm in (8)–(9) can be replaced by another vector norm, although it is a rather natural choice here. In Lempio and Veliov (1998), n SOSL constants $\mu_i \in \mathbb{R}$ were introduced separately for each coordinate. Here, we use $\mu := \max_{i=1,\dots,n} \mu_i$ for simplicity.

In Lempio (1993, 1995) the uniform version of the latter condition requires that (8) holds for *all* $\zeta \in F(y)$ with $x_i > y_i$ (we call this version S-UOSL as in Baier and Farkhi (2013)). Due to the symmetry for $\xi \in F(x)$ and $\zeta \in F(y)$, (9) is automatically fulfilled.

The *strengthened one-sided Lipschitz condition (SOSL)* essentially requires the OSL condition for each coordinate (in a given basis). Although it is stronger than the OSL condition, it does not imply continuity, but provides better stability than the OSL condition. It appears first in Lempio (1993, 1995) in an uniform form (S-UOSL, analogous to the UOSL condition). First order convergence of the Euler scheme for differential inclusions is derived for the S-UOSL right-hand side in Lempio (1995, Sec. 4). Lempio and Veliov (1998) formulated the weaker form as stated in Definition 2, analoguous to Definition 1, and proved that it ensures the first order convergence of the Euler scheme. The SOSL condition is stronger than the OSL condition and it has some interesting consequences which are not proved for general OSL maps as the order convergence $\mathcal{O}(h)$ of the Euler scheme instead of $\mathcal{O}(\sqrt{h})$ known for OSL maps. Also, the local existence of solutions of the differential inclusion (1) is shown in Farkhi et al. (2014), provided the negation $-F$ is SOSL with zero constant. The latter property of the negation $-F$ defines a special type of monotonicity of F.

Here we prove a Filippov-type stability theorem for the solutions of a discrete inclusion of the form (2) with perturbations in the right-hand side, both in the state and the set, for OSL and SOSL maps and present some applications. Similarly to the 'continuous' Filippov-type theorem for OSL map F (Donchev and Farkhi 2000), we show in the case of an OSL mapping F stability of the discrete solution set which is of order one half with respect to inner and outer perturbations and with respect to the time step h. For infinite time interval, we obtain stability (boundedness) of the discrete solutions if the OSL constant is negative. In the case of OSL map F we show first order of stability with respect to all perturbations and h. We apply these results to study the rate of convergence of the implicit Euler scheme considered in Beyn and Rieger (2010) for OSL (not necessarily continuous) maps F. In particular, we show that the iterates of the implicit Euler method are $\mathcal{O}(\sqrt{h})$-close to some iterates of the explicit one and even $\mathcal{O}(h)$-close for SOSL maps. An important possible application of the discrete Filippov-time theorems, together with the 'continuous' ones is to derive convergence rate of various discrete approximations of differential inclusions. Such discrete approximations like set-valued Euler and Runge-Kutta methods are studied in Veliov (1989, 1992, 2015), and may be useful also for investigation of discrete approximations of control systems. Detailed analysis of such discretizations may be found in Veliov (1997), Veliov (2005), Pietrus and Veliov (2009), Veliov (2010), Haunschmied et al. (2014).

Let us remark that in Chahma (2003, Proposition 2.2.3) a discrete Filippov-type theorem is proved in the Lipschitz case for the explicit Euler scheme and outer perturbations. In Beyn and Rieger (2010, Theorem 14) another discrete Filippov theorem is proved for the implicit Euler method and for outer perturbations in the case of time-dependent, jointly continuous and one-sided Lipschitz right-hand side.

We note that a discrete Filippov-type theorem can be deduced indirectly by the continuous one (Donchev and Farkhi 2000), and the approximation estimate of the continuous trajectories by the discrete ones (Donchev and Farkhi 1998, 2000). We prefer the direct proofs to obtain more precise approximation estimates. The presented discrete Filippov theorems may be useful for investigation of the stability also for discrete systems obtained by one-step set-valued Runge-Kutta methods or some multistep methods as the leap-frog scheme, as well as for infinite time behavior, in particular in the case of negative OSL constant.

The paper is organized as follows. The next section contains some preliminaries and the basic assumptions. In Sect. 3 we prove the stability in the case of OSL mappings. The case of SOSL maps is discussed in Sect. 4. The applications are given in the last subsection of each section.

2 Problem and Preliminaries

In this section the notation and some preliminary results used further in the text are stated. We also present the problem formulation, the continuous differential inclusion and its discretization, the discrete Euler iterates.

2.1 Preliminaries

We denote by $\mathbb{R}_+ := \{x \in \mathbb{R} \mid x \geq 0\}$ and vectors in $\mathbb{R}^n$ by $x := (x_1, x_2, \ldots, x_n)^\top \in \mathbb{R}^n$. The *closed unit ball* in $\mathbb{R}^n$ is denoted by $B_1(0)$, the usual *scalar product of two vectors* $x, y \in \mathbb{R}^n$ is denoted by $\langle x, y\rangle$. The corresponding *Euclidean norm* is denoted by $|x|_2$ or by $|x|$ for brevity, while the *sum norm* and the *maximum norm* of a vector $x \in \mathbb{R}^n$ are denoted by $|x|_1 := \sum_{i=1}^n |x_i|$ and $|x|_\infty := \max_{1\leq i\leq n} |x_i|$. For a real number μ we denote $\mu_+ := \max\{0, \mu\}$.

We denote by $\mathscr{K}(\mathbb{R}^n)$ the *set of compact, nonempty subsets* of $\mathbb{R}^n$. The *Hausdorff distance* between two sets $X, Y \in \mathscr{K}(\mathbb{R}^n)$ is

$$d_H(X, Y) := \max\{\operatorname{dist}(X, Y), \operatorname{dist}(Y, X)\},$$

where $\operatorname{dist}(X, Y) := \sup_{x\in X} \operatorname{dist}(x, Y)$ and the distance from a point to a set is $\operatorname{dist}(x, Y) := \inf_{y\in Y} |x - y|$. The *convex hull* of a set A is denoted by $\operatorname{co} A$, the *norm of a set* is defined by $\|A\| := d_H(A, \{0\})$.

For a L_p *function* $f : I \to \mathbb{R}^n$ we denote $\|f\|_{L_p}$ as its L_p-*norm* and for a grid function $\eta_h = \{\eta^j\}_{j=0}^{N-1}$ we define its *discrete* L_p-*norm* for $p \in \{1, 2, \infty\}$, by

$$\|\eta_h\|_1 := h \sum_{j=0}^{N-1} |\eta^j|, \quad \|\eta_h\|_2 := \sqrt{h \sum_{j=0}^{N-1} |\eta^j|^2}, \quad \|\eta_h\|_\infty := \sup_{0 \le j < N} |\eta^j|, \tag{10}$$

$$\lfloor\!\lfloor \eta_h \rfloor\!\rfloor_{\min,\mu} := \min\{\|\eta_h\|_1, \frac{1}{\sqrt{|\mu|}}\|\eta_h\|_2, \frac{1}{|\mu|}\|\eta_h\|_\infty\} \tag{11}$$

We summarize the equivalence of the discrete L_p-norms for later reference.

Lemma 1 *Let* $h = \frac{T-t_0}{N}$ *a given step size for* $N \in \mathbb{N}$ *and let* $\eta_h = \{\eta_h^j\}_{j=0}^{N-1}$ *be a grid function.*

Then,

$$\sqrt{h}\|\eta_h\|_2 \le \|\eta_h\|_1 \le \sqrt{T - t_0}\,\|\eta_h\|_2\,, \tag{12}$$

$$h\|\eta_h\|_\infty \le \|\eta_h\|_1 \le (T - t_0)\|\eta_h\|_\infty\,, \tag{13}$$

$$\sqrt{h}\|\eta_h\|_\infty \le \|\eta_h\|_2 \le \sqrt{T - t_0}\,\|\eta_h\|_\infty\,. \tag{14}$$

We denote the *Hausdorff distance between two sets* $\mathscr{S}, \widetilde{\mathscr{S}}$ *of grid functions*, using the $\|\cdot\|_\infty$ norm for the distance between the functions, by $\mathrm{d}_{\mathrm{H}}^{\infty}(\mathscr{S}, \widetilde{\mathscr{S}})$.

Next we present some notation and auxiliary inequalities used further in the text. Denote $\overline{1}_h := \{1\}_{j=0}^{N-1}$ and set for $\gamma_h = \{\gamma_k\}_{k=0}^{N-1} \subset \mathbb{R}_+$

$$g_h(\mu, j; \gamma_h) := h \sum_{k=0}^{j-1} (1 + \mu h)^{j-1-k} \gamma_k, \qquad g_h(\mu, j) := g_h(\mu, j; \overline{1}_h).$$

Note that $g_h(\mu, j; C'\gamma_h' + C''\gamma_h'') = C' g_h(\mu, j; \gamma_h') + C'' g_h(\mu, j; \gamma_h'')$.

Remark 1 Recall the simple claim that if $s_j \in \mathbb{R}$, $j = 0, 1, \ldots, k$, satisfy

$$s_{k+1} \le a \cdot s_k + \beta_k, \quad a \ge 0, \quad k = 0, 1, \ldots, j - 1,$$

then $s_j \le a^j s_0 + \sum_{k=0}^{j-1} a^{j-1-k} \beta_k$. Thus, for $a = 1 + \mu h$ and $\beta_k = h\gamma_k$, $k = 0, \ldots N - 1$, we get

$$s_j \le (1 + \mu h)^j s_0 + g_h(\mu, j; \gamma_h), \quad j = 0, 1, \ldots, N - 1. \tag{15}$$

We now estimate $g_h(\mu, j; \gamma_h)$ and $g_h(\mu, j)$.

Lemma 2 *Let $\mu \in \mathbb{R}$, $1 + \mu h > 0$, $\gamma_h = \{\gamma_k\}_{k=0}^{N-1} \subset \mathbb{R}_+$, $t_j = t_0 + jh$, $j = 0, \ldots, N$. Then for $j = 0, \ldots, N$,*

$$g_h(\mu, j; \gamma_h) \leq e^{\mu_+(t_j - t_0)} \llbracket \gamma_h \rrbracket_{\min,\mu} . \tag{16}$$

In particular,

$$g_h(\mu, j) \leq e^{\mu_+(t_j - t_0)} \min \left\{ t_j - t_0, \sqrt{\frac{t_j - t_0}{|\mu|}}, \frac{1}{|\mu|} \right\} . \tag{17}$$

Proof The case "$\mu = 0$" is trivial (we assume $\frac{1}{|\mu|} = \infty$ in the right-hand side). Let $\mu \neq 0$. Note that since $e^z \geq 1 + z$, we obtain for $\mu > 0$ that

$$(1 + \mu h)^j - 1 \leq (1 + \mu h)^j \leq e^{\mu j h} = e^{\mu_+ j h} . \tag{18}$$

For $\mu < 0$ we have

$$\left| (1 + \mu h)^j - 1 \right| = 1 - (1 + \mu h)^j < 1 = e^{\mu_+ j h} . \tag{19}$$

Thus we obtain

$$h \sum_{k=0}^{j-1} (1 + \mu h)^k = h \frac{(1 + \mu h)^j - 1}{\mu h} = \frac{|(1 + \mu h)^j - 1|}{|\mu|} \leq \frac{1}{|\mu|} e^{\mu_+ j h} . \tag{20}$$

To show that $g_h(\mu, j; \gamma_h) \leq e^{\mu_+(t_j - t_0)} \|\gamma_h\|_1$, we bound as in (18)–(19) for $k \leq j$,

$$(1 + \mu h)^{j-1-k} \leq e^{\mu_+(j-1-k)h} \leq e^{\mu_+ j h} = e^{\mu_+(t_j - t_0)} . \tag{21}$$

To show that $g_h(\mu, j; \gamma_h) \leq \frac{1}{\sqrt{|\mu|}} e^{\mu_+(t_j - t_0)} \|\gamma_h\|_2$, we use the Hölder inequality $h \sum_{k=0}^{j-1} \beta_k \gamma_k \leq \left(h \sum_{k=0}^{j-1} (\beta_k)^2 \right)^{\frac{1}{2}} \left(h \sum_{k=0}^{j-1} (\gamma_k)^2 \right)^{\frac{1}{2}}$ to get

$$h \sum_{k=0}^{j-1} (1 + \mu h)^{j-1-k} \gamma_k \leq \left(h \sum_{k=0}^{j-1} (1 + \mu h)^{2k} \right)^{\frac{1}{2}} \|\gamma_h\|_2 .$$

Then, using the formula for the geometric progression, the fact that $2 + \mu h > 1$ and (18)–(19), we get

$$\left(h \sum_{k=0}^{j-1} (1+\mu h)^{2k} \right)^{\frac{1}{2}} \le \left(h \, \frac{(1+\mu h)^{2j} - 1}{(1+\mu h)^2 - 1} \right)^{\frac{1}{2}} = \left(\frac{(1+\mu h)^{2j} - 1}{\mu(2+\mu h)} \right)^{\frac{1}{2}}$$

$$\le \left(\frac{|(1+\mu h)^{2j} - 1|}{|\mu| \cdot 1} \right)^{\frac{1}{2}} \le \left(\frac{e^{\mu_+ 2jh}}{|\mu|} \right)^{\frac{1}{2}} = \frac{1}{\sqrt{|\mu|}} \, e^{\mu_+ (t_j - t_0)} .$$

The inequality $g_h(\mu, j; \gamma_h) \le \frac{1}{|\mu|} \|\gamma_h\|_\infty$ follows directly from (20). □

2.2 Basic Assumptions

For definitions of notions as *upper semi-continuity (usc)* or *measurability* of set-valued maps and their properties which we do not define or formulate here in details, the reader may consult (Aubin and Frankowska 1990; Aubin and Cellina 1984) or Deimling (1992).

The *reachable set at time T* for the differential inclusion (1), starting from the set X_0, is denoted by $\mathscr{R}(T, t_0, X_0)$. For a given step-size $h := \frac{T-t_0}{N}$ and grid points $t_j := t_0 + jh$, $j = 0, \dots, N$, let $\eta_h := \{\eta^j\}_{j=0}^N$, $\eta^j \in \mathbb{R}^n$, be a *discrete solution of the Euler inclusion* (2). The *discrete reachable set at time T* for the Euler inclusion (2), called $\mathscr{R}_h(T, t_0, X_0)$, is defined as the set of all end points $\eta^N \in \mathbb{R}^n$ of admissible grid functions starting from points of the set X_0.

Additionally we allow outer and inner perturbations of the *discrete inclusion*

$$\eta^{j+1} \in \eta^j + h \left(F(\eta^j + \overline{\delta}^j) + \overline{\varepsilon}^j \right), \quad j = 0, \dots, N-1, \quad \eta^0 \in X_0, \qquad (22)$$

where the inner perturbations $\overline{\delta}_h := \left\{ \overline{\delta}^j \right\}_{j=0}^{N-1} \subset \mathbb{R}^n$ are uniformly bounded by a given constant K_δ, while the outer ones $\overline{\varepsilon}_h := \{\overline{\varepsilon}^j\}_{j=0}^{N-1} \subset \mathbb{R}^n$ are bounded in some discrete norm by a given constant K_ε.

For the rest of the paper we demand some of the following four assumptions:

(A1) $F(\cdot)$ has nonempty, compact images.
(A1') $F(\cdot)$ has convex images.
(A2) There are constants $C_B, C_F \ge 0$ such that all solutions of (22) satisfy

$$\|\eta_h\|_\infty \le C_B \qquad \max_{0 \le j \le N} \left\| F(\eta^j) \right\| \le C_F .$$

At places where the solutions of (1) are involved we assume also

(A2') There exist solutions of (1) on I and there are constants $C_B, C_F \geq 0$ such that all solutions of (1) satisfy

$$\|x\|_{L_\infty} \leq C_B\,, \quad \sup_{t\in I} \|F(x(t))\| \leq C_F\,.$$

Sufficient conditions for (A2') are discussed in the next remark.

Denote $S := C_B B_1(0)$ such that $x(t) \in S$ for $t \in I$ and $\eta^j \in S$ for each $j = 0, \ldots, N$.

(A3) $F(\cdot)$ is *one-sided Lipschitz (OSL)* with constant $\mu \in \mathbb{R}$, i.e., for all $x, y \in S$ and all $\xi \in F(x)$ there exists $\zeta \in F(y)$ with

$$\langle x - y, \xi - \zeta \rangle \leq \mu |x - y|^2\,.$$

(A3') $F(\cdot)$ is *strengthened one-sided Lipschitz (SOSL)* with a constant $\mu \in \mathbb{R}$, i.e., for all $x, y \in S$ and all $\xi \in F(x)$ there exists $\zeta \in F(y)$ such that if $x_i > y_i$ we have the inequality

$$\xi_i - \zeta_i \leq \mu |x - y|_\infty$$

and whenever $x_i < y_i$,

$$\zeta_i - \xi_i \leq \mu |x - y|_\infty\,.$$

Additionally, we sometimes require the assumption

(A0) $F : \mathbb{R}^n \rightrightarrows \mathbb{R}^n$ is *upper semi-continuous (usc)*, i.e., for all $x \in S$ and all $\varepsilon > 0$ there exists $\delta > 0$ such that for all $y \in \mathbb{R}^n$ with $|x - y| \leq \delta$, $F(y) \subset F(x) + \varepsilon B_1(0)$.

Remark 2 The assumption (A2') can be guaranteed by assuming a linear growth condition, i.e., $\|F(t, x)\| \leq C(1+|x|)$ for all $x \in \mathbb{R}^n$ (see Donchev and Farkhi 1998, Lemmas 3.1 and 4.1; Baier et al. 2007, Lemma 2.6), or by the weaker assumption of boundedness of $F(\cdot)$ on bounded sets together with the OSL condition on $\mathbb{R}^n$ (see Donchev and Farkhi 2000, Lemma 3.2 and Remark 3.1). To guarantee the existence in (A2') one can require additionally e.g., (A0), (A1) and (A1)', see Deimling (1992, Ch. 2). For simplicity we do not formulate the weakest possible assumptions.

3 Discrete Filippov-Type Theorems for One-Sided Lipschitz Maps

Here we discuss the stability with respect to inner and outer perturbations separately, for reader's convenience.

3.1 Outer Perturbations

We consider perturbed initial values and outer perturbations. A discrete counterpart of the continuous Filippov theorem for OSL maps in Donchev and Farkhi (2000) with $\overline{\delta}(t) \equiv 0$ is obtained.

Proposition 1 *Let the assumptions (A1)–(A3) and* $1 + 2\mu h > 0$ *be satisfied. Consider* $\overline{\varepsilon}_h = \{\overline{\varepsilon}^j\}_{j=0}^{N-1} \subset \mathbb{R}^n$ *with* $\|\overline{\varepsilon}_h\|_\infty \leq K_\varepsilon$ *and let* $\{y^j\}_{j=0}^N$ *be a discrete solution of the perturbed inclusion*

$$y^{j+1} \in y^j + h\big(F(y^j) + \overline{\varepsilon}^j\big), \quad j = 0, \dots, N-1, \quad y^0 \in X_0 \textit{ be given.} \tag{23}$$

Then there exists a discrete solution $\{x^j\}_{j=0}^N$ *of* (2) *with*

$$|y^{j+1} - x^{j+1}|^2 \leq (1 + 2\mu h)|y^j - x^j|^2 + 4C_B\varepsilon_j h + 2\big(\varepsilon_j^2 + 4C_F^2\big)h^2 , \tag{24}$$

$$|y^j - x^j| \leq \big(\sqrt{1+2\mu h}\,\big)^j |y^0 - x^0| + C_1\sqrt{g_h(2\mu, j, \varepsilon_h)} + C_2\sqrt{g_h(2\mu, j)}\sqrt{h}\,, \tag{25}$$

where $\varepsilon_h = \{\varepsilon_j\}_{j=0}^{N-1}$, $\varepsilon_j = |\overline{\varepsilon}^j|$, $C_1 = \sqrt{4C_B + 2C_\varepsilon}$, $C_2 = 2\sqrt{2}C_F$, *and* C_ε *is a bound of* hK_ε *and for finite* T *can be defined as* $C_\varepsilon = (T - t_0)K_\varepsilon$.

Proof By (A2), $y^j \in S$ for each $j = 0, \dots, N$. Given a solution $\{y^j\}_{j=0}^N$ of (23), there is $w^j \in F(y^j)$ with

$$y^{j+1} = y^j + h(w^j + \overline{\varepsilon}^j), \quad j = 0, \dots, N-1.$$

Suppose the iterates x^k are constructed for $0 \leq k \leq j$. By the OSL condition choose $v^j \in F(x^j)$ such that $\langle y^j - x^j, w^j - v^j\rangle \leq \mu|y^j - x^j|^2$. Then,

$$\begin{aligned}\langle y^j - x^j, w^j + \overline{\varepsilon}^j - v^j\rangle &= \langle y^j - x^j, w^j - v^j\rangle + \langle y^j - x^j, \overline{\varepsilon}^j\rangle \\ &\leq \mu|y^j - x^j|^2 + |y^j - x^j| \cdot \varepsilon_j\,. \end{aligned} \tag{26}$$

We set $x^{j+1} := x^j + hv^j$ which yields

$$y^{j+1} - x^{j+1} = (y^j - x^j) + h(w^j + \overline{\varepsilon}^j - v^j)\,.$$

Using the inequality $(a+b)^2 \le 2(a^2+b^2)$ and (A2), we get

$$|w^j + \overline{\varepsilon}^j - v^j|^2 \le 2(|w^j - v^j|^2 + |\overline{\varepsilon}^j|^2) \le 8C_F^2 + 2\varepsilon_j^2 .$$

We use this inequality and (26) in the estimate of the norm difference:

$$\begin{aligned}|y^{j+1} - x^{j+1}|^2 &= |y^j - x^j|^2 + 2h\langle y^j - x^j, w^j + \overline{\varepsilon}^j - v^j\rangle + |w^j + \overline{\varepsilon}^j - v^j|^2 h^2 \\ &\le (1+2\mu h)|y^j - x^j|^2 + 2\varepsilon_j h|y^j - x^j| + 2\varepsilon_j^2 h^2 + 8C_F^2 h^2\end{aligned}$$

We apply (A2) to estimate the second term in the right-hand side,

$$2\varepsilon_j h|y^j - x^j| \le 2\varepsilon_j h(|y^j| + |x^j|) \le 4C_B \varepsilon_j h .$$

The last two inequalities imply (24). Denoting $s_j := |y^j - x^j|^2$, $\varepsilon_h^2 := \{\varepsilon_j^2\}_{j=0}^{N-1}$ and using (15), we obtain

$$\begin{aligned}s_j &\le (1+2\mu h)^j s_0 + g_h\big(2\mu, j; 4C_B\varepsilon_h + 2h\varepsilon_h^2 + 8C_F^2 h \overline{1}_h\big) \\ &= (1+2\mu h)^j s_0 + 4C_B g_h(2\mu, j; \varepsilon_h) + 2h g_h(2\mu, j; \varepsilon_h^2) + 8C_F^2 h g_h(2\mu, j) .\end{aligned} \tag{27}$$

To simplify the estimate (27), we note that since by (A2) $y^j, y^{j+1}, F(y^j)$ are uniformly bounded, $h\varepsilon_j$ is uniformly bounded too. Let C_ε be a bound of $h\varepsilon_j$, then

$$g_h(2\mu, j; \varepsilon_h^2)\, h \le C_\varepsilon g_h(2\mu, j; \varepsilon_h)$$

and $C_\varepsilon \le 2C_B + hC_F$. Applied to (27), this yields

$$s_j \le (1+2\mu h)^j s_0 + (4C_B + 2C_\varepsilon) g_h(2\mu, j; \varepsilon_h) + 8C_F^2 h g_h(2\mu, j) . \tag{28}$$

Taking the square root we obtain the following estimate

$$|y^j - x^j| \le \big(\sqrt{1+2\mu h}\,\big)^j |y^0 - x^0| + C_1\sqrt{g_h(2\mu, j; \varepsilon_h)} + C_2\sqrt{g_h(2\mu, j)}\sqrt{h} \tag{29}$$

with the constants $C_1 := \sqrt{4C_B + 2C_\varepsilon}$, $C_2 := 2\sqrt{2}C_F$. □

Then, applying Lemma 2 to (25) we get

Theorem 1 (Discrete Filippov Theorem with Outer Perturbations) *Assuming the conditions of Proposition 1 with a step size* $h = \frac{T-t_0}{N}$ *such that* $1 + 2\mu h > 0$.

Then, for a discrete solution $y_h = \{y^j\}_{j=0}^N$ of the perturbed Euler inclusion (23), *there exists a discrete solution $x_h = \{x^j\}_{j=0}^N$ of the Euler inclusion* (2) *with*

$$|y^j - x^j| \le \left(\sqrt{1+2\mu h}\right)^j |y^0 - x^0| \\ + e^{\mu_+(t_j - t_0)} \left\{ C_1 \sqrt{\lfloor\lfloor \varepsilon_h \rfloor\rfloor_{\min,2\mu}} + C_2 \sqrt{\lfloor\lfloor \bar{1}_h \rfloor\rfloor_{\min,2\mu} \sqrt{h}} \right\} \quad (30)$$

with $\varepsilon_h = \{\varepsilon_j\}_{j=0}^{N-1}$, $\varepsilon_j = |\bar{\varepsilon}^j|$, $C_1 := \sqrt{4C_B + 2C_\varepsilon}$, $C_2 := 2\sqrt{2}C_F$, and C_ε is defined as in Proposition 1.

Remark 3 A similar but coarser estimate than (25) (with the norm $\|\varepsilon_h\|_1$ and bigger constants) follows in an indirect way, applying the continuous Filippov theorem for the OSL case (Theorem 3.1 in Donchev and Farkhi (2000)) and the error estimate for the Euler approximation (Theorem 4.1 for convex-valued maps in the same paper). In the theorem above we provide a direct proof with a refined estimate.

Note that $g_h(2\mu, j; \varepsilon_h)$ in (25) corresponds to the Riemann sum of the integral error term $e^{2\mu(t-\cdot)}\varepsilon(\cdot)$ in the above cited continuous-time Filippov theorem in Donchev and Farkhi (2000).

Corollary 1 *If we additionally assume in Theorem 1 that*

$$\lfloor\lfloor \varepsilon_h \rfloor\rfloor_{\min,\mu} \le C_\alpha h^\alpha$$

with $\alpha > 0$, then, since $1 + z \le e^z$, we may find a constant C such that

$$|y^j - x^j| \le e^{\mu(t_j - t_0)} |y^0 - x^0| + C h^{\frac{1}{2}\min\{\alpha,1\}}$$

with C depending on μ and, for $\mu \ge 0$ additionally, on $t_j - t_0$.

To obtain stability estimate for an infinite time (when $N \to \infty$, $h > 0$ is fixed), we define the *infinite-time discrete norms* for $\varepsilon_h^\infty := \{\varepsilon_j\}_{j=0}^\infty$ as in (10)–(11) by replacing $N-1, N$ by ∞ and suppose that $\mu < 0$. Then we get from (30) a discrete version of the stability result in Donchev and Farkhi (2000, Corollary 3.2), Frankowska and Rampazzo (2000).

Theorem 2 *Assuming the conditions of Proposition 1 on $I = [t_0, \infty)$, especially $\|\bar{\varepsilon}_h^\infty\|_\infty \le K_\varepsilon$, and let the OSL constant $\mu < 0$. For a fixed step size $h > 0$ with $1 + 2\mu h > 0$ we consider infinitely many steps with the Euler method.*

Then, $C_\varepsilon := \frac{1}{2|\mu|} K_\varepsilon \ge h \|\bar{\varepsilon}_h^\infty\|_\infty$ and for a discrete solution of the perturbed Euler inclusion $y_h^\infty = \{y^j\}_{j=0}^\infty$ there exists a discrete solution $x_h^\infty = \{x^j\}_{j=0}^\infty$ of the Euler inclusion (2) *with*

$$|y^j - x^j| \le e^{\mu(t_j - t_0)} |y^0 - x^0| + C_1 \sqrt{\lfloor\lfloor \{\varepsilon_\nu\}_{\nu=0}^{j-1} \rfloor\rfloor_{\min,\mu}} + C_2 \frac{1}{\sqrt{2|\mu|}} \sqrt{h}\,. \quad (31)$$

The Hausdorff distance between the original and the perturbed reachable sets is

$$\limsup_{j\to\infty} \mathrm{d_H}(\mathscr{R}_h(t_j, t_0, X_0),\ \mathscr{R}_h^{\varepsilon_h}(t_j, t_0, X_0)) \le C_1 \sqrt{\lfloor\!\lfloor \varepsilon_h^\infty \rfloor\!\rfloor_{\min,\mu}} + C_2 \frac{1}{\sqrt{|\mu|}} \sqrt{h}\,.$$

Thus, if $\lfloor\!\lfloor \varepsilon_h^\infty \rfloor\!\rfloor_{\min,\mu} = \mathscr{O}(h)$, *in particular if* $\sum_{j=0}^\infty \varepsilon_j < \infty$, *the estimate is*

$$\limsup_{j\to\infty} \mathrm{d_H}(\mathscr{R}_h(t_j, t_0, X_0),\ \mathscr{R}_h^{\varepsilon_h}(t_j, t_0, X_0)) = \mathscr{O}(\sqrt{h}).$$

3.2 Inner Perturbations

Here, the right-hand side contains only inner perturbation of the state variable.

Theorem 3 (Discrete Filippov Theorem with Inner Perturbations) *Let the assumptions (A1)–(A3) and* $1 + 2\mu h > 0$ *be satisfied and consider a grid function* $\overline{\delta}_h := \{\overline{\delta}^j\}_{j=0}^N \subset \mathbb{R}^n$ *with* $\delta_j := |\overline{\delta}^j|$, $j = 0, \ldots, N$, $\delta_h := \{\delta_j\}_{j=0}^N$ *and* $\|\overline{\delta}_h\|_\infty \le K_\delta$. *Let* $\{y^j\}_{j=0}^N$ *be a discrete solution of the perturbed Euler inclusion*

$$y^{j+1} \in y^j + hF(y^j + \overline{\delta}^j) \quad (j = 0, \ldots, N-1), \quad y^0 \in X_0 \ \textit{given}. \tag{32}$$

Then for every $x^0 \in X_0$ *there exists a discrete solution* $\{x^j\}_{j=0}^N$ *of* (2) *with*

$$|y^j - x^j| \le \left(\sqrt{1+2\mu h}\right)^j |y^0 - x^0| + C_1\sqrt{g_h(2\mu, j, \delta_h)} + C_2\sqrt{g_h(2\mu, j)}\sqrt{h}\,, \tag{33}$$

$$|y^j - x^j| \le \left(\sqrt{1+2\mu h}\right)^j |y^0 - x^0| + e^{\mu_+(t_j - t_0)} \left\{ C_1 \sqrt{\lfloor\!\lfloor \delta_h \rfloor\!\rfloor_{\min,2\mu}} + C_2 \sqrt{\lfloor\!\lfloor \bar{1}_h \rfloor\!\rfloor_{\min,2\mu}} \sqrt{h} \right\} \tag{34}$$

for $j = 0, \ldots, N$ *with constants* $C_1 := 2\sqrt{C_F + (2C_B + \frac{1}{2}K_\delta)|\mu|}$, *and* $C_2 := 2C_F$.

Proof Assume that we have constructed the sequence $\{x^k\}_k$ up to the index j. Let $y^{j+1} = y^j + hw^j$, $w^j \in F(y^j + \overline{\delta}^j)$. The OSL condition assures the existence of $v^j \in F(x^j)$ such that

$$\langle (y^j + \overline{\delta}^j) - x^j, w^j - v^j \rangle \le \mu \left| \left(y^j + \overline{\delta}^j\right) - x^j \right|^2$$

and we define x^{j+1} by

$$x^{j+1} := x^j + hv^j\,.$$

By assumption (A2) the sequence $\{v^j\}_{j=0}^{N-1}$, $\{w^j\}_{j=0}^{N-1}$ are uniformly bounded by C_F. Hence,

$$\begin{aligned}
|y^{j+1} - x^{j+1}|^2 &= |(y^j + hw^j) - (x^j + hv^j)|^2 = |(y^j - x^j) + h(w^j - v^j)|^2 \\
&\le |y^j - x^j|^2 + 2h\langle \left(y^j + \overline{\delta}^j\right) - x^j, w^j - v^j\rangle \\
&\quad - 2h\langle \overline{\delta}^j, w^j - v^j\rangle + 4C_F^2 h^2 \\
&\le |y^j - x^j|^2 + 2\mu h \left|\left(y^j + \overline{\delta}^j\right) - x^j\right|^2 + 2h|\overline{\delta}^j| \cdot |w^j - v^j| + 4C_F^2 h^2\,,
\end{aligned}$$

since the difference $|w^j - v^j|$ can be bounded by $2C_F$. We estimate

$$\begin{aligned}
\mu|(y^j + \overline{\delta}^j) - x^j|^2 &= \mu\left(|y^j - x^j|^2 + 2\langle y^j - x^j, \overline{\delta}^j\rangle + |\overline{\delta}^j|^2\right) \\
&\le \mu|y^j - x^j|^2 + 2|\mu| \cdot |y^j - x^j| \cdot |\overline{\delta}^j| + \mu|\overline{\delta}^j|^2 \qquad (35) \\
&\le \mu|y^j - x^j|^2 + 4C_B|\mu|\delta_j + |\mu| \cdot \delta_j^2\,.
\end{aligned}$$

The last inequalities and the bound $\delta_j^2 \le K_\delta \delta_j$ lead to

$$\begin{aligned}
|y^{j+1} - x^{j+1}|^2 &\le |y^j - x^j|^2 + 2\mu h|y^j - x^j|^2 + 2|\mu|h\delta_j^2 \\
&\quad + 8C_B|\mu|h\delta_j + 4C_F h\delta_j + 4C_F^2 h^2 \\
&\le (1 + 2\mu h)|y^j - x^j|^2 + 4\left(C_F + \left(2C_B + \frac{1}{2}K_\delta\right)|\mu|\right)h\delta_j + 4C_F^2 h^2\,.
\end{aligned}$$

We set $\Delta^j := |y^j - x^j|^2$ and as in Remark 1 get

$$\Delta^{j+1} \le (1 + 2\mu h)\Delta^j + 4\left(C_F + \left(2C_B + \frac{1}{2}K_\delta\right)|\mu|\right)h\delta_j + 4C_F^2 h^2\,,$$

$$\Delta^j \le (1 + 2\mu h)^j \Delta^0 + \sum_{k=0}^{j-1} (1 + 2\mu h)^{j-1-k}\left(C_1^2 h\delta_j + 4C_F^2 h^2\right)$$

Taking the square root yields

$$|y^j - x^j| \le \left(\sqrt{1 + 2\mu h}\right)^j |y^0 - x^0| + C_1\sqrt{g_h(2\mu, j, \delta_h)} + 2C_F\sqrt{g_h(2\mu, j)}\,\sqrt{h}\,.$$

We complete the proof applying Lemma 2. □

Corollary 2 *Let all assumptions of Theorem 3 are fulfilled and let*

$$\lfloor\lfloor \delta_h \rfloor\rfloor_{\min,\mu} \le C_\alpha h^\alpha$$

be fulfilled for the inner perturbation.

Then, for each solution $\{y^j\}_{j=0}^N$ *of the perturbed inclusion* (32) *there exists a discrete solution* $\{x^j\}_{j=0}^N$ *of* (2) *with*

$$|y^j - x^j| \le e^{\mu(t_j - t_0)}|y^0 - x^0| + \widetilde{C} e^{\mu_+(T-t_0)} h^{\frac{1}{2}\min\{\alpha,1\}},$$

where the constant $\widetilde{C}$ *may be easily estimated from* C_1, C_2 *in Theorem 3 and does not depend on the time length whenever* $\mu \neq 0$.

3.3 Both Perturbations and Applications

The general theorem for inner and outer perturbations may be obtained combining the last two theorems. In the estimate the square root of the discrete norms of the inner and outer perturbations as well as an error term $\mathcal{O}(\sqrt{h})$ will appear.

Theorem 4 (Discrete Filippov Theorem with Both Perturbations) *Let the assumptions (A1)–(A3) and* $1 + 2\mu h > 0$ *hold and consider grid functions* $\overline{\delta}_h = \left\{\overline{\delta}^j\right\}_{j=0}^{N-1} \subset \mathbb{R}^n$ *with* $\delta_j := \left|\overline{\delta}^j\right|$, $j = 0, \ldots, N-1$, *and* $\overline{\varepsilon}_h = \{\overline{\varepsilon}^j\}_{j=0}^{N-1} \subset \mathbb{R}^n$, $\varepsilon_j := \left|\overline{\varepsilon}^j\right|$ *satisfying* $\|\overline{\delta}_h\|_\infty \le K_\delta$, $\|\overline{\varepsilon}_h\|_\infty \le K_\varepsilon$.
Let $\{y^j\}_{j=0}^N$ *be a discrete solution of the perturbed Euler inclusion*

$$y^{j+1} \in y^j + h\big(F\big(y^j + \overline{\delta}^j\big) + \overline{\varepsilon}^j\big), \quad j = 0, \ldots, N-1, \quad y^0 \in X_0 \text{ is given.} \tag{36}$$

Then for every $x^0 \in X_0$ *there exists a discrete solution* $\{x^j\}_{j=0}^N$ *of* (2) *with*

$$\begin{aligned}|y^j - x^j| \le \big(\sqrt{1+2\mu h}\big)^j |y^0 - x^0| + C_1 \sqrt{g_h(2\mu, j, \delta_h)} \\ + C_2\sqrt{g_h(2\mu, j, \varepsilon_h)} + C_3\sqrt{g_h(2\mu, j)}\sqrt{h}\end{aligned}$$

for $j = 0, \ldots, N$ *with constants* $C_1 := 2\sqrt{C_F + \left(2C_B + \frac{1}{2}K_\delta\right)|\mu|}$, $C_2 := \sqrt{4C_B + 2C_\varepsilon}$, $C_3 := (2 + 2\sqrt{2})C_F$ *and* C_ε *is defined as in Proposition 1.*

Next we study the distance between the iterates of the explicit and the implicit set-valued Euler's method. The following proposition shows that each iterate of the second is close to some iterate of the first one, and thus provides convergence results for the implicit method whenever the corresponding convergence result for the explicit method is known. A more elaborated study for continuous right-hand sides can be found in Beyn and Rieger (2010). It is also shown in Beyn and Rieger (2010) that if F is usc and $1-\mu h > 0$, then the implicit inclusion (37) has a solution.

Proposition 2 *Let the step size h be so small that $hC_F \le K_\delta$, $1 - \mu h > 0$ and choose $x^0 \in X_0$. In addition to (A0)–(A3), assume that (A1') is fulfilled and (A2) also holds for the implicit Euler method.*

Then there is a constant C such that for each implicit Euler iterate $\{y^j\}_{j=0}^N$ of

$$y^{j+1} \in y^j + hF\big(y^{j+1}\big), \quad j = 0, 1, \dots, N-1, \quad y^0 = x^0, \tag{37}$$

there is an iterate $\{x^j\}_{j=0}^N$ of the explicit scheme (2) *with*

$$|y^j - x^j| \le C\sqrt{h}, \quad j = 0, \dots, N,$$

and the distance from the reachable set of (37) *to the one of* (2) *satisfies*

$$\operatorname{dist}\big(\mathscr{R}_h^{impl}(T, t_0, X_0), \mathscr{R}_h(t_j, t_0, X_0)\big) \le C\sqrt{h}.$$

Proof The restriction on the step size h and (A0), (A1'), (A3) guarantee the existence of iterates of the implicit Euler method by Beyn and Rieger (2010, Theorem 4).

Consider an iterate of the implicit scheme, i.e.,

$$y^{j+1} = y^j + hw^{j+1}, \quad w^{j+1} \in F\big(y^{j+1}\big).$$

We can rewrite it as perturbed Euler iteration with $\overline{\delta}^j := y^{j+1} - y^j$, since

$$y^{j+1} \in y^j + hF\big(y^{j+1}\big), \quad F\big(y^{j+1}\big) = F\big(y^j + \overline{\delta}^j\big).$$

The inner perturbations are bounded by $\mathscr{O}(h)$, since iterates of both schemes (and hence velocities) are bounded:

$$\big|\overline{\delta}^j\big| = \big|y^{j+1} - y^j\big| = h\big|w^{j+1}\big| \le C_F h \le K_\delta$$

Corollary 2 for the explicit Euler can be applied so that

$$|y^j - x^j| \le \widetilde{C}_1\sqrt{\|\delta_h\|_1} + \widetilde{C}_2\sqrt{h} = (\widetilde{C}_1\sqrt{C_F} + \widetilde{C}_2)\sqrt{h}. \qquad \square$$

Corollary 3 *Let the assumptions of Proposition 2 here and of Theorem 4.1 in Donchev and Farkhi (2000) be fulfilled. Then, there exists a constant C with*

$$\operatorname{dist}(\mathscr{R}_h^{impl}(T, t_0, X_0), \mathscr{R}(T, t_0, X_0)) \le C\sqrt{h}.$$

Proof Since F is convex-valued, we can apply the convergence result for the explicit Euler in Donchev and Farkhi (2000, Theorem 4.1) so that

$$\mathrm{d_H}\left(\mathscr{R}_h(T, t_0, X_0), \mathscr{R}(T, t_0, X_0)\right) \le C\sqrt{h}\,.$$

The rest follows by Proposition 2 and the triangle inequality. □

Remark 4 Similarly, $\mathscr{O}(\sqrt{h})$-estimates for other Runge-Kutta methods may be obtained applying Corollary 3 if F is OSL. As one example we mention the improved Euler scheme in Lempio (1993)

$$x^{j+1} \in x^j + hF\left(x^j + \frac{h}{2}v^j\right), \quad v^j \in F(x^j).$$

Here, $\left|\overline{\delta}^j\right| = \frac{h}{2}|v^j| \le \frac{C_F}{2}h$ and the order of the distance is $\mathscr{O}(\sqrt{h})$.

Remark 5 We can formulate Proposition 2 and Corollary 3 with the same assumptions and the Lipschitz condition replacing the OSL one provided that (A1') also holds. Then, explicit and implicit Euler iterates can be found that are $\mathscr{O}(h)$-close by applying the discrete Filippov theorem for the explicit Euler in Chahma (2003, Proposition 2.2.3) as well as the corresponding Filippov theorem for the implicit Euler in Beyn and Rieger (2010, Theorem 14). Therefore, the convergence of the implicit Euler is the same as for the explicit one, i.e., $\mathscr{O}(h)$ on a finite time interval. But, the OSL condition with a negative constant provides contractivity of the reachable set mapping and (exponential) stability at an infinite time interval.

Both mentioned convergence results for the implicit Euler method for OSL maps with additional continuous right-hand sides do not deliver the preferrable stability results for $\mu < 0$ as stated in Beyn and Rieger (2010).

3.4 Discrete Relaxation Stability Theorem

Consider the (set-valued) Euler discrete inclusion

$$\eta^{j+1} \in \eta^j + hF(\eta^j), \qquad j = 0, 1, \ldots, N-1, \quad \eta^0 = x^0 \in X_0\,, \tag{38}$$

and its convexified counterpart

$$\eta^{j+1} \in \eta^j + h\,\mathrm{co}\,F(\eta^j), \quad j = 0, 1, \ldots, N-1, \quad \eta^0 = x^0\,. \tag{39}$$

In Grammel (2003), an estimate of order $\mathscr{O}(\sqrt{h})$ is obtained for the Hausdorff distance between the solutions sets of the relaxed differential inclusion (with co F at the right-hand side in (1)) and the Euler difference inclusion (38).

We denote by $\mathscr{S}_h$ the set of solutions of (38) and by $\mathscr{S}_h^{co}$ the set of solutions of (39). Here, these solutions are considered in the space of grid functions $\eta_h = \{\eta^j\}_{j=0}^N$ and are studied under the weaker OSL condition.

Theorem 5 (Discrete Relaxation Stability) *Let the assumptions (A1)–(A3) hold. Then, there is a constant C such that*

$$\mathrm{d_H}(\mathscr{S}_h, \mathscr{S}_h^{co}) \le C\sqrt{h}.$$

Proof For a solution $\{y^j\}_{j=0}^N$ of (39) and all $0 \le j \le N-1$ there is $w^j \in \operatorname{co} F(y^j)$ with

$$y^{j+1} = y^j + hw^j\,, \quad j = 0, \ldots, N-1.$$

We construct a solution $\{x^j\}_{j=0}^N$ of (38) which is at the required distance from $(y^j)_{j=0}^N$. Suppose x^k are constructed for $0 \le k \le j$. We recall that since $F(x)$ is OSL, then also the map $\operatorname{co} F(x)$ is OSL with the same constant (this can be easily verified by the definition).

By the OSL condition there is $\widetilde{v}^j \in \operatorname{co} F(x^j)$ such that

$$\langle y^j - x^j, w^j - \widetilde{v}^j \rangle \le \mu |y^j - x^j|^2. \tag{40}$$

Then,

$$|y^{j+1} - \big(x^j + h\widetilde{v}^j\big)|^2 \le |y^j - x^j|^2 + 2h\langle y^j - x^j, w^j - \widetilde{v}^j \rangle + h^2 |w^j - \widetilde{v}^j|^2\,. \tag{41}$$

We note that the linear function $\varphi(v) = \langle y^j - x^j, w^j - v \rangle$ achieves its minimum on the convex compact set $\operatorname{co} F(x^j)$ at some extremal point $v^j \in F(x^j)$, since the compact $F(x^j)$ contains all extremal points of its convex hull (see Marti 1977, Sec. III.2, Lemma 1). Hence, we may choose $v^j \in F(x^j)$ to replace $\widetilde{v}^j$ in (40).

We set $x^{j+1} := x^j + hv^j$ and obtain from (41), (40) and (A2)

$$|y^{j+1} - x^{j+1}|^2 \le (1 + 2\mu h)|y^j - x^j|^2 + 4C_F^2 h^2\,. \tag{42}$$

Applying (15) and Lemma 2, we get the estimates:

$$|y^j - x^j|^2 \le 4C_F^2 e^{2\mu_+(T-t_0)} \min\big\{T - t_0, \frac{1}{\mu}\big\} h,$$

$$|y^j - x^j| \le 2C_F e^{\mu_+(T-t_0)} \min\left\{\sqrt{T - t_0}, \frac{1}{\sqrt{|\mu|}}\right\} \sqrt{h}\,. \qquad \square$$

Denote by $\mathscr{S}^{co}$ the set of solutions of the convexified differential inclusion (1) in which $F(x)$ is replaced by co $F(x)$. The following corollary extends a theorem of Grammel (2003) from Lipschitz to OSL mappings F.

Corollary 4 *Under the assumptions of Theorem 5 and of Theorem 4.1 in Donchev and Farkhi (2000), there is a constant C such that*

$$\mathrm{d_H}(\mathscr{S}_h, \mathscr{S}^{co}) \le C\sqrt{h}.$$

Proof The convergence result for the explicit Euler (Donchev and Farkhi 2000, Theorem 4.1) yields

$$\mathrm{d_H}(\mathscr{S}_h^{co}, \mathscr{S}^{co}) \le C\sqrt{h}\,.$$

The rest follows by Theorem 5 and the triangle inequality. □

Let us mention the conjecture of Veliov in (2015) that for the Lipschitz map F the above rate is $\mathscr{O}(h)$. This conjecture is proved in some important special cases.

4 Discrete Filippov-Type Theorems for Strengthened One-Sided Lipschitz Maps

Let us recall that the SOSL condition is stronger than the OSL, but it also provides stronger stability. In its earlier uniform version (with "for all" instead of "there exist" in its definition) in Lempio (1993, 1995), Lempio and Silin (1997) it is implemented to gain the order of convergence 1 for the Euler method (instead of $\frac{1}{2}$ for UOSL). Several classes of discontinuous right-hand sides in applications (see Mannshardt 1978/1979, Model 1988 and Kastner-Maresch 1990a,b, 1992, Kastner-Maresch and Lempio 1993 as well as references in Kastner-Maresch 1990a) fulfill the SOSL condition (see Lempio 1995, Lempio and Veliov 1998).

4.1 Both Perturbations

The analysis of the convergence of the Euler scheme made in Lempio (1995) and Lempio and Veliov (1998) lies in the basis of our proofs here. Let us stress that the uniform condition of Lempio (1995), as the UOSL condition, implies uniqueness of the solution of the differential inclusion (1) which does not hold in general if the right-hand side is OSL or SOSL. In Lempio and Veliov (1998, Remark 2.1 and Theorem 2.4) convergence order 1 is proved for the Euler method. In the following we state discrete Filippov theorems and stability results for infinite time with estimates of order 1 improving the estimates obtained for the OSL case in Sect. 3.

Proposition 3 (Local Estimate, Discrete Filippov Theorem for SOSL with Both Perturbations) *Let the assumptions (A1)–(A2), (A3') be satisfied with the SOSL constant* $\mu \in \mathbb{R}$ *and choose a step size with* $1 + \mu h > 0$.
Consider $\{\overline{\delta}^j\}_{j=0}^{N-1} \subset \mathbb{R}^n$ *with* $\delta_j := |\overline{\delta}^j|_\infty$, $j = 0, \dots, N$, $\overline{\varepsilon}_h = \{\overline{\varepsilon}^j\}_{j=0}^{N-1} \subset \mathbb{R}^n$, $\varepsilon_j := |\overline{\varepsilon}^j|$, $\{\varepsilon^j\}_{j=0}^{N-1} \subset \mathbb{R}_+$ *satisfying* $\|\overline{\delta}_h\|_\infty \le K_\delta$, $\|\overline{\varepsilon}_h\|_\infty \le K_\varepsilon$ *and let* $\{y^j\}_{j=0}^N$ *be a discrete solution of the perturbed Euler inclusion*

$$y^{j+1} \in y^j + h\big(F\big(y^j + \overline{\delta}^j\big) + \overline{\varepsilon}^j\big), \quad j = 0, \dots, N-1, \quad y^0 \in X_0 \text{ be given.} \tag{43}$$

Then for every $x^0 \in X_0$ *there exists a discrete solution* $\{x^j\}_{j=0}^N$ *of* (2) *with*

$$|y^{j+1} - x^{j+1}|_\infty \le \max\big\{(1+\mu h)|y^j - x^j|_\infty + |\mu| h \delta_j + h\varepsilon_j \,,\; 2C_F h + \delta_j + h\varepsilon_j\big\}. \tag{44}$$

Proof By (A2), we know that all discrete Euler solutions are bounded. We denote the selections by $\{w^j\}_{j=0}^{N-1}$ such that

$$y^{j+1} = y^j + hw^j + h\overline{\varepsilon}^j\,, \quad w^j \in F\big(y^j + \overline{\delta}^j\big).$$

For one iterate $y^j \in \mathbb{R}^n$ or selection $w^j \in \mathbb{R}^n$, we denote y_i^j resp. w_i^j as the i-th coordinate, where $i = 1, \dots, n$.

By assumption (A2) (the boundedness condition), the sequence $\{w^j\}_{j=0}^{N-1}$ is also uniformly bounded by C_F.

(i) construction of the sequence $\{x^j\}_{j=0}^{N-1}$
Assume that we have constructed the sequence $\{x^j\}_{j=0}^k$ up to the time step k. We denote the corresponding selections by $\{v^j\}_{j=0}^{k-1}$, i.e.,

$$x^{j+1} = x^j + hv^j\,, \quad v^j \in F(x^j)\,.$$

By the SOSL condition (assumption (A3')) there exists $v^k \in F(x^k)$ such that we have the SOSL inequalities as stated in (A3') in the two case: $y_i^k + \overline{\delta}_i^k > x_i^k$ (case a1 below) and $y_i^k + \overline{\delta}_i^k < x_i^k$ (case a2 below).
Other cases have to be dealt separately, e.g., case c).

(ii) local error estimate with previous error term
We set $x^{k+1} := x^k + hv^k$ with suitable $v^k \in F(x^k)$ and consider the following cases:
case a) $\operatorname{sign}((y_i^k + hw_i^k) - x_i^{k+1}) = \operatorname{sign}\big((y_i^k + \overline{\delta}_i^k) - x_i^k\big) \ne 0$
case a1) $(y_i^k + hw_i^k) - x_i^{k+1} > 0$, $\big(y_i^k + \overline{\delta}_i^k\big) - x_i^k > 0$
By the SOSL condition (assumption (A3')), we find $v^k \in F(x^k)$ with

$$w_i^k - v_i^k \le \mu\big|\big(y^k + \overline{\delta}^k\big) - x^k\big|_\infty\,.$$

In this case we distinguish two subcases, $\mu \geq 0$ and $\mu < 0$. If $\mu \geq 0$, we estimate

$$w_i^k - v_i^k \leq \mu \left|(y^k + \overline{\delta}^k) - x^k\right|_\infty \leq \mu |y^k - x^k|_\infty + \mu \left|\overline{\delta}^k\right|_\infty.$$

If $\mu < 0$, we use the estimate

$$\left|(y^k - x^k) + \overline{\delta}^k\right|_\infty \geq |y^k - x^k|_\infty - \left|\overline{\delta}^k\right|_\infty$$

so that

$$w_i^k - v_i^k \leq \mu \left|(y^k + \overline{\delta}^k) - x^k\right|_\infty \leq \mu |y^k - x^k|_\infty - \mu \left|\overline{\delta}^k\right|_\infty.$$

Hence, we have in both subcases the common estimate

$$w_i^k - v_i^k \leq \mu |y^k - x^k|_\infty + |\mu| \cdot \left|\overline{\delta}^k\right|_\infty$$

so that

$$\begin{aligned}
|y_i^{k+1} - x_i^{k+1}| &\leq |(y_i^k + hw_i^k) - x_i^{k+1}| + h|\overline{\varepsilon}^k| \\
&\leq \left((y_i^k + hw_i^k) - (x_i^k + hv_i^k)\right) + h\varepsilon_k \\
&= (y_i^k - x_i^k) + h(w_i^k - v_i^k) + h\varepsilon_k \\
&\leq |y_i^k - x_i^k| + h\left(\mu |y^k - x^k|_\infty + |\mu| \cdot \left|\overline{\delta}^k\right|_\infty + \varepsilon_k\right) \\
&\leq (1 + \mu h)|y^k - x^k|_\infty + |\mu| h\delta_k + h\varepsilon_k
\end{aligned}$$

a2) $(y_i^k + hw_i^k) - x_i^{k+1} < 0$, $\left(y_i^k + \overline{\delta}_i^k\right) - x_i^k < 0$
By the SOSL condition in (A3') and, as above, we have the other inequality

$$v_i^k - w_i^k \leq \mu |x^k - y^k|_\infty + |\mu| \cdot \left|\overline{\delta}^k\right|_\infty.$$

Similarly to subcase a1), we get similarly

$$\begin{aligned}
|y_i^{k+1} - x_i^{k+1}| &\leq |(y_i^k + hw_i^k) - x_i^{k+1}| + h|\overline{\varepsilon}^k| \\
&= (x_i^k - y_i^k) + h(v_i^k - w_i^k) + h\varepsilon_k \\
&\leq |x_i^k - y_i^k| + h\left(\mu |y^k - x^k|_\infty + |\mu| \cdot \left|\overline{\delta}^k\right|_\infty + \varepsilon_k\right) \\
&\leq (1 + \mu h)|y^k - x^k|_\infty + |\mu| h\delta_k + h\varepsilon_k
\end{aligned}$$

case b) $\operatorname{sign}((y_i^k + hw_i^k) - x_i^{k+1}) = -\operatorname{sign}\left((y_i^k + \overline{\delta}_i^k) - x_i^k\right) \neq 0$

In these cases, we have an error reset, since the past estimates are not used.
b1) $(y_i^k + hw_i^k) - x_i^{k+1} > 0$, $\left(y_i^k + \overline{\delta}_i^k\right) - x_i^k < 0$
Here, we first proceed as in subcase a1) but do not use the SOSL condition and simply neglect negative terms:

$$
\begin{aligned}
|y_i^{k+1} - x_i^{k+1}| &\le |(y_i^k + hw_i^k) - x_i^{k+1}| + h|\overline{\varepsilon}^k| \\
&\le (y_i^k - x_i^k) + h(w_i^k - v_i^k) + h\varepsilon_k \\
&= \underbrace{\left(y_i^k + \overline{\delta}_i^k\right) - x_i^k}_{<0} + h(w_i^k - v_i^k) - \overline{\delta}_i^k + h\varepsilon_k \\
&< \left|\overline{\delta}_i^k\right| + h|w_i^k - v_i^k| + h\varepsilon_k \le 2C_F h + \delta_k + h\varepsilon_k
\end{aligned}
$$

b2) $(y_i^k + hw_i^k) - x_i^{k+1} < 0$, $\left(y_i^k + \overline{\delta}_i^k\right) - x_i^k > 0$
Again, we first proceed as in subcase a2) and then neglect negative terms:

$$
\begin{aligned}
|y_i^{k+1} - x_i^{k+1}| &\le |(y_i^k + hw_i^k) - x_i^{k+1}| + h|\overline{\varepsilon}^k| \\
&= (x_i^k - y_i^k) + h(v_i^k - w_i^k) + h\varepsilon_k \\
&= \underbrace{x_i^k - \left(y_i^k + \overline{\delta}_i^k\right)}_{<0} + h(v_i^k - w_i^k) + \overline{\delta}_i^k + h\varepsilon_k \\
&< \left|\overline{\delta}_i^k\right| + h|w_i^k - v_i^k| + h\varepsilon_k \le 2C_F h + \delta_k + h\varepsilon_k
\end{aligned}
$$

case c) $(y_i^k + hw_i^k) - x_i^{k+1} = 0$ or $\left(y_i^k + \overline{\delta}_i^k\right) - x_i^k = 0$
c1) $(y_i^k + hw_i^k) - x_i^{k+1} = 0$
This is the simplest case, since the essential term is zero and simply disappears.

$$
|y_i^{k+1} - x_i^{k+1}| \le |\underbrace{(y_i^k + hw_i^k) - x_i^{k+1}}_{=0}| + h|\overline{\varepsilon}^k| \le h\varepsilon_k
$$

c2) $\left(y_i^k + \overline{\delta}_i^k\right) - x_i^k = 0$

Here we have

$$
\begin{aligned}
|y_i^{k+1} - x_i^{k+1}| &\le |(y_i^k + hw_i^k) - x_i^{k+1}| + h|\overline{\varepsilon}^k| \\
&= \left|\underbrace{\left(y_i^k + \overline{\delta}_i^k\right) - x_i^k}_{=0} - \overline{\delta}_i^k + h(w_i^k - v_i^k)\right| + h\varepsilon_k \\
&\le \left|\overline{\delta}_i^k\right| + h|w_i^k - v_i^k| + h\varepsilon_k \le 2C_F h + \delta_k + h\varepsilon_k \,.
\end{aligned}
$$

To summarize, in the subcases a1) and a2) we have

$$|y_i^{k+1} - x_i^{k+1}| \le (1 + \mu h)|y^k - x^k|_\infty + |\mu| h\delta_k + h\varepsilon_k \,, \tag{45}$$

while in all other cases we have

$$|y_i^{k+1} - x_i^{k+1}| \le 2C_F h + \delta_k + h\varepsilon_k \,. \tag{46}$$

Hence, (44) holds. □

We now deduce a global error estimate from the local one of Proposition 3.

Proposition 4 (Global Estimate, Discrete Filippov Theorem for SOSL with Both Perturbations) *Assume the conditions of Proposition 3.*
Let $\{y^j\}_{j=0}^N$ be a discrete solution of the perturbed Euler inclusion

$$y^{j+1} \in y^j + h\big(F\big(y^j + \overline{\delta}^j\big) + \overline{\varepsilon}^j\big), \quad j = 0, \ldots, N-1\,, \quad y^0 \in X_0 \text{ be given.} \tag{47}$$

Then for every $x^0 \in X_0$ there exists a discrete solution $\{x^j\}_{j=0}^N$ of (2) *with*

$$|y^j - x^j|_\infty \le \max\Big\{(1+\mu h)^j |y^0 - x^0|_\infty,\ (1+\mu h)^j\big(2C_F h + \|\delta_h\|_\infty\big) + g_h(\mu, j, \varepsilon_h)\Big\}$$
$$+ |\mu|\, g_h(\mu, j, \delta_h) + g_h(\mu, j, \varepsilon_h) \tag{48}$$

for $j = 0, \ldots, N$ and a step size $h > 0$ with $1 + \mu h > 0$.

Proof The sequence $\{x^j\}_{j=0}^N$ is constructed as in Proposition 3.
We consider an index set $J_\Delta \subset \{0, \ldots, N-1\}$ for which the following holds:

- J_Δ consists of subsequent numbers, i.e., there are $k' \le k''$, $k', k'' \in J_\Delta$ such that
$$J_\Delta = \{k', k'+1, \ldots, k''\}\,,$$
- for all indices $k \in J_\Delta$ we require that case a) in Proposition 3 holds,
- the index set J_Δ is maximal with respect to inclusion within the set of numbers $\{0, \ldots, N\}$.

Whenever $k \in J_\Delta$, we never encounter the estimate (46) and can use (45). Hence, it follows by Remark 1 that

$$|y_i^k - x_i^k| \le (1+\mu h)^{k-k'} |y^{k'} - x^{k'}|_\infty + h \sum_{\nu=k'}^{k-1} (1+\mu h)^{k-1-\nu}\big(|\mu| \cdot \delta_\nu + \varepsilon_\nu\big)\,.$$

If this index set J_Δ contains the element $k' = 0$, we can rewrite the term $|y^{k'} - x^{k'}|_\infty$ as $|y^0 - x^0|_\infty$ and we have the estimate

$$|y^k - x^k|_\infty \le (1 + \mu h)^k |y^0 - x^0|_\infty + |\mu|\, g_h(\mu, k, \delta_h) + g_h(\mu, k, \varepsilon_h)\,. \tag{49}$$

If otherwise $k' > 0$, then we encounter an error reset due to (46) so that the maximality of J_Δ yields with a suitable index $i_0 \in \{1, \dots, n\}$:

$$|y^{k'} - x^{k'}|_\infty = \left|y^{k'}_{i_0} - x^{k'}_{i_0}\right| \le 2C_F h + \delta_{k'-1} + h\varepsilon_{k'-1}$$

Thus, if $k' > 0$, we have for $k' \le k \le k''$

$$\begin{aligned}
|y^k - x^k|_\infty \le\, & (1 + \mu h)^{k-k'} \big(2C_F h + \delta_{k'-1} + h\varepsilon_{k'-1}\big) \\
& + |\mu|\, g_h\left(\mu, k - k', \{\delta_\nu\}_{\nu=k'}^{k-1}\right) + g_h\left(\mu, k - k', \{\varepsilon_\nu\}_{\nu=k'}^{k-1}\right) \\
\le\, & (1 + \mu h)^{k-k'} \big(2C_F h + \delta_{k'-1}\big) + h(1 + \mu h)^{k-k'} \varepsilon_{k'-1} \\
& + |\mu|\, g_h\left(\mu, k - k', \{\delta_\nu\}_{\nu=k'}^{k-1}\right) + g_h\left(\mu, k - k', \{\varepsilon_\nu\}_{\nu=k'}^{k-1}\right)
\end{aligned}$$

Thus, we obtain in case $k' > 0$, $k' \le k \le k''$,

$$\begin{aligned}
|y^k - x^k|_\infty \le\, & (1 + \mu h)^k \big(2C_F h + \|\delta_h\|_\infty\big) + g_h\left(\mu, k - k', \{\varepsilon_\nu\}_{\nu=k'-1}^{k-2}\right) \\
& + |\mu|\, g_h\left(\mu, k - k', \{\delta_\nu\}_{\nu=k'}^{k-1}\right) + g_h\left(\mu, k - k', \{\varepsilon_\nu\}_{\nu=k'}^{k-1}\right) \\
\le\, & (1 + \mu h)^k \big(2C_F h + \|\delta_h\|_\infty\big) + g_h(\mu, k, \varepsilon_h) \\
& + |\mu|\, g_h(\mu, k, \delta_h) + g_h(\mu, k, \varepsilon_h).
\end{aligned} \tag{50}$$

The maximum of the estimates (49)–(50) implies (48). □

An immediate consequence is the convergence order 1 with respect to the step size, if both perturbations are $\mathcal{O}(h)$ (measured in different norms).

Theorem 6 *Assume the conditions of Proposition 4. Then,*

$$\begin{aligned}
\max_{j=0,\dots,N} |y^j - x^j| \le e^{\mu_+(T-t_0)} \Bigg(& \max\left\{|y^0 - x^0|_\infty,\ 2C_F h + \|\delta_h\|_\infty + \lfloor\!\lfloor \varepsilon_h \rfloor\!\rfloor_{\min,\mu}\right\} \\
& + |\mu| \min\left\{T - t_0, \frac{1}{|\mu|}\right\} \|\delta_h\|_\infty + \lfloor\!\lfloor \varepsilon_h \rfloor\!\rfloor_{\min,\mu} \Bigg).
\end{aligned}$$

If $\max\left\{|y^0 - x^0|_\infty,\ \|\delta_h\|_\infty,\ \|\varepsilon_h\|_1\right\} \le C_e h$ *also holds, then there is a constant* C *with*

$$|y^j - x^j| \le Ch, \quad j = 0, \dots, N.$$

A similar result as in Theorem 2 with $\mathscr{O}(h)$-perturbations holds for infinite time in the SOSL case.

Theorem 7 *Assume the conditions of Proposition 4 on* $I = [t_0, \infty)$ *especially that* $\|\overline{\delta}_h^\infty\|_\infty \le K_\delta$ *and* $\|\overline{\varepsilon}_h^\infty\|_\infty \le K_\varepsilon$, *and let the OSL constant satisfy* $\mu < 0$. *For a fixed step size* $h > 0$ *with* $1 + \mu h > 0$ *we consider infinitely many steps with the Euler method.*

Then, for a discrete solution $y_h^\infty = \{y^j\}_{j=0}^\infty$ *of the perturbed Euler inclusion* (47) *there exists a discrete solution* $x_h^\infty = \{x^j\}_{j=0}^\infty$ *of the Euler inclusion* (2) *with*

$$\begin{aligned} |y^j - x^j| \le \max\Big\{ & e^{\mu(t_j - t_0)} |y^0 - x^0|_\infty,\ e^{\mu(t_j - t_0)} \big(2C_F h + \|\{\delta_\nu\}_{\nu=0}^{j-1}\|_\infty\big) \\ & + e^{\mu_+(t_j - t_0)} \|\{\varepsilon_\nu\}_{\nu=0}^{j-1}\|_1 \Big\} \\ & + e^{\mu_+(t_j - t_0)} \big(\|\{\delta_\nu\}_{\nu=0}^{j-1}\|_\infty + \|\{\varepsilon_\nu\}_{\nu=0}^{j-1}\|_1\big). \end{aligned} \tag{51}$$

Hence, the Hausdorff distance between the original and the perturbed reachable sets satisfies

$$\limsup_{j \to \infty} \mathrm{d_H}(\mathscr{R}_h(t_j, t_0, X_0),\ \mathscr{R}_h^{\varepsilon_h}(t_j, t_0, X_0)) \le 2\big(\|\delta_h^\infty\|_\infty + \|\varepsilon_h^\infty\|_1\big).$$

Thus, if $\max\{\|\delta_h^\infty\|_\infty,\ \|\varepsilon_h^\infty\|_1\} \le C_e h$, *the estimate is*

$$\limsup_{j \to \infty} \mathrm{d_H}(\mathscr{R}_h(t_j, t_0, X_0),\ \mathscr{R}_h^{\varepsilon_h}(t_j, t_0, X_0)) \le 4 C_e h,$$

where we have used Lemma 2 and the inequalities

$$|\mu|\, e^{\mu_+(t_j - t_0)} \llbracket \delta_h^\infty \rrbracket_{\min,\mu} \le \|\delta_h^\infty\|_\infty, \quad e^{\mu_+(t_j - t_0)} \llbracket \varepsilon_h^\infty \rrbracket_{\min,\mu} \le \|\varepsilon_h^\infty\|_1.$$

4.2 Application

For SOSL right-hand side and for the implicit Euler method we show an analogous result to Proposition 2, but with first order estimate replacing the $\mathscr{O}(\sqrt{h})$ order in the OSL case.

Proposition 5 *Let the step size h be so small that $hC_F \le K_\delta$, $1 + 2\mu h > 0$ and the assumptions (A0)–(A1), (A1'), (A3') as well as (A2) also for the implicit Euler method hold.*

Then for all $x^0 \in X_0$ there exists a constant C such that for each iterate $\{y^j\}_{j=0}^N$ of the implicit Euler scheme

$$y^{j+1} \in y^j + hF(y^{j+1}), \quad j = 0, 1, \dots, N-1, \quad y^0 = x^0$$

there is one iterate $\{x^j\}_{j=0}^N$ of the explicit scheme (2) *with*

$$|y^j - x^j| \le Ch, \quad j = 0, \dots, N.$$

The distance from the reachable set of the implicit Euler method to the one of the explicit Euler and to the reachable set of (1) *respectively, can be estimated by*

$$\operatorname{dist}\left(\mathscr{R}_h^{impl}(T, t_0, X_0),\ \mathscr{R}_h(T, t_0, X_0)\right) \le Ch\,,$$
$$\operatorname{dist}\left(\mathscr{R}_h^{impl}(T, t_0, X_0),\ \mathscr{R}(T, t_0, X_0)\right) \le Ch\,.$$

Proof The proof is almost identical to the one of Proposition 2 and Corollary 3. Only Lempio and Veliov (1998, Theorem 2.4) and Theorem 6 replace Donchev and Farkhi (2000, Theorem 4.1) and Corollary 2 which guarantee

$$|y^j - x^j| \le Ch\,. \qquad \square$$

This result is similar to the convergence order $\mathscr{O}(h)$ attained for the explicit Euler scheme in Lempio (1995), but the SOSL property for the right-hand side replaces the S-UOSL property. Also the implicit Euler converges with $\mathscr{O}(h)$, if F is S-UOSL.

Corollary 5 *Let all assumptions of Proposition 5 hold except that (A3') is replaced by the S-UOSL property of F, then all implicit Euler iterates converges to the (single) solution of* (1) *with convergence order $\mathscr{O}(h)$.*
Similarly, if all assumptions of Proposition 2 hold except that (A3) is replaced by the UOSL property of F, then the convergence order for the same method to the (single) solution of (1) *is $\mathscr{O}(\sqrt{h})$ by Donchev and Farkhi (2000, Theorem 4.1).*

Proof The UOSL property enforces the uniqueness of the solution $x(\cdot)$ starting from a given point such that Proposition 5 yields

$$\operatorname{dist}\left(\mathscr{R}_h^{\text{impl}}(T, t_0, \{x_0\}),\ \{x(T)\}\right) \le Ch,$$
$$\operatorname{dist}\left(\{x(T)\},\ \mathscr{R}_h^{\text{impl}}(T, t_0, \{x_0\})\right) = \inf_{\eta^N \in \mathscr{R}_h^{\text{impl}}(T, t_0, \{x_0\})} |x(T) - \eta^N|$$
$$\le \sup_{\eta^N \in \mathscr{R}_h^{\text{impl}}(T, t_0, \{x_0\})} |x(T) - \eta^N| = \operatorname{dist}\left(\mathscr{R}_h^{\text{impl}}(T, t_0, \{x_0\}),\ \{x(T)\}\right) \le Ch. \qquad \square$$

Remark 6 The same $\mathcal{O}(h)$-estimate may also be obtained for the improved Euler (cp. Remark 4) in the case if F is SOSL.

Acknowledgements The authors would like to thank Frank Lempio and Janosch Rieger for the helpful discussions on the explicit and implicit Euler method. The authors also acknowledge the partial support by The Hermann Minkowski Center for Geometry at Tel Aviv University, Israel.

References

J.-P. Aubin, A. Cellina, *Differential Inclusions*. Grundlehren der Mathematischen Wissenschaften, vol. 264 (Springer, Berlin-Heidelberg-New York-Tokyo, 1984)

J.-P. Aubin, H. Frankowska, *Set-Valued Analysis*. Systems & Control: Foundations & Applications, vol. 2 (Birkhäuser, Boston, MA, 1990)

W. Auzinger, R. Frank, F. Macsek, Asymptotic error expansions for stiff equations: the implicit Euler scheme. SIAM J. Numer. Anal. **27**(1), 67–104 (1990)

R. Baier, I.A. Chahma, F. Lempio, Stability and convergence of Euler's method for state-constrained differential inclusions. SIAM J. Optim. **18**(3), 1004–1026 (2007). (electronic). D. Dentcheva, J. Revalski (eds.), special issue on "Variational Analysis and Optimization"

R. Baier, E. Farkhi, Regularity of set-valued maps and their selections through set differences. Part 2: one-sided Lipschitz properties. Serdica Math. J. **39**(3–4), 391–422 (2013). Special issue dedicated to the 65th anniversary of Professor Asen L. Dontchev and to the 60th anniversary of Professor Vladimir M. Veliov

W.-J. Beyn, J. Rieger, The implicit Euler scheme for one-sided Lipschitz differential inclusions. Dyn. Contin. Discrete Impuls. Syst. Ser. B Appl. Algorithms **14**(2), 409–428 (2010)

F.E. Browder, Nonlinear accretive operators in Banach spaces. Bull. Am. Math. Soc. **73**, 470–476 (1967a)

F.E. Browder, Nonlinear mappings of nonexpansive and accretive type in Banach spaces. Bull. Am. Math. Soc. **73**, 875–882 (1967b)

I.A. Chahma, Set-valued discrete approximation of state-constrained differential inclusions. Bayreuth. Math. Schr. **67**, 3–162 (2003)

K. Deimling, *Multivalued Differential Equations*. de Gruyter Series in Nonlinear Analysis and Applications, vol. 1 (de Gruyter, Berlin-New York, 1992)

K. Dekker, J.G. Verwer, *Stability of Runge-Kutta Methods for Stiff Nonlinear Differential Equations*. CWI Monographs, vol. 2 (North-Holland, Amsterdam, 1984)

T.D. Donchev, Functional-differential inclusion with monotone right-hand side. Nonlinear Anal. **16**(6), 533–542 (1991)

T.D. Donchev, Properties of one-sided Lipschitz multivalued maps. Nonlinear Anal. **49**(1), 13–20 (2002)

T.D. Donchev, One sided Lipschitz multifunctions and applications, in *Optimal Control, Stabilization and Nonsmooth Analysis*. Lecture Notes in Control and Information Science (Springer, Berlin, 2004), pp. 333–341

T.D. Donchev, E. Farkhi, Stability and Euler approximation of one-sided Lipschitz differential inclusions. SIAM J. Control Optim. **36**(2), 780–796 (1998) (electronic)

T.D. Donchev, E. Farkhi, Approximations of one-sided Lipschitz differential inclusions with discontinuous right-hand sides, in *Calculus of Variations and Differential Equations (Haifa, 1998)*. Chapman & Hall/CRC Research Notes in Mathematics, vol. 410 (Chapman & Hall/CRC, Boca Raton, FL, 2000), pp. 101–118

T.D. Donchev, E. Farkhi, On the theorem of Filippov-Pliś and some applications. Control and Cybern. **38**(4A), 1251–1271 (2009)

T.D. Donchev, R. Ivanov, On the existence of solutions of differential inclusions in uniformly convex Banach space. Math. Balkanica (N.S.) **6**(1), 13–24 (1992)

E. Farkhi, T.D. Donchev, R. Baier, Existence of solutions for nonconvex differential inclusions of monotone type. C. R. Acad. Bulg. Sci. **67**(3), 323–330 (2014)

A.F. Filippov, Classical solutions of differential equations with multi-valued right-hand side. SIAM J. Control **5**, 609–621 (1967)

H. Frankowska, F. Rampazzo, Filippov's and Filippov-Wazewski's theorems on closed domains. J. Differ. Equ. **161**(2), 449–478 (2000)

G. Grammel, Towards fully discretized differential inclusions. Set-Valued Anal. **11**(1), 1–8 (2003)

E. Hairer, G. Wanner, *Solving Ordinary Differential Equations. II Stiff and Differential-Algebraic Problems*. Springer Series in Computational Mathematics, vol. 14, 2nd edn. (Springer, Berlin, 1996)

J.L. Haunschmied, A. Pietrus, V.M. Veliov, The Euler method for linear control systems revisited, in *Large-Scale Scientific Computing. Revised Selected Papers from the 9th International Conference (LSSC '13), Sozopol, June 3–7, 2013*, ed. by I. Lirkov, S.D. Margenov, J. Waśniewski. Lecture Notes in Computer Science, vol. 8353 (Springer, Heidelberg, 2014), pp. 90–97

A. Kastner-Maresch, Diskretisierungsverfahren zur Lösung von Differentialinklusionen [Discretization Methods for the Solution of Differential Inclusions]. Ph.D. thesis, Department of Mathematics, University of Bayreuth, Bayreuth, (1990a)

A. Kastner-Maresch, Implicit Runge-Kutta methods for differential inclusions. Numer. Funct. Anal. Optim. **11**(9–10), 937–958 (1990b)

A. Kastner-Maresch, The implicit midpoint rule applied to discontinuous differential equations. Computing **49**(1), 45–62 (1992)

A. Kastner-Maresch, F. Lempio, Difference methods with selection strategies for differential inclusions. Numer. Funct. Anal. Optim. **14**(5–6), 555–572 (1993)

F. Lempio, Difference methods for differential inclusions, in *Modern Methods of Optimization.* Proceedings of a Summer School at the Schloß Thurnau of the University of Bayreuth (Germany), FRG, October 1–6, 1990. Lecture Notes in Economics and Mathematical Systems, vol. 378 (Springer, Berlin-Heidelberg-New York, 1992), pp. 236–273

F. Lempio, Modified Euler methods for differential inclusions, in *Set-Valued Analysis and Differential Inclusions. A Collection of Papers resulting from a Workshop held in Pamporovo, September 17–21, 1990*. Progress in Systems and Control Theory, vol. 16 (Birkhäuser, Boston, MA-Basel-Berlin, 1993), pp. 131–148

F. Lempio, Euler's method revisited. Proc. Steklov Inst. Math. **211**, 429–449 (1995)

F. Lempio, D.B. Silin, Differential inclusions with strongly one-sided-Lipschitz right-hand sides. Differ. Equ. **32**(11), 1485–1491 (1997)

F. Lempio, V.M. Veliov, Discrete approximations of differential inclusions. Bayreuth. Math. Schr. **54**, 149–232 (1998)

G. Lumer, R.S. Phillips, Dissipative operators in a Banach space. Pac. J. Math. **11**, 679–698 (1961)

R. Mannshardt, One-step methods of any order for ordinary differential equations with discontinuous right-hand sides. Numer. Math. **31**(2), 131–152 (1978/1979)

J.T. Marti, *Konvexe Analysis*. Lehrbücher und Monographien aus dem Gebiet der Exakten Wissenschaften, Mathematische Reihe, vol. 54 (Birkhäuser, Basel-Stuttgart, 1977)

R.H. Martin Jr., A global existence theorem for autonomous differential equations in a Banach space. Proc. Am. Math. Soc. **26**, 307–314 (1970)

R. Model, Zur Integration über Unstetigkeiten in gewöhnlichen Differentialgleichungen. Z. Angew. Math. Mech. **68**(3), 161–169 (1988)

A. Pietrus, V.M. Veliov, On the discretization of switched linear systems. Syst. Control Lett. **58**(6), 395–399 (2009)

V.M. Veliov, Second order discrete approximations to strongly convex differential inclusions. Syst. Control Lett. **13**(3), 263–269 (1989)

V.M. Veliov, Second order discrete approximation to linear differential inclusions. SIAM J. Numer. Anal. **29**(2), 439–451 (1992)

V.M. Veliov, On the time-discretization of control systems. SIAM J. Control Optim. **35**(5), 1470–1486 (1997)

V.M. Veliov, Error analysis of discrete approximations to bang-bang optimal control problems: the linear case. Control Cybern. **34**(3), 967–982 (2005)

V.M. Veliov, On the relationship between continuous- and discrete-time control systems. Cent. Eur. J. Oper. Res. **18**(4), 511–523 (2010)

V.M. Veliov, Relaxation of Euler-type discrete-time control system, in *Large-Scale Scientific Computing. Revised Selected Papers from the 10th International Conference (LSSC '15), Sozopol, June 8–12, 2015*, ed. by I. Lirkov, S. D. Margenov, and J. Waśniewski. Lecture Notes in Computer Science, vol. 9374 (Springer, Cham, 2015), pp. 134–141

Solution Stability and Path-Following for a Class of Generalized Equations

Radek Cibulka and Tomáš Roubal

Abstract We study strong metric (sub)regularity of a special non-monotone generalized equation with either smooth or locally Lipschitz single-valued part. The existence of a Lipschitz selection of a solution mapping associated with a parametric generalized equation is proved. An inexact Euler-Newton continuation method for tracking a solution trajectory is introduced and demonstrated to have an accuracy of order $O(h^4)$. The theoretical results are applied in the study of non-regular electrical circuits involving devices like diodes and transistors.

1 Introduction

Given matrices $B \in \mathbb{R}^{n \times m}$, $C \in \mathbb{R}^{m \times n}$ with $m \leq n$, a vector $p \in \mathbb{R}^n$, a single-valued mapping $f : \mathbb{R}^n \to \mathbb{R}^n$, and a set-valued mapping $F : \mathbb{R}^m \rightrightarrows \mathbb{R}^m$, we consider the problem of finding a solution $z \in \mathbb{R}^n$ to the generalized equation

$$p \in f(z) + BF(Cz). \tag{1}$$

In Adly and Outrata (2013), the authors considered a special case of the above inclusion with the linear single-valued part f, with $B := C^T$, and F being the

R. Cibulka (✉)
NTIS - New Technologies for the Information Society and Department of Mathematics, Faculty of Applied Sciences, University of West Bohemia, Pilsen, Czech Republic
e-mail: cibi@kma.zcu.cz

T. Roubal
Department of Mathematics, Faculty of Applied Sciences, University of West Bohemia, Pilsen, Czech Republic
e-mail: roubalt@students.zcu.cz

G. Feichtinger et al. (eds.), *Control Systems and Mathematical Methods in Economics*, Lecture Notes in Economics and Mathematical Systems 687,
https://doi.org/10.1007/978-3-319-75169-6_4

Clarke's subdifferential Clarke (1983) of the super-potential j defined by

$$j(x) := j_1(x_1) + j_2(x_2) + \cdots + j_m(x_m) \quad \text{whenever} \quad x = (x_1, \ldots, x_m)^T \in \mathbb{R}^m, \tag{2}$$

where $j_i : \mathbb{R} \to \mathbb{R}$ is a locally Lipschitz continuous function for each $i \in \{1, \ldots, m\}$. They investigated two important stability properties called Aubin (pseudo Lipschitz or Lipschitz-like) property and the isolated calmness of the solution mapping corresponding to (1). A generalization to the present setting with a smooth single-valued part can be found in Adly and Cibulka (2014), where also the calmness of the solution mapping is considered. In the second section, we derive conditions guaranteeing the strong metric (sub)regularity of the mapping $\Phi = f + BF(C\cdot)$ when f is either smooth or non-smooth but locally Lipschitz continuous.

In the third and fourth section, we study the case when the parameter p in (1) depends on time. More precisely, given a constant $\varepsilon > 0$ and a function $p : [0, \varepsilon] \to \mathbb{R}^n$ we want to find a function $z : [0, \varepsilon] \to \mathbb{R}^n$ such that

$$p(t) \in f(z(t)) + BF(Cz(t)) \quad \text{for all} \quad t \in [0, \varepsilon]. \tag{3}$$

Based on a generalization of Dontchev et al. (2013, Theorem 2.4) we prove the existence of a Lipschitz continuous response $z(\cdot)$ to the Lipschitz continuous input signal $p(\cdot)$.

The last section is devoted to numerical simulations and applications in electronics. In Adly et al. (2013), simulations are performed by using Xcos (a component of Scilab). In order to use this software package, the set-valued part F is approximated by a single-valued one. In Dontchev et al. (2013), the authors considered an Euler-Newton continuation method for tracking solution trajectories of parametric variational inequalities. We derive an inexact method for tracking solution trajectories of parametric generalized equations under point-wise [strong] metric regularity of the "partial linearizations" of Φ at every point of the solution trajectory and under slightly weaker assumptions on differentiability of the corresponding single-valued part f. Finally, implementing this method (in Matlab) we provide a simulation of the behavior of some basic non-regular circuits, that is, the circuits where various types of diodes are present. Unlike Adly et al. (2013), we work directly with a set-valued model here.

1.1 Notation

In $\mathbb{R}^d$, the norm, the scalar product, the closed and the open ball with a center $x \in \mathbb{R}^d$ and a radius $r \geq 0$, are denoted by $\|\cdot\|$, $\langle \cdot, \cdot \rangle$, $\mathbb{B}[x, r]$, and $\mathbb{B}(x, r)$, respectively. We set $\mathbb{B} = \mathbb{B}[0, 1]$. Fix a non-empty subset Ω of $\mathbb{R}^d$ containing a point $\bar{x}$; the

Fréchet/regular normal cone to Ω at $\bar{x}$ is the set

$$\widehat{N}(\bar{x}; \Omega) := \left\{ \xi \in \mathbb{R}^d : \limsup_{x \to \bar{x}, x \in \Omega \setminus \{\bar{x}\}} \frac{\langle \xi, x - \bar{x} \rangle}{\|x - \bar{x}\|} \leq 0 \right\};$$

the *general/limiting normal cone* $N(\bar{x}; \Omega)$ to Ω at $\bar{x}$ contains all $\xi \in \mathbb{R}^d$ for which there are sequences $(x^k)_{k \in \mathbb{N}}$ in Ω and $(\xi^k)_{k \in \mathbb{N}}$ in $\mathbb{R}^d$ converging to $\bar{x}$ and ξ, respectively, such that $\xi^k \in \widehat{N}(x^k; \Omega)$ for each $k \in \mathbb{N}$; the *Bouligand-Severi tangent cone* $T(\bar{x}; \Omega)$ to Ω at $\bar{x}$ contains those $v \in \mathbb{R}^d$ for which there are sequences $(t^k)_{k \in \mathbb{N}}$ in $(0, \infty)$ and $(v^k)_{k \in \mathbb{N}}$ in $\mathbb{R}^d$ converging to 0 and v, respectively, such that $\bar{x} + t^k v^k \in \Omega$ for each $k \in \mathbb{N}$; and finally the *Bouligand paratingent cone* $\widetilde{T}(\bar{x}; \Omega)$ to Ω at $\bar{x}$ contains those $v \in \mathbb{R}^d$ for which there are sequences $(t^k)_{k \in \mathbb{N}}$ in $(0, \infty)$, $(v^k)_{k \in \mathbb{N}}$ in $\mathbb{R}^d$, and $(x^k)_{k \in \mathbb{N}}$ in Ω converging to 0, v, and $\bar{x}$, respectively, such that $x^k + t^k v^k \in \Omega$ for each $k \in \mathbb{N}$. The *distance* from a point $x \in \mathbb{R}^d$ to Ω is denoted by $d(x, \Omega)$ with convention that $d(x, \emptyset) = \infty$. Throughout $s : \mathbb{R}^d \to \mathbb{R}^l$ means that s is single-valued while $S : \mathbb{R}^d \rightrightarrows \mathbb{R}^l$ denotes a general mapping which may be set-valued. For such a mapping S, the domain, the graph, and the range are denoted by $\operatorname{dom} S$, $\operatorname{gph} S$ and $\operatorname{rge} S$. Fix a point $(\bar{x}, \bar{y}) \in \operatorname{gph} S$. A *selection for S around $\bar{x}$ for $\bar{y}$* is any single-valued mapping s defined on a neighborhood U of $\bar{x}$ such that $s(\bar{x}) = \bar{y}$ and $s(x) \in S(x)$ for each $x \in U$; a *(graphical) localization of S around $\bar{x}$ for $\bar{y}$* is any set-valued mapping $\widetilde{S}$ such that for some neighborhoods U of $\bar{x}$ and V of $\bar{y}$ we have $\operatorname{gph} \widetilde{S} = \operatorname{gph} S \cap (U \times V)$ and $\operatorname{dom} \widetilde{S} = U$. A mapping $\Phi : \mathbb{R}^l \rightrightarrows \mathbb{R}^d$ is *strongly (metrically) regular* at $\bar{y}$ for $\bar{x}$ provided that $S := \Phi^{-1}$ has a Lipschitz continuous single-valued localization around $\bar{x}$ for $\bar{y}$; the mapping Φ is called *(metrically) regular at $\bar{y}$ for $\bar{x}$* if there is a constant $\kappa > 0$ along with neighborhoods V of $\bar{y}$ and U of $\bar{x}$ such that

$$d\big(y, \Phi^{-1}(x)\big) \leq \kappa\, d(x, \Phi(y)) \quad \text{for each} \quad (y, x) \in V \times U;$$

and Φ is *strongly (metrically) subregular at $\bar{y}$ for $\bar{x}$* provided that there is a constant $\kappa > 0$ along with a neighborhood V of $\bar{y}$ such that

$$\|y - \bar{y}\| \leq \kappa\, d(\bar{x}, \Phi(y)) \quad \text{for each} \quad y \in V.$$

As a rule we omit the part "metrically" in the whole text. We also use global versions of [strong] regularity. Given a constant $\kappa > 0$ and sets $V \subset \mathbb{R}^l$ and $U \subset \mathbb{R}^d$, the mapping Φ is *regular on V for U with the constant κ* provided that

$$d\big(y, \Phi^{-1}(x)\big) \leq \kappa\, d(x, \Phi(y) \cap U) \quad \text{for each} \quad (y, x) \in V \times U;$$

and Φ is said to be *strongly regular on V for U with the constant κ* if the mapping $U \ni x \longmapsto \Phi^{-1}(x) \cap V$ is single-valued and Lipschitz continuous on U with the constant κ. The *paratingent/strict graphical derivative* of Φ at $(\bar{y}, \bar{x}) \in \operatorname{gph} \Phi$ is the mapping $\widetilde{D}\Phi(\bar{y},\bar{x}) : \mathbb{R}^l \rightrightarrows \mathbb{R}^d$ defined by

$$\widetilde{D}\Phi(\bar{y},\bar{x})(v) := \{u \in \mathbb{R}^d : (v,u) \in \widetilde{T}((\bar{y},\bar{x}); \operatorname{gph}\Phi)\}, \quad v \in \mathbb{R}^l.$$

Consider a locally Lipschitz continuous function $h : \mathbb{R}^l \to \mathbb{R}^d$ and a point $\bar{u} \in \mathbb{R}^l$. The *Bouligand's limiting Jacobian of h at $\bar{u}$* is the (non-empty compact) set $\partial_B h(\bar{u})$ consisting of all matrices $A \in \mathbb{R}^{d\times l}$ for which there is a sequence $(u^k)_{k\in\mathbb{N}}$ converging to $\bar{u}$ such that h is differentiable at each u^k and $\nabla h(u^k) \to A$ as $k \to \infty$. The *Clarke's generalized Jacobian of h at $\bar{u}$*, denoted by $\partial h(\bar{u})$, is the convex hull of $\partial_B h(\bar{u})$.

1.2 Standing Assumptions

Consider the generalized equation (1). Assume that f is continuous, F has a closed graph, B is injective, and C is surjective. Denote by Φ and Q the set-valued mappings from $\mathbb{R}^n$ into itself defined by $\Phi(z) := f(z) + BF(Cz)$ and $Q(z) := BF(Cz)$ for all $z \in \mathbb{R}^n$, respectively. We also suppose that we have in hand a point $(\bar{z},\bar{p}) \in \operatorname{gph}\Phi$. Finally, put $\bar{v} := (B^TB)^{-1}B^T(\bar{p} - f(\bar{z}))$.

2 Regularity Properties of Φ at $\bar{z}$ for $\bar{p}$

We start with a geometric lemma which is an analogue of Adly and Cibulka (2014, Lemma 4.1), where the classical Bouligand-Severi tangent cone was considered.

Lemma 1 *Let $E \in \mathbb{R}^{r\times d}$ be any matrix, let $G \in \mathbb{R}^{l\times d}$ be injective, and let Γ be a subset of* rge E. *Put $\Xi := E^{-1}(\Gamma)$ and $\Lambda := G(\Xi)$. For $\bar{x} \in \Lambda$ denote by $\bar{y}$ the (unique) point in Ξ with $G\bar{y} = \bar{x}$. Then*

$$\widetilde{T}(\bar{x};\Lambda) = \{u \in \mathbb{R}^l : \exists w \in \mathbb{R}^d \text{ such that } u = Gw \text{ and } Ew \in \widetilde{T}(E\bar{y};\Gamma)\}.$$

Proof We claim that $\widetilde{T}(\bar{y};\Xi) = \{w \in \mathbb{R}^d : Ew \in \widetilde{T}(E\bar{y},\Gamma)\}$. First, take any $w \in \widetilde{T}(\bar{y};\Xi)$. Find sequences (t^k) in $(0,\infty)$, (y^k) in Ξ, and (w^k) in $\mathbb{R}^d$ converging to 0, $\bar{y}$ and w, respectively, such that $y^k + t^kw^k \in \Xi$ for all $k \in \mathbb{N}$. Therefore $Ey^k + t^kEw^k = E(y^k + t^kw^k) \in \Gamma$ for each $k \in \mathbb{N}$. Hence $Ew \in \widetilde{T}(E\bar{y};\Gamma)$. On the other hand, let $w \in \mathbb{R}^d$ be such that $Ew \in \widetilde{T}(E\bar{y},\Gamma)$. Pick sequences (t^k) in $(0,\infty)$, (u^k) in Γ, and (v^k) in $\mathbb{R}^r$ converging to 0, $E\bar{y}$, and Ew, respectively, such that $u^k + t^kv^k \in \Gamma$ for each $k \in \mathbb{N}$. As $\Gamma \subset \operatorname{rge} E$, where the latter set is a closed subspace of $\mathbb{R}^r$, we get that $v^k \in \operatorname{rge} E$ for each $k \in \mathbb{N}$. By Banach open mapping theorem, there are sequences (y^k) converging to $\bar{y}$ and (w^k) converging to w, both in $\mathbb{R}^d$, such that

$$Ey^k = u^k \quad \text{and} \quad Ew^k = v^k \quad \text{for each} \quad k \in \mathbb{N}.$$

Thus, for an arbitrary index k, we have $Ey^k \in \Gamma$ and $E(y^k + t^k w^k) \in \Gamma$, hence both y^k and $y^k + t^k w^k$ are in $E^{-1}(\Gamma) = \Xi$. So $w \in \widetilde{T}(\bar{y}; \Xi)$. The claim is proved.

Second, we show that $\widetilde{T}(\bar{x}; \Lambda) = G\big(\widetilde{T}(\bar{y}; \Xi)\big)$. Pick any $w \in G\big(\widetilde{T}(\bar{y}; \Xi)\big)$. Find $v \in \widetilde{T}(\bar{y}; \Xi)$ with $Gv = w$. Thus there are sequences (t^k) in $(0, \infty)$ converging to 0, (y^k) in Ξ converging to $\bar{y}$, and (v^k) in $\mathbb{R}^d$ converging to v such that $y^k + t^k v^k \in \Xi$ for all $k \in \mathbb{N}$. For each $k \in \mathbb{N}$, let $u^k := Gy^k$ and $w^k := Gv^k$. Clearly, (u^k) converges to $G\bar{y} = \bar{x}$ and (w^k) converges to w. Moreover,

$$u^k + t^k w^k = G(y^k + t^k v^k) \in G(\Xi) = \Lambda \quad \text{for all} \quad k \in \mathbb{N}.$$

So $w \in \widetilde{T}(\bar{x}; \Lambda)$. Thus $G\big(\widetilde{T}(\bar{y}; \Xi)\big) \subset \widetilde{T}(\bar{x}; \Lambda)$. To see the opposite inclusion, pick any $w \in \widetilde{T}(\bar{x}; \Lambda)$. As $\Lambda = G(\Xi)$, we find sequences (t^k) in $(0, \infty)$ converging to 0, (x^k) in $G(\Xi)$ converging to $\bar{x}$, and (w^k) in $\mathbb{R}^l$ converging to w such that

$$x^k + t^k w^k \in G(\Xi) \quad \text{for all} \quad k \in \mathbb{N}.$$

For each $k \in \mathbb{N}$, find $y^k \in \Xi$ such that $x^k = Gy^k$. Then (y^k) is bounded. Indeed, if this is not the case, find a cluster point $\bar{h}$ of $(y^k/\|y^k\|)$. Let N be an infinite subset of $\mathbb{N}$ such that $\lim_{N \ni k \to \infty} y^k/\|y^k\| = \bar{h}$. Then

$$0 = \lim_{N \ni k \to \infty} \frac{x^k}{\|y^k\|} = \lim_{N \ni k \to \infty} G\left(\frac{y^k}{\|y^k\|}\right) = G\bar{h}.$$

This contradicts the injectivity of G because $\|\bar{h}\| = 1$. Therefore there is an infinite subset N of $\mathbb{N}$ such that $(y^k)_{k \in N}$ converges, to $\tilde{y} \in \mathbb{R}^d$ say. Then

$$G\bar{y} = \bar{x} = \lim_{N \ni k \to \infty} x^k = \lim_{N \ni k \to \infty} Gy^k = G\tilde{y}.$$

Employing, the injectivity once more, we get $\bar{y} = \tilde{y}$. For each $k \in N$, find v^k in Ξ such that $w^k = G\big((v^k - y^k)/t^k\big)$, and let $u^k := (v^k - y^k)/t^k$. Similar argument as in the case of (y^k) shows that $(u^k)_{k \in N}$ is bounded. Therefore there is an infinite subset N' of N such that $(u^k)_{k \in N'}$ converges to some $u \in \mathbb{R}^d$. For each $k \in N'$, we have $y^k + t^k u^k = v^k \in \Xi$, therefore $u \in \widetilde{T}(\bar{y}; \Xi)$. Moreover, $w = G(u)$. Thus $G\big(\widetilde{T}(\bar{y}; \Xi)\big) \supset \widetilde{T}(\bar{x}; \Lambda)$. We showed that $G\big(\widetilde{T}(\bar{y}; \Xi)\big) = \widetilde{T}(\bar{x}; \Lambda)$, which in combination with the claim yields the assertion. ■

We also need a slight modification of the condition by B. Kummer guaranteeing the strong regularity of a set-valued mapping (Dontchev and Rockafellar 2014, Theorem 4D.1).

Proposition 1 *Consider a set-valued mapping $H : \mathbb{R}^n \rightrightarrows \mathbb{R}^n$ and a point $(\bar{x}, \bar{y}) \in$ gph H. Then H is strongly regular at $\bar{x}$ for $\bar{y}$ if and only if it satisfies the following three conditions:*

(a) for each neighborhood U of $\bar{x}$ there is a neighborhood V of $\bar{y}$ such that $H^{-1}(y) \cap U \neq \emptyset$ whenever $y \in V$;

(b) *the set* $\operatorname{gph} H \cap \big(\mathbb{B}[\bar{x}, r] \times \mathbb{B}[\bar{y}, r]\big)$ *is closed for some* $r > 0$;
(c) $0 \in \widetilde{D}H(\bar{x}, \bar{y})(u) \implies u = 0.$

Proof Suppose that H is strongly regular at $\bar{x}$ for $\bar{y}$. Then (c) holds by Dontchev and Rockafellar (2014, Theorem 4D.1). Observe also that H has necessarily locally closed graph at the reference point (cf., Dontchev and Rockafellar 2014, Proposition 3G.6). Finally, (a) is satisfied since H is open at $(\bar{x}, \bar{y})$, that is, for any neighborhood U of $\bar{x}$ the set $V := H(U)$ is a neighborhood of $\bar{y}$. The converse implication follows from Dontchev and Rockafellar (2014, Theorem 4D.1) provided that conditions (1) and (2) therein hold. The first one follows from (c). Suppose that the latter condition fails. Then there is a sequence $(y^k)_{k\in\mathbb{N}}$ converging to $\bar{y}$ along with an open neighborhood U of $\bar{x}$ such that for each $l \in \mathbb{N}$ there is $k > l$ such that $H^{-1}(y^k) \cap U = \emptyset$. This contradicts (a). ∎

Trivial examples show that the premise (a) cannot be omitted. Indeed, define $h : \mathbb{R} \to \mathbb{R}$ by $h(x) = x$, $x \geq 0$. Then both (b) and (c) are valid, (a) fails and h is not strongly regular at 0 for 0.

Now, we are in position to formulate the main statement of this section.

Theorem 1 *In addition to the standing assumptions, suppose that f is continuously differentiable on $\mathbb{R}^n$. Then*

(i) *Φ is regular at $\bar{z}$ for $\bar{p}$ if and only if*

$$\left.\begin{array}{c}\Big(\big(CC^T\big)^{-1}C\nabla f(\bar{z})^T\xi,\, B^T\xi\Big) \in -N\big((C\bar{z}, \bar{v}); \operatorname{gph} F\big)\\ \nabla f(\bar{z})^T\xi \in \operatorname{rge} C^T\end{array}\right\} \implies \xi = 0;$$

(ii) *Φ is strongly subregular at $\bar{z}$ for $\bar{p}$ if and only if*

$$\left.\begin{array}{c}\Big(Cb, -\big(B^TB\big)^{-1}B^T\nabla f(\bar{z})b\Big) \in T\big((C\bar{z}, \bar{v}); \operatorname{gph} F\big)\\ \nabla f(\bar{z})b \in \operatorname{rge} B\end{array}\right\} \implies b = 0;$$

(iii) *Φ is strongly regular at $\bar{z}$ for $\bar{p}$ if and only if*

(a) *for each neighborhood U of $\bar{z}$ there is a neighborhood V of $\bar{p}$ such that $\Phi^{-1}(p) \cap U \neq \emptyset$ whenever $p \in V$; and*

(b)

$$\left.\begin{array}{c}\Big(Cb, -\big(B^TB\big)^{-1}B^T\nabla f(\bar{z})b\Big) \in \widetilde{T}\big((C\bar{z}, \bar{v}); \operatorname{gph} F\big)\\ \nabla f(\bar{z})b \in \operatorname{rge} B\end{array}\right\} \implies b = 0.$$

Proof The statement (i) is Adly and Cibulka (2014, Corollary 3.1), whereas (ii) is Adly and Cibulka (2014, Corollary 4.1).

(iii) First, we show that

$$\widetilde{D}\Phi(\bar{z}, \bar{p})(b) = \nabla f(\bar{z})b + B\,\widetilde{D}F(C\bar{z}, \bar{v})(Cb) \quad \text{for all} \quad b \in \mathbb{R}^n. \tag{4}$$

As in Dontchev and Rockafellar (2014, Proposition 4A.2), it is elementary to show that

$$\widetilde{D}\Phi(\bar{z}, \bar{p})(b) = \nabla f(\bar{z})b + \widetilde{D}Q(\bar{z}, \bar{p} - f(\bar{z}))(b) \quad \text{for each} \quad b \in \mathbb{R}^n.$$

Further, observe that

$$\operatorname{gph} Q = \left\{ \begin{pmatrix} u \\ v \end{pmatrix} \in \mathbb{R}^{2n} : \exists \begin{pmatrix} b \\ c \end{pmatrix} \in \mathbb{R}^{n+m} : \begin{pmatrix} u \\ v \end{pmatrix} = G \begin{pmatrix} b \\ c \end{pmatrix} \text{ and } E \begin{pmatrix} b \\ c \end{pmatrix} \in \operatorname{gph} F \right\},$$

with

$$G := \begin{pmatrix} I_n & 0 \\ 0 & B \end{pmatrix} \quad \text{and} \quad E := \begin{pmatrix} C & 0 \\ 0 & I_m \end{pmatrix}.$$

As B is injective, so is G. Lemma 1 (with $r := 2m$, $l := 2n$, $d := n + m$, $\Gamma :=$ gph F, $\bar{x} := (\bar{z}, \bar{p} - f(\bar{z}))^T$, and $\bar{y} := (\bar{z}, \bar{v})^T$) reveals that

$$\widetilde{T}\big((\bar{z}, \bar{p} - f(\bar{z})); \operatorname{gph} Q\big) = \left\{ \begin{pmatrix} b \\ Bc \end{pmatrix} : \begin{pmatrix} Cb \\ c \end{pmatrix} \in \widetilde{T}\big((C\bar{z}, \bar{v}); \operatorname{gph} F\big) \right\}.$$

Thus $\widetilde{D}Q(\bar{z}, \bar{p} - f(\bar{z}))(b) = B\, \widetilde{D}F(C\bar{z}, \bar{v})(Cb)$ for all $b \in \mathbb{R}^n$, which proves (4).

As B is injective, the matrix $B^T B \in \mathbb{R}^{m \times m}$ is non-singular. The graph of Φ is closed. It suffices to show that (b) is equivalent to Proposition 1(c) with $H := \Phi$. First, let $b \in \mathbb{R}^n$ be such that $0 \in \nabla f(\bar{z})b + B\widetilde{D}F(C\bar{z}, \bar{v})(Cb)$. Find a point $w \in \widetilde{D}F(C\bar{z}, \bar{v})(Cb)$ with $\nabla f(\bar{z})b + Bw = 0$. Thus $-(B^T B)^{-1} B^T \nabla f(\bar{z})b$ is in $\widetilde{D}F(C\bar{z}, \bar{v})(Cb)$. Clearly, we have $\nabla f(\bar{z})b \in \operatorname{rge} B$. On the other hand, pick any $b \in \mathbb{R}^n$ with $\big(Cb, -(B^T B)^{-1} B^T \nabla f(\bar{z})b\big)$ in $\widetilde{T}\big((C\bar{z}, \bar{v}); \operatorname{gph} F\big)$ and $\nabla f(\bar{z})b \in \operatorname{rge} B$. Then $w := -(B^T B)^{-1} B^T \nabla f(\bar{z})b \in \widetilde{D}F(C\bar{z}, \bar{v})(Cb)$. Thus $B^T Bw = -B^T \nabla f(\bar{z})b$. So $Bw + \nabla f(\bar{z})b \in \ker B^T \cap \operatorname{rge} B = \{0\}$. Then $0 \in \nabla f(\bar{z})b + B\widetilde{D}F(C\bar{z}, \bar{v})(Cb)$. ■

We derive a simple sufficient condition guaranteeing the strong regularity of Φ which will be applied in Sect. 4. This statement provides a stronger conclusion than Adly et al. (2013, Corollary 1) where regularity of Φ is proved. Recall that a matrix $M \in \mathbb{R}^{n \times n}$ is called a *P-matrix* provided that all its k-by-k principal minors are positive whenever $k \in \{1, \ldots, n\}$. It is well known, that M is a P-matrix if and only if for any non-zero $x \in \mathbb{R}^n$ there is $j \in \{1, \ldots, n\}$ such that $x_j(Mx)_j > 0$. A mapping $S : \mathbb{R}^n \rightrightarrows \mathbb{R}^n$ is *monotone* provided that $\langle \hat{y} - \tilde{y}, \hat{x} - \tilde{x} \rangle \geq 0$ whenever $(\hat{y}, \hat{x}), (\tilde{y}, \tilde{x}) \in \operatorname{gph} S$. A monotone mapping $S : \mathbb{R}^n \rightrightarrows \mathbb{R}^n$ is called *maximal*, if there does not exist other monotone mapping whose graph strictly contains the graph of S.

Corollary 1 *In addition to the standing assumptions, suppose that $n = m$, $B = C = I_n$, $\nabla f(\bar{z})$ is a P-matrix, and there are maximal monotone mappings $F_i : \mathbb{R} \rightrightarrows \mathbb{R}$, $i \in \{1, \dots, n\}$ such that $F(x) = \prod_{i=1}^{n} F_i(x_i)$ whenever $x = (x_1, \dots, x_n)^T \in \mathbb{R}^n$. Then Φ is strongly regular at $\bar{z}$ for $\bar{p}$.*

Proof Note that $\prod_{i=1}^n \operatorname{gph} F_i = \varphi(\operatorname{gph} F)$, where

$$\varphi(x, y) = \big((x_1, y_1), \dots, (x_n, y_n)\big), \quad x = (x_1, \dots, x_n)^T, y = (y_1, \dots, y_n)^T \in \mathbb{R}^n.$$

Clearly, φ is linear and one-to-one. The definition of the paratingent cone and Lemma 1 imply that

$$\prod_{i=1}^{n} \widetilde{T}\big((\bar{z}_i, \bar{v}_i); \operatorname{gph} F_i\big) \supset \widetilde{T}\Big(\varphi(\bar{z}, \bar{v}); \prod_{i=1}^{n} \operatorname{gph} F_i\Big) = \varphi\Big(\widetilde{T}\big((\bar{z}, \bar{v}); \operatorname{gph} F\big)\Big).$$

Also, it is well-known that

$$\prod_{i=1}^{n} N\big((\bar{z}_i, \bar{v}_i); \operatorname{gph} F_i\big) = N\Big(\varphi(\bar{z}, \bar{v}); \prod_{i=1}^{n} \operatorname{gph} F_i\Big) = \varphi\Big(N\big((\bar{z}, \bar{v}); \operatorname{gph} F\big)\Big).$$

As all F_i's are maximal monotone, we have $N\big((\bar{z}_i, \bar{v}_i); \operatorname{gph} F_i\big) \subset \{(a, b) \in \mathbb{R}^2 : ab \le 0\}$ and $\widetilde{T}\big((\bar{z}_i, \bar{v}_i); \operatorname{gph} F_i\big) \subset \{(a, b) \in \mathbb{R}^2 : ab \ge 0\}$ for each $i \in \{1, \dots, n\}$. Fix any non-zero $\eta \in \mathbb{R}^n$. Since $\nabla f(\bar{z})$ is a P-matrix, so is $\nabla f(\bar{z})^T$. There are $k, l \in \{1, \dots, n\}$ such that $\eta_k(\nabla f(\bar{z})\eta)_k > 0$ and $\eta_l(\nabla f(\bar{z})^T \eta)_l > 0$, which means that $\big(\eta_k, -(\nabla f(\bar{z})\eta)_k\big) \notin \widetilde{T}\big((\bar{z}_k, \bar{v}_k); \operatorname{gph} F_k\big)$ and $\big((\nabla f(\bar{z})^T \eta)_l, \eta_l\big) \notin -N\big((\bar{z}_l, \bar{v}_l); \operatorname{gph} F_l\big)$. The above relations for the normal and parantingent cone and the fact that φ is one-to-one imply that both the conditions in Theorem 1(iii) hold (the first one thanks to the statement (i) of this theorem). ■

Remark 1 Goeleven (2008) considered the case when $f(z) := Az$, $z \in \mathbb{R}^n$, with a given P-matrix $A \in \mathbb{R}^{n \times n}$, and F is the Fenchel-Moreau-Rockafellar subdifferential of the super-potential j defined by (2) with $m := n$ and $j_i : \mathbb{R} \to \mathbb{R} \cup \{\infty\}$, $i \in \{1, \dots, n\}$, being a proper, lower semi-continuous convex function such that

$$\lambda j_i(x) = j_i(\lambda x) \quad \text{for each} \quad \lambda \ge 0 \quad \text{and each} \quad x \in \operatorname{dom} j_i.$$

Then, for each $i \in \{1, \dots, n\}$, the mapping $F_i := \partial j_i$ is "piecewise constant", that is, the value $F_i(x)$ equals either

$$\begin{cases} \{\alpha\}, & x < 0, \\ \{\beta\}, & x > 0, \\ [\alpha, \beta], & x = 0; \end{cases} \quad \text{or} \quad \begin{cases} \{\alpha\}, & x < 0, \\ [\alpha, \infty), & x = 0; \end{cases} \quad \text{or} \quad \begin{cases} \{\beta\}, & x > 0, \\ (-\infty, \beta], & x = 0, \end{cases}$$

with $\alpha := -j_i(-1) \leq j_i(1) =: \beta$ provided that the corresponding value is finite. In this case, Goeleven (2008, Theorem 2.1) says that Φ^{-1} is single-valued with the whole of $\mathbb{R}^n$ as its domain. By Corollary 1, we get a generalization of Goeleven (2008, Proposition 2.1) proving that the solution mapping is not only continuous but locally Lipschitz (and therefore Lipschitz on any compact set).

At the end of this section, we state sufficient conditions for strong (sub)regularity when the single-valued part f in (1) is locally Lipschitz continuous only.

Theorem 2 *In addition to the standing assumptions, suppose that f is locally Lipschitz continuous on $\mathbb{R}^n$. For any $A \in \partial f(\bar{z})$, define the mapping $J_A : \mathbb{R}^n \rightrightarrows \mathbb{R}^n$ by $J_A(z) := f(\bar{z}) + A(z - \bar{z}) + Q(z)$, $z \in \mathbb{R}^n$. Then*

(i) Φ is strongly subregular at $\bar{z}$ for $\bar{p}$ if for each $A \in \partial f(\bar{z})$, we have

$$\left.\begin{array}{c}\left(Cb, -\left(B^T B\right)^{-1} B^T Ab\right) \in T\left((C\bar{z}, \bar{v}); \operatorname{gph} F\right) \\ Ab \in \operatorname{rge} B\end{array}\right\} \Longrightarrow b = 0;$$

(ii) Φ is strongly regular at $\bar{z}$ for $\bar{p}$, if for each $A \in \partial f(\bar{z})$, we have

(a) for each neighborhood U of $\bar{z}$ there is a neighborhood V of $\bar{p}$ such that $J_A^{-1}(p) \cap U \neq \emptyset$ whenever $p \in V$;

(b)

$$\left.\begin{array}{c}\left(Cb, -\left(B^T B\right)^{-1} B^T Ab\right) \in \widetilde{T}\left((C\bar{z}, \bar{v}); \operatorname{gph} F\right) \\ Ab \in \operatorname{rge} B\end{array}\right\} \Longrightarrow b = 0.$$

Proof

(i) For each $A \in \partial f(\bar{z})$, the mapping J_A is strongly subregular at $\bar{z}$ for $\bar{p}$ by Theorem 1(ii) with $\Phi := J_A$. Apply Cibulka et al. (2016, Theorem 3.1) to get the conclusion.
(ii) Conditions (a) and (b) guarantee that, for each $A \in \partial f(\bar{z})$, the mapping J_A is strongly regular at $\bar{z}$ for $\bar{p}$. Izmailov's theorem (Izmailov 2014) implies the statement. ■

3 Existence of a Lipschitz Continuous Selection

Given $h : \mathbb{R}^d \times \mathbb{R}^n \to \mathbb{R}^n$ and $G : \mathbb{R}^n \rightrightarrows \mathbb{R}^n$, consider the parametric generalized equation

$$\text{For} \quad p \in \mathbb{R}^d \quad \text{find} \quad z \in \mathbb{R}^n \quad \text{such that} \qquad 0 \in h(p, z) + G(z). \tag{5}$$

The *solution mapping* corresponding to (5) is the mapping $S : \mathbb{R}^d \ni p \longmapsto \{z \in \mathbb{R}^n : 0 \in h(p, z) + G(z)\}$. Solving problem (3) with a fixed $p(\cdot)$ means to find $z(\cdot)$ such that $z(t) \in S(t)$ for each $t \in [0, \varepsilon]$ with $d := 1$, $G := Q$, and $h(t, z) := f(z) - p(t)$, $(t, z) \in [0, \varepsilon] \times \mathbb{R}^n$. Although the next three statements are valid in general Banach spaces we state them in finite dimensions. The first result generalizes a parametric version of Dontchev et al. (2013, Theorem 2.4), where a Lipschitz localization instead of a Lipschitz selection is considered.

Theorem 3 *Given $G : \mathbb{R}^n \rightrightarrows \mathbb{R}^n$ and $h : \mathbb{R}^d \times \mathbb{R}^n \to \mathbb{R}^n$, let S be a solution mapping for* (5). *Suppose that a point $(\bar{p}, \bar{z}) \in \mathbb{R}^d \times \mathbb{R}^n$ and positive constants α, β, δ, κ, μ and ν are such that*

(i) there is a function $s : \mathbb{B}[0, \delta] \to \mathbb{R}^n$ which is Lispchitz continuous with the constant κ and such that $s(0) = \bar{z}$ and $s(y) \in G^{-1}(y)$ for each $y \in \mathbb{B}[0, \delta]$;
(ii) $\|h(\bar{p}, \bar{z})\| \leq \beta$;
(iii) $\|h(p, \hat{z}) - h(p, \tilde{z})\| \leq \mu\|\hat{z} - \tilde{z}\|$ whenever $p \in \mathbb{B}[\bar{p}, \delta]$ and $\hat{z}, \tilde{z} \in \mathbb{B}[\bar{z}, \delta]$;
(iv) $\|h(\hat{p}, z) - h(\tilde{p}, z)\| \leq \nu\|\hat{p} - \tilde{p}\|$ whenever $z \in \mathbb{B}[\bar{z}, \delta]$ and $\hat{p}, \tilde{p} \in \mathbb{B}[\bar{p}, \delta]$;
(v) $\kappa\mu < 1$, $\beta\kappa'(1 + \nu) \leq \alpha$, and $\alpha \leq \delta \min\{1, \kappa\}$, where $\kappa' := \kappa/(1 - \mu\kappa)$.

Then there is a function $\sigma : \mathbb{R}^d \to \mathbb{R}^n$ which is Lipschitz continuous on $\mathbb{B}[\bar{p}, \beta]$ with the constant $\kappa'\nu$ and such that

$$\|\bar{z} - \sigma(\bar{p})\| \leq \kappa'\|h(\bar{p}, \bar{z})\| \quad \textit{and} \quad \sigma(p) \in S(p) \cap \mathbb{B}[\bar{z}, \alpha] \quad \textit{whenever } p \in \mathbb{B}[\bar{p}, \beta].$$

Proof First, we observe that

$$\|h(p, z)\| \leq \alpha/\kappa \leq \delta \quad \text{for each} \quad (p, z) \in \mathbb{B}[\bar{p}, \beta] \times \mathbb{B}[\bar{z}, \alpha]. \tag{6}$$

Indeed, fix any such (p, z). As $\alpha \leq \delta$ and $\beta < \delta$ due to (v), using (ii)–(v), we get

$$\begin{aligned} \|h(p, z)\| &\leq \|h(p, z) - h(p, \bar{z})\| + \|h(p, \bar{z}) - h(\bar{p}, \bar{z})\| + \|h(\bar{p}, \bar{z})\| \\ &\leq \mu\|z - \bar{z}\| + \nu\|p - \bar{p}\| + \beta \leq \mu\alpha + \beta(1 + \nu) \leq \alpha\mu + \alpha(1 - \kappa\mu)/\kappa \\ &= \alpha/\kappa \leq \delta. \end{aligned}$$

Fix any $p \in \mathbb{B}[\bar{p}, \beta]$. Consider the function $\varphi_p : \mathbb{B}[\bar{z}, \alpha] \ni z \longmapsto \varphi_p(z) = s(-h(p, z))$, where $s : \mathbb{B}[0, \delta] \to \mathbb{R}^n$ satisfies (i). By (6), φ_p is well-defined. For any $z \in \mathbb{B}[\bar{z}, \alpha]$, (i) and (6) imply that

$$\|\bar{z} - \varphi_p(z)\| = \|s(0) - s(-h(p, z))\| \leq \kappa\|h(p, z)\| \leq \alpha.$$

Moreover, for any $\hat{z}, \tilde{z} \in \mathbb{B}[\bar{z}, \alpha]$, the Lipschitz continuity of s and (iii) reveal that

$$\|\varphi_p(\hat{z}) - \varphi_p(\tilde{z})\| = \|s(-h(p, \hat{z})) - s(-h(p, \tilde{z}))\| \leq \kappa\|h(p, \hat{z}) - h(p, \tilde{z})\| \leq \kappa\mu\|\hat{z} - \tilde{z}\|.$$

As $\kappa\mu < 1$, φ_p is a contraction from $\mathbb{B}[\bar{z}, \alpha]$ into itself, hence it has a unique fixed point in $\mathbb{B}[\bar{z}, \alpha]$.

For each $p \in \mathbb{B}[\bar{p}, \beta]$, denote by $\sigma(p)$ the unique fixed point of φ_p in $\mathbb{B}[\bar{z}, \alpha]$. For each $p \in \mathbb{B}[\bar{p}, \beta]$, we have that

$$\sigma(p) = z \quad \Leftrightarrow \quad z = \varphi_p(z) \quad \Rightarrow \quad 0 \in h(p, z) + G(z) \quad \Rightarrow \quad \sigma(p) \in S(p). \tag{7}$$

To show that σ is Lipschitz continuous on $\mathbb{B}[\bar{p}, \beta]$ with the constant $\kappa'\nu$, fix arbitrary $\hat{p}, \tilde{p} \in \mathbb{B}[\bar{p}, \beta]$. The first equivalence in (7) together with the definitions of $\varphi_{\hat{p}}$ and $\varphi_{\tilde{p}}$ yields that

$$\begin{aligned}\|\sigma(\hat{p}) - \sigma(\tilde{p})\| &= \|s(-h(\hat{p}, \sigma(\hat{p}))) - s(-h(\tilde{p}, \sigma(\tilde{p})))\| \le \kappa\|h(\hat{p}, \sigma(\hat{p})) - h(\tilde{p}, \sigma(\tilde{p}))\| \\ &\le \kappa\|h(\hat{p}, \sigma(\hat{p})) - h(\tilde{p}, \sigma(\hat{p}))\| + \kappa\|h(\tilde{p}, \sigma(\hat{p})) - h(\tilde{p}, \sigma(\tilde{p}))\| \\ &\le \kappa\nu\|\hat{p} - \tilde{p}\| + \kappa\mu\|\sigma(\hat{p}) - \sigma(\tilde{p})\|.\end{aligned}$$

Consequently, $\|\sigma(\hat{p}) - \sigma(\tilde{p})\| \le \kappa\nu/(1 - \kappa\mu)\|\hat{p} - \tilde{p}\|$ as claimed. Since

$$\begin{aligned}\|\bar{z} - \sigma(\bar{p})\| &= \|s(0) - s(-h(\bar{p}, \sigma(\bar{p})))\| \le \kappa\|h(\bar{p}, \sigma(\bar{p}))\| \\ &\le \kappa\|h(\bar{p}, \sigma(\bar{p})) - h(\bar{p}, \bar{z})\| + \kappa\|h(\bar{p}, \bar{z})\| \le \kappa\mu\|\sigma(\bar{p}) - \bar{z}\| + \kappa\|h(\bar{p}, \bar{z})\|,\end{aligned}$$

we conclude that $\|\bar{z} - \sigma(\bar{p})\| \le \kappa'\|h(\bar{p}, \bar{z})\|$. The proof is finished. ■

The above statement (with $d := n$ and $h(p, z) = -p + g(z) + \bar{p}$), immediately implies the following non-parametric version.

Corollary 2 *Let $G : \mathbb{R}^n \rightrightarrows \mathbb{R}^n$ and $(\bar{z}, \bar{y}) \in \operatorname{gph} G$ be given. Assume that there exist constants $\delta > 0$ and $\kappa > 0$ along with a function $s : \mathbb{B}[\bar{y}, \delta] \to \mathbb{R}^n$, which is Lipschitz continuous on $\mathbb{B}[\bar{y}, \delta]$ with the constant κ and such that $s(\bar{y}) = \bar{z}$ and $s(y) \in G^{-1}(y)$ for each $y \in \mathbb{B}[\bar{y}, \delta]$. Let $\mu > 0$ be such that $\kappa\mu < 1$. Then for every positive α and β such that*

$$2\beta\kappa \le \alpha(1 - \mu\kappa) \quad \textit{and} \quad \alpha \le \delta \min\{1, \kappa\},$$

and for every function $g : \mathbb{R}^n \to \mathbb{R}^n$ satisfying

$$\|g(\bar{z})\| \le \beta \quad \textit{and} \quad \|g(\hat{z}) - g(\tilde{z})\| \le \mu\|\hat{z} - \tilde{z}\| \quad \textit{for all} \quad \hat{z}, \tilde{z} \in \mathbb{B}[\bar{z}, \delta],$$

there is a function $\sigma : \mathbb{R}^n \to \mathbb{R}^n$ which is Lipschitz continuous on $\mathbb{B}[\bar{y}, \beta]$ with the constant $\kappa' := \kappa/(1 - \kappa\mu)$ and such that

$$\|\bar{z} - \sigma(\bar{y})\| \le \kappa'\|g(\bar{z})\| \quad \textit{and} \quad \sigma(y) \in (g+G)^{-1}(y) \cap \mathbb{B}[\bar{z}, \alpha] \quad \textit{whenever} \quad y \in \mathbb{B}[\bar{y}, \beta].$$

If there is a single-valued Lipschitz localization of G^{-1} around $\bar{y}$ for $\bar{z}$ then this also is the selection for G^{-1} around $\bar{y}$ for $\bar{z}$ and all such selections coincide locally.

Therefore the function φ_p, for $p := y$, from the proof of Theorem 3 is uniquely determined by y and the uniqueness of the fixed point in $\mathbb{B}[\bar{z}, \alpha]$ implies that the mapping $\mathbb{B}[\bar{y}, \beta] \ni y \longmapsto (g + G)^{-1}(y) \cap \mathbb{B}[\bar{z}, \alpha]$ is single-valued. Also the implications in (7) become equivalences. Therefore one arrives at a strong regularity part of Dontchev et al. (2013, Theorem 2.4). Let us present a slightly modified version of this statement (cf. Cibulka and Fabian 2016, Theorem 2.3) which will be used later.

Theorem 4 *Let $G : \mathbb{R}^n \rightrightarrows \mathbb{R}^n$ and $(\bar{z}, \bar{y}) \in \operatorname{gph} G$ be given. Assume that there are positive constants a, b, and κ such that the set $\operatorname{gph} G \cap \big(\mathbb{B}[\bar{z}, a] \times \mathbb{B}[\bar{y}, b]\big)$ is closed and G is [strongly] regular on $\mathbb{B}[\bar{z}, a]$ for $\mathbb{B}[\bar{y}, b]$ with the constant κ. Let $\mu > 0$ be such that $\kappa\mu < 1$ and let $\kappa' > \kappa/(1 - \kappa\mu)$. Then for every positive α and β such that*

$$2\kappa'\beta + \alpha \leq a \quad \text{and} \quad \mu(2\kappa'\beta + \alpha) + 2\beta \leq b$$

and for every function $g : \mathbb{R}^n \to \mathbb{R}^n$ satisfying

$$\|g(\bar{z})\| \leq \beta \text{ and } \|g(\hat{z}) - g(\tilde{z})\| \leq \mu\|\hat{z} - \tilde{z}\| \quad \text{for all} \quad \hat{z}, \tilde{z} \in \mathbb{B}[\bar{z}, 2\kappa'\beta + \alpha],$$

the mapping $g + G$ has the following property: for every $y, y' \in \mathbb{B}[\bar{y}, \beta]$ and every $z \in (g + G)^{-1}(y) \cap \mathbb{B}[\bar{z}, \alpha]$ there exists a [unique] point $z' \in \mathbb{B}[\bar{z}, 2\kappa'\beta + \alpha]$ such that

$$y' \in g(z') + G(z') \quad \text{and} \quad \|z - z'\| \leq \kappa'\|y - y'\|.$$

Under strong regularity of G, the mapping $\mathbb{B}[\bar{y}, \beta] \ni y \longmapsto (g+G)^{-1}(y) \cap \mathbb{B}[\bar{z}, \alpha]$ is single-valued and Lipschitz continuous with the constant κ'.

If the mapping in question is (locally) monotone then the assumption on the existence of a single-valued Lipschitz localization is equivalent to the existence of a Lipschitz selection. Recall that $S : \mathbb{R}^n \rightrightarrows \mathbb{R}^n$ is *locally monotone* at $(\bar{y}, \bar{z}) \in \operatorname{gph} S$ if there is a neighborhood W of $(\bar{y}, \bar{z})$ such that

$$\langle \hat{y} - \tilde{y}, \hat{z} - \tilde{z} \rangle \geq 0 \quad \text{whenever} \quad (\hat{y}, \hat{z}), (\tilde{y}, \tilde{z}) \in \operatorname{gph} S \cap W. \tag{8}$$

Lemma 2 *A set-valued mapping $S : \mathbb{R}^n \rightrightarrows \mathbb{R}^n$, which is locally monotone at $(\bar{y}, \bar{z}) \in \operatorname{gph} S$, has a single-valued Lipschitz continuous localization around $\bar{y}$ for $\bar{z}$ if and only if it has a Lipschitz continuous selection around $\bar{y}$ for $\bar{z}$.*

Proof We shall imitate (and simplify) the proof of Dontchev and Rockafellar (2014, Theorem 3G.5). Find a neighborhood W of $(\bar{y}, \bar{z})$ such that (8) holds. Let s be a local selection for S which is both defined and Lipschitz continuous on $\mathbb{B}(\bar{y}, r)$ for some $r > 0$ such that $\mathbb{B}(\bar{y}, r) \times \mathbb{B}(\bar{z}, \kappa r) \subset W$, where $\kappa > 0$ is the Lipschitz constant of s. As $s(\bar{y}) = \bar{z}$, we have $s(\mathbb{B}(\bar{y}, r)) \subset \mathbb{B}(\bar{z}, \kappa r)$. Fix any $y \in \mathbb{B}(\bar{y}, r)$. Then $s(y) \in S(y) \cap \mathbb{B}(\bar{z}, \kappa r)$. It suffices to show that the latter set is singleton.

Assume, on the contrary, that there is $z \in S(y) \cap \mathbb{B}(\bar{z}, \kappa r)$ distinct from $s(y)$. Let $b := \|z - s(y)\|$ and $c := (z - s(y))/b$. Then

$$b > 0, \quad \|c\| = 1, \quad \text{and} \quad \langle z, c\rangle = b + \langle s(y), c\rangle. \tag{9}$$

Find $\tau > 0$ such that $\kappa\tau < b$ and that $y + \tau c \in \mathbb{B}(\bar{y}, r)$. Since $\|c\| = 1$, the Cauchy-Schwartz inequality and the Lipschitz continuity of s imply that

$$\langle s(y + \tau c) - s(y), c\rangle \;\leq\; \|s(y + \tau c) - s(y)\|\,\|c\| \;\leq\; \kappa\tau. \tag{10}$$

As $(y + \tau c, s(y + \tau c))$ and (y, z) are in gph $S \cap W$, inequality (8) reveals that

$$0 \;\leq\; \langle s(y + \tau c) - z, y + \tau c - y\rangle \;=\; \tau\langle s(y + \tau c) - z, c\rangle. \tag{11}$$

Using (9), (10), and (11), we get that

$$b + \langle s(y), c\rangle \;=\; \langle z, c\rangle \;\leq\; \langle s(y + \tau c), c\rangle \;\leq\; \langle s(y), c\rangle + \kappa\tau < \langle s(y), c\rangle + b,$$

a contradiction. Hence $S(y) \cap \mathbb{B}(\bar{z}, \kappa r) = \{s(y)\}$ for each $y \in \mathbb{B}(\bar{y}, r)$. The opposite implication is trivial. ∎

In particular, for problem (3) we get the following consequence.

Theorem 5 *Under the standing assumptions, consider problem* (3) *along with its solution mapping* $S : [0, \varepsilon] \ni t \longmapsto S(t) := \{z \in \mathbb{R}^n : p(t) \in f(z) + Q(z)\}$, *where* $p : [0, \varepsilon] \to \mathbb{R}^n$ *is a given Lipschitz continuous function and* f *is continuously differentiable on* $\mathbb{R}^n$. *Suppose that for each* $(t, z) \in$ gph S *the inverse of the mapping* $\mathbb{R}^n \ni v \longmapsto H_{t,z}(v) := f(z) - p(t) + \nabla f(z)(v - z) + Q(v)$ *has a Lipschitz continuous selection around* 0 *for* z. *Then for any* $(t, z) \in$ gph S *there are neighborhoods* T *of* t *in* $[0, \varepsilon]$ *and* U *of* z *in* $\mathbb{R}^n$ *along with a Lipschitz continuous function* $u : T \to \mathbb{R}^n$ *such that* $u(\tau) \in S(\tau) \cap U$ *for each* $\tau \in T$ *and* $u(t) = z$.

Assume, in addition, that there is $r > 0$ *such that* $S(t) \subset r\mathbb{B}$ *for all* $t \in [0, \varepsilon]$ *and*

$$\sup_{(t,z)\in \text{gph}\, S,\, t\in(0,\varepsilon)} \{\text{lip}\,(s; t) : s \text{ is a Lipschitz selection for } S \text{ around } t \text{ for } z\} \tag{12}$$

is finite. Then, for each $z^0 \in S(0)$, *there exists a Lipschitz continuous function* $s : [0, \varepsilon] \to \mathbb{R}^n$ *such that* $s(0) = z^0$ *and* $s(t) \in S(t)$ *for each* $t \in (0, \varepsilon]$.

Proof Fix any $(t, z) \in$ gph S. Note that for any $(\tau, v) \in [0, \varepsilon] \times \mathbb{R}^n$ we have

$$h(\tau, v) + H_{t,z}(v) = f(v) - p(\tau) + Q(v),$$

where $h(\tau, v) := f(v) - f(z) - \nabla f(z)(v - z) + p(t) - p(\tau)$. By Theorem 3 (with $G := H_{t,z}$, the reference point (t, z), μ arbitrary small, ν being the Lipschitz constant of $p(\cdot)$, and $h(t, z) = 0$), there exists a closed neighborhood $T_{t,z}$ of t in $[0, \varepsilon]$, a closed neighborhood $U_{t,z}$ of z in $\mathbb{R}^n$, and a Lipschitz continuous function $u_{t,z} : T_{t,z} \to \mathbb{R}^n$ such that $u_{t,z}(\tau) \in S(\tau) \cap U_{t,z}$ for each $\tau \in T_{t,z}$ and $u_{t,z}(t) = z$.

To prove the rest, pick any $z^0 \in S(0)$. As the quantity in (12) is finite, by the first part of the proof, there is a strictly increasing sequence (t^k) in $(0, \varepsilon)$ along with a function u which is Lipschitz continuous on $[0, t^k]$ for each $k \in \mathbb{N}$ with a constant $\ell_1 > 0$ (independent of k), and such that $u(0) = z^0$ and $u(\tau) \in S(\tau)$ for each $\tau \in [0, \bar{t})$ where $\bar{t} := \lim_{k\to\infty} t^k$. Assume that $\bar{t} < \varepsilon$ for all such sequences (t^k), that is, for any $\hat{t} > \bar{t}$ we cannot extend u to be Lipschitz continuous on $[0, \hat{t}]$. Let $z^k := u(t^k)$, $k \in \mathbb{N}$. Then (z^k) is bounded and $z^k \in S(t^k)$ for each $k \in \mathbb{N}$. Choose an infinite set $N \subset \mathbb{N}$ such that $\bar{z} := \lim_{N \ni k\to\infty} z^k$ exists. As gph S is closed, we have $\bar{z} \in S(\bar{t})$. Set $u(\bar{t}) = \bar{z}$. Find $\hat{t} > \bar{t}$ and $u_{\bar{t},\bar{z}} : [\bar{t}, \hat{t}] \to \mathbb{R}^n$ such that $u_{\bar{t},\bar{z}}$ is Lipschitz continuous on $[\bar{t}, \hat{t}]$ with a constant $\ell_2 > 0$, that $u_{\bar{t},\bar{z}}(\bar{t}) = \bar{z}$ and $u_{\bar{t},\bar{z}}(\tau) \in S(\tau)$ for each $\tau \in [\bar{t}, \hat{t}]$. Let $u(\tau) := u_{\bar{t},\bar{z}}(\tau)$, $\tau \in [\bar{t}, \hat{t}]$.

Fix any $\hat{\tau} \in [0, \bar{t})$ and $\tilde{\tau} \in [\bar{t}, \hat{t}]$. Find $k_0 \in \mathbb{N}$ such that $t^{k_0} > \hat{\tau}$. Then, for any $k > k_0$, we have

$$\begin{aligned} \|u(\hat{\tau}) - u(\tilde{\tau})\| &\le \|u(\hat{\tau}) - u(t^k)\| + \|u(t^k) - u(\bar{t})\| + \|u(\bar{t}) - u(\tilde{\tau})\| \\ &\le \ell_1|\hat{\tau} - t^k| + \|z^k - \bar{z}\| + \ell_2|\bar{t} - \tilde{\tau}|. \end{aligned}$$

Passing to the limit as $k \to \infty$, we get that $\|u(\hat{\tau}) - u(\tilde{\tau})\| \le \ell|\hat{\tau} - \tilde{\tau}|$ where $\ell := \max\{\ell_1, \ell_2\}$. Thus u is Lipschitz on $[0, \hat{t}]$, a contradiction. ■

Contrary to Dontchev et al. (2013, Theorem 2.4) where all the points of gph S are supposed to be strongly regular in the sense of Robinson (1980), there does not exist a Lipschitz selection for S on the whole interval $[0, \varepsilon]$, in general. Indeed, it suffices to consider $[0, \varepsilon] := [0, 3]$, and $S(t) := \sqrt{-t+2}$ if $t \in [0, 1]$, $S(t) := \{0, \sqrt{-t+2}\}$ if $t \in (1, 2]$, and $S(t) = 0$ when $t \in (2, 4]$.

Remark 2 Let $S : \mathbb{R}^d \rightrightarrows \mathbb{R}^l$ be a mapping with $(\bar{p}, \bar{u}) \in \operatorname{gph} S$. Assume that there is $\kappa > 0$ along with closed convex neighborhoods U of $\bar{u}$ and V of $\bar{p}$ such that $S(p) \cap U$ is a closed convex set for each $p \in V$ and $S(\tilde{p}) \cap U \subset S(\hat{p}) + \kappa\|\tilde{p} - \hat{p}\|\,\mathbb{B}$ for each $\tilde{p}, \hat{p} \in V$. Note that the last inclusion holds, for some U and V, provided that S^{-1} is regular at $\bar{u}$ for $\bar{p}$. By Dontchev and Rockafellar (2014, Theorem 3E.3), there is $\kappa' > 0$ together with closed convex neighborhoods U' of $\bar{u}$ and V' of $\bar{p}$ such that $S(p) \cap U'$ is a closed convex set for each $p \in V'$ and $S(\tilde{p}) \cap U' \subset S(\hat{p}) \cap U' + \kappa'\|\tilde{p} - \hat{p}\|\,\mathbb{B}$ for each $\tilde{p}, \hat{p} \in V'$. Using Steiner selection, the remark following Aubin and Frankowska (1990, Theorem 9.4.3) implies that S has a Lipschitz continuous selection around $\bar{p}$ for $\bar{u}$.

4 Numerical Simulation and Applications in Electronics

For the input-output simulation, we derive an extension of the Euler-Newton path-following method from Dontchev and Rockafellar (2014, Section 6G) which, for variational inequalities, was introduced in Dontchev et al. (2013). We will apply this method to a generalized equation (3) with $B = C = I_n$, that is,

$$p(t) \in f(z(t)) + F(z(t)) \quad \text{for all} \quad t \in [0, \varepsilon]. \tag{13}$$

A general case with $B, C \neq I_n$, which we can also obtain in electronics (see, e.g., Adly and Cibulka (2014, Example 6.2) and Adly et al. (2013, Example 5)), is beyond the scope of this note. Let us just mention that often we have $B = C^T$ or the problem can be transformed, under suitable conditions, to ensure this (Adly and Outrata 2013, Section 5).

For an integer $N > 1$, define the uniform grid $t^i := ih$, $i \in \{0, 1, \dots, N\}$, with a step size $h := \varepsilon/N$. Given $\Delta > 0$ and points $(e^i)_{i=0}^{N-1}$ in $\mathbb{B}[p(t^{i+1}), \Delta h^2]$, we study a predictor-corrector scheme in the form

$$\begin{cases} e^i & \in \quad f(z^i) + \nabla f(z^i)(v^{i+1} - z^i) + F(v^{i+1}), \\ p(t^{i+1}) & \in \quad f(v^{i+1}) + \nabla f(v^{i+1})(z^{i+1} - v^{i+1}) + F(z^{i+1}), \end{cases} \tag{14}$$

where z^0 is sufficiently close to the exact solution of (13) at time $t := 0$.

Theorem 6 *Let $\bar{z} : [0, \varepsilon] \to \mathbb{R}^n$ be a Lipschitz continuous solution of (13), where $p : [0, \varepsilon] \to \mathbb{R}^n$ is Lipschitz continuous, $f : \mathbb{R}^n \to \mathbb{R}^n$ is differentiable on whole of $\mathbb{R}^n$ and its derivative mapping is locally Lipschitz continuous, and $F : \mathbb{R}^n \rightrightarrows \mathbb{R}^n$ has a closed graph. Suppose that for each $t \in [0, \varepsilon]$ the mapping*

$$\mathbb{R}^n \ni v \longmapsto \mathcal{G}_t(v) := f(\bar{z}(t)) + \nabla f(\bar{z}(t))(v - \bar{z}(t)) + F(v) \subset \mathbb{R}^n$$

is [strongly] regular at $\bar{z}(t)$ for $p(t)$. Then there is $\alpha > 0$ such that for any $\Delta > 0$ there are constants $N_0 \in \mathbb{N}$ and $c > 0$ such that for each $N > N_0$ and each $z^0 \in \mathbb{B}[\bar{z}(t^0), \Delta h^4]$, where $h := \varepsilon/N$, there are [uniquely determined] points $(z^i)_{i=1}^N$ generated by the iteration (14), with the initial point z^0 and arbitrarily chosen points e^i in $\mathbb{B}[p(t^{i+1}), \Delta h^2]$, $i \in \{0, 1, \dots, N-1\}$, such that $z^i \in \mathbb{B}[\bar{z}(t^i), \alpha]$ for each $i \in \{0, \dots, N\}$ and

$$\max_{0<i<N} \|z^i - \bar{z}(t^i)\| \le ch^4. \tag{15}$$

Proof We divide the proof into two steps.

STEP 1. *There are positive constants a, b, and κ such that, for each $t \in [0, \varepsilon]$, the mapping $\mathcal{G}_t$ is [strongly] regular on $\mathbb{B}[\bar{z}(t), a]$ for $\mathbb{B}[p(t), b]$ with the constant κ.*

Fix any $t \in [0, \varepsilon]$. The mapping $\mathcal{G}_t$ has a closed graph and there are positive constants a_t, b_t, and κ_t such that $\mathcal{G}_t$ is [strongly] regular on $\mathbb{B}[\bar{z}(t), a_t]$ for $\mathbb{B}[p(t), b]$ with the constant κ_t. Let $\mu_t := 1/(2\kappa_t)$ and $\kappa_t' := 3\kappa_t$. Then $\kappa_t \mu_t < 1$ and $\kappa_t' > 2\kappa_t = \kappa_t/(1 - \kappa_t \mu_t)$. The continuity of the function $s \longmapsto \nabla f(\bar{z}(s))$ yields $\alpha_t \in (0, \min\{a_t/2, 3\kappa_t b_t/4\})$ such that

$$\|\nabla f(\bar{z}(\tau)) - \nabla f(\bar{z}(t))\| < \mu_t \quad \text{whenever} \quad \tau \in (t - \alpha_t, t + \alpha_t) \cap [0, \varepsilon]. \tag{16}$$

Let $\beta_t := \alpha_t/(2\kappa_t')$. Then

$$2\kappa_t'\beta_t + \alpha_t = 2\alpha_t < a_t \quad \text{and} \quad \mu_t(2\kappa_t'\beta_t + \alpha_t) + 2\beta_t = \frac{\alpha_t}{\kappa_t} + \frac{\alpha_t}{3\kappa_t} = \frac{4\alpha_t}{3\kappa_t} < b_t. \tag{17}$$

The continuity of the functions $s \longmapsto f(\bar{z}(s))$, $s \longmapsto \nabla f(\bar{z}(s))$, $p(\cdot)$, and $\bar{z}(\cdot)$ implies that there is $r_t \in (0, \alpha_t/2)$ such that for each $\tau \in (t - r_t, t + r_t) \cap [0, \varepsilon]$ we have

$$\|f(\bar{z}(t)) - f(\bar{z}(\tau)) - \nabla f\big(\bar{z}(\tau)\big)\big(\bar{z}(t) - \bar{z}(\tau)\big)\| < \beta_t, \tag{18}$$

and also

$$\|\bar{z}(\tau) - \bar{z}(t)\| < \frac{\alpha_t}{2} \quad \text{and} \quad \|p(\tau) - p(t)\| < \frac{\beta_t}{2}. \tag{19}$$

Fix any $\tau \in [0, \varepsilon] \cap (t - r_t, t + r_t)$. Define the function $g_{t,\tau} : \mathbb{R}^n \to \mathbb{R}^n$, for each $v \in \mathbb{R}^n$, by

$$g_{t,\tau}(v) := f(\bar{z}(t)) - f(\bar{z}(\tau)) + \nabla f(\bar{z}(t))(v - \bar{z}(t)) - \nabla f\big(\bar{z}(\tau)\big)\big(v - \bar{z}(\tau)\big).$$

Then $\mathcal{G}_t = \mathcal{G}_\tau + g_{t,\tau}$. Using (16), we get that for every $\hat{v}, \tilde{v} \in \mathbb{R}^n$, we have

$$\|g_{t,\tau}(\hat{v}) - g_{t,\tau}(\tilde{v})\| = \big\|\big(\nabla f(\bar{z}(t)) - \nabla f\big(\bar{z}(\tau)\big)\big)(\hat{v} - \tilde{v})\big\| \le \mu_t \|\hat{v} - \tilde{v}\|.$$

Also $\big\|g_{t,\tau}\big(\bar{z}(t)\big)\big\| = \big\|f(\bar{z}(t)) - f(\bar{z}(\tau)) - \nabla f\big(\bar{z}(\tau)\big)\big(\bar{z}(t) - \bar{z}(\tau)\big)\big\| < \beta_t$ by (18). Using (19), we conclude that

$$\mathbb{B}[\bar{z}(\tau), \kappa_t'\beta_t] = \mathbb{B}[\bar{z}(\tau), \alpha_t/2] \subset \mathbb{B}(\bar{z}(t), \alpha_t) \quad \text{and} \quad \mathbb{B}[p(\tau), \beta_t/2] \subset \mathbb{B}(p(t), \beta_t).$$

First, assume that $\mathcal{G}_t$ is regular (not strongly). Applying Theorem 4, with $G := \mathcal{G}_t$, $g := -g_{t,\tau}$, $\bar{z} := \bar{z}(t)$, and $\bar{y} := p(t)$, we conclude that the following `claim` holds: *for every y, $y' \in \mathbb{B}[p(t), \beta_t]$ and every $v \in \mathcal{G}_\tau^{-1}(y') \cap \mathbb{B}[\bar{z}(t), \alpha_t]$ there exists a point $v' \in \mathbb{B}[\bar{z}(t), 2\kappa_t'\beta_t + \alpha_t]$ such that*

$$y \in \mathcal{G}_\tau(v') \quad \text{and} \quad \|v - v'\| \le \kappa_t' \|y - y'\|.$$

Consequently, for all $(v, y) \in \mathbb{B}[\bar{z}(\tau), \kappa_t'\beta_t] \times \mathbb{B}[p(\tau), \beta_t/2]$ we have

$$d\big(v, \mathcal{G}_\tau^{-1}(y)\big) \le \kappa_t' \, d\big(y, \mathcal{G}_\tau(v) \cap \mathbb{B}[p(\tau), \beta_t/2]\big). \tag{20}$$

Indeed, fix any such a pair (v, y). Pick an arbitrary $y' \in \mathcal{G}_\tau(v) \cap \mathbb{B}[p(\tau), \beta_t/2]$ (if there is any). Then $v \in \mathbb{B}(\bar{z}(t), \alpha_t)$ and $y' \in \mathbb{B}(p(t), \beta_t)$. The claim yields $v' \in \mathcal{G}_\tau^{-1}(y)$ with $\|v - v'\| \le \kappa_t' \|y - y'\|$. Hence $d(v, \mathcal{G}_\tau^{-1}(y)) \le \|v - v'\| \le \kappa_t' \|y - y'\|$. As y' was an arbitrary point in $\mathcal{G}_\tau(v) \cap \mathbb{B}[p(\tau), \beta_t/2]$, inequality (20) is proved.

Second, assume that $\mathcal{G}_t$ is strongly regular. Again, applying Theorem 4, we get that the mapping $\mathbb{B}[p(t), \beta_t] \ni y \longmapsto \sigma_{t,\tau}(y) := \mathcal{G}_\tau^{-1}(y) \cap \mathbb{B}[\bar{z}(t), \alpha_t]$ is single-valued and Lipschitz continuous with the constant κ'_t. Note that $p(\tau) \in \mathbb{B}(p(t), \beta_t/2)$ and $\bar{z}(\tau) \in \mathcal{G}_\tau^{-1}(p(\tau)) \cap \mathbb{B}(\bar{z}(t), \alpha_t/2) = \{\sigma_{t,\tau}(p(\tau))\}$. Fix any $y \in \mathbb{B}[p(\tau), \beta_t/2] \subset \mathbb{B}(p(t), \beta_t)$. Then

$$\|\sigma_{t,\tau}(y) - \bar{z}(\tau)\| = \|\sigma_{t,\tau}(y) - \sigma_{t,\tau}(p(\tau))\| \le \kappa'_t \|y - p(\tau)\| \le \kappa'_t \beta_t/2.$$

Hence $\sigma_{t,\tau}\big(\mathbb{B}[p(\tau), \beta_t/2]\big) \subset \mathbb{B}(\bar{z}(\tau), \kappa'_t\beta_t) \subset \mathbb{B}(\bar{z}(t), \alpha_t)$. Consequently, the mapping $\mathbb{B}[p(\tau), \beta_t/2] \ni y \longmapsto \mathcal{G}_\tau^{-1}(y) \cap \mathbb{B}[\bar{z}(\tau), \kappa'_t\beta_t]$ is single-valued and Lipschitz continuous with the constant κ'_t.

Summarizing, for each $\tau \in [0, \varepsilon] \cap (t - r_t, t + r_t)$, the mapping $\mathcal{G}_\tau$ is [strongly] regular on $\mathbb{B}[\bar{z}(\tau), \kappa'_t\beta_t]$ for $\mathbb{B}[p(\tau), \beta_t/2]$ with the constant κ'_t, that is, the size of neighborhoods and the constant of regularity are independent of τ in a vicinity of t. From the open covering $\cup_{t\in[0,\varepsilon]}\big((t - r_t, t + r_t) \cap [0, \varepsilon]\big)$ of $[0, \varepsilon]$ choose a finite subcovering $\mathcal{O}^i := [0, \varepsilon] \cap (t^i - r_{t^i}, t^i + r_{t^i})$, $i \in \{1, 2, \dots, k\}$. Let $\kappa := \max\{\kappa'_{t^i} : i = 1, \dots, k\}$, $b := \min\{\beta_{t^i}/2 : i = 1, \dots, k\}$, and $a := \min\{\kappa b, \min\{\kappa'_{t^i}\beta_{t^i} : i = 1, \dots, k\}\}$. STEP 1 is finished.

STEP 2. Let a, b, and κ be the constants found in STEP 1. Denote by ℓ_1 and ℓ_2 the Lipschitz constant of $\bar{z}(\cdot)$ and $p(\cdot)$, respectively. Let $r > 0$ be such that $\bar{z}([0, \varepsilon]) + a\mathbb{B} \subset r\mathbb{B}$. As $\nabla f(\cdot)$ is locally Lipschitz, it is Lipschitz on the (compact) set $r\mathbb{B}$ with the constant $\ell_3 > 0$, say. Let $\ell := \max\{1, \ell_1, \ell_2, \ell_3\}$. Then, for any $\hat{z}$, $\tilde{z} \in r\mathbb{B}$, we have

$$\begin{aligned}\|f(\tilde{z}) - f(\hat{z}) - \nabla f(\hat{z})(\tilde{z} - \hat{z})\| &= \left\| \int_0^1 \big(\nabla f(\hat{z} + t(\tilde{z} - \hat{z})) - \nabla f(\hat{z})\big)(\tilde{z} - \hat{z})\, \mathrm{d}t \right\| \\ &\le \ell \left\|\tilde{z} - \hat{z}\right\|^2 \int_0^1 t\, \mathrm{d}t = \frac{\ell}{2} \left\|\tilde{z} - \hat{z}\right\|^2 . \end{aligned} \tag{21}$$

Let

$$\kappa' := 2\kappa, \quad \mu := 1/(3\kappa), \quad \alpha := \min\{a/2, 1/(3\ell\kappa), b\kappa\}, \quad \text{and} \quad \beta := \frac{1}{2}\ell\alpha^2. \tag{22}$$

We show the following claim: *For any $(t, y, z) \in [0, \varepsilon] \times \mathbb{B}[p(t), \beta] \times \mathbb{B}[\bar{z}(t), \alpha]$, there is a [unique] $\tilde{z} \in \mathbb{B}[\bar{z}(t), \alpha]$ such that*

$$y \in f(z) + \nabla f(z)(\tilde{z} - z) + F(\tilde{z}) \quad \text{and} \quad \|\tilde{z} - \bar{z}(t)\| \le \kappa'\ell\big(\|z - \bar{z}(t)\|^2 + \|y - p(t)\|\big).$$

Indeed, fix any such (t, y, z). Consider a function $g : \mathbb{R}^n \to \mathbb{R}^n$ defined by

$$g(v) = g_{t,y,z}(v) := f(z) + \nabla f(z)(v - z) - f(\bar{z}(t)) - \nabla f(\bar{z}(t))(v - \bar{z}(t)), \quad v \in \mathbb{R}^n.$$

We are going to check the assumptions of Theorem 4 with $(\bar{z}, \bar{y}, G, z, y, y')$ replaced by $(\bar{z}(t), p(t), \mathcal{G}_t, \bar{z}(t), g(\bar{z}(t)) + p(t), y)$. Note that $\mathcal{G}_t$ has a closed graph. Clearly, (22) implies that $\kappa\mu < 1$ and $\kappa' > 3\kappa/2 = \kappa/(1-\mu\kappa)$. Moreover,

$$2\kappa'\beta + \alpha = (2\alpha\ell\kappa)\alpha + \alpha \leq 5\alpha/3 < 2\alpha \leq a$$

and, consequently,

$$\mu(2\kappa'\beta + \alpha) + 2\beta \leq \frac{5\alpha}{9\kappa} + (\ell\alpha)\alpha \leq \frac{5\alpha}{9\kappa} + \frac{\alpha}{3\kappa} = \frac{8\alpha}{9\kappa} < b.$$

Noting that $z \in \mathbb{B}[\bar{z}(t), \alpha] \subset \mathbb{B}(\bar{z}(t), a) \subset r\mathbb{B}$ and using (21) we get

$$\|g(\bar{z}(t))\| = \|f(\bar{z}(t)) - f(z) - \nabla f(z)(\bar{z}(t) - z)\| \leq \frac{\ell}{2}\|\bar{z}(t) - z\|^2 \leq \beta. \quad (23)$$

Since $\ell\alpha \leq 1/(3\kappa) = \mu$, for arbitrary $\hat{v}, \tilde{v} \in \mathbb{R}^n$, we have

$$\|g(\hat{v}) - g(\tilde{v})\| = \|\big(\nabla f(z) - \nabla f(\bar{z}(t))\big)(\hat{v} - \tilde{v})\| \leq \ell\alpha\,\|\hat{v} - \tilde{v}\| \leq \mu\|\hat{v} - \tilde{v}\|.$$

Moreover, observing that $g + \mathcal{G}_t = f(z) + \nabla f(z)(\cdot - z) + F$, we get

$$\begin{aligned} g(\bar{z}(t)) + p(t) &= f(z) + \nabla f(z)(\bar{z}(t) - z) - f(\bar{z}(t)) + p(t) \\ &\in f(z) + \nabla f(z)(\bar{z}(t) - z) + F(\bar{z}(t)) = (g + \mathcal{G}_t)(\bar{z}(t)). \end{aligned}$$

Hence $\bar{z}(t) \in (g + \mathcal{G}_t)^{-1}(g(\bar{z}(t)) + p(t))$, and, by (23), also $g(\bar{z}(t)) + p(t) \in \mathbb{B}[p(t), \beta]$. Remembering that $y \in \mathbb{B}[p(t), \beta]$, Theorem 4 implies that there exists a [unique] point $\tilde{z} \in (g + \mathcal{G}_t)^{-1}(y)$ such that $\|\tilde{z} - \bar{z}(t)\| \leq \kappa'\|y - g(\bar{z}(t)) - p(t)\|$. Then $y \in f(z) + \nabla f(z)(\tilde{z} - z) + F(\tilde{z})$ and (23) implies that

$$\|\tilde{z} - \bar{z}(t)\| \leq \kappa'\big(\ell\|z - \bar{z}(t)\|^2 + \|y - p(t)\|\big),$$

which establishes the claim because $1 \leq \ell$.

Pick any $\Delta > 0$. Let

$$m := \max\{\ell, \Delta, \kappa'\ell, \varepsilon\} \quad \text{and} \quad c := 25m^7. \quad (24)$$

Choose $N_0 \in \mathbb{N}$ such that $c < N_0$ and $m\varepsilon \leq N_0 \min\{\alpha/2, \beta\}$. Fix any $N > N_0$ and let $h := \varepsilon/N$. Then

$$m \geq 1, \quad h < \varepsilon/N_0 \leq m/N_0 < m/c < 1, \text{ and } mh < m\varepsilon/N_0 \leq \min\{\alpha/2, \beta\}. \quad (25)$$

Pick any $z^0 \in \mathbb{B}[\bar{z}(t^0), \Delta h^4]$. By (25) we have $\|z^0 - \bar{z}(t^0)\| \leq mh^4 < mh < \alpha/2$. We proceed by induction. Suppose that z^i verifies $\|z^i - \bar{z}(t^i)\| \leq ch^4$ for some

$i \in \{0, 1, \ldots, N-1\}$. Pick any $e^i \in \mathbb{B}[p(t^{i+1}), \Delta h^2]$. We will show that there is a [unique] point $z^{i+1} \in \mathbb{B}[\bar{z}(t^{i+1}), \alpha]$ generated by (14) such that $\|z^{i+1} - \bar{z}(t^{i+1})\| \leq ch^4$. Inequalities (25) imply that

$$\|e^i - p(t^{i+1})\| \leq mh^2 < mh < \beta \tag{26}$$

and

$$\begin{aligned}\|z^i - \bar{z}(t^{i+1})\| &\leq \|z^i - \bar{z}(t^i)\| + \|\bar{z}(t^i) - \bar{z}(t^{i+1})\| \leq ch^4 + \ell h < (ch)h + mh \\ &< 2mh < \alpha. \end{aligned} \tag{27}$$

The claim with $t := t^{i+1}$, $y := e^i$, and $z := z^i$ yields a [unique] point $v^{i+1} \in \mathbb{B}[\bar{z}(t^{i+1}), \alpha]$ such that

$$e^i \in f(z^i) + \nabla f(z^i)(v^{i+1} - z^i) + F(v^{i+1})$$

and, taking into account (25), (26), and (27), we also have

$$\begin{aligned}\|v^{i+1} - \bar{z}(t^{i+1})\| &\leq m\big(\|z^i - \bar{z}(t^{i+1})\|^2 + \|e^i - p(t^{i+1})\|\big) \leq m(4m^2h^2 + mh^2) \\ &\leq m(5m^2h^2) = 5m^3h^2 < mh(5m^3/c) < mh < \alpha/2. \end{aligned} \tag{28}$$

The claim with $t := t^{i+1}$, $y := p(t^{i+1})$, and $z := v^{i+1}$ yields a [unique] point $z^{i+1} \in \mathbb{B}[\bar{z}(t^{i+1}), \alpha]$ such that

$$p(t^{i+1}) \in f(v^{i+1}) + \nabla f(v^{i+1})(z^{i+1} - v^{i+1}) + F(z^{i+1})$$

and, taking into account (28) and (24), we also have

$$\|z^{i+1} - \bar{z}(t^{i+1})\| \leq m\|v^{i+1} - \bar{z}(t^{i+1})\|^2 \leq 25m^7h^4 = ch^4.$$

The induction step is finished, and hence so is the proof. ∎

The point e^i appearing in (14) can be interpreted as a sufficiently precise prediction at time t^i of the (possibly unknown) value of $p(t^{i+1})$. Then we wait until the precise value of $p(t^{i+1})$ is known and compute a correction z^{i+1}. On the other hand, taking $e^i := p(t^i) + hp'(t^i)$, $i \in \{0, 1, \ldots, N-1\}$, then we have $\|e^i - p(t^{i+1})\| \leq \Delta h^2$ provided that $p'(\cdot)$ exists and is Lipschitz on $[0, \varepsilon]$ with the constant 2Δ. Hence the algorithm proposed in Dontchev and Rockafellar (2014, Section 6G) is a particular case of (14). Finally, it is clear from the proof, that instead of $p(t^{i+1})$ in the latter inclusion of (14) one can take any $\tilde{e}^i \in \mathbb{B}[p(t^{i+1}), \Delta h^4]$, that is, the corrector step can be done via an inexact method (which is always the case in practice).

In the remaining part, we discuss two elementary examples from electronics.

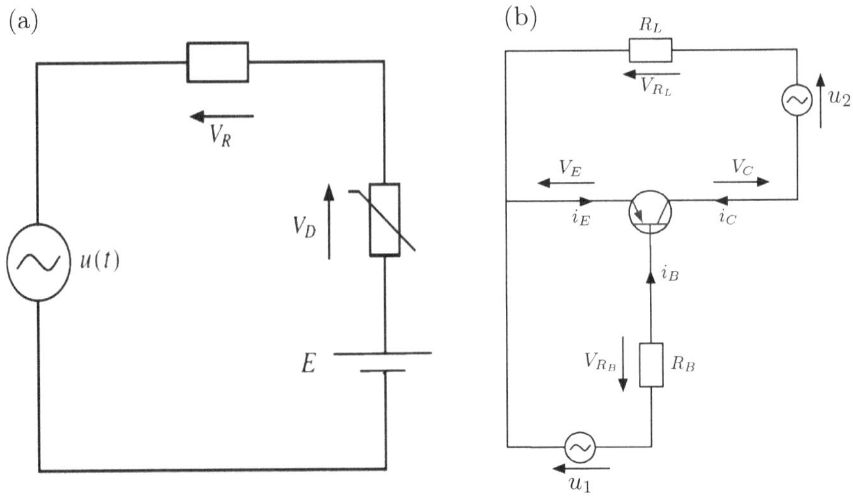

Fig. 1 The circuits considered. (**a**) Example 1. (**b**) Example 2

Example 1 Consider the circuit in Fig. 1a involving a non-linear resistor with current-voltage (i–v) characteristic given by $f(z) := \operatorname{argsinh}(z)$, $z \in \mathbb{R}$, a source $E > 0$, an input-signal source u with the corresponding instantaneous current i, and a practical diode with i–v characteristic

$$F(z) := \begin{cases} [V_1, V_2], & \text{if } z = 0, \\ V_2, & \text{if } z > 0, \\ V_1, & \text{if } z < 0, \end{cases}$$

where $V_1 < 0 < V_2$ are given constants. Letting $p := u - E$ and $z := i$, Kirchhoff's voltage law reveals that, during a fixed time interval $[0, \varepsilon]$, problem (3) reads as

$$p(t) \in \underbrace{\operatorname{argsinh}(z(t))}_{V_R} + \underbrace{F(z(t))}_{V_D} \quad \text{for} \quad t \in [0, \varepsilon].$$

Corollary 1 and Remark 1 imply that Φ^{-1} is a Lipschitz function on any compact interval in $\mathbb{R}$, where $\Phi := f + F$ is the mapping corresponding to static problem (1). For a particular choice of parameters, the solution of the time-dependent problem along with the absolute error of the Euler-Newton path-following method (14) with $e^i = 0$, $i \geq 0$, and $p(\cdot) \in C^\infty(\mathbb{R})$ can be found in Fig. 2a, b, respectively. In this setting, the precision of our method (14) and the method from Dontchev and Rockafellar (2014, Section 6G) is compared Table 1. The input-output simulation for a non-smooth $p(\cdot)$ along with the corresponding errors is in Fig. 3.

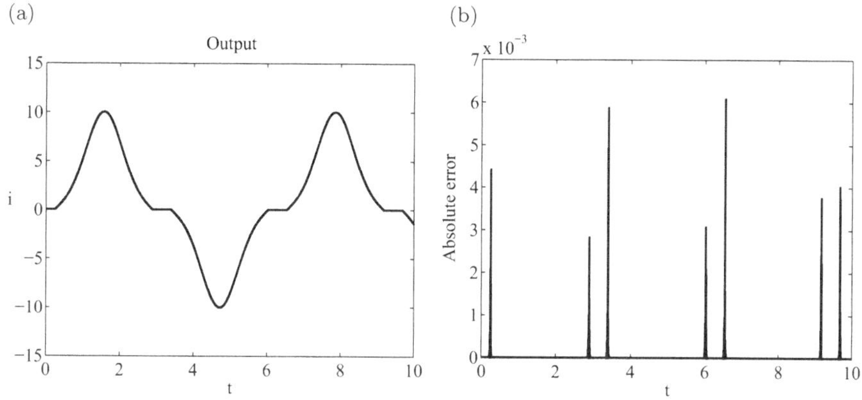

Fig. 2 Example 1 with $[0, \varepsilon] = [0, 10]$, $V_1 = -1$, $V_2 = 1$, $p(t) = 4\sin t$, $h = 0.01$, $E = 0$. (**a**) The output current. (**b**) The absolute error

Table 1 The grid error for different step sizes in Example 1

h	Method from Dontchev and Rockafellar (2014, Section 6G)	Method (14) with $e^i = 0$
0.1	0.0020	0.0027
0.01	2.0035e−07	2.2500e−07
0.001	2.0197e−11	2.2057e−11
0.0001	2.4514e−13	2.4869e−13

Example 2 Consider the circuit in Fig. 1b involving load resistances $R_B > 0$ and $R_L > 0$, two input-signal sources u_1 and u_2, and a P-N-P transistor (see Fig. 4) having three terminals labeled emitter, base and collector. Its behavior can be described by the Ebers-Moll model (Sedra and Smith 2004, p. 409) involving two diodes placed back to back and two dependent current-controlled sources $\alpha_I I'$ and $\alpha_N I$ shunting the diodes. Here $\alpha_N \in [0, 1)$ is known as the current gain in normal operation and $\alpha_I \in (0, 1]$ is known as the inverted common-base current gain. Therefore $i_E = I - \alpha_I I'$ and $i_C = I' - \alpha_N I$. This means that

$$\begin{pmatrix} i_E \\ i_C \end{pmatrix} = \begin{pmatrix} 1 & -\alpha_I \\ -\alpha_N & 1 \end{pmatrix} \begin{pmatrix} I \\ I' \end{pmatrix}.$$

Kirchhoff's laws also reveal that $i_B = -(i_E + i_C)$, so $R_B(-i_C - i_E) + u_1 - V_E = 0$ and $0 = V_C + u_2 + R_L i_C - V_E = V_C + u_2 + R_L i_C + R_B(i_C + i_E) - u_1$. Given $V_{E1} < 0 < V_{E2}$, $V_{C1} < 0 < V_{C2}$, $\alpha > 0$, and $\beta > 0$, assume that the characteristics

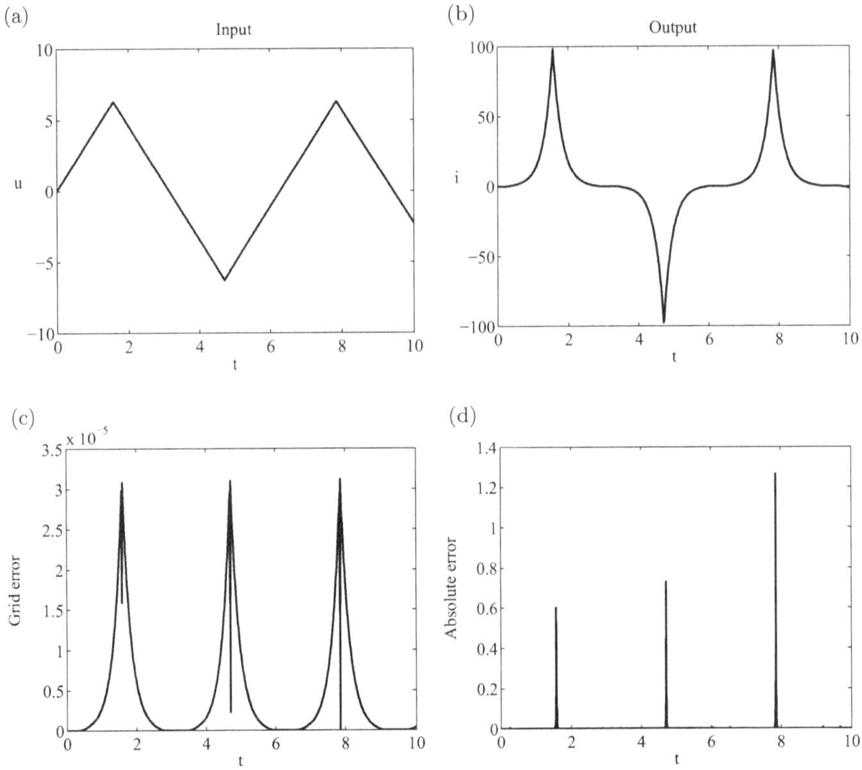

Fig. 3 Example 1 with $[0, \varepsilon] = [0, 10]$, $V_1 = -1$, $V_2 = 1$, $h = 0.01$, $E = 0$. (**a**) The input signal. (**b**) The output current. (**c**) The grid error. (**d**) The absolute error

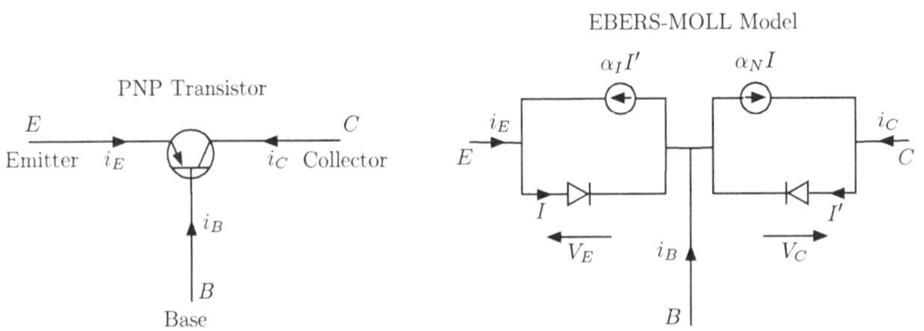

Fig. 4 The P-N-P transistor and its Ebers-Moll model

of the diodes involved in Ebers-Moll model are defined by

$$G_1(x) := \begin{cases} [V_{E1}, V_{E2}], & x = 0, \\ V_{E1} - \alpha \arctan(x), & x < 0, \\ V_{E2} - \alpha \arctan(x), & x > 0, \end{cases} \quad \text{and}$$

$$G_2(x) := \begin{cases} [V_{C1}, V_{C2}], & x = 0, \\ V_{C1} - \beta \arctan(x), & x < 0, \\ V_{C2} - \beta \arctan(x), & x > 0. \end{cases}$$

Then $V_E \in G_1(I) = -\alpha \arctan(I) + F_1(I)$ and $V_C \in G_2(I') = -\beta \arctan(I') + F_2(I')$, where

$$F_1(x) := \begin{cases} [V_{E1}, V_{E2}], & x = 0, \\ V_{E1}, & x < 0, \\ V_{E2}, & x > 0, \end{cases} \quad \text{and} \quad F_2(x) := \begin{cases} [V_{C1}, V_{C2}], & x = 0, \\ V_{C1}, & x < 0, \\ V_{C2}, & x > 0. \end{cases}$$

To sum up, we obtained that

$$\begin{pmatrix} u_1 \\ u_1 - u_2 \end{pmatrix} \in \begin{pmatrix} R_B & R_B \\ R_B & R_B + R_L \end{pmatrix} \begin{pmatrix} 1 & -\alpha_I \\ -\alpha_N & 1 \end{pmatrix} \begin{pmatrix} I \\ I' \end{pmatrix} - \begin{pmatrix} \alpha \arctan(I) \\ \beta \arctan(I') \end{pmatrix} + \begin{pmatrix} F_1(I) \\ F_2(I') \end{pmatrix}.$$

So we arrived at (1) with $n = m = 2$, $B = C = I_2$, $p := (u_1, u_1 - u_2)^T$, $z := (I, I')^T$. Fix any $\bar{z} \in \mathbb{R}^2$. Then

$$\nabla f(\bar{z}) = \begin{pmatrix} (1-\alpha_N)R_B - \alpha/(1+\bar{z}_1^2) & (1-\alpha_I)R_B \\ (1-\alpha_N)R_B - \alpha_N R_L & (1-\alpha_I)R_B + R_L - \beta/(1+\bar{z}_2^2) \end{pmatrix}.$$

Then $A := \nabla f(\bar{z})$ is a P-matrix provided that $a_{11} > 0$, $a_{22} > 0$, and $a_{11}a_{22} - a_{12}a_{21} > 0$, which is the case, for example, when α and β are sufficiently small. As both F_1 and F_2 are maximal monotone, Corollary 1 says that Φ is strongly regular at any reference point. The solution obtained by the Euler-Newton method (14), with $e^i = 0$, $i \geq 0$, is in Fig. 5a, b.

Acknowledgements We would like to thank an anonymous referee for her/his helpful comments.

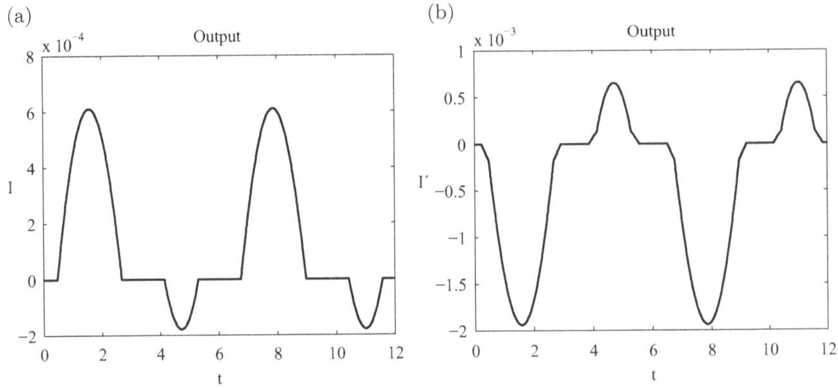

Fig. 5 Example 2 with $[0, \varepsilon] = [0, 12]$, $V_{E1} = -2$, $V_{E2} = 2$, $V_{C1} = -4$, $V_{C2} = 4$, $\alpha = 2/\pi$, $\beta = 2$, $u_1(t) = \sin(t)$, $u_2(t) = 10\sin(t)$, $R_L = 3000$, $R_B = 30000$, $\alpha_I = 0.7$, $\alpha_N = 0.1$, and $h = 0.01$. (**a**) The first component of the solution. (**b**) The second component of the solution

References

S. Adly, R. Cibulka, Quantitative stability of a generalized equation. Application to non-regular electrical circuits. J. Optim. Theory Appl. **160**, 90–110 (2014)

S. Adly, J.V. Outrata, Qualitative stability of a class of non-monotone variational inclusions. Application in electronics. J. Convex Anal. **20**, 43–66 (2013)

S. Adly, R. Cibulka, H. Massias, Variational analysis and generalized equations in electronics. Stability and simulation issues. Set-Valued Var. Anal. **21**, 333–358 (2013)

J-P. Aubin, H. Frankowska, *Set-Valued Analysis*. Systems & Control: Foundations & Applications (Birkhäuser, Boston, 1990)

R. Cibulka, M. Fabian, On primal regularity estimates for set-valued mappings. J. Math. Anal. Appl. **438**, 444–464 (2016)

R. Cibulka, A.L. Dontchev, A.Y. Kruger, Strong metric subregularity of mappings in variational analysis and optimization. J. Math. Anal. Appl. **457**, 1247–1282 (2018)

F.H. Clarke, *Optimization and Nonsmooth Analysis* (Wiley, New York, 1983)

A.L. Dontchev, R.T. Rockafellar, *Implicit Functions and Solution Mappings. A View from Variational Analysis*, 2nd edn. (Springer, New York, 2014)

A.L. Dontchev, M.I. Krastanov, R.T. Rockafellar, V.M. Veliov, An Euler–Newton continuation method for tracking solution trajectories of parametric variational inequalities. SIAM J. Control Optim. **51**, 1823–1840 (2013)

D. Goeleven, Existence and uniqueness for a linear mixed variational inequality arising in electrical circuits with transistors. J. Optim. Theory Appl. **138**, 397–406 (2008)

A.F. Izmailov, Strongly regular nonsmooth generalized equations. Math. Program. **147**, 581–590 (2014)

S.M. Robinson, Strongly regular generalized equations. Math. Oper. Res. **5**, 43–62 (1980)

A.S. Sedra, K.C. Smith, *Microelectronic Circuits*, 5th edn. (Oxford University Press, New York, Oxford, 2004)

On the Dynamic Programming Approach to Incentive Constraint Problems

Fausto Gozzi, Roberto Monte, and M. Elisabetta Tessitore

Abstract In this paper, we study a class of infinite horizon optimal control problems with incentive constraints in the discrete time case. More specifically, we establish sufficient conditions under which the value function satisfies the Dynamic Programming Principle.

1 Introduction

We study a family of discrete time deterministic dynamic optimization problems with an infinite horizon and an *incentive compatibility constraint*, in the sequel referred to as *incentive constrained problems* (ICP). These problems (see Sect. 2 for the detailed formulation) are classical infinite horizon optimal control problems with a constraint on the continuation value of the payoff function: the continuation value of the payoff function at any current date must be larger than some prescribed function of the current state and control.[1]

[1] From an economic point of view, the constraint can be interpreted as an incentive to respect some contract. In this sense we are in a *normative perspective*. A social planner seeks an optimal policy among those not including the termination of the process. The goal is the definition of a social contract, whose breach is prevented by the incentive compatibility constraint.

F. Gozzi (✉)
Dipartimento di Economia e Finanza, Università Luiss - Roma, Rome, Italy
e-mail: fgozzi@luiss.it

R. Monte
Dipartimento di Ingegneria Informatica e Ingegneria Civilei, Macroarea di Ingegneria, Università di Roma "Tor Vergata", Rome, Italy
e-mail: monte@uniroma2.it

M. E. Tessitore
Dipartimento di Economia e Finanza, Macroarea di Economia, Università di Roma "Tor Vergata", Rome, Italy
e-mail: tessitore@uniroma2.it

G. Feichtinger et al. (eds.), *Control Systems and Mathematical Methods in Economics*, Lecture Notes in Economics and Mathematical Systems 687,
https://doi.org/10.1007/978-3-319-75169-6_5

ICP arise in many economic applications (see e.g. Dilip et al. 1990, Marcet and Marimon 2011, Rustichini 1998a, Rustichini 1998b, Pavan et al. 2014, for the discrete time case, and Barucci et al. 2000, Von Thadden 2002, Cvitanic and Zhang 2013, for the continuous time case). ICP are not easily manageable due to the specific nature of the incentive constraint. In fact, this type of constraints binds the future of optimal strategies, differently than the more commonly used constraints which concern the states of the system or the range of admissible strategies. For this reason, in general, the Dynamic Programming (DP) approach cannot be employed.

In Marcet and Marimon (2011) a method is proposed to deal with stochastic incentive constrained problems in discrete time when the DP method cannot be applied. This method is based on embedding the problem in a wider family by introducing the function defining the incentive constraint in the objective function and thereby solving it by a Lagrangean approach, which allows to circumvent the DP perspective. In general, this result is surely useful, though we suspect that some assumptions, like the boundedness and the concavity of the functions defining the incentive constraints, might be weakened.

A natural question arise: although, in general, the DP approach cannot be exploited, is it still possible to determine a class of incentive constrained problems such that the DP procedure can be successfully employed?

In this direction, we are aware of two results in the literature (see Rustichini 1998a and Barucci et al. 2000) showing the applicability of DP approach to a class of ICP, under the assumption that the intertemporal incentive constraint at a future time t is equal to the intertemporal utility of the maximizing agent from t on. More precisely, Rustichini (1998a) deals with a discrete time case, and Barucci et al. (2000) is concerned with the deterministic case in continuous time. Clearly, such a result has a quite narrow domain of applicability since it is not very natural to require that the incentive constraint is equal to the future intertemporal utility of the agent (see examples in Section 3 of Marcet and Marimon (2011)).

The main contribution of this paper is to show that the DP approach can be applied to a much wider family of incentive constraints and present the assumption which makes this possible: a weak form of comonotonicity assumption between the future of the intertemporal utility and the incentive constraint (Assumption 2 below).

Using this assumption we extend the domain of applicability of the DP approach to a larger class of ICP, though dealing only with the discrete deterministic case. This is just our preliminary step to tackle the discrete stochastic case and the continuous time cases. Our main result is that the Dynamic Programming Principle (DPP), formally the equation

$$V(x) = \sup_{c \in \mathcal{C}_g(x)} \sum_{t=0}^{T} \beta^t r\left(x_t(x, c), c_t\right) + \beta^T V\left(x_T(x, c)\right), \qquad (1)$$

where $\mathcal{C}_g(x)$ is an admissible set of controls associated to the initial point x, holds true under the above mentioned comonotonicity assumption presented in Sect. 2.

The paper has the following structure: in Sect. 2 we formulate the problem; in Sect. 3 we prove the DPP (1) for ICP problems via two Lemmas, which are related to some properties of the trajectories: stability under shift (SS) (see Lemma 4) and stability under concatenation (SC) (see Lemma 6).

2 Incentive Constrained Problems in the Discrete-Time Deterministic Case

We first give a general formulation.

2.1 The General Model

Let $(x_t)_{t\in\mathbb{N}}$ be a controlled discrete-time stationary dynamical system with states in the real Euclidean space $\mathbb{R}^n$. Given a map $f : \mathbb{R}^n \times \mathbb{R}^m \to \mathbb{R}^n$, the *transition function* of the dynamical system, a nonempty subset $\mathbb{X}_0$ of $\mathbb{R}^n$, the *state constraint* on the dynamical system, and a nonempty-valued correspondence $\Gamma : \mathbb{R}^n \to \mathfrak{P}(\mathbb{R}^m)$, the *technological constraint* on the control, we assume that $(x_t)_{t\in\mathbb{N}}$ is subject to the difference equation

$$\begin{cases} x_{t+1} = f(x_t, c_t), & t \in \mathbb{N}, \\ x_0 = x \in \mathbb{X}_0, \end{cases} \tag{2}$$

where x is the *initial state* of the dynamical system, and $(c_t)_{t\in\mathbb{N}}$ is a discrete-time control process such that $c_t \in \Gamma(x_t)$ and $f(x_t, c_t) \in \mathbb{X}_0$, for every $t \in \mathbb{N}$. Such a process $(c_t)_{t\in\mathbb{N}}$ is called an *admissible control* for the dynamical system with initial state x. We denote by $\mathcal{C}(x)$ the set of all admissible control processes with initial state x and we write $x_t(x, c)$ for the solution of (2) corresponding to the choice of a specific control process $c \equiv (c_t)_{t\in\mathbb{N}}$ in $\mathcal{C}(x)$.

Since in this paper we are not concerned in dealing with the minimal hypotheses that assure the existence of solutions of (2), we assume that for every $x \in \mathbb{X}_0$ the set $\mathcal{C}(x)$ is nonempty.

Given a function $r : \mathbb{R}^n \times \mathbb{R}^m \to \mathbb{R}$, known as *intertemporal utility*, and $\beta \in (0, 1)$, the *discount factor*, for any $x \in \mathbb{X}_0$ and any $c \in \mathcal{C}(x)$, we introduce the *objective functional* $J_r : \mathbb{X}_0 \times \mathcal{C}(x) \to \mathbb{R}$, given by

$$J_r(x; c) \stackrel{\text{def}}{=} \sum_{t=0}^{+\infty} \beta^t r(x_t(x, c), c_t).$$

Then, the standard optimization problem for the functional $J_r : \mathbb{X}_0 \times \mathcal{C}(x) \to \mathbb{R}$ is to compute the value function $V_0 : \mathbb{X}_0 \to \tilde{\mathbb{R}}$, where $\tilde{\mathbb{R}} \equiv \mathbb{R} \cup \{-\infty, +\infty\}$, given by

$$V_0(x) \stackrel{\text{def}}{=} \sup_{c\in\mathcal{C}(x)} J_r(x; c),$$

and try to determine an optimal control $c^* \in \mathcal{C}(x)$ such that

$$V_0(x) = J_r\left(x; c^*\right),$$

for every $x \in \mathbb{X}_0$.

Now, given the functions $g_1 : \mathbb{R}^n \times \mathbb{R}^m \to \mathbb{R}$, the *initial incentive constraint*, and $g_2 : \mathbb{R}^n \times \mathbb{R}^m \to \mathbb{R}$, known as *intertemporal incentive constraint*, and given $N \in \mathbb{N} \cup \{+\infty\}$, for any $x \in \mathbb{X}_0$, we introduce the set $\mathcal{C}_g(x)$ of all controls $c \in \mathcal{C}(x)$ fulfilling the incentive constraint

$$g_1\left(x_t\left(x, c\right), c_t\right) + \sum_{n=1}^{N} \beta^n g_2\left(x_{t+n}\left(x, c\right), c_{t+n}\right) \geq 0, \qquad \forall t \in \mathbb{N}. \tag{3}$$

We assume that the set $\mathbb{X} \equiv \{x \in \mathbb{X}_0 \mid \mathcal{C}_g(x) \neq \emptyset\}$ is nonempty.[2] The incentive constraint problem for the functional $J_r : \mathbb{X} \times \mathcal{C}_g(x) \to \mathbb{R}$ is to compute the value function $V : \mathbb{X} \to \tilde{\mathbb{R}}$ given by

$$V(x) \stackrel{\text{def}}{=} \sup_{c \in \mathcal{C}_g(x)} J_r\left(x; c\right), \tag{4}$$

and try to determine an optimal control $c^* \in \mathcal{C}_g(x)$ such that

$$V(x) = J_r\left(x; c^*\right),$$

for every $x \in \mathbb{X}$.

In the sequel, for simplicity, we set

$$\sum_{t=0}^{+\infty} \beta^t g_2\left(x_t\left(x, c\right), c_t\right) \equiv J_{g_2}\left(x; c\right),$$

and

$$\sum_{t=0}^{N-h} \beta^t g_2\left(x_t\left(x, c\right), c_t\right) \equiv J_{g_2}^h\left(x; c\right),$$

for every $N \in \mathbb{N}$, and $h = 1, \ldots, N$.

[2] We recall that in Barucci et al. (2000) the set $\mathbb{X}$ is one of the unknown to be found.

2.2 Main Assumptions and Statement of DPP

To tackle the constrained optimization problem (4) via the DP approach, we begin with assuming:

Assumption 1 *For every $x \in \mathbb{X}$ we have $V(x) \in \mathbb{R}$.*

This is a clear restriction. Sometimes one would like to consider problems where V may be infinite. On the other hand, the latter case is not considered in most of the quoted papers treating ICP. We formulate this assumption for sake of simplicity. Indeed, sometimes the finiteness of V arises from compactness or just boundedness, some other times from the choice of the state-control constraints. In this context, the treatment of all these different topics would result in excessively technical complications.

Having assumed the finiteness of the value function, we still need to introduce an assumption on the structure of the family of all admissible controls, which will turn out to be crucial to exploit a the DP approach.

Assumption 2 *For any $c \in \mathcal{C}_g(x)$, any $T > 0$, and any $\varepsilon > 0$ there exists a ε-optimal control at $x_T(x, c)$, denoted by c^ε, such that either*

1.

$$J_{g_2}(x_T(x,c); c^\varepsilon) \geq J_{g_2}(x_T(x,c); c^T),$$

when $N = +\infty$, or

2.

$$J^h_{g_2}(x_T(x,c); c^\varepsilon) \geq J^h_{g_2}(x_T(x,c); c^T), \qquad h = 1, \ldots, N,$$

when $N = +\infty$,

where c^T is the control at $x_T(x, c)$ given by

$$c^T_t \stackrel{def}{=} c_{T+t}, \quad t \in \mathbb{N}$$

Some comments are in order

First, we recall that c^ε is an *ε-optimal control* at $x_T(x, c)$ when

$$V(x_T(x,c)) < J_r(x_T(x,c); c^\varepsilon) + \varepsilon.$$

The possibility of finding an ε-optimal control is assured by the finiteness of the value function. Assumption 2 just prescribes the existence of at least one of such controls fulfilling also either 1 or 2.

Second, observe that, as required by 1 and 2, the control c^T actually belongs to $\mathcal{C}_g(x_T(x, c))$. This will be shown in subsequent Lemma 4.

Third, Assumption 2 may be seen as an hypothesis of *comonotonicity* of the functionals $J_{g_2}(x;\cdot)$ (or $J^h_{g_2}(x;\cdot)$) with respect to $J_r(x;\cdot)$.[3] Actually, if this is the case, 1 and 2 clearly hold true. This has a plain economic explanation. Our "comonotonicity" means that r and g_2 are compatible, which means that the social planner and the agent maximizing $J_{g_2}(x;\cdot)$ (or $J^h_{g_2}(x;\cdot)$) and $J_r(x;\cdot)$ have the same structure of preferences along optimal (or almost optimal) strategies. In this case, the incentive does not affect the recursivity.

Note that in Rustichini (1998a) and Barucci et al. (2000), one has $g_2 = r$. Hence Assumption 2 is clearly satisfied. Our point is that Assumption 2:

- strongly extends the applicability of the DP approach to cases in which $g_2 \neq r$;
- clarifies what is deeply needed to preserve the applicability of the DP approach, i.e. a "weak" comonotonicity between the objective function and the incentive constraints.

We are now in a position to state:

Theorem 3 *For any $x \in \mathbb{X}$, any $c \in \mathcal{C}_g(x)$, and any $T \in \mathbb{N}$, write*

$$J_{r,T}(x;c) \equiv \sum_{t=0}^{T} \beta^t r\left(x_t(x,c), c_t\right) + \beta^T V\left(x_T(x,c)\right),$$

Then, under Assumptions 1 and 2, we have

$$V(x) = \sup_{c \in \mathcal{C}_g(x)} J_{r,T}(x;c).$$

The proof follows by applying the two inequalities proven in Propositions 5 and 7.

[3]The functional $J_{g_2}(x;\cdot)$ is comonotone with respect to $J_r(x;\cdot)$ when for all controls $c_1, c_2 \in \mathcal{C}_g(x)$ the condition

$$J_r(x;c_1) \leq J_r(x;c_2)$$

implies

$$J_{g_2}(x;c_1) \leq J_{g_2}(x;c_2).$$

Clearly, the same holds for $J^h_{g_2}(x;\cdot)$

3 Dynamic Programming: Statement and Proof

The main idea beyond our result is the following (contained also "in nuce" in Barucci et al. (2000, Section 3)).

The classical proof of the Dynamic Programming Principle (1) depends on two key properties of the family of the sets of admissible strategies $\{\mathcal{C}_g(x)\}_{x \in \mathbb{X}}$ The first is the "stability under shift" (SS), while the second is the "stability under concatenation" (SC).

Property (SS) states that $c \in \mathcal{C}_g(x)$ implies $c^T \in \mathcal{C}_g(x_T(x, c))$. Starting with an admissible control c at $t = 0$ from $x_0 = x$, shifting it at T then c^T is admissible when starting at $t = 0$ from $x_0 = x_T(x, c)$. Even in cases more general than ours, (SS) is true as it is shown in Lemma 4 and it yields Inequality (6) presented below and proven in Sect. 3.1.

Property (SC) states that picking two admissible controls $c \in \mathcal{C}_g(x)$ and $\hat{c} \in \mathcal{C}_g(x_T(x, c))$ then the control $\tilde{c}$ given by

$$\tilde{c}_t \stackrel{def}{=} \begin{cases} c_t, & \text{if } t < T \\ \hat{c}_{t-T} & \text{if } t \geq T \end{cases}, \tag{5}$$

belongs to $\mathcal{C}_g(x)$, hence the concatenated control $\tilde{c}$ is admissible starting at $t = 0$ from $x_0 = x$. Unfortunately, even in simple cases (SC) is not true. On the other hand, the failure of (SC) does not mean that DPP is false since (SC) is a sufficient condition. In Sect. 3.2 we show (see Lemma 6) that (SC) holds if $\hat{c}$ is "better" than c along the functional J_{g_2}. This is done by analyzing the family $\{\mathcal{C}_g(x)\}_{x \in \mathbb{X}}$. Given this fact, Inequality (7) presented below is an easy consequence of Assumption 2.

In next two subsection we consider the following inequalities

$$V(x) \leq \sup_{c \in \mathcal{C}_g(x)} \sum_{t=0}^{T} \beta^t r\left(x_t(x, c), c_t\right) + \beta^T V\left(x_T(x, c)\right), \tag{6}$$

$$V(x) \geq \sup_{c \in \mathcal{C}_g(x)} \sum_{t=0}^{T} \beta^t r\left(x_t(x, c), c_t\right) + \beta^T V\left(x_T(x, c)\right). \tag{7}$$

3.1 *The Easy Inequality*

We begin by proving Property (SS).

Lemma 4 *For any $x \in \mathbb{X}$, any $c \in \mathcal{C}_g(x)$, and any $T > 0$, the control c^T given by*

$$c_t^T \stackrel{def}{=} c_{T+t}, \quad t \in \mathbb{N}$$

belongs to $\mathcal{C}_g\left(x_T(x, c)\right)$.

Proof Since the transition function $f : \mathbb{R}^n \times \mathbb{R}^m \to \mathbb{R}^n$ of the dynamical system is stationary, for any $x \in \mathbb{X}$, any $c \in \mathcal{C}(x)$, and any $T > 0$, we clearly have

$$x_{T+t}(x, c) = x_t(x_T(x, c), c^T) \tag{8}$$

for every $t \in \mathbb{N}$. Hence, $c_T \in \mathcal{C}(x_T(x, c))$. Assuming in addition $c \in \mathcal{C}_g(x)$, we have also

$$g_1(x_s(x, c), c_s) + \sum_{n=1}^{N} \beta^n g_2(x_{s+n}(x, c), c_{s+n}) \geq 0,$$

for every $s \in \mathbb{N}$. Thus, setting $s \equiv t + T$, it follows

$$g_1(x_{T+t}(x, c), c_{T+t}) + \sum_{n=1}^{N} \beta^n g_2(x_{T+t+n}(x, c), c_{T+t+n}) \geq 0$$

for every $t \in \mathbb{N}$. Therefore, on account of (8) and of the definition of c^T, we obtain

$$g_1(x_t(x_T(x, c), c^T), c_t^T) + \sum_{n=1}^{N} \beta^n g_2(x_{t+n}(x_T(x, c), c^T), c_{t+n}^T) \geq 0,$$

which completes the proof. □

From (SS) it easily follows

Proposition 5 *For any $x \in \mathbb{X}$, any $c \in \mathcal{C}_g(x)$, and any $T \in \mathbb{N}$ we have*

$$V(x) \leq \sup_{c \in \mathcal{C}_g(x)} J_{r,T}(x; c). \tag{9}$$

Proof To prove (9), which is an abbreviation for (6), it is sufficient to observe that, for every $\varepsilon > 0$, there exists $c^\varepsilon \in \mathcal{C}_g(x)$ such that

$$J_r(x; c^\varepsilon) + \varepsilon > V(x). \tag{10}$$

On the other hand, on account of Lemma 4, for any $T \in \mathbb{N}$, we can write

$$\begin{aligned} J_r(x; c^\varepsilon) &= \sum_{t=0}^{T} \beta^t r(x_t(x; c^\varepsilon), c_t^\varepsilon) + \beta^T J(x_T(x; c^\varepsilon); c^{\varepsilon,T}) \\ &\leq \sum_{t=0}^{T} \beta^t r(x_t(x; c^\varepsilon), c_t^\varepsilon) + \beta^T V(x_T(x; c^\varepsilon)) \\ &= J_{r,T}(x; c^\varepsilon). \end{aligned} \tag{11}$$

Combining (10) and (11), it then follows

$$J_{r,T}(x; c^{\varepsilon}) + \varepsilon > V(x),$$

and the latter clearly implies the desired result. □

3.2 The Hard Inequality

Now, we prove Property (SC)

Lemma 6 *For any $x \in \mathbb{X}$, any $c \in \mathcal{C}_g(x)$ and any $T > 0$, let $\hat{c} \in \mathcal{C}_g(x_T(x, c))$ satisfying either the condition*

$$J_{g2}\left(x_T(x, c); \hat{c}\right) \geq J_{g2}\left(x_T(x, c); c^T\right) \tag{12}$$

when $N = +\infty$, or

$$J_{g_2}^{h}\left(x_T(x, c); \hat{c}\right) \geq J_{g_2}^{h}\left(x_T(x, c); c^T\right), \qquad \forall h = 1, \ldots, N, \tag{13}$$

when $N = +\infty$. Then the control $\tilde{c}$ given by

$$\tilde{c}_t \overset{def}{=} \begin{cases} c_t, & if\, t < T \\ \hat{c}_{t-T} & if\, t \geq T \end{cases}, \tag{14}$$

belongs to $\mathcal{C}_g(x)$.

Proof For any $x \in \mathbb{X}$, any $c \in \mathcal{C}_g(x)$ and any $T > 0$, by (14), we have

$$x_t(x, \tilde{c}) = \begin{cases} x_t(x, c), & \text{if } t < T \\ x_{t-T}(x_T(x, c), \hat{c}), & \text{if } t \geq T \end{cases}. \tag{15}$$

Hence, $\tilde{c} \in \mathcal{C}(x)$. Thus, to prove that $\tilde{c} \in \mathcal{C}_g(x)$, we are left with the task of showing that the incentive constraint (3) is satisfied for every $t \geq 0$. To this, we start to consider the case $N = +\infty$.

For $0 \leq t < T$, we can write

$$\begin{aligned} & g_1(x_t(x, \tilde{c}), \tilde{c}_t) + \sum_{n=1}^{+\infty} \beta^n g_2(x_{t+n}(x, \tilde{c}), \tilde{c}_{t+n}) \\ & \quad = g_1(x_t(x, \tilde{c}), \tilde{c}_t) + \sum_{n=1}^{T-t-1} \beta^n g_2(x_{t+n}(x, \tilde{c}), \tilde{c}_{t+n}) \\ & \qquad + \sum_{n=T-t}^{+\infty} \beta^n g_2(x_{t+n}(x, \tilde{c}), \tilde{c}_{t+n}) \end{aligned}$$

$$= g_1(x_t(x,c),c_t) + \sum_{n=1}^{T-t-1} \beta^n g_2(x_{t+n}(x,c),c_{t+n})$$

$$+ \sum_{n=T-t}^{+\infty} \beta^n g_2(x_{t-T+n}(x_T(x,c),\hat{c}),\hat{c}_{t-T+n}).$$

Hence, adding and subtracting the term

$$\sum_{n=T-t}^{+\infty} \beta^n g_2(x_{t+n}(x,c),c_{t+n}) = \sum_{n=T-t}^{+\infty} \beta^n g_2(x_{t-T+n}(x_T(x,c),c^T),c^T_{t-T+n}),$$

we obtain

$$g_1(x_t(x,\tilde{c}),\tilde{c}_t) + \sum_{n=1}^{+\infty} \beta^n g_2(x_{t+n}(x,\tilde{c}),\tilde{c}_{t+n})$$

$$= g_1(x_t(x,c),c_t) + \sum_{n=1}^{+\infty} \beta^n g_2(x_{t+n}(x,c),c_{t+n})$$

$$+ \sum_{n=T-t}^{+\infty} \beta^n g_2(x_{t-T+n}(x_T(x,c),\hat{c}),\hat{c}_{t-T+n})$$

$$- \sum_{n=T-t}^{+\infty} \beta^n g_2(x_{t-T+n}(x_T(x,c),c^T),c^T_{t-T+n}).$$

On the other hand, observe that $c \in \mathcal{C}_g(x)$ implies that

$$g_1(x_t(x,c),c_t) + \sum_{n=1}^{+\infty} \beta^n g_2(x_{t+n}(x,c),c_{t+n}) \geq 0.$$

Moreover, setting $k \equiv n - (T - t)$, we have

$$\sum_{n=T-t}^{+\infty} \beta^n g_2(x_{t-T+n}(x_T(x,c),\hat{c}),\hat{c}_{t-T+n}) \tag{16}$$

$$= \beta^{T-t} \sum_{k=0}^{+\infty} \beta^k g_2(x_k(x_T(x,c),\hat{c}),\hat{c}_k)$$

$$= \beta^{T-t} J_{g_2}(x_T(x,c);\hat{c}), \tag{17}$$

and

$$\sum_{n=T-t}^{+\infty} \beta^n g_2(x_{t-T+n}(x_T(x,c),c^T),c^T_{t-T+n}) \tag{18}$$

$$= \beta^{T-t} \sum_{k=0}^{+\infty} \beta^k g_2(x_k(x_T(x,c),c^T),c^T_k)$$

$$= \beta^{T-t} J_{g_2}(x_T(x,c);c^T). \tag{19}$$

Therefore, combining (17) and (19) with (12), thanks to (14), it follows that, for every $0 \le t < T$,

$$g_1(x_t(x,\tilde{c}),\tilde{c}_t) + \sum_{n=1}^{+\infty} \beta^n g_2(x_{t+n}(x,\tilde{c}),\tilde{c}_{t+n}) \ge 0.$$

To complete the proof when $N = +\infty$, it is sufficient to observe that, on account of (14) and (15), for every $t \ge T$ we have

$$g_1(x_t(x,\tilde{c}),\tilde{c}_t) + \sum_{n=1}^{+\infty} \beta^n g_2(x_{t+n}(x,\tilde{c}),\tilde{c}_{t+n})$$

$$= g_1(x_{t-T}(x_T(x,c),\hat{c}),\hat{c}_{t-T}) + \sum_{n=1}^{+\infty} \beta^n g_2(x_{t-T+n}(x_T(x,c),\hat{c}),\hat{c}_{t-T+n})$$

and that

$$g_1(x_{t-T}(x_T(x,c),\hat{c}),\hat{c}_{t-T}) + \sum_{n=1}^{+\infty} \beta^n g_2(x_{t-T+n}(x_T(x,c),\hat{c}),\hat{c}_{t-T+n}) \ge 0$$

for $\hat{c} \in \mathcal{C}_g(x_T(x,c))$.

Now, with regard to the case $N \in \mathbb{N}$, let us observe first that for every $0 \le t < T - N$ we clearly have

$$g_1(x_t(x,\tilde{c}),\tilde{c}_t) + \sum_{n=1}^{N} \beta^n g_2(x_{t+n}(x,\tilde{c}),\tilde{c}_{t+n})$$

$$= g_1(x_t(x,c),c_t) + \sum_{n=1}^{N} \beta^n g_2(x_{t+n}(x,c),c_{t+n}) \ge 0.$$

Second, for every $T - N \leq t < T$, namely $T - t \leq N$,

$$
\begin{aligned}
& g_1(x_t(x, \tilde{c}), \tilde{c}_t) + \sum_{n=1}^{N} \beta^n g_2(x_{t+n}(x, \tilde{c}), \tilde{c}_{t+n}) \\
&\quad = g_1(x_t(x, \tilde{c}), \tilde{c}_t) + \sum_{n=1}^{T-t-1} \beta^n g_2(x_{t+n}(x, \tilde{c}^T), \tilde{c}^T_{t+n}) \\
&\qquad + \sum_{n=T-t}^{N} \beta^n g_2(x_{t+n}(x, \tilde{c}^T), \tilde{c}^T_{t+n}) \\
&\quad = g_1(x_t(x, c), c_t) + \sum_{n=1}^{T-t-1} \beta^n g_2(x_{t+n}(x, c), c_{t+n}) \\
&\qquad + \sum_{n=T-t}^{N} \beta^n g_2(x_{t-T+n}(x_T(x, c), \hat{c}), \hat{c}_{t-T+n}),
\end{aligned}
$$

Hence, adding and subtracting the term

$$
\sum_{n=T-t}^{N} \beta^n g_2(x_{t+n}(x, c), c_{t+n}) = \sum_{n=T-t}^{N} \beta^n g_2(x_{t-T+n}(x_T(x, c), c^T), c^T_{t-T+n})
$$

we obtain

$$
\begin{aligned}
& g_1(x_t(x, \tilde{c}), \tilde{c}_t) + \sum_{n=1}^{N} \beta^n g_2(x_{t+n}(x, \tilde{c}), \tilde{c}_{t+n}) \\
&\quad = g_1(x_t(x, c), c_t) + \sum_{n=1}^{N} \beta^n g_2(x_{t+n}(x, c), c_{t+n}) \\
&\qquad \sum_{n=T-t}^{N} \beta^n g_2(x_{t-T+n}(x_T(x, c), \hat{c}), \hat{c}_{t-T+n}) \\
&\qquad - \sum_{n=T-t}^{N} \beta^n g_2(x_{t-T+n}(x_T(x, c), c^T), c^T_{t-T+n}).
\end{aligned}
$$

On the other hand, observe that $c \in \mathcal{C}_g(x)$ implies that

$$
g_1(x_t(x, c), c_t) + \sum_{n=1}^{N} \beta^n g_2(x_{t+n}(x, c), c_{t+n}) \geq 0.
$$

Moreover, setting $k = n - T - t$, we have

$$\sum_{n=T-t}^{N} \beta^n g_2(x_{t-T+n}(x_T(x,c),\hat{c}),\hat{c}_{t-T+n})$$
$$= \beta^{T-t} \sum_{k=0}^{N-(T-t)} \beta^k g_2(x_k(x_T(x,c),\hat{c}),\hat{c}_k)$$
$$= \beta^{T-t} J_{g_2}^{T-t}(x_T(x,c);\hat{c}) \tag{20}$$

and

$$\sum_{n=T-t}^{N} \beta^n g_2(x_{t-T+n}(x_T(x,c),c^T),c^T_{t-T+n})$$
$$= \beta^{T-t} \sum_{k=0}^{N-(T-t)} \beta^k g_2(x_k(x_T(x,c),c^T),c^T_k)$$
$$= \beta^{T-t} J_{g_2}^{T-t}(x_T(x,c);c^T) \tag{21}$$

Therefore, combining (20) and (21) with (13), and observing that $T - N \le t < T$ means $T - t = 1, \ldots, N$, thanks to (14), we obtain again

$$g_1(x_t(x,\tilde{c}^T),\tilde{c}^T_t) + \sum_{n=1}^{N} \beta^n g_2(x_{t+n}(x,\tilde{c}^T),\tilde{c}^T_{t+n}) \ge 0.$$

Finally, for $t \ge T$ we can argue exactly as in the case $N = +\infty$. This completes the proof. □

We are now in a position to prove

Proposition 7 *Under Assumption 2, for any $x \in \mathbb{X}$ and any $T \in \mathbb{N}$, we have*

$$V(x) \ge \sup_{c \in \mathcal{C}_g(x)} J_{r,T}(x;c). \tag{22}$$

Proof To prove (22), which is an abbreviation for (7), fix any control $c \in \mathcal{C}_g(x)$ and consider

$$J_{r,T}(x;c) = \sum_{t=0}^{T} \beta^t r(x_t(x,c),c_t) + \beta^T V(x_T(x,c)).$$

Now, consider the corresponding control $c^T \in \mathcal{C}_g(x_T(x,c))$ (see Lemma 4). If c^T turns out to be optimal at $x_T(x,c)$, we have

$$\begin{aligned} V(x_T(x,c)) &= \sup_{\hat{c}\in\mathcal{C}_g(x_T(x,c))} \sum_{t=0}^{+\infty} \beta^t r(x_t(x_T(x,c),\hat{c}),\hat{c}_t) \\ &= \sum_{t=0}^{+\infty} \beta^t r(x_t(x_T(x,c),c^T),c_t^T). \end{aligned}$$

Thus,

$$\begin{aligned} \beta^T V(x_T(x,c)) &= \sum_{t=0}^{+\infty} \beta^{T+t} r(x_t(x_T(x,c),c^T),c_t^T) \\ &= \sum_{t=T}^{+\infty} \beta^t r(x_{t-T}(x_T(x,c),c^T),c_{t-T}^T), \end{aligned}$$

and

$$J_{r,T}(x;c) = \sum_{t=0}^{T} \beta^t r(x_t(x,c),c_t) + \sum_{t=T}^{+\infty} \beta^t r(x_{t-T}(x_T(x,c),c^T),c_{t-T}^T).$$

Therefore, by (8), we can write

$$J_{r,T}(x;c) = \sum_{t=0}^{+\infty} \beta^t r(x_t(x,c),c_t) = J_r(x;c),$$

and the latter clearly implies

$$J_{r,T}(x;c) \leq V(x),$$

On the other hand, if c^T is not optimal at $x_T(x,c)$, then for any $\varepsilon > 0$ we can find an ε-optimal control $c^\varepsilon \in \mathcal{C}_g(x_T(x,c))$ satisfying Assumption 2. Then, we have

$$V(x_T(x,c)) < J_r(x_T(x,c);c^\varepsilon) + \varepsilon. \tag{23}$$

$$J_r(x_T(x,c);c^T) \leq J_r(x_T(x,c);c^\varepsilon). \tag{24}$$

This implies that

$$\begin{aligned} &J_{r,T}(x;c) \\ &= \sum_{t=0}^{T} \beta^t r(x_t(x,c),c_t) + \beta^T V(x_T(x,c)) \end{aligned}$$

$$\leq \sum_{t=0}^{T} \beta^t r(x_t(x,c), c_t) + \beta^T J_r(x_T(x,c); c^\varepsilon) + \beta^T \varepsilon$$

$$= \sum_{t=0}^{T} \beta^t r(x_t(x,c), c_t) + \sum_{t=0}^{+\infty} \beta^{T+t} r(x_t(x_T(x,c), c^\varepsilon), c_t^\varepsilon) + \beta^T \varepsilon$$

$$= \sum_{t=0}^{T} \beta^t r(x_t(x,c), c_t) + \sum_{t=T}^{+\infty} \beta^t r(x_{t-T}(x_T(x,c), c^\varepsilon), c_{t-T}^\varepsilon) + \beta^T \varepsilon \quad (25)$$

Now, thanks to Lemma 6, the concatenation of c and c^ε, i.e. the control

$$\tilde{c}_t^\varepsilon \stackrel{def}{=} \begin{cases} c_t, & \text{if } t < T \\ c_{t-T}^\varepsilon & \text{if } t \geq T \end{cases},$$

belongs to $\mathcal{C}_g(x)$. Hence, by (8),

$$\sum_{t=0}^{T} \beta^t r(x_t(x,c), c_t) + \sum_{t=T}^{+\infty} \beta^t r(x_{t-T}(x_T(x,c), c^\varepsilon), c_{t-T}^\varepsilon) = J_r\left(x; \tilde{c}^\varepsilon\right),$$

Thus, substituting the latter into (25), we obtain

$$J_{r,T}(x; c) \leq J\left(x; \tilde{c}^\varepsilon\right) + \beta^T \varepsilon \leq V(x) + \beta^T \varepsilon,$$

and for the arbitrariness of ε, it follows

$$J_T(x; c) \leq V(x).$$

Finally, the arbitrariness of $c \in \mathcal{C}_g(x)$ yields the desired result. □

References

E. Barucci, F. Gozzi, A. Swiech, Incentive compatibility constraints and dynamic programming in continuous time. J. Math. Econ. **34**(4), 471–508 (2000)

J. Cvitanic, J. Zhang, Contract Theory in Continuous-Time Models, in *Springer Finance* (Springer, Berlin Heidelberg, 2013)

A. Dilip, D. Pearce, E. Stacchetti, Towards a theory of discounted repeated games with imperfect monitoring. Econometrica **58**(5), 1041–1064 (1990)

A. Marcet, R. Marimon, Recursive Contracts, EUI Working Papers mwp 2011/03, Florence (2011)

A. Pavan, I. Segal, J. Toikka, Dynamic mechanism design: a myersonian approach. Econometrica **82**, 601–653 (2014)

A. Rustichini, Dynamic programming solutions of incentive constrained problems. J. Econ. Theory **78** (2), 329–354 (1998)

A. Rustichini, Lagrange multipliers in incentive–constrained problems. J. Math. Econ. **29**(4), 365–380 (1998)

E.L. Von Thadden, An incentive problem in the dynamic theory of banking. J. Math. Econ. **38**(1–2), 271–292 (2002)

A Functional Analytic Approach to a Bolza Problem

Mikhail I. Krastanov and Nadezhda K. Ribarska

To Vlado, with gratitude and friendship

Abstract The classical problem of the calculus of variations is studied under the assumption that the integrand is a continuous function. A non-smooth variant of the classical du Bois-Reymond lemma is presented. Under suitable additional assumptions, a non-smooth version of the classical Euler equation is proved.

1 Introduction

One of the recent directions of the scientific activity of Vladimir Veliov is closely related to the heterogeneity. The reason is that heterogeneity can play a substantial role in the evolution of populations, economic systems, epidemic diseases, physical systems, social models, and etc. Optimal control problems for heterogeneous systems attracted attention relatively recently, and the obtained results applied mainly to age-structured systems. Many of the published papers presented optimality conditions for particular models, usually in the form of a maximum principle of Pontryagin's type. A general maximum principle for nonlinear McKendrick-type systems is obtained by Brokate in (1985). However, a number of extensions of the McKendrick and Gurtin-MacCamy models arose, where the existing optimality conditions were not applicable (cf., for example, Gurtin and MacCamy (1974)).

M. I. Krastanov (✉) · N. K. Ribarska
Faculty of Mathematics and Informatics, University of Sofia, Sofia, Bulgaria

Institute of Mathematics and Informatics, Bulgarian Academy of Sciences, Sofia, Bulgaria
e-mail: krast@math.bas.bg; ribarska@fmi.uni-sofia.bg

G. Feichtinger et al. (eds.), *Control Systems and Mathematical Methods in Economics*, Lecture Notes in Economics and Mathematical Systems 687,
https://doi.org/10.1007/978-3-319-75169-6_6

A nontrivial extension of the result of Brokate, proved by Feichtinger, Tragler and Veliov in (2003) for an age structured control system with nonlocal dynamics and nonlocal boundary conditions, was able to cope with such situations and found numerous applications.

At that time Vladimir Veliov posed to Tsvetomir Tsachev the problem of generalization the results in Feichtinger et al. (2003) for the case of presence of point-wise terminal state constraints. This was the starting point of a long-term investigation of optimal control problems in infinite-dimensional state spaces. We are very grateful to Vladimir Veliov for attracting our attention to this problem. Our first result was a Pontryagin maximum principle for optimal control problems with terminal constraints in arbitrary Banach state space (cf. Krastanov et al. 2011). During the years in the process of further investigations of optimal control problems in infinite-dimensional state space we realized the importance of the so called "uniform approximating cones" to closed sets in Banach spaces. Each uniform approximating cone is contained in the corresponding Clarke tangent cone and has the following property (which holds true for Clarke tangent cones only in finite-dimensional spaces): roughly speaking, the strong transversality of two uniform approximating cones at a common point of two closed sets implies local nonseparation of these sets. Based on this nonseparation property of closed sets, we obtained an abstract Lagrange multiplier rule and a necessary optimality condition of Pontryagin maximum principle type. These results will appear in Krastanov (2017). A natural next step is to apply the obtained Lagrange multiplier rule to optimal control problems in finite-dimensional state space considered as optimization problems on an infinite-dimensional space of the corresponding trajectories. There already exist very general nonsmooth necessary optimality conditions (cf., for example, Clarke 2013, Ioffe 2017, and the references therein) for such problems. Here we consider the most simple, but non trivial control problem: the so called problem of the calculus of variations. The main result is Theorem 3.10 which is very close to Theorem 18.1 of Clarke (2013). Our approach is completely different from the techniques presented in Clarke (2013). We do not use any variational principles. Our proofs heavily rely on the specific functional-analytic properties of the considered problem and on our previous results from Krastanov et al. (2011, 2015) and Krastanov (2017). We think that the proposed approach could be extended to cope with more general problems.

The organization of this paper is as follows: In Sect. 1 we introduce some notations, pose the problem, prove a variant of the classical du Bois-Reymond lemma and prove measurability of a suitable multivalued map. Uniform tangent sets and cones are defined in Sect. 2. A uniform tangent set to the epigraph of an integral functional is explicitly constructed and the corresponding tangent cone as well as its polar and prepolar are studied. The main result is formulated and proved at the end of the same section.

2 Statement of the Problem and Preliminaries

Let Z be a Banach space. Throughout the paper $\mathbf{B}_Z$ ($\bar{\mathbf{B}}_Z$) will denote the open (closed) unit ball of Z centered at the origin. If A is a subset of Z, its polar is the set $A^0 := \{z^* \in Z^* : z^*(z) \leq 1 \text{ for every } z \in A\}$. If B is a subset of Z^*, its pre-polar is the set $B_0 := \{z \in Z : z^*(z) \leq 1 \text{ for every } z^* \in B\}$.

We study the classical problem of the calculus of variations:

$$\varphi(x) = \int_a^b L(x(t), \dot{x}(t))\, dt \to \min \text{ subject to } x(a) = x_a \text{ and } x(b) = x_b,$$

where $x : [a, b] \to \mathbb{R}^n$ is an absolutely continuous curve. This classical problem is considered under different assumptions imposed on the integrand L and various necessary conditions are already obtained. Here we are going to consider the continuous case, i.e. we assume that $L : \mathbb{R}^n \times \mathbb{R}^n \to \mathbb{R}$ is continuous.

To rigorously formulate the variational problem we are interested in, we introduce some notations. Let X be the Banach space $L^1([a, b]; \mathbb{R}^n)$ and X^* be its dual $L^\infty([a, b]; \mathbb{R}^n)$. We consider the integral functional $\varphi : X^* \times X^* \to \mathbb{R}$ defined by

$$\varphi(x, u) = \int_a^b L(x(t), u(t))\, dt.$$

Note that the continuity of L and the fact that both arguments belong to X^* imply that φ is well defined.

We set

$$P := \{(x, u) \in X^* \times X^* : x(t) = x_a + \int_a^t u(s)\, ds,\ t \in [a, b]\}$$

and

$$Q := \{(x, u) \in X^* \times X^* : \int_u^b u(s)\, ds = x_b - x_a\},$$

where x_a and x_b are fixed points of $\mathbb{R}^n$. We now consider the variational problem

$$\textbf{(VP)} \qquad \varphi(x, u) \to \min \qquad \text{subject to } (x, u) \in P \cap Q.$$

Let $(\bar{x}, \bar{u}) \in X^* \times X^*$ be a solution of the problem (**VP**) and let

$$A := P \cap Q - (\bar{x}, \bar{u}).$$

Let us remind that the pre-polar Y_0 of a linear space $Y \subset X^*$ is the set

$$Y_0 = \{x \in X : y(x) = 0 \text{ for each } y \in Y\}.$$

The following variant of the classical du Bois-Reymond lemma holds true:

Lemma 2.1 *The set A is a w^*-closed linear subspace of $X^* \times X^*$ and its pre-polar is the set*

$$A_0 = \{(y, v) \in X \times X : v \textit{ absolutely continuous, } \dot{v}(t) = y(t) \textit{a.e. in [a,b]}\}.$$

Proof Indeed, the definition of A implies that

$$A = \{(x, u) \in X^* \times X^* : \int_a^b u(s)ds = 0,\ x(t) = \int_a^t u(s)\, ds,\ t \in [a, b]\}$$

and it is clear that A is a linear subspace of $X^* \times X^*$. To prove that A is weak star closed we take an arbitrary generalized sequence $\{(x_\alpha, u_\alpha)\}_{\alpha \in I} \subset A$ which tends to (x_0, u_0) in the weak star topology, i.e.

$$\int_a^b ((x_\alpha(t) - x_0(t))y(t) + (u_\alpha(t) - u_0(t))v(t))\, dt \to_\alpha 0 \text{ for each } (y, v) \in X \times X.$$

We have to prove that $(x_0, u_0) \in A$. Indeed, we have $\int_a^b u_0(s)ds = 0$ taking y and v to be identically zero and one, respectively, in the above expression. It remains to prove that $x_0(t) = \int_a^t u_0(s)ds$. Let us assume the contrary. Then there exists a subset σ of $[a, b]$ with positive measure and $r > 0$ such that $\left| x_0(t) - \int_a^t u_0(s)ds \right|_1 \geq r$ for each $t \in \sigma$ (here by $|z|_1$, where $z = (z_1, \ldots, z_n)^T$ is a vector in $\mathbb{R}^n$, we mean $\sum_{i=1}^n |z_i|$). We choose $y \in X$ to be such that

$$\left(x_0(t) - \int_a^t u_0(s)ds \right) y(t) = \chi_\sigma(t) \left| x_0(t) - \int_a^t u_0(s)ds \right|_1,$$

where $\chi_\sigma(t) = 0$ for $t \notin \sigma$ and $\chi_\sigma(t) = 1$ for $t \in \sigma$. Then

$$0 < r.\text{meas}\,(\sigma) \leq \int_a^b \left(x_0(t) - \int_a^t u_0(s)ds \right) y(t)dt =$$

$$= \int_a^b (x_0(t) - x_\alpha(t))\, y(t)dt + \int_a^b \left(\int_a^t u_\alpha(s)ds - \int_a^t u_0(s)ds \right) y(t)dt. \tag{1}$$

It is clear that

$$\int_a^b (x_0(t) - x_\alpha(t))\, y(t)dt \to_\alpha 0 \tag{2}$$

because $\{x_\alpha\}_{\alpha\in I}$ tends to x_0 in the weak star topology. Also,

$$\int_a^b \left(\int_a^t u_\alpha(s)ds - \int_a^t u_0(s)ds\right) y(t)dt =$$
$$= \iint_{[a,b]\times[a,b]} (u_\alpha(s) - u_0(s))\chi_{[a,t]}(s)y(t)dsdt$$

and the last integral exists because $u_\alpha - u_0$ and y belong to X^*, and $\chi_{[a,\cdot]} \in L^\infty([a,b]\times[a,b],\mathbb{R})$. Therefore, Fubini's theorem is applicable and

$$\int_a^b \left(\int_a^t u_\alpha(s)ds - \int_a^t u_0(s)ds\right) y(t)dt =$$
$$= \int_a^b (u_\alpha(s) - u_0(s)) \left(\int_a^b y(t)\chi_{[a,t]}(s)dt\right) ds \to_\alpha 0 \tag{3}$$

because $\{u_\alpha\}_{\alpha\in I}$ tends to u_0 in the weak star topology and

$$h(s) := \int_a^b y(t)\chi_{[a,t]}(s)dt = \int_s^b y(t)dt,\ s \in [a,b],$$

belongs to X. Thus (2) and (3) contradict to (1). This completes the proof that A is weak star closed.

Let (y, v) be an arbitrary element of the pre-polar $A_0 \subset X \times X$. Then for each $(x, u) \in A$ we have that

$$0 = \int_a^b (x(t)y(t) + u(t)v(t))dt = \int_a^b \left(\left(\int_a^t u(s)ds\right) y(t) + u(t)v(t)\right) dt =$$
$$= \int_a^b \left(\int_a^t u(s)ds\right) d\left(\int_a^t y(s)ds\right) + \int_a^b u(t)v(t)dt =$$
$$= \left(\int_a^t u(s)ds\right)\left(\int_a^t y(s)ds\right)\Big|_a^b + \int_a^b u(t)\left(-\int_a^t y(s)ds + v(t)\right) dt.$$

Because $(x, u) \in A$ we have that $\int_a^b u(s)ds = 0$, and therefore,

$$0 = \int_a^b u(t)\left(-\int_a^t y(s)ds + v(t)\right) dt = \int_a^b u(t)\left(-\int_a^t y(s)ds + v(t) + c\right) dt.$$

for every constant vector c of $\mathbb{R}^n$. We set $w(t) = -\int_a^t y(s)ds + v(t) + c$, where the constant vector c is chosen in such a way that $\int_a^b w(t)dt = 0$. Thus we have that $\int_a^b u(t)w(t)dt = 0$ for each $u \in X^*$ with $\int_a^b u(s)ds = 0$. We set $w^*(t) :=$ $\mathbf{sgn}(w(t)) + d \in X^*$, where the constant vector $d^T \in \mathbb{R}^n$ is chosen in such a way that $\int_a^b w^*(t)dt = 0$ and $\mathbf{sgn}(z_1, z_2, \dots, z_n) = (\operatorname{sgn} z_1, \operatorname{sgn} z_2, \dots, \operatorname{sgn} z_n)$ (here $\operatorname{sgn} r = -1$ for $r < 0$, $\operatorname{sgn} r = 0$ for $r = 0$ and $\operatorname{sgn} r = 1$ for $r > 0$). Hence,

$$0 = \int_a^b w^*(t)w(t)dt = \int_a^b (\mathbf{sgn}\,(w(t)) + d)\, w(t)dt =$$

$$= \int_a^b (\mathbf{sgn}(w(t)))\, w(t)dt = \int_a^b |w(t)|_1 dt.$$

This implies that $w(t) = 0$ for almost all $t \in [a, b]$, i.e. $v(t) = -c + \int_a^t y(s)ds$ a.e. in $[a, b]$. Hence for every element $(y, v) \in A_0$ we have that v is absolutely continuous on $[a, b]$ and $\dot{v} = y$ a.e. in $[a, b]$.

Now let $(y, v) \in X \times X$ be such that v is absolutely continuous on $[a, b]$ and $\dot{v} = y$ a.e. in $[a, b]$ and let (x, u) be an arbitrary element of A. Then one can check that

$$\int_a^b (x(t)y(t) + u(t)v(t))dt = \int_a^b x(t)dv(t) + \int_a^b u(t)v(t)dt =$$

$$= -\int_a^b u(t)v(t)dt + x(t)v(t)\Big|_a^b + \int_a^b u(t)v(t)dt = 0$$

because $(x, u) \in A$. This completes the proof. ◇

Lemma 2.2 *Let Z be a separable Banach space, S be a closed subset of Z, K be a closed subset of $\mathbb{R}^n$ and $F : K \to S$ be continuous. Then the multivalued mapping $x \to \hat{T}_S(F(x))$ which assigns to each point $x \in K$ the Clarke tangent cone $\hat{T}_S(F(x))$ to S at the point $F(x)$ is Lebesgue measurable.*

Proof Let $\mathcal{B}_n = \{U_1^n, U_2^n, \dots\}$ (for $n = 1, 2, \dots$) be the family of the open balls in Z of radius $1/n$ centered at some of the elements of a countable dense subset of Z. We fix an arbitrary open set $U \subset Z$. Let $(n_1, n_2, \dots, n_k)$ be a finite sequence of positive integers and let us define the set

$$W_{n_1 n_2 \dots n_k} := \left\{ z \in S : \begin{array}{l} \exists\, \delta_z > 0\; \forall\, y \in B_{\delta_z}(z) \cap S\; \forall \lambda \in (0, \delta_z) \text{ it is true} \\ \text{that } \left(y + \lambda \left(U \cap U_1^{n_1} \cap U_2^{n_2} \cap \dots U_k^{n_k}\right)\right) \cap S \neq \emptyset \end{array} \right\}.$$

We put $W_{n_1 n_2 \dots n_k}$ to be the empty set if $U \cap U_1^{n_1} \cap U_2^{n_2} \cap \dots U_k^{n_k} = \emptyset$. It is straightforward to check that the sets $W_{n_1 n_2 \dots n_k}$, where $(n_1, n_2, \dots, n_k)$ is a finite sequence of positive integers, are relatively open in S. Moreover, $v \in \hat{T}_S(z)$ if and only if there exists an infinite sequence of positive integers $(n_1, n_2, \dots)$ such that

$$\{v\} = \bigcap_{k=1}^{\infty} \left(U \cap U_1^{n_1} \cap U_2^{n_2} \cap \dots U_k^{n_k}\right) \text{ and } z \in \bigcap_{k=1}^{\infty} W_{n_1 n_2 \dots n_k}$$

because $\operatorname{diam}\left(U \cap U_1^{n_1} \cap U_2^{n_2} \cap \dots U_k^{n_k}\right) \longrightarrow_{k\to\infty} 0$. Equivalently, we have that $v \in \hat{T}_S(F(x))$ if and only if there exists an infinite sequence of positive integers $(n_1, n_2, \dots)$ such that

$$\{v\} = \bigcap_{k=1}^{\infty} \left(U \cap U_1^{n_1} \cap U_2^{n_2} \cap \dots U_k^{n_k}\right) \text{ and } x \in \bigcap_{k=1}^{\infty} F^{-1}\left(W_{n_1 n_2 \dots n_k}\right) .$$

As for every infinite sequence of positive integers $\zeta = (n_1, n_2, \dots)$ the intersection $\bigcap_{k=1}^{\infty} \left(U \cap U_1^{n_1} \cap U_2^{n_2} \cap \dots U_k^{n_k}\right)$ is either empty or an one-point set, we have

$$\left(\hat{T}_S(F)\right)^{-1}(U) := \{x \in K : \hat{T}_S(F(x)) \cap U \neq \emptyset\} = \bigcup_{\zeta \in \mathbf{N}^{\mathbf{N}}} \left(\bigcap_{k=1}^{\infty} F^{-1}\left(W_{\zeta|_k}\right)\right).$$

Continuity of F implies that the sets $F^{-1}\left(W_{\zeta|_k}\right)$ are relatively open in K, hence Borel in $\mathbb{R}^n$. Therefore $\left(\hat{T}_S(F)\right)^{-1}(U)$ is a Suslin set and thus it is Lebesgue measurable. This completes the proof. ◇

3 Uniform Tangent Sets and Prepolars

We set $Y := X \times X \equiv L^1([a, b]; \mathbb{R}^{2n})$ and $Y^* := X^* \times X^* \equiv L^\infty([a, b]; \mathbb{R}^{2n})$, and consider the integral functional $\varphi(y) - \int_a^b L(y(t))dt$.

Remind that $L : \mathbb{R}^{2n} \to \mathbb{R}$ is assumed to be continuous. Next we are going to study some tangent cones to the epigraph

$$\text{Epi } (\varphi) := \{(y, r) \in Y^* \times \mathbb{R} : \varphi(y) \le r, \ y \in Y\}$$

at the point $(\bar{y}, \varphi(\bar{y}))$. We consider $Y^* \times \mathbb{R}$ equipped with the uniform norm $\|(y, r)\| := \max\{\|y\|, |r|\}$. We denote by $|\cdot|$ the usual Euclidian norm on $\mathbb{R}^n$. The following definitions are taken from Krastanov (2017):

Definition 3.1 Let S be a closed subset of Y and y_0 belong to S. We say that the bounded set D is a uniform tangent set to S at the point y_0 if for each $\varepsilon > 0$ there exists $\delta > 0$ such that for each $v \in D$ and for each point $y \in S \cap (y_0 + \delta\bar{\mathbf{B}})$ there exists $\bar{\lambda} > 0$ such that for each $\lambda \in [0, \bar{\lambda}]$ the set $S \cap (y + \lambda(v + \varepsilon\bar{\mathbf{B}}))$ is non empty.

Definition 3.2 Let S be a closed subset of Y and y_0 belong to S. We say that the cone C is a uniform tangent cone to S at the point y_0 if $C \cap \bar{\mathbf{B}}$ is a uniform tangent set to S at the point y_0.

Let $\hat{T}_{epi\ L}(\bar{y}(t), L(\bar{y}(t)))$ denote the Clarke tangent cone to the epigraph $epi\ L \subset \mathbb{R}^{2n+1}$ of L at the point $(\bar{y}(t), L(\bar{y}(t)))$. Let $R : [a, b] \longrightarrow \mathbb{R}$ be a nonnegative summable function. We are going to assume that $R \geq 1$ on $[a, b]$. We introduce the multivalued map

$$t \to G_R(t) := \hat{T}_{epi\ L}(\bar{y}(t), L(\bar{y}(t))) \cap \left(\bar{\mathbf{B}}_{\mathbb{R}^{2n}} \times [-R(t), R(t)]\right), \quad t \in [a, b].$$

Lemma 3.3 *The multivalued map* $G_R : [a, b] \longrightarrow \mathbb{R}^{2n+1}$ *is measurable.*

Proof Let us fix an arbitrary positive integer n. Applying the Lusin's theorem to the measurable function $\bar{y}$, there exists a compact subset E_n of $[a, b]$ with meas $([a, b] \setminus E_n) < \frac{1}{n}$ such that the restriction of $\bar{y}$ on E_n is continuous. Because E_n is compact, we can apply Lemma 2.2 for $Z \equiv \mathbb{R}^{2n+1}$, $S = epi\ L$, $K = E_n$ and $F : K \to S$ defined by $F(t) := (\bar{y}(t), L(\bar{y}(t)))$, and to conclude that the multivalued map $t \to \hat{T}_{epi\ L}(\bar{y}(t), L(\bar{y}(t)))$, $t \in E_n$, is measurable. As the multivalued map $t \to \bar{\mathbf{B}}_{\mathbb{R}^{2n}} \times [-R(t), R(t)]$, $t \in E_n$ is measurable and has compact values, the intersection map $t \to G_R(t)$ is measurable as well (cf. Proposition 3.4 of Deimling 1992). Clearly,

$$E := [a, b] \setminus \left(\bigcup_{n=1}^{\infty} E_n\right) = \bigcap_{n=1}^{\infty} ([a, b] \setminus E_n)$$

has Lebesgue measure zero. Let U be an arbitrary open subset of $\mathbb{R}^{2n+1}$. Then

$$G_R^{-1}(U) := \{t \in [a, b] : \ G_R(t) \cap U \neq \emptyset\} = \bigcup_{n=1}^{\infty} \left(G_R^{-1}(U) \cap E_n\right) \cup \left(G_R^{-1}(U) \cap E\right).$$

Each set $G_R^{-1}(U) \cap E_n, n = 1, 2, \ldots,$ is measurable because $(G_R)|_{E_n}$ is measurable. Also, $G_R^{-1}(U) \cap E$ is measurable because of the completeness of the Lebesgue measure. Therefore the set $G_R^{-1}(U)$ is measurable, and so the map G_R is measurable on $[a, b]$. $\diamond$

We set

$$B_R := \left\{ (v, r) \in Y^* \times \mathbb{R} : \ \begin{array}{c} \text{there exist measurable functions } v \text{ and } r_v \text{ with} \\ r = \displaystyle\int_a^b r_v(t)dt \text{ and for almost all } t \in [a, b] \\ \text{it is true that } (v(t), r_v(t)) \in G_R(t) \end{array} \right\}.$$

Also, we set C to be the cone generated by B_R, i.e.

$$C = \{\lambda y : \ y \in B_R, \ \lambda \geq 0\}.$$

Definition 3.4 It is said that the set $\mathcal{F}$ of summable functions $f : [a, b] \to \mathbb{R}^n$ is uniformly integrable if for each $\varepsilon > 0$ there exists $\delta > 0$ such that for every measurable subset E of $[a, b]$ with meas $(E) < \delta$ it holds true $\int_E |f(t)|dt < \varepsilon$ for every $f \in \mathcal{F}$.

In order to ensure that the above defined set B_R is a uniform tangent set to the epigraph of φ at the point $\bar{y}$, we need the following standing assumption (**SA**):

There exists a positive real $\bar{\delta}$ such that the family of summable functions

$$\mathcal{F} := \left\{ \frac{L(y + \lambda v) - L(y)}{\lambda} - r_v : \begin{array}{l} y \in Y^* \text{ with } \|y - \bar{y}\| < \bar{\delta}, \lambda \in (0, \bar{\delta}), \\ (v, r) \in B_R \text{ with } r = \displaystyle\int_a^b r_v(t)dt \\ \text{and for almost all } t \in [a, b] \\ \text{it is true that } (v(t), r_v(t)) \in G_R(t) \end{array} \right\}$$

is uniformly integrable.

Remark 3.5 Clearly, the existence of a summable function $\mu : [a, b] \to \mathbb{R}$ such that $|f(t)| \leq \mu(t)$ almost everywhere on $[a, b]$ for every $f \in \mathcal{F}$ implies uniform integrability of $\mathcal{F}$.

Lemma 3.6 *If the condition* (**SA**) *holds true, then the set B_R is a uniform tangent set to Epi (φ) at the point $\bar{y}$.*

Proof Let us fix an arbitrary $\varepsilon \in (0, 1)$.

Step 1: Regularization on a set with big measure. According to the standing assumption (**SA**), there exists $\delta_0 > 0$ such that for each measurable subset E of $[a, b]$ with meas $E < \delta_0$ and for each element $f \in \mathcal{F}$ it holds true that $\int_E |f(t)|dt < \frac{\varepsilon}{2}$. Applying the Lusin's theorem to the measurable function $\bar{y}$, there exists an open subset E_1 of $[a, b]$ with meas $E_1 < \delta_0/3$ such that the restriction of $\bar{y}$ on $[a, b] \setminus E_1$ is continuous. Applying the Lusin's theorem to the measurable function R, there exists an open subset E_2 of $[a, b]$ with meas $E_2 < \delta_0/3$ such that the restriction of R on $[a, b] \setminus E_2$ is continuous. Because G_R is measurable on $[a, b]$ (cf. Lemma 3.3) we can apply Corollary 1 (cf. Lojasiewicz 1985), Proposition 1.2 and Proposition 2.1 (cf. Deimling 1992) to obtain existence of an open subset E_3 of $[a, b] \setminus (E_1 \cup E_2)$ with meas $E_3 < \delta_0/3$ such that the map G_R is continuous with respect to the Hausdorff metric on $T := [a, b] \setminus (E_1 \cup E_2 \cup E_3)$.

Step 2: Definition of δ. Let t be an arbitrary element of T. Since $\hat{T}_{epi\ L}(\bar{y}(t), L(\bar{y}(t)))$ is a uniform tangent cone to *epi* L (cf. Theorem 1, Rockafellar

1979), there exists $\delta_t > 0$ such that for every element $(y, p) \in epi\ L$ which is δ_t-close to $(\bar{y}(t), L(\bar{y}(t))$, for every direction $(v, r) \in G_R(t)$ and for every $\lambda \in (0, \delta_t)$ it is true that

$$\left[(y, p) + \lambda\left((v, r) + \frac{\varepsilon}{4(b-a)}\left(\bar{\mathbf{B}}_{\mathbb{R}^{2n}} \times [-R(t), R(t)]\right)\right)\right] \bigcap epi\ L \neq \emptyset\,. \tag{4}$$

This means that whenever $(y, p) \in epi\ L$ is δ_t-close to $(\bar{y}(t), L(\bar{y}(t)), (v, r) \in G_R(t)$ and $\lambda \in (0, \delta_t)$ are fixed, there exists

$$(w, s) \in (v, r) + \frac{\varepsilon}{4(b-a)}\left(\bar{\mathbf{B}}_{\mathbb{R}^{2n}} \times [-R(t), R(t)]\right)$$

such that $L(y + \lambda w) \leq p + \lambda s$.

The continuity of $\bar{y}$, $L \circ \bar{y}$ and G_R (with respect to the Hausdorff metric) at t implies the existence of $\eta_t \in (0, \delta_t)$ such that

$$\|\bar{y}(\tau) - \bar{y}(t)\| < \frac{\delta_t}{2},\ |L(\bar{y}(\tau)) - L(\bar{y}(t))| < \frac{\delta_t}{2},\ \text{dist}\,(G_R(\tau), G_R(t)) < \frac{\varepsilon}{4(b-a)}$$

whenever $\tau \in T$ and $|\tau - t| < \eta_t$.

Because $\{(t - \eta_t, t + \eta_t)\}_{t \in T}$ is an open covering of the compact set T, there exists a finite set $\{t_i\}_{i=1}^{k} \subset T$ such that $T \subset \bigcup_{i=1}^{k}(t_i - \eta_{t_i}, t_i + \eta_{t_i})$. We set $\eta := \min\{\eta_{t_1}, \eta_{t_2}, \ldots, \eta_{t_k}\} > 0$.

The set $D := \{\bar{y}(t) : t \in [a, b]\} + \eta\bar{\mathbf{B}}_{\mathbb{R}^{2n}}$ is compact. The continuity of L on D implies that L is uniformly continuous on D. Hence there exists $\delta > 0$ such that $|L(y) - L(x)| < \eta/2$ whenever $\|y - x\| < \delta$. Without loss of generality we may assume that $\delta < \eta/2$ and $\delta < \bar{\delta}$, where $\bar{\delta}$ is the constant from (SA).

Step 3: Verification of uniformity of the set B_R. Now we turn to the proof that for the so defined δ we have

$$\left[(y(\cdot), q) + \lambda\left((v(\cdot), r) + \varepsilon\bar{\mathbf{B}}_{Y^* \times \mathbb{R}}\right)\right] \bigcap Epi\ \varphi \neq \emptyset$$

whenever $(v(\cdot), r) \in B_R$, $(y(\cdot), q) \in \left[(\bar{y}(\cdot), \varphi(\bar{y}(\cdot))) + \delta\bar{\mathbf{B}}_{Y^* \times \mathbb{R}}\right] \bigcap Epi\ \varphi$ and $\lambda \in (0, \delta)$.

Indeed, let (y, q) be an arbitrary fixed element of $Epi\ \varphi \subset Y^* \times \mathbb{R}$ with $\|y - \bar{y}\| < \delta$ and $|q - \varphi(\bar{y})| < \delta$, $(v, r) \in B_R$ and $\lambda \in (0, \delta)$. Note that $(v, r) \in B_R$ implies that there exists a summable function $r_v : [a, b] \to \mathbb{R}$ with $r = \int_a^b r_v(t)dt$ and for almost all $t \in [a, b]$ we have $(v(t), r_v(t)) \in \hat{T}_{epi\ L}(\bar{y}(t), L(\bar{y}(t))) \cap \left(\bar{\mathbf{B}}_{\mathbb{R}^{2n}} \times [-R(t), R(t)]\right)$.

We define the multi-valued map $H : [a, b] \to \mathbb{R}^{2n+1}$ as follows: $H(t) :=$

$$= \begin{cases} \left[(y(t), L(y(t))) + \lambda \left((v(t), r_v(t)) + \frac{\varepsilon}{2(b-a)} \left(\bar{\mathbf{B}}_{\mathbb{R}^{2n}} \times [-R(t), R(t)] \right) \right) \right] \bigcap \\ epi\ L, \qquad \text{for } t \in T; \\ \\ (y(t), L(y(t))) + \lambda(v(t), r_v(t)), \qquad \text{for } t \in [a, b] \setminus T. \end{cases}$$

It is straightforward to check that H is measurable (cf., for example, Deimling 1992 and Castaing and Valadier 1977). We are going to prove that $H(t)$ is nonempty whenever $t \in T$.

Indeed, let us now fix an arbitrary $t \in T$. Then there exists an index $i \in \{1, 2, \ldots, k\}$ for which $t \in (t_i - \eta_{t_i}, t_i + \eta_{t_i})$. We have

$$\|y(t) - \bar{y}(t_i)\| \le \|y(t) - \bar{y}(t)\| + \|\bar{y}(t) - \bar{y}(t_i)\| \le \|y - \bar{y}\|_{Y^*} + \frac{\delta_{t_i}}{2} < \delta + \frac{\delta_{t_i}}{2} \le \delta_{t_i}$$

because $\delta < \frac{\eta}{2} \le \frac{\eta_{t_i}}{2} < \frac{\delta_{t_i}}{2}$. Using that $\|y(t) - \bar{y}(t)\| < \delta$ and $|t - t_i| < \eta_{t_i}$ we obtain

$$|L(y(t)) - L(\bar{y}(t_i))| \le |L(y(t)) - L(\bar{y}(t))| + |L(\bar{y}(t)) - L(\bar{y}(t_i))| \le$$
$$\le \frac{\eta}{2} + \frac{\delta_{t_i}}{2} < \frac{\delta_{t_i}}{2} + \frac{\delta_{t_i}}{2} \le \delta_{t_i}.$$

The inequality $|t - t_i| < \eta_{t_i}$ implies that dist $(G_R(t), G_R(t_i)) < \frac{\varepsilon}{4(b-a)}$. Since $(v(t), r_v(t)) \in G_R(t)$, the above written inequality implies that there exists $(v_i, r_i) \in G_R(t_i)$ such that dist $((v(t), r_v(t)), (v_i, r_i)) < \frac{\varepsilon}{4(b-a)}$. As we have already proved that $(y(t), L(y(t)))$ is δ_{t_i}-close to $(\bar{y}(t_i), L(\bar{y}(t_i)))$, we can apply (4) for $t := t_i$, $y := y(t)$, $p := L(y(t))$, $v := v_i$, $r := r_i$ and obtain that

$$(y(t), L(y(t))) + \lambda \left((v_i, r_i) + \frac{\varepsilon}{4(b-a)} \left(\bar{\mathbf{B}}_{\mathbb{R}^{2n}} \times [-R(t), R(t)] \right) \right) \bigcap \text{epi } L \neq \emptyset.$$

On the other hand the inclusion

$$(v_i, r_i) \in (v(t), r_v(t)) + \frac{\varepsilon}{4(b-a)} \left(\bar{\mathbf{B}}_{\mathbb{R}^{2n}} \times [-R(t), R(t)] \right)$$

implies

$$(y(t), L(y(t))) + \lambda \left((v_i, r_i) + \frac{\varepsilon}{4(b-a)} \left(\bar{\mathbf{B}}_{\mathbb{R}^{2n}} \times [-R(t), R(t)] \right) \right) \subset$$

$$\subset (y(t), L(y(t))) + \lambda \left((v(t), r_v(t)) + \frac{\varepsilon}{4(b-a)} \left(\bar{\mathbf{B}}_{\mathbb{R}^{2n}} \times [-R(t), R(t)] \right) + \right.$$

$$\left. + \frac{\varepsilon}{4(b-a)} \left(\bar{\mathbf{B}}_{\mathbb{R}^{2n}} \times [-R(t), R(t)] \right) \right) \subset$$

$$\subset (y(t), L(y(t))) + \lambda \left((v(t), r_v(t)) + \frac{\varepsilon}{2(b-a)} \left(\bar{\mathbf{B}}_{\mathbb{R}^{2n}} \times [-R(t), R(t)] \right) \right).$$

Therefore

$$H(t) = \left[(y(t), L(y(t))) + \lambda \left((v(t), r_v(t)) + \tfrac{\varepsilon}{2(b-a)} \left(\bar{\mathbf{B}}_{\mathbb{R}^{2n}} \times [-R(t), R(t)] \right) \right) \right] \bigcap \operatorname{epi} L \neq \emptyset$$

for every $t \in T$. Remind that the multi-valued map H is measurable, so we apply Theorem III.6 from Castaing and Valadier (1977) to obtain that there exists a measurable selection $h : [a, b] \to \mathbb{R}^{2n+1}$ of H. Next we define $(w(t), r_w(t))$ for each $t \in [a, b]$ from the equality

$$(y(t) + \lambda w(t), L(y(t)) + \lambda r_w(t)) = h(t).$$

Hence

$$\|(w(t), r_w(t)) - (v(t), r_v(t))\| \leq \frac{\varepsilon}{2(b-a)} \text{ for } t \in T$$

and $(w(t), r_w(t)) = (v(t), r_v(t))$ for $t \in [a, b] \setminus T$. In particular, $\|v - w\|_{Y^*} \leq \varepsilon$ (without loss of generality we may think that $b - a \geq 1$). Further we estimate

$$\frac{\varphi(y + \lambda w) - \varphi(y)}{\lambda} - r = \int_a^b \left[\frac{L(y(t) + \lambda w(t)) - L(y(t))}{\lambda} - r_v(t) \right] dt =$$

$$= \int_T \left[\frac{L(y(t) + \lambda w(t)) - L(y(t))}{\lambda} - r_v(t) \right] dt +$$

$$+ \int_{[a,b] \setminus T} \left[\frac{L(y(t) + \lambda w(t)) - L(y(t))}{\lambda} - r_v(t) \right] dt.$$

Since $(y(t) + \lambda w(t), L(y(t)) + \lambda r_w(t)) \in H(t) \subset \operatorname{epi} L$ for each $t \in T$, we have

$$L(y(t) + \lambda w(t)) \leq L(y(t)) + \lambda r_w(t),$$

and therefore

$$\frac{L(y(t)+\lambda w(t))-L(y(t))}{\lambda}-r_v(t)\leq r_w(t)-r_v(t)\leq\frac{\varepsilon}{2(b-a)}.$$

After integrating this inequality, we obtain

$$\int_T\left[\frac{L(y(t)+\lambda w(t))-L(y(t))}{\lambda}-r_v(t)\right]dt\leq\frac{\varepsilon}{2(b-a)}\cdot\text{ meas }(T)\leq\frac{\varepsilon}{2}.$$

On the other hand

$$\int_{[a,b]\setminus T}\left[\frac{L(y(t)+\lambda w(t))-L(y(t))}{\lambda}-r_v(t)\right]dt=$$

$$=\int_{[a,b]\setminus T}\left[\frac{L(y(t)+\lambda v(t))-L(y(t))}{\lambda}-r_v(t)\right]dt<\frac{\varepsilon}{2}$$

The last inequality follows from the standing assumption **(SA)**, the choice of δ_0 and the definition of T (meas $[a,b]\setminus T<\delta_0$).

Then

$$\frac{\varphi(y+\lambda w)-\varphi(y)}{\lambda}-r<\varepsilon$$

which implies that

$$(y,\varphi(y))+\lambda(w,r+\varepsilon)=(y+\lambda w,\varphi(y)+\lambda(r+\varepsilon))\in\text{Epi }\varphi.$$

Because $(w,r+\varepsilon)\in(v,r)+\varepsilon\bar{\mathbf{B}}_{Y^*\times\mathbb{R}}$ we have

$$\left[(y,\varphi(y))+\lambda\left((v,r)+\varepsilon\bar{\mathbf{B}}_{Y^*\times\mathbb{R}}\right)\right]\bigcap\text{Epi }\varphi\neq\emptyset.$$

Taking into account that $q\geq\varphi(y)$, we obtain

$$\left[(y,q)+\lambda\left((v,r)+\varepsilon\bar{\mathbf{B}}_{Y^*\times\mathbb{R}}\right)\right]\bigcap\text{Epi }\varphi\neq\emptyset$$

which completes the proof. ◇

Let us introduce the set of all summable selections of the Clarke subdifferential of L along the optimal couple $\bar{y}(t)$, $t\in[a,b]$, i.e.

$$K:=\{w\in Y:\ w(t)\in\partial_C L(\bar{y}(t))\text{ a.e. in }[a,b]\}.$$

Lemma 3.7 *Let the set $\partial_C L(\bar{y}(t))$ be nonempty, bounded for almost every $t \in [a,b]$ and let all its measurable selections be summable. Then the prepolar C_0 of the cone C coincides with the set*

$$\tilde{K} := \{\lambda(w,-1) : \lambda \geq 0,\ w \in K\}.$$

Proof Recall that

$$C_0 = \{(y,s) \in Y \times \mathbb{R} : \langle v, y\rangle + rs \leq 1 \text{ whenever } (v,r) \in C\} =$$
$$= \{(y,s) \in Y \times \mathbb{R} : \langle v, y\rangle + rs \leq 0 \text{ whenever } (v,r) \in B_R\}$$

because the set C is a cone. As C_0 is a cone as well, it is enough to check that $(w,-1) \in C_0$ for each $w \in K$ to conclude that $\tilde{K} \subset C_0$. Indeed, let us fix an arbitrary $(w,-1)$ with $w \in K$. Let $(v,r) \in B_R$, i.e. there exists a summable function r_v with $r = \displaystyle\int_a^b r_v(t)dt$ and $(v(t), r_v(t)) \in G_R(t)$ for almost all $t \in [a,b]$. Then for almost each $t \in [a,b]$ we have that $w(t) \in \partial_C L(\bar{y}(t))$ and $(v(t), r_v(t)) \in G_R(t)$, hence

$$\langle w(t), v(t)\rangle - r_v(t) = \langle (w(t),-1), (v(t), r_v(t))\rangle \leq 0.$$

Integrating this inequality, we obtain that

$$\langle w, v\rangle - r = \int_a^b (\langle w(t), v(t)\rangle - r_v(t))\ dt \leq 0,$$

and therefore $(w,-1) \in C_0$.

Next we show the reverse inclusion $C_0 \subset \tilde{K}$. Let us assume the contrary, i.e. there exists $(w,s) \in C_0 \setminus \tilde{K}$. As $(\mathbf{0}, 1) \in G_R(t)$ for all $t \in [a,b]$, then $(\mathbf{0}, b-a) \in B_R$, and hence $\langle (w,s), (\mathbf{0}, b-a)\rangle \leq 0$. Therefore $s \leq 0$. If $s < 0$, then $w_0 := \frac{w}{|s|} \notin K$ (because $K \times \{-1\}$ generates the cone $\tilde{K}$). Hence meas $(T) > 0$ where

$$T := \{t \in [a,b] : w_0(t) \notin \partial L(\bar{y}(t))\}.$$

Then for each $t \in T$ there exists $(v, r_v) \in G_R(t)$ with $\langle w_0(t), v\rangle - r_v > 0$. For each positive integer n we set

$$E_n := \left\{ t \in [a,b] : \text{ there exists } (v, r_v) \in G_R(t) \text{ with } \langle w_0(t), v\rangle - r_v > \frac{1}{n} \right\}.$$

As $T = \bigcup_{n=1}^{\infty} E_n$ we can find a positive integer n_0 such that meas $(E_{n_0}) > 0$. We consider the following multi-valued map $\Gamma : [a, b] \to \mathbb{R}^{2n+1}$ defined by

$$\Gamma(t) := \begin{cases} \left\{ (v, r_v) \in \mathbb{R}^{2n+1} : \langle w_0(t), v \rangle - r_v > \dfrac{1}{n_0} \right\} \bigcap G_R(t), & \text{if } t \in E_{n_0}; \\ (\mathbf{0}, 0), & \text{if } t \notin E_{n_0}. \end{cases}$$

The measurability of Γ implies the existence of a measurable selection γ of Γ, i.e. $\gamma(t) := (v(t), r_v(t))$, where $(v(t), r_v(t)) \in \Gamma(t) \subset G_R(t)$ for $t \in E_{n_0}$ and $(v(t), r_v(t)) = (\mathbf{0}, 0) \in G_R(t)$ for $t \notin E_{n_0}$.

Since $\left(v(\cdot), \int_a^b r_v(t)\, dt \right) \in B_R$, we have

$$0 \geq \int_a^b (\langle w_0(t), v(t) \rangle - r_v(t))\, dt =$$

$$= \int_{[a.b]\setminus E_{n_0}} (\langle w_0(t), \mathbf{0} \rangle - 0)\, dt + \int_{E_{n_0}} (\langle w_0(t), v(t) \rangle - r_v(t))\, dt \geq$$

$$\geq 0 + \text{ meas } E_{n_0} . \frac{1}{n_0} > 0.$$

The obtained contradiction shows that $s = 0$. Then

$$\int_a^b \langle w(t), v(t) \rangle dt \leq 0 \text{ for every } (v, r) \in B_R.$$

This inequality and the same reasoning as above imply that $\langle w(t), \zeta \rangle \leq 0$ whenever $(\zeta, r) \in \hat{T}_{epi\ L}(\bar{y}(t), L(\bar{y}(t))$ for almost every $t \in [a, b]$.

Let us assume that for some $t \in [a, b]$ we have that $w(t) \neq 0$, $\partial_C L(\bar{y}(t))$ is nonempty and bounded and $\langle w(t), \zeta \rangle \leq 0$ whenever $(\zeta, r) \in \hat{T}_{epi\ L}(\bar{y}(t), L(\bar{y}(t))$. Let $\xi \in \partial_C L(\bar{y}(t))$ (i.e. $(\xi, -1) \in \left(\hat{T}_{epi\ L}(\bar{y}(t), L(\bar{y}(t)) \right)_0$). Then the convexity of the Clarke normal cone yields that for each $\lambda \in (0, 1]$ we have that $(\lambda\xi + (1 - \lambda)w(t), -\lambda) \in \left(\hat{T}_{epi\ L}(\bar{y}(t), L(\bar{y}(t)) \right)_0$, hence

$$\xi + \left(\frac{1}{\lambda} - 1 \right) w(t) \in \partial_C L(\bar{y}(t)).$$

Clearly, for all $\lambda > 0$ small enough the norm of the above written element of $\partial_C L(\bar{y}(t))$ can be made as large as we want. This contradicts the boundedness of

$\partial_C L(\bar{y}(t))$. Therefore, $w(t) = 0$ almost everywhere on $[a, b]$, and so $(\mathbf{0}, 0) \in C_0 \setminus \tilde{K}$ which is impossible. ◇

Lemma 3.8 *Let us assume that* $\partial_C L(\bar{y}(t)) \subset R(t)\bar{\mathbf{B}}_{\mathbb{R}^n}$ *for almost every* $t \in [a, b]$. *Then* C *is weak star closed and* $C = \tilde{K}^0$. *Moreover, if the standing assumption (SA) holds true, then* C *is a uniform tangent cone to* $Epi\ \varphi$.

Proof Let us fix $(\bar{v}, \bar{r})$ in the weak star closure of $C \subset Y^* \times \mathbb{R}$.

Step 1: Definition of $\bar{r}_v$. Because $\overline{C}^{w^*}$ is convex, Y is separable, it is enough to assume that $(\bar{v}, \bar{r})$ is weak star limit of a sequence of elements of C (cf., for example, see Corollary 4.45 from Fabian et al. 2001), i.e. there exists a sequence $\{(v_n, r_n)\}_{n=1}^{\infty} \subset C$, where $r_n = \int_a^b r_{v_n}(t)dt$, which is weak star convergent and its limit in the weak star topology is $(\bar{v}, \bar{r})$.

The sequence $\{v_n\}_{n=1}^{\infty}$ is weak star convergent, and hence it is bounded. Let $\|v_n\| \leq M$ for each positive integer n. Then we put

$$\bar{r}_{v_n}(t) := \min \left\{ s \in \mathbb{R} : \ (v_n(t), s) \in \hat{T}_{epi\ L}(\bar{y}(t), L(\bar{y}(t))) \right\}.$$

It is clear that

$$\bar{r}_{v_n}(t) \in [-\|v_n\| R(t), \|v_n\| R(t)] \subset [-M R(t), M R(t)]$$

almost everywhere and that the so defined mapping is measurable. Because

$$\{\bar{r}_{v_n}\}_{n=1}^{\infty} \subset W := \{w \in L^1([a, b], \mathbb{R}) : \ |w(t)| \leq M R(t) \text{ a. e. in } [a, b]\}$$

and the set W is weakly compact, there exists a subsequence $\{\bar{r}_{v_{n_k}}\}_{k=1}^{\infty}$ which is weakly convergent to some $\bar{r}_v \in W$.

Step 2: Proof that $(v(\cdot), \bar{r}_v(\cdot))$ **is a selection of the Clarke tangent cone, i.e.**

$$(\bar{v}(t), \bar{r}_v(t)) \in \hat{T}_{epi\ L}(\bar{y}(t), L(\bar{y}(t))) \ \text{ a. e. in } [a, b].$$

Let us assume the contrary, i.e. there exists $T \subset [a, b]$ with positive measure where the above written inclusion is violated, i.e. there exists $(\xi, \eta) \in \left(\hat{T}_{epi\ L}(\bar{y}(t), L(\bar{y}(t)))\right)_0 \bigcap \bar{\mathbf{B}}_{\mathbb{R}^{2n+1}}$ such that $\langle \xi, \bar{v}(t) \rangle + \eta \bar{r}_v(t) > 0$ for each $t \in T$. Because $T = \bigcup_{m=1}^{\infty} T_m$, where

$$T_m := \left\{ t \in [a, b] : \begin{array}{l} \text{there exists } (\xi, \eta) \in \left(\hat{T}_{epi\ L}(\bar{y}(t), L(\bar{y}(t)))\right)_0 \bigcap \bar{\mathbf{B}}_{\mathbb{R}^{2n+1}} \\ \\ \text{such that } \langle \xi, \bar{v}(t) \rangle + \eta \bar{r}_v(t) > \dfrac{1}{m} \end{array} \right\},$$

there exists a positive integer m_0 such that meas $(T_{m_0}) > 0$.

We consider the following multi-valued map $\Gamma : [a, b] \to \mathbb{R}^{2n+1}$ defined by

$$\Gamma(t) := \begin{cases} \left\{(\xi, \eta) \in \left(\hat{T}_{epi\ L}(\bar{y}(t), L(\bar{y}(t)))\right)_0 : \langle \xi, \bar{v}(t)\rangle + \eta \bar{r}_v(t) > \dfrac{1}{m_0}\right\} \\ \bigcap \bar{\mathbf{B}}_{\mathbb{R}^{2n+1}}, & \text{if } t \in T_{m_0}; \\ (\mathbf{0}, 0), & \text{if } t \notin T_{m_0}. \end{cases}$$

The measurability of Γ implies the existence of a measurable selection γ of Γ, i.e. $\gamma(t) := (\xi(t), \eta(t))$, where $(\xi(t), \eta(t)) \in \Gamma(t) \subset \left(\hat{T}_{epi\ L}(\bar{y}(t), L(\bar{y}(t)))\right)_0$ for $t \in T_{m_0}$ and $(\xi(t), \eta(t)) = (\mathbf{0}, 0)$ for $t \notin T_{m_0}$.

Since $\left(v_{n_k}(t),\ \bar{r}_{v_{n_k}}(t)\right) \in \hat{T}_{epi\ L}(\bar{y}(t), L(\bar{y}(t)))$ a.e. in $[a, b]$, we have

$$\int_a^b \left(\langle \xi(t), v_{n_k}(t)\rangle + \eta(t)\bar{r}_{v_{n_k}}(t)\right) dt \le 0\,.$$

Because $\bar{v}$ is a weak star limit of $\{v_{n_k}\}_{k=1}^{\infty}$, $\bar{r}_v$ is a weak limit of $\{r_{v_{n_k}}\}_{k=1}^{\infty}$ and clearly $\xi \in L^{\infty}([a, b], \mathbb{R}^{2n}) \subset Y$, $\eta \in L^{\infty}([a, b], \mathbb{R})$, we have that

$$\int_a^b \left(\langle \xi(t), v_{n_k}(t)\rangle + \eta(t)\bar{r}_{v_{n_k}}(t)\right) dt \longrightarrow_{k\to\infty} \int_a^b (\langle \xi(t), \bar{v}(t)\rangle + \eta(t)\bar{r}_v(t))\ dt.$$

Therefore

$$\int_a^b (\langle \xi(t), \bar{v}(t)\rangle + \eta(t)\bar{r}_v(t))\ dt \le 0.$$

On the other hand-side

$$\int_a^b (\langle \xi(t), \bar{v}(t)\rangle + \eta(t)\bar{r}_v(t))\ dt \ge \frac{1}{m_0} \cdot \text{meas}\,(T_{m_0}) > 0,$$

which is a contradiction. Hence $(\bar{v}(t), \bar{r}_v(t)) \in \hat{T}_{epi\ L}(\bar{y}(t), L(\bar{y}(t)))$ almost everywhere in $[a, b]$.

Step 3: Weak star closedness of C. According to Step 2 together with the estimate for the values of $\bar{r}_v$, we obtain that

$$\left(\bar{v}, \int_a^b \bar{r}_v(t)dt\right) \in MB_R \subset C\,.$$

Also, we have

$$\bar{r} = \lim_{k\to\infty} r_{n_k} = \lim_{k\to\infty} \int_a^b r_{v_{n_k}}(t)dt \ge \lim_{k\to\infty} \int_a^b \bar{r}_{v_{n_k}}(t)dt = \int_a^b \bar{r}_v(t)dt.$$

Now the above estimate, convexity of C and the fact that $(\mathbf{0}, b-a) \in C$ imply $(\bar{v}, \bar{r}) \in C$, which completes the proof that C is weak star closed.

Step 4: Uniformity of C and its relation to $\tilde{K}^0$. Let $(\bar{v}, \bar{r}) \in C \cap \bar{\mathbf{B}}_{Y^* \times \mathbb{R}}$. Setting (as above)

$$\bar{r}_{\bar{v}}(t) := \min \left\{ s \in \mathbb{R} : \ (\bar{v}(t), s) \in \hat{T}_{epi\ L}(\bar{y}(t), L(\bar{y}(t))) \right\},$$

we obtain that $\bar{r}_{\bar{v}}$ is a summable function satisfying $\bar{r}_{\bar{v}}(t) \in [-R(t) R(t)]$ almost everywhere in $[a, b]$, therefore $\left(\bar{v}, \int_a^b \bar{r}_{\bar{v}}(t)\, dt\right) \in B_R$. It is clear that

$$\bar{r} - \int_a^b \bar{r}_{\bar{v}}(t)\, dt \leq 1 - \left(- \int_a^b R(t)\, dt \right) = 1 + \int_a^b R(t)\, dt =: N .$$

Then $\frac{\bar{r} - \int_a^b \bar{r}_{\bar{v}}(t)\, dt}{b-a} (\mathbf{0}, b-a) \in N \cdot B_R$ and

$$(\bar{v}, \bar{r}) = \left(\bar{v}, \int_a^b \bar{r}_{\bar{v}}(t)\, dt \right) + \frac{\bar{r} - \int_a^b \bar{r}_{\bar{v}}(t)\, dt}{b-a} (\mathbf{0}, b-a)$$

together with the convexity of B_R imply that $C \cap \bar{\mathbf{B}}_{Y^* \times \mathbb{R}} \subset 2N \cdot B_R$. The last inclusion and the fact that B_R is a uniform tangent set to $Epi\ \varphi$ prove that C is a uniform tangent cone to $Epi\ \varphi$.

It remained to show that $C = \tilde{K}^0$. This is a direct consequence of the fact that $\tilde{K} = C_0$, C is a weak star closed convex set containing the origin, and the bipolar theorem. ◇

Lemma 3.9 *Let us assume that $\partial_C L(\bar{y}(t)) \subset R(t) \bar{\mathbf{B}}_{\mathbb{R}^n}$ for almost every $t \in [a, b]$. Then C has a nonempty interior and $C^0 = J(C_0) = J(\tilde{K})$, where J is the canonical embedding of $Y \times \mathbb{R}$ in $Y^{**} \times \mathbb{R}$.*

Proof Let us first note that under the assumptions of the lemma the set

$$K = \{ w \in Y : \ w(t) \in \partial_C L(\bar{y}(t)) \text{ a.e. in } [a, b] \}$$

is weakly compact in Y as all its elements are dominated by the summable function R. Hence the set

$$K^{-1} := \{ (w, -1) \in Y \times \mathbb{R} : \ w \in K \}$$

is weakly compact in $Y \times \mathbb{R}$ as well. Therefore $J(K^{-1})$ is weakly star compact in $Y^{**} \times \mathbb{R}$, hence it is weakly star closed. Let us prove that the set

$$J(\tilde{K}) = J \left(\bigcup_{\lambda \geq 0} \lambda K^{-1} \right) = \bigcup_{\lambda \geq 0} \lambda J \left(K^{-1} \right) \quad \text{is weak star closed.}$$

Indeed, let $(\bar{w}, \bar{\lambda}) \in \overline{J(\tilde{K})}^{w^*}$, that is there exist nets (generalised sequences)

$$\{\lambda_\alpha\}_{\alpha \in I} \subset [0, +\infty),\ \{w_\alpha\}_{\alpha \in I} \subset K \ \text{ with } \ (\bar{w}, \bar{\lambda}) = w^* - \lim_{\alpha \in I} \lambda_\alpha (\bar{J}(w_\alpha), -1),$$

where $J = (\bar{J}, id_{\mathbb{R}})$. First we note that the above implies that $\bar{\lambda} \leq 0$. Also, the weak compactness of K yields the existence of a weakly convergent subnet $\{w_{\alpha_\beta}\}_{\beta \in \bar{I}}$ of $\{w_\alpha\}_{\alpha \in I}$. Let $w_0 = w - \lim_{\beta \in \bar{I}} w_{\alpha_\beta}$. Then $w_0 \in K$ and $(-\bar{\lambda})\bar{J}(w_0) = w^* - \lim_{\beta \in \bar{I}} \lambda_{\alpha_\beta} \bar{J}\left(w_{\alpha_\beta}\right)$. Therefore

$$(\bar{w}, \bar{\lambda}) = (-\bar{\lambda}) \left(\bar{J}(w_0), -1\right) \in (-\bar{\lambda}) J\left(K^{-1}\right) \subset J(\tilde{K}).$$

We have from the previous lemma that $\left[J(\tilde{K})\right]_0 = \tilde{K}^0 = C$. Therefore the weak star closedness of $J(\tilde{K})$ and the bipolar theorem imply

$$C^0 = \left(\left[J(\tilde{K})\right]_0\right)^0 = \overline{\text{conv}}^{w^*}\left(\{\mathbf{0}\} \cup J(\tilde{K})\right) = \overline{J(\tilde{K})}^{w^*} = J(\tilde{K}).$$

It remains to show that C has nonempty interior. As weak compacts are bounded, there exists $M > 0$ such that $K \subset M \cdot \bar{\mathbf{B}}_Y$. Then

$$\tilde{K} \subset \{\lambda(w, -1) \in Y \times \mathbb{R} :\ \lambda \geq 0,\ \|w\| \leq M\}.$$

Therefore

$$C = \tilde{K}^0 \supset \{\lambda(w, -1) \in Y \times \mathbb{R} :\ \lambda \geq 0,\ \|w\| \leq M\}^0 =$$

$$= \left\{(v, r) \in Y^* \times \mathbb{R} :\ \langle v, w \rangle - r \leq 0 \ \text{ whenever } w \in Y,\ \|w\| \leq M\right\} =$$

$$= \left\{(v, r) \in Y^* \times \mathbb{R} :\ \|v\| \leq \frac{r}{M}\right\}$$

and it is clear that the set written above has nonempty interior. ◇

Theorem 3.10 *Let $L : \mathbb{R}^n \times \mathbb{R}^n \longrightarrow \mathbb{R}$ be a continuous mapping. Let $\bar{y} = (\bar{x}, \bar{u})$ be a solution of* **(VP)**. *Let us assume that there exists a nonnegative summable function R on $[a, b]$ such that $\partial_C L(\bar{y}(t)) \subset R(t)\bar{\mathbf{B}}_{\mathbb{R}^n}$ for almost every $t \in [a, b]$. Let the standing assumption (SA) hold true. Then there exists an absolutely continuous function $p : [a, b] \longrightarrow \mathbb{R}^n$ such that*

$$(\dot{p}(t), p(t)) \in \partial_C L(\bar{y}(t)) \text{ a.e. in } [a, b].$$

Proof We are going to apply the Lagrange multipliers theorem (Theorem 3.8) from Krastanov (2017). Without loss of generality we may assume that $\bar{y} = (\bar{x}, \bar{u})$

is a strict minimum of (VP). As C is a uniform tangent cone to $Epi\ \varphi$ (see Lemma 3.8) and (trivially) $A = P \cap Q - (\bar{x}, \bar{u})$ is a uniform tangent cone to $P \cap Q$, the only assumption to be checked is the quasisolidity of the set $D := \big((A \times (-\infty, 0]) \cap \bar{\mathbf{B}}_{Y^* \times \mathbb{R}}\big) - \big(C \cap \bar{\mathbf{B}}_{Y^* \times \mathbb{R}}\big)$. The convex cone C has a nonempty interior according to Lemma 3.9, thus D has nonempty interior as well, hence it is quasisolid.

Now the Lagrange multipliers theorem yields the existence of a non-trivial pair $(\xi, \eta) \in Y^{**} \times R$ such that $\eta \in \{0, 1\}, \xi \in A^0$ and $-(\xi, \eta) \in C^0$. Lemma 3.9 implies $C^0 = J(\tilde{K})$. If we assume that $\eta = 0$, then $\xi = \mathbf{0}$ as well because of the definition of $\tilde{K}$, which is a contradiction. Therefore $\eta = 1$ and $-(\xi, \eta) = J(w, -1)$, where $w \in K$. On the other hand side, $\xi \in A^0$ now implies $w \in -A_0 = A_0$. Lemma 2.1 gives that

$$A_0 = \{(y, v) \in X \times X : v \text{ absolutely continuous}, \dot{v}(t) = y(t) \text{a.e. in [a,b]}\}.$$

Therefore $w \in K \cap A_0$ implies the existence of an absolutely continuous function $p : [a, b] \longrightarrow \mathbb{R}^n$ such that

$$(\dot{p}(t), p(t)) \in \partial_C L(\bar{y}(t)) \text{ a.e. in } [a, b]\,.$$

◇

Acknowledgements This work was supported by the Bulgarian Ministry of Science and Higher Education—National Fund for Science Research under grant DFNI-IO2/10 and by the Sofia University "St. Kliment Ohridski" under contract No. 80-10-220/24.04.2017.

References

M. Brokate, Pontryagin's principle for control problems in age-dependent population dynamics. J. Math. Biol. **23**, 75–101 (1985)

C. Castaing, M. Valadier, Convex Analysis and Measurable Multifunctions, in *Lecture Notes in Mathematics*, vol. 580 (Springer, Berlin, 1977)

F. Clarke, Functional Analysis, in *Calculus of Variations and Optimal Control* (Springer, Berlin, 2013)

K. Deimling, *Multivalued Differential Equations* (Walter de Gruyter, Berlin, 1992)

M. Fabian, P. Habala, P. Hajek, V. Montesinos, J. Pelant, V. Zizler, *Functional Analysis and Infinite-Dimensional Geometry* (Springer, Berlin, 2001)

G. Feichtinger, G. Tragler, V. Veliov, Optimality conditions for age-structured control systems. J. Math. Anal. Appl. **288**, 47–68 (2003)

M.E. Gurtin, R.C. MacCamy, Nonlinear age-dependent population dynamics. Arch. Rational Mech. Anal. **54**, 281–300 (1974)

A.D. Ioffe, Variational Analysis of Regular Mappings Theory and Applications (Springer, 2017)

M.I. Krastanov, N.K. Ribarska, Nonseparation of sets and optimality conditions. SIAM J. Control. Optim. **55**(3):1598–1618 (2017)

M.I. Krastanov, N.K. Ribarska, Ts.Y. Tsachev, A pontryagin maximum principle for infinite-dimensional problems. SIAM J. Control. Optim. **49**(5), 2155–2182 (2011)

M.I. Krastanov, N.K. Ribarska, Ts.Y. Tsachev, On the geometry of the pontryagin maximum principle in banach spaces. Set-Valued Var. Anal. **25**(3), 443–463 (2015). https://doi.org/10.1007/s11228-015-0316-9

St. Lojasiewicz, jr., Some theorems of Scorza Dragoni type for multifunctions with application to the problem of existence of solutions for differential multivalued equations. Math. Control Theory, Banach Cent. Publ. **14**, 625–643 (1985)

R.T. Rockafellar, Clarke's tangent cones and the boundaries of closed sets in R^n. Nonlinear Anal. Theory Methods Appl. **3**(1), 145–154 (1979)

On the Generalized Duality Principle for State-Constrained Control and State Estimation Under Impulsive Inputs

Alexander B. Kurzhanski

Abstract This paper compares solutions to the problem of optimal control under scale of inputs that range from impulses of higher order to differentiable functions with that of guaranteed state estimation under unknown but bounded inputs taken within the same range. Indicated is a similarity between duality of these problems in the systems sense and in the sense of mathematical programming.

1 Introduction

In this paper we deal with the problem of optimal control under state constraints subject to controls that range from impulses of higher order to differentiable input functions. These are compared with the problem of optimal state estimation for systems with unknown but bounded inputs within the same range (Kurzhanski and Veliov 1993; Kurzhanski and Varaiya 2014).We indicate the interrelations between their solutions, emphasizing a *system duality* between these problems. We then formulate a *generalized duality principle* indicating in particular the analogies between primal and dual variables for solving the problem of optimal state-constrained control in the mathematical sense of convex analysis, (Rockafellar 1999; Rockafellar and Wets 2005) with similar variables for the problem of optimal state estimation (observation) treated in the same sense. There thus exists a clear analogy of *system duality* between problems of optimal state-constrained control and those of optimal state estimation, apart from the *duality in mathematical sense* between primal and dual variables for each type of the considered problems. The important issue is the indication of functional spaces for solving such problems

A. B. Kurzhanski (✉)
Moscow State University, Moscow, Russia

University of California at Berkeley, Berkeley, CA, USA
e-mail: kurzhans@mail.ru

G. Feichtinger et al. (eds.), *Control Systems and Mathematical Methods in Economics*, Lecture Notes in Economics and Mathematical Systems 687,
https://doi.org/10.1007/978-3-319-75169-6_7

within the mentioned range. Such results are produced here and summarized in tabular form.

This paper continues further developments of topics motivated earlier by publications (Kalman 1960; Krasovski 1964; Aubin 1991; Bensoussan and Lions 1982; Kurzhanski 1983; Krasovski and Subbotin 1988; Veliov 1997).

1.1 Duality in the Mathematical Sense

The typical **Primal Problem of Optimal Control** for linear-convex systems considered in this paper consists of *minimizing a given cost functional* along the trajectories of *a given controlled system* under constraints that may be

(i) given class of controls and constraints on the controls,
(ii) given boundary conditions,
(iii) given state constraints,
(iv) restrictions that depend on available advanced or on-line information on the parameters of the controlled process which may be (a) complete or (b) incomplete due to *uncertainty in the system model and system inputs*.

These may be selected within various types of functional spaces that depend

(a) on the physical nature of the problem and related properties of the system,
(b) the mathematical tools selected for the solution.

The solution requires to select *an open loop control* (a function of time) or *a closed loop (feedback) control*—a function of both time and the *generalized system state*.[1]

This is typically reached by solving a **Dual Problem of Optimization** that consists of

(1) specifying a generalized Lagrangian by introducing appropriate multipliers used to form a *dual cost functional.*
(2) reducing the solution to a *maximization* of the dual cost functional *over the Lagrangian multipliers*along the solutions of an appropriate *adjoint system*, without any additional constraints.
(3) The Dual Problem of Optimization may be interpreted as one of optimal maximizing control for the adjoint system where the functional Lagrange multipliers are interpreted as the controls.

[1]The present paper deals with open-loop control. The problem of closed-loop impulse control that depends on information of type (iv) is the topic of paper (Kurzhanski and Daryin 2008).

2 Duality Scale in State-Constrained Impulse Control

The problems of optimal control under state constraints may be posed in different functional spaces leading to solutions of dual problems which are presented accordingly in terms of respective conjugate spaces. Arrays of such problems form a scale discussed below. Here we consider stationary systems. (A similar situation arises in the problem of dynamic state estimation which is treated below following the present one).

In this paper the collection of functional spaces within which we treat the problems of state-constrained control vary from those where controls are selected from the space of generalized functions that include high order impulses to those taken from the space of finite times continuously differentiable, smooth functions. In between these two emphasized classes lies a broad range of other functional spaces. Such range is indicated in two tabular scales that follows further.

We begin by introducing some definitions and notations.

2.1 *The Space of Generalized Functions: Some Definitions and Notations*

Consider the linear space $D_k^n[\alpha, \beta] = D_k$ of finite, k-times continuously differentiable n-vector functions $\varphi(\cdot)$, each of which is such that $\varphi(t) = 0$ beyond $[\alpha, \beta]$. A sequence $\varphi_j(t), j = 1, \dots, \infty$, tends to zero in interval $[\alpha, \beta]$ if it uniformly converges to zero together with its derivatives $d^i\varphi_j(t)/dt^i$ of order "i" up to "k" on this interval. Here k is a nonnegative integer and D_0 is taken continuous. This is the space D_k^n of n-dimensional *basic functions* with norm

$$\|\varphi(\cdot)\|_{D_k} = \max_{i,t}\{\|d^i\varphi(t)/dt^i\| \mid i = 0, \dots, k, \quad t \in [t_\alpha, t_\beta]\}.$$

A *generalized function* or (*distribution*) f (over D_k^n) is a linear, continuous functional $< f, \varphi >$ over D_k^n. The collection of all such functionals f form the space D_k^{n*} conjugate to D_k^n. The properties of such spaces are indicated in references (Schwartz 1950–1951, 1966; Gelfand and Shilov 1991; Agranovich 2008). A differential equation of type $dx(t)/dt = F(t)x(t), +H(t)u(t))$ is called *an equation in distributions* if the products of type $< f(\cdot), \varphi(\cdot) >$ are equal for both sides of the equation and for each basic function $\varphi(\cdot) \in D_k^n$.

A definite integral of type $\int_{t_\alpha}^t ds$, with varying upper limit t, for a generalized function h, given at interval $t \in [t_\alpha, t_\beta + 0] \subset [\alpha, \beta]$, is defined as

$$\left\langle \int_{t_\alpha}^t h d\tau, \varphi \right\rangle = \left\langle h, \int_t^\beta \left[\varphi_0(\tau) \int_\alpha^\beta \varphi(\xi)d\xi - \varphi(\tau)\right] d\tau \right\rangle, \; t \le t_\beta + 0,$$

where φ is any basic function from $D_k^n[\alpha, \beta]$ and $\varphi_0 \in D_k^n[\alpha, \beta]$ is concentrated on $[\alpha, t_\alpha]$ with $\int_\alpha^{t_\alpha} \varphi_0(\tau)d\tau = 1$. The value of such integral does not depend on φ_0.

Note that a linear operation given by product of type $< \mathsf{W}(\cdot), \varphi(\cdot) >$ for a generalized function $\mathsf{W}(\cdot)$ of order k may be presented in terms of function of bounded variation as

$$< \mathsf{W}(\cdot), \varphi(\cdot) >= \sum_{i=0}^{k-1} \int_\alpha^{\beta+0} (d^i\varphi(t)/dt^i)^T d\mathcal{W}_i(t), \ \{\mathcal{W}_0(\cdot), \ldots, \mathcal{W}_{k-1}(\cdot)\} = \mathsf{W}(\cdot), \tag{1}$$

where $\mathcal{W}_i(\cdot) \in \mathbf{BV}[\alpha, \beta]$ belong to the space of functions with bounded variation on $[\alpha, \beta + 0]$.

Systems which control inputs in the class of high-order distributions are treated here along the lines of paper (Kurzhanski and Osipov 1969).

2.2 High Order Impulse Control

We begin with the problem of *state-constrained impulse control under higher impulses* (SCCHI).

Consider a differential equation in distributions

$$dx/dt = Ax + Bu + x^{(\alpha)}\delta^{(k)}(t - t_\alpha), \quad t \in [t_\alpha, t_\beta], \tag{2}$$

with admissible control $u(\cdot) \in D_k^{p*}[\alpha, \beta], \ u(t) \in \mathbb{R}_k^p \ k \geq 0$, defined within interval $[t_\alpha, t_\beta], \ \alpha < t_\alpha \leq t_\beta < \beta$, being a distribution of higher order that includes delta-functions and their higher derivatives. It generates a solution $x(\cdot) \in D_{k-1}^{n*}[\alpha, \beta]$ described as

$$x(t) = \mathcal{G}(t, t_\alpha) \int_{t_\alpha}^t \mathcal{G}(t_\alpha, \tau)\Big[Bu + x^{(\alpha)}\delta^{(k)}(t - t_\alpha)\Big]d\tau$$

with $\mathcal{G}(t, t_\alpha)$ being the fundamental matrix for equation $dx/dt = Ax$ and the integral taken as defined in previous subsection.

Also introduced is a *state constraint* which is here applied to an output $y(\cdot) \in D_0^m[\alpha, \beta]$ which is an "ordinary" function. A terminal condition is given accordingly. Namely, we have

$$y(t) = N\mathbf{x}(t), \ \ \mathbf{x}(t) = (x * \zeta_+^{k-1})(t, t_\alpha), \ \ \|y(t)\| \leq \nu, \ \ t \in [t_\alpha, t_\beta], \tag{3}$$

$$\mathbf{x}^\alpha = \sum_{j=0}^{k-1} (-1)^j A^j x^{(\alpha)}, \ \ \mathbf{x}(t_\beta) = \mathbf{x}^\beta \in \mathcal{M}.$$

Here $N \in \mathbb{R}^{m\times n}$, $\mathcal{M} \subseteq \mathbb{R}^n$ is a given convex compact set and $\mathbf{x}(\cdot) \in D_0^n[\alpha, \beta]$, $(D_0 = \mathcal{C})$

Symbol $(x * \zeta_+^{(k-1)})(t, t_\alpha)$ stands for the operation of *convolution*, namely

$$(x * \zeta_+^{(k-1)})(t, t_\alpha) = \int_{t_\alpha}^{t} x(\tau)\zeta_+^{k-1}(t-\tau)d\tau,$$

where $\zeta^{(l)} = t^l/l!$ if $l \geq 1$, $\zeta^{(0)}(t) = 1$, $\zeta^{(-1)} = \delta(t)$, and $f_+(t) \equiv f(t)$ if $t \geq 0$, $f_+(t) = 0$ if $t < 0$.

Problem 2.1 (Existence-SCCHI) *Given $\mu > 0$, specify conditions that ensure the existence of control $u(\cdot) \in D_k^{p*}$ which transfers the system (2) from $\mathbf{x}^\alpha$ to $\mathbf{x}^\beta \in \mathcal{M}$ under given state constraint (3), with $\|u(\cdot)\|_{D_k^{p*}} \leq \mu$. Here*

$$\|u(\cdot)\|_{D_k^{p*}} = \Big\{\sum_{i=0}^{k} \max < U_i(\cdot), d^i\varphi(\cdot)/dt^i > \mid \| d^i\varphi(\cdot)/dt^i(\cdot)\|_C \leq 1\Big\},$$

Problem 2.2 (Primal-SCCHI) *Among solutions $u(\cdot)$ to Problem 2.1 find the optimal $u^0(\cdot)$, that ensures*

$$\mu = \mu^0 = \|u^0(\cdot)\|_{D_k^{p*}} = \min.$$

The conditions of Problem 2.1 imply inequalities

$$\langle l, x(t_\beta)\rangle \leq \rho(l \mid \mathcal{M}), \quad \langle\lambda^\sharp(\cdot), Nx(\cdot)\rangle = \langle\lambda(\cdot), N\mathbf{x}(\cdot)\rangle \leq \nu\|\lambda^\sharp(\cdot)\|D_{k-1}$$

for all $l \in \mathbb{R}^n$, $\lambda^\sharp(\cdot) \in D_{k-1}$, where

$$\lambda^\sharp(t) = \int_t^{t_\beta} \lambda(\xi)\zeta_+^{(k-1)}(t-\xi)d\xi = (\lambda(\cdot)*\zeta_+^{(k-1)})(t, t_\beta), \quad \lambda^\sharp(\cdot) \in D_{k-1} \subset \mathbf{BV}^m[t_\alpha, t_\beta].$$

Following schemes of paper (Kurzhanski 2016), we further consider equation

$$ds/dt = -sA - \lambda^\sharp(t)N, \quad s(t_\beta) - l^T, \tag{4}$$

where function $\lambda^\sharp(\cdot)$ is $(k-1)$ times differentiable, so same times "smoother" than λ. Its solution, that depends on l, $\lambda^\sharp(\cdot)$, is denoted as $s^\sharp[\cdot] = s(\cdot; l, \lambda^\sharp(\cdot)) \in D_k^n$.

In view of previous relations and assuming $\|u(\cdot)\|_{D_k^*} \leq \mu$, we observe that solvability of the Primal Problem 2.2 depends on inequality

$$\max_{u\in\mathcal{U}}\left\{\int_{t_\alpha}^{t_\beta} s(t; l, \alpha\lambda^\sharp(\cdot))Bu(t)dt\right\} - \int_{t_\alpha}^{t_\beta} s(t; l, \alpha\lambda^\sharp(\cdot))(f(t) + f^{(\alpha)})dt +$$

$$+ \rho(l \mid \mathcal{M}) + \nu\|\lambda^\sharp(\cdot)\|_{D_{k-1}} \geq 0, \quad l \in \mathbb{R}^n, \lambda^\sharp(\cdot) \in D_{k-1}. \tag{5}$$

which yields

Theorem 2.1 *The "Existence SCCHI" Problem 2.1 is solvable iff conditions (4), (5) are fulfilled for all $l \in \mathbb{R}^q$, $\lambda^\sharp(\cdot) \in D_{k-1}[t_\alpha, t_\beta]$.*

This theorem is true for all integers $k \geq 1$. With $k = 1$ the multiplier $\lambda^\sharp \in D_0$ where $D_0 \subset C$ is the space of continuous functions that are zero-valued beyond $[\alpha, \beta] \supset [t_\alpha, t_\beta]$.

The given theorem forestalls the next one—to find the optimal control $u^0(\cdot)$. This is achieved P by solving an adjoint optimization problem. Namely, the optimal, norm-minimal control $u^0 = \mu^0$ that solves Problem 2.2 (the Primal-SCCHI) is found by solving a dual problem of maximization over multipliers $\{l, \lambda^\sharp\}$ which is as follows.

Denoting

$$\mathbf{H}(l, \lambda^\sharp)$$
$$= \int_{t_\alpha}^{t_\beta} s(t \mid l, \lambda^\sharp(\cdot))(f(t) + f^{(\alpha)})dt - \rho(l \mid \mathcal{M}) - \nu\|\lambda^\sharp(\cdot)\|_{D_{k-1}},$$
$$\lambda(\cdot) = d^{k-1}\lambda^\sharp(\cdot)/dt^{k-1},$$

and using relation

$$\mathbf{H}(l, \lambda^\sharp) \leq \max_{u \in \mathcal{U}} \left\{ \int_{t_\alpha}^{t_\beta} s(t; \mid l\lambda^\sharp(\cdot))Bu(t)dt \right\} \leq \mu\|s(\cdot \mid l, \lambda^\sharp(\cdot))B(\cdot)\|_{D_k^n}$$

we arrive at the next formulation

Problem 2.3 (Dual SCCHI) *Find maximizer*

$$\mu^0 = \sup_{l,\lambda} \left\{ \frac{\mathbf{H}(l, \lambda^\sharp)}{\|s(\cdot \mid l, \lambda^\sharp(\cdot))B\|} \right\} \tag{6}$$

along the solutions to (4).

This yields

Theorem 2.2

(i) The optimal norm-minimal control with $\|u^0(\cdot)\| = \mu^0$ that solves the Primal Problem 2.2 is determined by the maximizer $\{l = l^0, \lambda^\sharp(\cdot) = \lambda_0^\sharp(\cdot)\}$ for the Dual SCCHI Problem 2.3.

(ii) The optimal control $u^0(\cdot)$ is determined from a Maximum Principle of type indicated in Kurzhanski and Varaiya (2014), Section 6.1.1 (pp. 255–257) and Section 7.1.1 (pp. 289–291).

Remark 2.1

(i) By reversing relation (6) we have an equivalent minimization problem

$$\gamma^0 = (\mu^0)^{-1} = \inf_{l,\lambda^\sharp}\left\{\|s(\cdot \mid l, \lambda^\sharp(\cdot))B\|_{D_k} \,\middle|\, \mathbf{H}(l, \lambda^\sharp) = 1\right\} \tag{7}$$

which may be interpreted as a problem of optimal control for the adjoint system (4), where $\lambda^\sharp(\cdot)$ may be treated as the control.

(ii) From the previous lines one may observe that the high-order impulse control $u(\cdot) \in D_k^{p*}$ generates a related adjoint Lagrange-type multiplier $\lambda^\sharp(\cdot)$—a smooth function, with corresponding level of smoothness.

(iii) The maximum principle for control problems of this paper is considered within a finite time interval. For an infinite time interval such principle was introduced in paper (Aseev and Veliov 2012) and related investigations.

2.3 Smooth Controls

We now pass to Problem SCCSM of *state-constrained control in the class of smooth, (continuously differentiable) functions*. Consider system

$$dx/dt = Ax + BU(t), \quad t \in [t_\alpha, t_\beta], \tag{8}$$

where the vector -valued control inputs $U(t) \in \mathbb{R}^p$ and the solution outputs $x(t) \in \mathbb{R}^n$ belong to related classes of smooth functions: belong to related classes of smooth functions $U(\cdot) \in D^p_{k-1}$, $x(\cdot) \in D^n_k$, assuming these are concentrated on $[t_\alpha, t_\beta] \subset [\alpha, \beta]$.

The input $U(t)$ of system (8) is the output of a multiple integrator $U(t) = U^{(0)}(t)$, where

$$dU^{(0)}(t)/dt = U^{(1)}(t), \ldots, dU^{(k-1)}(t)/dt = \mathsf{U}(t), \quad \mathsf{U}(\cdot) \in D_0^p, \tag{9}$$

so that

$$U(t) = U^{(0)}(t) = \int_{t_\alpha}^{t} \frac{(t-\xi)^{k-1}}{(k-1)!}\mathsf{U}(\xi)d\xi = (\mathsf{U} * \zeta_+^{(k-1)})(t, t_\alpha).$$

The state constraint is

$$y(t) = Nx(t) = k^*(t) + \xi(t), \quad \|\xi(\cdot)\|_{D_k} \le \nu, \quad t \in [t_\alpha, t_\beta], \tag{10}$$

where $k^*(\cdot)$, $\xi(\cdot) \in D_k^m$ are smooth m-vector functions. Hence the output $y(t)$ is an ordinary function.

Problem 2.4

(i) *(Existence—SCCSM) Given Eq. (8) and time interval* $t \in [t_\alpha, t_\beta]$, *specify bounded input* $\|\mathsf{U}(t)\| \leq \gamma,\ \gamma > 0,\ t \in [t_\alpha, t_\beta]$, *that transfers* $x(t)$ *from given* $x^{(\alpha)} = x(t_\alpha)$ *to* $x^{(\beta)} = x(t_\beta) \in \mathcal{M}$ *(the given target set) under state constraint (10).*

(ii) *(Primal—SCCSM) Among solutions to the previous point (i) find the optimal one* $\mathsf{U}^0(\cdot)$, *for which the bound* γ *on control* $\|\mathsf{U}^0(\cdot)\|_{D_0}$ *will be minimal:* $\gamma = \gamma^0 = \min$.

The control input $U(\cdot)$ in (8) now belongs to space $D_{k-1}^{(m)}$ of (k-1)-times continuously differentiable functions, being generated by continuous function $\mathsf{U}(\cdot)$ in D_0.

To formulate conditions of solvability for Problem 2.4 (i) we apply procedures similar to those of previous section, but now dealing with smooth inputs. We have

$$\int_{t_\alpha}^{t_\beta} l^T \mathcal{G}(t_\beta, \tau) BU(\tau) d\tau \leq \rho(l \mid \mathcal{M}) - l^T c^{(1)} = \mathbf{h}(l),$$

and

$$\int_{t_\alpha}^{t_\beta} \lambda^T(t) N\mathcal{G}(t, t_\alpha) dt x^{(\alpha)} + \int_{t_\alpha}^{t_\beta} \Big(\int_{\tau}^{t_\beta} \lambda^T(t) N G(t, \tau) dt \Big) BU(\tau) d\tau \leq$$

$$\int_{t_\alpha}^{t_\beta} \lambda^T(t)(k^*(t) - c_2(t)) dt + \nu \|\lambda(\cdot)\|_{D_k^*},$$

where

$$c^{(1)} = G(t_\beta, t_\alpha) x^{(\alpha)}, \quad c^{(2)}(t) = NG(t, t_\alpha) x^{(\alpha)}.$$

Denoting $s[\cdot] = s(\cdot \mid l, \lambda) \in D_{k-1}^*$ as the solution to equation

$$ds/d\tau = -sA - \lambda^T N, \quad s(t_\beta) = l, \tag{11}$$

we come to inequalities

$$0 \leq -s(t_\alpha \mid l, \lambda) x^\alpha + \mathbf{h}(l), \tag{12}$$

and

$$\langle s[\cdot], BU(\cdot) \rangle = \int_{t_\alpha}^{t_\beta} s(\tau \mid l, \lambda) BU(\tau) d\tau \leq \langle \lambda(\cdot), k^*(\cdot) - c^{(2)}(\cdot) \rangle + \nu \|\lambda(\cdot)\|_{D_k^*} = +\mathbf{H}(\lambda). \tag{13}$$

Assuming

$$\mathbf{s}[\xi] = \int_{\xi}^{t_\beta} s(\tau)\frac{(\tau - \xi)^{k-1}}{(k-1)!}d\tau,$$

adding relation (12) with (13), and applying equality

$$\langle s[\cdot], BU(\cdot)\rangle = \langle \mathbf{s}(\cdot), B\mathsf{U}(\cdot)\rangle,$$

we get

$$\langle \mathbf{s}[\cdot], B\mathsf{U}(\cdot)\rangle \leq \mathbf{h}(l) - s(t_\alpha \mid l, \lambda)x^\alpha + \mathbf{H}(\lambda). \tag{14}$$

Remark 2.2 The relations of the above involve an integration of generalized functions from D_k^*, D_{k-1}^* multiplied by ordinary functions. In this case the notation for the related integrals is symbolic and is understood in the sense of the theory of distributions, as mentioned above, in Sect. 2.1 (see Schwartz 1950–1951, 1966; Gelfand and Shilov 1991; Agranovich 2008) □

Theorem 2.3 *Problem 2.4 (i) is solvable if and only if the inequality (14) is true for all* $l \in \mathbb{R}^n$, $\lambda(\cdot) \in D_{k-1}^{m*}$, *under* ***some*** *control function* $\mathsf{U}(\cdot) \in D_0[t_\alpha, t_\beta]$.

From here, as before, we find the solution $\mathsf{U}^0(\cdot)$ to Problem (2.4)(ii) with minimal norm $\|\mathsf{U}^0(\cdot)\|_{D_0}$. Since

$$\|\mathsf{U}(\cdot)\|_{D_0}\|B^T\mathbf{s}^T[\cdot]\|_{D_0^*} \geq -\langle \mathbf{s}[\cdot], B\mathsf{U}(\cdot)\rangle$$

this gives

$$\|\mathsf{U}(\cdot)\|_{D_0} \geq \sup_{l,\lambda}\left\{\frac{s(t_\alpha \mid l, \lambda)x^\alpha - \mathbf{H}(\lambda) - \mathbf{h}(l)}{\|B^T\mathbf{s}^T[\cdot]\|_{D_0^*}}\right\} = \gamma^0 \tag{15}$$

over all $l \in \mathbb{R}^n$, $\lambda(\cdot) \in D_k^*[t_\alpha, t_\beta]$.

Problem 2.5 (Dual—SCCSM) *Solve the maximization problem (15) along the solutions to the adjoint equation (11).*

Under conditions of Problem 2.5 the supremum in the last relation (15) is actually a maximum. This relation yields the next conclusion.

Theorem 2.4 *The minimal norm* $\gamma^0 = \|\mathsf{U}^0(\cdot)\|_{D_0}$ *of the control* $\mathsf{U}^0(\cdot)$ *that solves the Primal Problem 2.4(ii) is found by solving the Dual SCCSM Problem 2.5 through relation (15).*

2.4 Duality Scale in Mathematics of State Constrained Control: Table SCC

Summarizing the items of the above, we now collect these results in tabular form. Indicated here are the functional spaces used for solving Problem 2.2 (Primal SCCHI) of state constrained control under high order impulses, in a receding scale. This is followed further by Problem 2.4(ii) (Primal SCCSM) of state constrained control under smooth controls. Indicated here are also the classes of functional spaces. for generalized Lagrange multipliers $\lambda^{\sharp}(\cdot)$ in the first case and $\lambda(\cdot)$ in the second, both used to treat the state constraints, with related adjoint equations for the dual problems of optimization. Note that for all the types of control inputs the state constraint is applied *to an ordinary function*.

Table of Functional Spaces for Problems of State-Constrained Control (SCC)

(SCC)	PRIMALS-x, $\mathbf{x}$	CONT-u, U	$y = Nx$, $N\mathbf{x}$,	MLTP-$\lambda^{\sharp}$, λ	DUALS-$s^{\sharp}$, s
(1)	D^{n*}_{k-1}, D^n_0	D^{q*}_k, D^*_0	D^{m*}_{k-1}, D^m_0	$D^m_{k-1}., D^{m*}_0$	D^m_k, D^n_0
...	...	...	...	...	
(2)	D^{n*}_1	D^{q*}_2	D^m_0	D^{m*}_0	D^n_0
(3)	D^{n*}_0	D^{q*}_1	D^m_0	D^{m*}_0	D^n_0
(4)	D^n_0	D^{q*}_0	D^m_0	D^{m*}_0	D^n_0
(5)	D^n_1	D^q_0	D^m_1	D^{m*}_1	$D^{n^*}_0$
...	...	...	...	...	
(6)	D^n_{k+1}	D^q_k	D^m_{k+1}	D^{m*}_{k+1}	D^{n*}_k

Here above, in lines (1)–(3), the state is x or $\mathbf{x}$, the control is u, the constrained trajectory is $y(t) = Nx(t)$ or $\mathbf{y}(t) = N\mathbf{x}(t)$, the Lagrange-type multiplier is $\lambda^{\sharp}(t)$ or $\lambda(t)$ and the dual (adjoint) variable is either $s^{\sharp}$ or s. In lines (4)–(6) these are $x, U, y(t), \lambda(t)$ and $s(t)$.

In the given table

- the first column PRIMS-x indicates the space in which lies solution $x(\cdot)$ of the Primal System,
- the second column CONTROL-u,U indicates the space to which belong the controls $u(\cdot), U(\cdot)$,
- the third column STATC-y=**Nx**, Nx indicates the space within which the state constraint on $y(\cdot)$ is placed,
- the fourth column MLTP-$\lambda^{\sharp}$, λ indicate the functional spaces to which belong the generalized Lagrange-type multipliers $\lambda^{\sharp}$and $\lambda(\cdot)$ responsible for treating the state constraint,
- the fifth column DUAL -s, $s^{\sharp}$ indicates the spaces to which belong the solutions $s^{\sharp}[\cdot]$ and $s[\cdot]$ of adjoint equations that solve the related Dual Problems of optimization.

Remark 2.3 One may observe that the range of spaces within which we pose the problem of state constrained control varies from generalized functions that include high order derivatives of δ-functions.to very smooth types of controls. □

3 Duality Scale in Problems of Guaranteed State Estimation

In this section we deal with problems of guaranteed state estimation for systems that operate under unknown inputs, emphasizing two classes of these—those described by high order distributions (Problem GSEHI) and those by smooth controls (Problem GSESM), as well as by those that lie in between. Indicated are related functional spaces within which the problems of state estimation are to be solved correctly.

3.1 *Guaranteed State Estimation: High Order Impulsive Disturbances*

We consider equation in distributions

$$dx/dt = Ax + Cv + p\delta^{(k)}(t - t_\alpha), \quad t \in [t_\alpha, t_\beta], \tag{16}$$

Here $x(t) \in \mathbb{R}^n$, $v(t) \in \mathbb{R}^q$, with norm $\|v(\cdot)\|_{D_k^*} \leq \nu$ in D_k^{q*}, and integer $k \geq 0$. Matrices $A \in \mathbb{R}^{n\times n}$, $C \in \mathbb{R}^{n\times q}$ are assumed constant.

The given system is complemented by a measurement equation

$$y(t) = H(x * \zeta_+^{(k-1)})(t, t_\alpha), \tag{17}$$

where $y(\cdot) \in D_0^{m*}$ is an ordinary function

Problem 3.1 (Solvability—GSEHI) *Given system (16), with measurement Eq. (17), estimate vector "p" from available observation* $y(t)$, $t \in [t_\alpha, t_\beta]$.

This problem is solved within the class of linear operators

$$W(\cdot) = \{w^{(1)}(\cdot), \ldots, w^{(n)}(\cdot)\}, \ w^{(i)}(\cdot) \in D_0^m[t_\alpha, t_\beta], \ W(t) \in \mathbb{R}^{m\times n},$$

that produce the non-biased guaranteed error

$$\Upsilon_E[W] = \|\langle W(\cdot), y(\cdot)\rangle^T - p\| = \max_{v(\cdot)}\{\|\langle W(\cdot), y(\cdot)\rangle^T - p\| \tag{18}$$

under $\|v(\cdot))\|_{D_k^*} \leq \nu,\ y(\cdot) \in D_0^{m*}$, and the "non-bias" condition

$$\int_{t_\alpha}^{t_\beta} W^T(t) H \mathbf{G}(t, t_\alpha) p dt = p, \quad \forall p. \tag{19}$$

The last relation ensures that the estimate of p under $v(\cdot) \equiv 0$ is exact. Here, since the input $p\delta^{(k)}(t - t_\alpha)$ to system (16) is a generalized function, we have

$$\mathbf{G}(t, t_\alpha) = (-1)^k \Big[\Big(\frac{\partial}{\partial \tau} \mathcal{G}(t, \tau) \Big) * \zeta_+^{k-1} \Big](t, t_\alpha),$$

and $(\partial^0 \mathcal{G}(t, \tau)/\partial \tau^0) = \mathcal{G}(t, \tau)$.

Taking $p = \{\mathbf{e}_1, \ldots, \mathbf{e}_n\}$ implies that for (19) we have

$$\int_{t_\alpha}^{t_\beta} W^T(t) H \mathbf{G}(t, t_\alpha) dt = I, \quad W(\cdot) \in D_0[t_\alpha, t_\beta]. \tag{20}$$

Calculating (18), (19) involves further relations

$$\int_{t_\alpha}^{t_\beta} W^T(t) y(t) dt = \int_{t_\alpha}^{t_\beta} \Big(\int_{\tau}^{t_\beta} W^T(t) \frac{(t-\tau)^{k-1}}{(k-1)!} dt \Big) H x(\tau) d\tau$$

$$\int_{\tau}^{t_\beta} W(t) \frac{(t-\tau)^{k-1}}{(k-1)!} dt = \mathbf{W}[\tau], \quad \mathbf{W}[\cdot] \in D_{k-1}[t_\alpha, t_\beta].$$

Then, denoting

$$x(\cdot) = x(\cdot \mid t_\alpha, p, v(\cdot)) = x(\cdot \mid t_\alpha, 0, v(\cdot)) + x(\cdot \mid t_\alpha, p, 0) = x_v[\cdot] + x_0[\cdot],$$

we have, in view of (20),

$$\langle W(\cdot), y(\cdot) \rangle - p = \int_{t_\alpha}^{t_\beta} W^T(t) H (x_v * \zeta_+^{(k-1)})(t, t_\alpha) dt =$$

$$\int_{t_\alpha}^{t_\beta} \mathbf{W}^T(t) H C_v(t) dt = \Psi[\mathbf{W}(\cdot), v(\cdot)].$$

Problem 3.2 (Of Guaranteed State Estimation (Primal GSEHI)) *Find solution operator* $\mathbf{W}^{(0)}(\cdot) = \mathbf{W}(\cdot)$ *to Problem 3.1 GSEHI (Solvability) for the optimal estimate* $\Upsilon[\mathbf{W}^{(0)}]$ *of the worst-case error as*

$$\Upsilon[\mathbf{W}^0] = \min_{\mathbf{W}} \{\Upsilon_E[\mathbf{W}]\},$$

under condition (20).

Then, after a legal interchange of operations min and max, (see Fan 1953), we come to relations for the optimal operation $\mathbf{W}^0[\cdot]$,

$$\Upsilon_E[\mathbf{W^0}(\cdot)] = \min_{\mathbf{W}} \max_{v}\{\|\Psi[\mathbf{W}(\cdot), v(\cdot)] \mid \|v(\cdot))\|_{D_k^*} \leq \mu\},$$

$$\max_{v(\cdot)} \|\Psi[\mathbf{W}(\cdot), v(\cdot)] \mid \|v(\cdot))\|_{D_k^*} \leq \nu\} =$$

$$= \max_{v}\{< s_w[\cdot], Cv(\cdot) > \mid \|v(\cdot)\|_{D^{k*}} \leq \nu\} = \nu\|s_w[\cdot]C\|_{D_k}, \tag{21}$$

under condition (20).

Here

$$ds_w/dt = -s_w A - \mathbf{W}^T(t)H(t), \quad s_w[t_\alpha] = l^T, \tag{22}$$

$$\|s_w[\cdot]C\|_{D^k} = \sum_{i=0}^{k-1} \max_t \left(\frac{d^i s_w[t]}{dt^i}C \;\Big|\, t \in [t_\alpha, t_\beta]\right).$$

Theorem 3.1 *The solution of Primal GSEHI Problem 3.2 is reduced to the next optimization problem: find*

$$\Upsilon_0[\mathbf{W}^0(\cdot)] = \min_{\mathbf{W}}\{\|s_w[\cdot]C\|_{D^k} \mid (20)\} \tag{23}$$

under condition (20). Then $\Upsilon_E[\mathbf{W^0}(\cdot)] = \nu\Upsilon_0[\mathbf{W}^0(\cdot)]$.

Problem (23) may be interpreted as one of optimal control for system (22) with $\mathbf{W}(t)$ treated as the control.

3.2 State Estimation Under Smooth Disturbances

Now we deal with input disturbances given by differentiable functions. Consider the equation

$$dx/dt = Ax + CV(t), \tag{24}$$

with $x(t) \in \mathbb{R}^n$ and unknown vector input $V(t) \in \mathbb{R}^q$ is such that $V(\cdot) \in D_{k-1}^q$, being concentrated on $[t_\alpha, t_\beta] \subset [\alpha, \beta]$ and norm-bounded as $\|V(\cdot)\|_{D_{k-1}} \leq \mu$. Matrices $A \in \mathbb{R}^{n\times n}$, $C \in \mathbb{R}^{n\times q}$ are constant and such that they ensure system (24) to be *dissipative* (see Willems Jan 2007).

The input $V(t)$ of system (24) is the output of a multiple integrator $V(t) = V^{(0)}(t)$, where

$$dV^{(0)}(t)/dt = V^{(1)}(t), \ldots, dV^{(k-1)}(t)/dt = \mathbf{V}(t), \quad \mathbf{V}(\cdot) \in D_0^q[t_\alpha, t_\beta], \tag{25}$$

so that

$$V^{(0)}(t) = \int_{t_0}^{t} \frac{(t-\xi)^{k-1}}{(k-1)!}\mathbf{V}(s)ds,$$

$$V^{(i)}(t) = \int_{t_0}^{t} \frac{(t-\xi)^{k-i-1}}{(k-i-1)!}\mathbf{V}(s)ds,\, V^{(k-1)}(t) = \int_{t_0}^{t} \mathbf{V}(s)ds.$$

The available observation of system (24) is modeled by the scalar output of noise-free *measurement equation*

$$y(t) = h^T x(t), \quad t \in [t_\alpha, t_\beta], \quad y(\cdot) \in D_k[\alpha, \beta], \tag{26}$$

which generates

$$\mathbf{y}^T = \left\{y, \frac{dy}{dt}, \ldots, \frac{d^{k-1}y}{dt^{k-1}}\right\}^T,$$

Here $y(\cdot) \in D_k^1 = D_k$ is a physically realizable k-times differentiable scalar function.

The objective is *to estimate the coordinate* $\mathbf{e}_j^T x^\beta = x_j^\beta$ through measurement of $y(t)$. Here, treating $y(\cdot) \in D_k$, we introduce the estimating operation as $\langle \mathsf{W}^{(j)}(\cdot), y(\cdot)\rangle = x_{j\star}^\beta$ and look for the best estimate

$$\|x_{j\star}^\beta - x_j^\beta\| = \min_{\mathsf{W}(\cdot)}$$

under nonbias condition $x_{est}^\beta = x^\beta$ when $V(\cdot) = 0$.

For further considerations we shall present the high order distribution $\mathsf{W}^{(j)}(\cdot)$ in terms of space **BV** of functions **W** with bounded variation as

$$< \mathsf{W}^{(j)}(\cdot), y(\cdot) >= \langle \mathbf{W}^{(j)}(\cdot), \mathbf{y}(\cdot)\rangle = \sum_{i=0}^{k-1} \int_{t_\alpha}^{t_\beta} \frac{d^i y(t)}{dt^i} d\mathbf{W}_i^{(j)}(t), \tag{27}$$

where $\mathbf{W}^{(j)}(\cdot) = \{\mathbf{W}_0^{(j)}(\cdot), \ldots, \mathbf{W}_{k-1}^{(j)}(\cdot)\}$ and for all i $\mathbf{W}_i^{(j)}(\cdot) \in \mathbf{BV}[t_\alpha, t_\beta + 0]$ are scalar functions with $k \le n$ and possibility of jump at $t = t_\beta$. We also denote $d^0 y(t)/dt^0 = y(t)$. Note that the role of $y(t)$ is similar to that of basic function $\varphi(\cdot)$ in the definition of generalized functions (see Schwartz 1966; Agranovich 2008 and also (1)).

Calculating derivatives of the above we have $d^i y(t)/dt^i = h^T d^i x(t)/dt^i, i = 0, \ldots, k-1$, and

$$d^i x(t)/dt^i = A^i G(t, t_\beta)x^\beta + \sum_{j=0}^{i} A^{j-1} C V^{(j-1)}(t) - A^i x^{(0)}(t),$$

$$x^{(0)}(t) = \int_t^{t_\beta} G(t, \tau) C V^{(0)}(t) dt.$$

Here $A^j = 0$, $V^{(j)} = 0$, when $j < 0$. Since a non-biased estimation of x^β requires that with disturbance $\mathbf{V}(t) \equiv 0$ the estimate would be exact, this gives

$$\int_{t_\alpha}^{t_\beta} \sum_{i=0}^{k-1} h^T A^i G(t, t_\beta) d\mathbf{W}_i^{(j)}(t) = \mathbf{e}_j^T, \quad j = 1, \ldots, n; \tag{28}$$

Denoting $\mathcal{V} = \{V^{(0)}, \ldots, V^{(k-1)}\}$, $\mathbf{W} = \{\mathbf{W}^{(1)}, \ldots, \mathbf{W}^{(n)}\}$, we are now able to formulate the required problem.

Problem 3.3 (Primal GSESM) *Find the linear operations $\langle \mathbf{W}^{(j)}(\cdot), \mathbf{y}(\cdot)\rangle$ that jointly ensure the minimum of guaranteed estimation error for vector x^β, namely, introducing*

$$\Upsilon[\mathbf{W}(\cdot), \mathbf{y}(\cdot)] = \max_{\mathcal{V}} \sum_{j=1}^{n} \{\|\langle \mathbf{W}^{(j)}(\cdot), \mathbf{y}(\cdot)\rangle - x_j^\beta\|^2 \mid \|\mathcal{V}^{(j)}\|_{D_{k-j}} \le \mu\}. \tag{29}$$

find operation $\mathbf{W}(\cdot)$ that ensures

$$\Upsilon[\mathbf{W}(\cdot), \mathbf{y}(\cdot)] = \min_{\mathbf{W}}, \tag{30}$$

under condition (28), in the class of functions $\mathbf{W}^{(j)}(\cdot) \in \mathbf{BV}$.

Theorem 3.2 *The optimal solution of Problem 3.3 (the Primal GSESM) under smooth disturbances $V(\cdot)$ is to find the operator $\mathbf{W}^0$ that ensures*

$$\Upsilon^0 = \Upsilon[\mathbf{W}^0(\cdot), \mathbf{y}(\cdot)] = \min_{\mathbf{W}}\{\Upsilon[\mathbf{W}, \mathbf{y}], \tag{31}$$

This is achieved by finding Primal Υ^0 through solving maximization problem (29)—the related Dual GSESM—under condition (28).

3.3 Duality Scale in Mathematics of Guaranteed State Estimation

Summarizing the sections of the above on state estimation we now collect the results in tabular form, indicating the functional spaces for the solution elements of the problems.

Table GSE of Functional Spaces for Problems of Guaranteed State Estimation Here in lines (1)–(3) the inputs are v, with phase state x, measurement y, solution operator W, adjoint system variable s, in lines (4)–(6) they are V, x, y, W, s.

(GSE)	PRIMS-x, x	DSTRB-v, V	MEASNT-y, y	SOLOPR-W, W	DUALS-s, s
(1)	D^{n*}_{k-1}	D^{q*}_{k}	D^{m}_{0}	D^{m*}_{0}	D^{n}_{0}
...	...	...	...	...	...
(2)	D^{n*}_{1}	D^{q*}_{2}	D^{m}_{0}	D^{m*}_{0}	D^{n}_{0}
(3)	D^{n}_{0}	D^{q*}_{0}	D^{m}_{0}	D^{m*}_{0}	D^{n}_{0}
(4)	D^{n}_{1}	D^{q}_{0}	D^{m}_{1}	D^{m*}_{1}	D^{n*}_{0}
...	...	...	...	...	...
(5)	D^{n}_{k}	D^{q}_{k-1}	D^{m}_{k}	D^{m*}_{k}	D^{n*}_{k-1}
(6)	D^{n}_{k+1}	D^{q}_{k}	D^{m}_{k+1}	D^{m*}_{k+1}	D^{n*}_{k}

In the given table

- the first column PRIMS-x indicates the space to which belong the trajectories of the Primal system,
- the second column DSTRB v, V indicates the space to which belong the unknown disturbances $v(\cdot)$ (impulsive, of higher order) or V (smooth inputs),
- the third column MEASNT-y indicates the spaces within which lie the available measurements $y(t) = H(t)(x * \zeta_{+}^{k-1})(t, t_{\alpha})$ and $y(t) = h^T x(t)$.
- the fourth column SOLOPR indicates the functional space to which belongs the solution operator $W(t)$ (the estimator),
- the fifth column DUALW $s, s^{\sharp}$ indicates the spaces for solutions $s(\cdot)$ of the adjoint equations generated by inputs $W(\cdot)$ which define a Dual optimization problem in the system sense.

4 Comparative System Duality in State Constrained Control and Guaranteed State Estimation

Here we emphasize *two types of duality*. The first is the property of duality *in the mathematical sense* within the pairs of primal and dual optimization variables in state constrained control and in guaranteed state estimation. The second type is the property of duality *in the system sense* which is due to the similarity between the problem of control and that of the of state estimation.

4.1 Duality Under Ordinary Impulses

The case of ordinary input functions has some specificity that is clarified here in terms of standard spaces. In this section we thus interpret the pair $D_0^*[t_\alpha, t_\beta]$, $D_0[t_\alpha, t_\beta]$ as $\mathbf{BV}[t_\alpha, t_\beta]$, $\mathcal{C}[t_\alpha, t_\beta]$.

4.1.1 State Constrained Control

For future discussion on the comparison between this problem and that of state estimation we need to deal with an array of controlled systems. To formulate the Primal Problem (SCCOI) we proceed as follows.

Given is an array of identical systems

$$dx_\star^{(i)}(t)/dt = A_\star x_\star^{(i)}(t) + H_\star(t)u^{(i)},\ x_\star(t_\alpha) = \sum_{i=1}^{n} x_\star^{(i)}(t_\alpha) = p^\star,\quad i = 1, \dots, n, \tag{32}$$

where $\|u^{(i)}(\cdot)\|_{D_0^*} \leq r_\star$, $x_\star^{(i)}(\cdot) \in \mathcal{C}$, $u^{(i)}(t) \in \mathbb{R}^m$, $x_\star^{(i)}(t_\alpha) = x_{\star\alpha}^{(i)}$, $H_\star(t) \in \mathbb{R}^{n\times m}$ and state constraint

$$y_\star^{(i)}(t) = C_\star x_\star^{(i)}(t),\ \|y_\star^{(i)}(t)\| \leq \nu_i,\quad C_\star \in \mathbb{R}^{q\times n},\quad t \in [t_\alpha, t_\beta]. \tag{33}$$

with terminal set $\mathcal{M}$ similar to Sect. 2.1.

Problem 4.1 (Primal SCCOI) *Find controls* $u(\cdot) = \{u^{(1)}(t), \dots, u^{(n)}\}$ *that ensure*

$$\begin{aligned}\varepsilon_\star^0 &= \varepsilon_\star[u^0(\cdot)] = \min_u \varepsilon_\star[u(\cdot)] \\ &= \min_u \Big\{ \sum_{i=1}^{n} \|C_\star x_\star^{(i)}(\cdot \mid t_\alpha, x_\alpha^{(i)}, H_\star u^{(i)})\|_{D_0} \mid \|u^{(i)}(\cdot)\|_{D_0^*} \leq r_\star\},\end{aligned} \tag{34}$$

under state constraint (33) and $x(t_\beta) \in \mathcal{M}$.

The conditions of this problem imply the next inequalities

$$\langle l^{(i)}, x_*^{(i)}(t_\beta)\rangle \leq \rho(l^{(i)} \mid \mathcal{M}),\quad \langle \Lambda_i(\cdot), Nx(\cdot)\rangle \leq \|\Lambda_i(\cdot)\|_{D_0^*}. \tag{35}$$

for all $l^{(i)} \in \mathbb{R}^n$, $\Lambda_i(\cdot) \in D_0^* \subset \mathbf{BV}$.

Passing to the *the Dual Problem (SCCOI)* , denote

$$\Phi[u^{(i)}, l^{(i)}, \Lambda_i] =$$

$$\left\langle H_\star(\cdot)u^{(i)}(\cdot), s_\star^{(i)}(\cdot \mid t_\alpha, l^{(i)}, C_\star\Lambda_i)\right\rangle = \left\langle x_\star(\cdot \mid t_\beta, 0, H_\star u^{(i)}), \Lambda_i(\cdot)^T C_\star)\right\rangle$$

where

$$ds_\star^{(i)}(t)/dt = -s_\star^{(i)}(t)A_\star - \Lambda_i^T(t)C_\star, \quad s^{(i)}(t_\beta) = l^{(i)}. \tag{36}$$

We now formulate

Problem 4.2 (Dual SCCOI) *Find*

$$\max_i \max_{l^{(i)},\Lambda_i} \{\Phi^{(i)}[u^{(i)}, l^{(i)}, \Lambda_i] \mid \|l^{(i)}\|^2 + \|\Lambda_i(\cdot)\|_{BV}^2 \le \mu_\star^2\}$$

$$= \max_i \{r_\star \|s_*^{(i)}(\cdot \mid t_\beta, l^{(i)0}, C_\star\Lambda_i^0)H_\star^t\|\},$$

through maximizers $l^0 = \{l^{(0)1}, \dots, l^{(0)n}\}, \; \Lambda^0(\cdot) = \{\Lambda_1^0(\cdot), \dots, \Lambda_n^0(\cdot)\}.$

The solution $\{l^0, \Lambda^0(\cdot)\}$ to this Dual Problem allows to calculate cost functions $\Phi^{(i)}[u^{(i)}, l^{(i)}, \Lambda_i]$, $(i = 1, \dots, n)$, thus solving the main part of the Primal Problem. *The Optimal Control* $u^0(\cdot)$ is then found using maximizers l^0, $\Lambda^0(\cdot)$, by applying a standard open-loop Maxmin Principle. (See those of the type given in Kurzhanski and Varaiya (2014, Section 7.2) and also Kurzhanski (2016)).

The parameters of Primal Problem SCCOI are indicated at line 4 of table SCC in the above.

4.1.2 Guaranteed State Estimation: Ordinary Impulses

We have *the Primal Problem (GSEOI)* for system

$$dx = Ax(t)dt + CdV, \; x^\alpha = x(t_\alpha) = p, \quad V(t_\alpha) = V(t_\alpha + 0). \tag{37}$$

where $V(\cdot) \in \mathbf{BV}^q[t_\alpha, t_\beta]$, $\|V(\cdot)\|_{\mathbf{BV}} \le 1$, $V(t) \in \mathbb{R}^q$, and p is unknown, but given is a noise-free measurement

$$y(t) = Hx(t), \quad x(t) \equiv 0, \text{ if } t < t_\alpha, \text{and } t > t_\beta. \tag{38}$$

Problem 4.3 *Estimate each coordinate* x_i^α *of vector* $x^\alpha = p$ *under nonbias condition*

$$\int_{t_\alpha}^{t_\beta} W^T(t)Hx(t \mid t_\beta + 0, p, 0)dt = I, \quad (V(\cdot) \equiv 0), \tag{39}$$

where $W(\cdot)$ *is the observation operator.*

Under (39) the estimation error will be

$$\varepsilon[W] = \max_V \sum_{i=1}^{n} | < w^{(i)}(\cdot), y(\cdot) > -p_i |,$$

with minimal error achieved through minimizer $W(\cdot) = W^0(\cdot)$ whose rows would be $W^0(\cdot) = \{w^{(1)0}(\cdot), \ldots, w^{(n)0}(\cdot)\}$, where

$$< w^{(i)0}(\cdot), y(\cdot) >= p_i, \quad V(\cdot) \equiv 0. \tag{40}$$

Hence we are to find the guaranteed state estimate $p^* = \{p_1^*, \ldots, p_n^*\}$ and the optimal guaranteed estimation error in view of relations

$$| < w^{(i)}(\cdot), y(\cdot) > -p_i | = |\langle w^{(i)}(\cdot), Hx(\cdot \mid t_\alpha, 0, V(\cdot))\rangle| = \langle s^{(i)T}(\cdot \mid t_\beta, \mathbf{e}^{(i)}, Hw^{(i)}), CV(\cdot)\rangle$$

where $s^{(i)}[t] = s(t \mid t_\beta, \mathbf{e}^{(i)}, Hw^{(i)})$ is the solution to adjoint equation

$$ds^{(i)}/dt = -s^{(i)}(t)A(t) - w^{(i)}(t)H, \quad s^{(i)T}(t_\beta) = \mathbf{e}^{(i)}. \; i = 1, \ldots, n. \tag{41}$$

We thus come to the next procedure.

Problem 4.4 (Primal GSEOI) *Given system (37), with measurement constraint (38), find*

$$\varepsilon[W^0] = \min_W \sum_{i=1}^{n} \left\{ \|s^{(i)}(\cdot \mid t_\beta, \mathbf{e}_i, Hw^{(i)})C\|_{\mathbf{BV}} \mid \|w^{(i)}(\cdot)\| \leq r \right\} \tag{42}$$

along solutions of equation (41) under condition (40).

This is reached through solving the Dual Problem (GSEOI)of calculating the cost function for the related Primal Problem, namely, by dealing with the next optimization.

Problem 4.5 (Dual GSEOI) *Find*

$$\|s(\cdot \mid t_\beta, I, Hw)C\|_{\mathbf{BV}}$$

$$= \max_V \sum_{i=1}^{n} \left\{ < s^{(i)T}(\cdot \mid t_\beta, \mathbf{e}^{(i)}, Hw^{(i)}), CV(\cdot) > \;\middle|\; \mathrm{Var}\{V\} \leq 1, \; \forall j \right\} \tag{43}$$

along Eq. (41) under condition (40).

The *duality properties in between Primal and Dual Problems* for SCCOI and for GSEOI *are of mathematical nature*, since their solutions are achieved through methods of Convex Analysis with generalizations treated through broader techniques of Nonlinear Analysis. Another duality type is as follows.

4.1.3 The System Duality Under Ordinary Impulses (SDOI)

Considering Problems 4.1 and 4.4 assume that the parameters of Problem 4.1 are changed as

$$A_* = -A,\ H_\star^T = -H,\ C_\star^T = C,\ U(\cdot) = W(\cdot),\ \Lambda(\cdot) = V(\cdot),\ \ p = q,\ r_* = r. \tag{44}$$

Then one may observe that under the new notations the solution formulas for Problems 4.1, 4.2 will coincide with those for Problems 4.4, 4.5, while (34) coincides with (42).

Theorem 4.1 (SDOI) *Assume parameters of Problem 4.1 have been changed as in (44). Then this problem will coincide with Problem 4.4 demonstrating a* ***system duality*** *between problems of state-constrained ordinary impulse control and guaranteed estimation under ordinary impulsive disturbances.*

4.2 Duality Under Impulses of Higher Order

State-Constrained Control (SCC) Consider system

$$dx/dt = A_\star x + B_\star(t)u + f + f^{(\alpha)},\quad i = \{1, \ldots, n\}, \tag{45}$$

in terms of distributions. Here vector $x \in \mathbb{R}^n$, and the generalized control $u(\cdot) \in D_k^{p*}$. Vectors $f(\cdot) \in D_k^{n*}$ are the n-dimensional disturbances and $f_i^{(\alpha)} = x^\alpha \delta^{(k)}(t - t_\alpha)$.

The state constraints are

$$\|\mathbf{y}(\cdot)\|_{D_0^{(m)}} \leq \kappa,\ \text{ where } \mathbf{y}(t) = N\mathbf{x}(t),\quad \mathbf{x}(t) = (x(\cdot) * \zeta^{(k-1)})(t, t_\alpha), \tag{46}$$

so that $x(\cdot) \in D_{k-1}^{n*},\ \mathbf{x}(t) \in D_0^n,\ \mathbf{y}(\cdot) \in D_0^m$.

Within the next schemes we have *the Primal Problem (SCCHI)*, preceded by

Problem 4.6 (Solvability) *Specify control $u(\cdot)$ that transfers the system (45) from $\mathbf{x}^\alpha$ to $\mathbf{x}^{(\beta)} \in \mathcal{M}$ under given state constraint $\|y(t)\|_{D_0} \leq \nu$, ensuring*

$$\|u(\cdot)\|_{D_k^*} = min,$$

where

$$\|u(\cdot)\|_{D_k^m} = \max_i \left\{ \sum_{i=0}^{k} < u(\cdot), d^i\varphi(\cdot)/dt^i > \ \mid \ \| d^i\varphi(\cdot)/dt^i\|_C \leq 1 \right\},\quad \varphi(\cdot) \in D_k^p.$$

This leads to

Problem 4.7 (Primal SCCHI) *Find controls $u = u^0(\cdot)$ that achieves*

$$\nu_0 = \nu[u^0(\cdot)] =$$

$$\min_{\{u(\cdot)\}} \left\{ \sum_{i=1}^{n} \|Nx_\star(\cdot \mid \cdot, Bu(\cdot))\|_{D^*_{k-1}} \mid \|u(\cdot)\|_{D^*_k} \leq r_h \right\} \quad (47)$$

along solutions to equation (45), given $x^\beta \in \mathcal{M}$, and state constraint (46).

Here we may treat the bounding state constraint as applied either to $\mathbf{x}(\cdot) \in D^n_0 = C^n$—through multiplier $\lambda(\cdot) \in D^{m*}_0$, or to $x(t)$—through multiplier

$$\lambda^\sharp(t) = \int_t^{t_\beta} \lambda(\xi)\zeta_+^{(k-1)}(t-\xi)d\xi = (\lambda(\cdot) * \zeta_+^{(k-1)})(\cdot, t), \;\; \lambda(\cdot) \in D_0 = C, \;\; \lambda^\sharp(\cdot) \in D^m_{k-1}.$$

Function $\lambda^\sharp(\cdot)$ is $(k-1)$ times differentiable, so same times "smoother" than λ. The solution of Problem Primal SCCHI 4.6 depends on $\{l, \lambda^\sharp(\cdot)\}$, through solution $s(\cdot; l, \lambda^\sharp)$ of adjoint equation

$$ds/dt = -sA_\star + \lambda^{\sharp T}(t)N, \quad s(t_\beta) = l^T. \quad (48)$$

This yields the next *the Dual Problem* (SCCHI) of maximization. Namely, denoting

$$\Phi^{(i)}[u, \lambda^\sharp] =$$

$$\left\langle B_\star u(\cdot), s_\star^T(\cdot \mid t_\alpha, 0, \lambda^T N)\right\rangle = \left\langle x(\cdot \mid t_\beta, 0), B_\star u, N^T\lambda_i(\cdot)\right\rangle,$$

with

$$\mathbf{H}(l, \lambda^\sharp) = \int_{t_\alpha}^{t_\beta} s(t \mid l, \lambda^\sharp(\cdot))(f(t) + f^{(\alpha)})dt - \rho(l \mid \mathcal{M}) - \nu\|\lambda^\sharp(\cdot)\|_{D_{k-1}},$$

and using relation

$$\max_{u(\cdot)} \left\{ \left| \int_{t_\alpha}^{t_\beta} s(t; \mid l, \lambda^\sharp(\cdot))B_\star(t)u(t)d \right| \|u(\cdot)\| \leq r \right\} = r\|s(\cdot \mid l, \lambda^\sharp(\cdot))B\|_{D_k} \quad (49)$$

we formulate

Problem 4.8 (Dual SCHI) *Find*

$$\nu^0 = \max_{l,\lambda^\sharp} \left\{ \frac{\mathbf{H}(l,\lambda^\sharp)}{r\|s(\cdot \mid l,\lambda^\sharp)B_\star(\cdot)\|_{D_k}} \right\} \tag{50}$$

over solutions to (48).

This is equivalent to a minimization

Problem 4.9 *Find*

$$\nu_0^{-1} = \min_{l,\lambda} \sum_{i=1}^{n} r\|s(\cdot \mid t_\alpha, 0, N^T\lambda^\sharp)\| \tag{51}$$

under condition $\mathbf{H}(l,\lambda^\sharp) = 1$ *over solutions to (48).*

The minimizers $\{l_0, \lambda_0^\sharp\}$ of this problem are then used to figure out the respective control solution u^0 according to a maximum principle that follows from (49). The functional spaces for solving this problem are indicated above, in table SCC, line 6.

4.2.1 Guaranteed State Estimation (GSE)

Consider system similar to (16) under disturbances being impulsive inputs $v(\cdot) \in D_k^{q*}$ of higher order, where $x(t) \in \mathbb{R}^n$, $x(\cdot) \in D_{k-1}^{n*}$. The noise free measurement $y(t) \in \mathbb{R}^m$, is similar to (17).

Now the *Primal Problem (GSEHI)* will be to estimate input vector "p" of Eq. (16) on the basis of measurement $y(t)$, which reduces to the next item

Problem 4.10 (Primal-GSEHI) *For the identification of input "p", due to system (16), (17), minimize the unbiased guaranteed estimation error, namely, find the optimal solution operator $W(\cdot)$ that realizes*

$$\Psi[W^0(\cdot)] = \min_W \{\Upsilon_E[W(\cdot)] \mid \|W(\cdot)\|_{D_0} \leq r\}, \tag{52}$$

$$\Upsilon_E[W(\cdot)] = \|\langle W(\cdot), y(\cdot)\rangle - p\| = \max_{v(\cdot)}\{\|\langle W(\cdot), y(\cdot)\rangle - p\| \mid \|v(\cdot)\|_{D_k^{q*}} \leq \nu\}, \tag{53}$$

under nonbias condition (20).

The solution to maximization (53) of Problem 4.10 is to be treated within the next scheme: solve *Dual Problem (GSEHI)* which is to define the functional $\Upsilon_E[W(\cdot)]$. We have

Problem 4.11 (Dual GSEHI) *For a fixed $W(\cdot)$ find $\Upsilon_E[W(\cdot)]$ of (53) under nonbias condition (20).*

This yields

Theorem 4.2 *The solution to Problem 4.11 is reduced to the following: find*

$$\Psi[W(\cdot)] = \max_{v}\{< s_w[\cdot], Cv(\cdot) > \mid \|v(\cdot)\|_{D_k^{q*}} \leq \nu\} = \nu \|s_w[\cdot]C\|_{D_k}, \tag{54}$$

under condition (20).

Here $s_w[\cdot]$ is the solution to equation

$$ds_w/dt = -s_w A(t) - \mathbf{W}(t)H(t), \quad s_w[t_\alpha] = l^T, \tag{55}$$

where

$$\mathbf{W}[t] = \int_{\tau}^{t_\beta} W(t)\frac{(\xi - t)^{k-1}}{(k-1)!}d\xi, \ \mathbf{W}[\cdot] \in D_{k-1}[t_\alpha, t_\beta].$$

$$\|s_w[\cdot]C\|_{D_k^n} = \sum_{i=0}^{k-1} \max_t \left(\frac{d^i s_w[t]}{dt^i}C \;\middle|\; t \in [t_\alpha, t_\beta]\right).$$

The functional spaces for the elements of these problems are indicated in table EHI line 1.

4.2.2 The System Duality Under Higher Impulses

Considering Problems 4.6 and 4.10, assume that the parameters of Problem 4.6 are changed as

$$A_* = -A, \ B_*^T = -H, \ N^T = C, \ U(\cdot) = W(\cdot), \ \Lambda(\cdot) = V(\cdot), \ \ p = q. \tag{56}$$

Then one may observe that under the new notations the solution to Problem 4.6 will coincide with the one for Problem 4.10, so that relations (47), (50) and (52), (53) are similar.

Theorem 4.3 *Assume parameters of Problem 4.6 have been changed as in (56). Then the solution to this problem will coincide with the one for Problem 4.10, (54) demonstrating a* ***system duality*** *between problems of state-constrained higher impulse control and guaranteed estimation under higher impulsive disturbances.*

4.3 Duality Under Smooth Inputs

State-Constrained Control (SCC) Consider the equation

$$dx/dt = A_* x + B_* U(t), \quad t \in [t_\alpha, t_\beta], \tag{57}$$

where p-vector controls $U(\cdot) \in D^p_{k-1}[\alpha, \beta]$, and related trajectories $x(\cdot) \in D^n_k[\alpha, \beta]$, are smooth functions, as indicated above.

The input $U(t)$ of system (57) is the output of a multiple integrator

$$U(t) = U^{(0)}(t) = \int_{t_\alpha}^{t} \frac{(t-\xi)^{k-1}}{(k-1)!} \mathsf{U}(\xi) d\xi$$
$$= (\mathsf{U}(\cdot) * \zeta^{(k-1)})(t, t_\alpha), \quad \mathsf{U}(\xi) \in \mathbb{R}^p, \ \mathsf{U}(\cdot) \in D^p_0.$$

and such that $\|U(\cdot)\|_{D_{k-1}} \le \gamma$. The state constraint is taken as

$$y(t) = Nx(t), \quad \|y(\cdot) - \kappa(t)\|_{D^m_k} \le \nu, \tag{58}$$

with $\kappa(\cdot) \in D^m_k$ and ν given.

A related *Primal Problem* now sounds as follows.

Problem 4.12 (Primal SCCSM)

(i) Given system (57), with state constraint (58), and bounded control input $\|U(\cdot)\|_{D_{k-1}} \le \gamma$, indicate conditions for the existence of such control $U(\cdot)$ that transfers $x(t)$ from given $x^{(\alpha)} = x(t_\alpha)$ to given $x^{(\beta)} = x(t_\beta) \in \mathcal{M}$.

(ii) Among such existing controls $U(\cdot)$ indicate the optimal one $U^0(\cdot)$ which produces the minimal bound $\nu = \nu^0$ in the state constraint (58).

Applying the previous types of schemes used above for solvability, we observe that Problem 4.12 is solvable iff under adjoint equation (11) the inequality (12) yields, together with condition

$$-\gamma \|s(\cdot \ |l, \lambda)B(\tau)\|_{D^*_{k-1}} \le \langle s(\cdot \mid l, \lambda)BU(\cdot)\rangle \le \gamma \|s(\cdot) \ |l, \lambda)B\|_{D^*_{k-1}},$$

the relation

$$\Gamma(l, \lambda(\cdot)) - \gamma \|s(\cdot \mid l, \lambda)B\| \le \nu \|\lambda(\cdot)\|_{D^*_k}, \tag{59}$$

for all $l \in \mathbb{R}^n$, $\lambda(\cdot) \in D^{m*}_k$ that generate $s(\tau \ |l, \lambda) \in D^{n*}_{k-1}$, while

$$\Gamma(l, \lambda(\cdot)) = \langle l, c^{(1)}\rangle + \langle \lambda(\cdot), c^{(2)}(\cdot)\rangle - \rho(l \mid \mathcal{M}) - \langle \lambda(\cdot), \kappa(\cdot)\rangle.$$

Here $s(\cdot \ l, \lambda(\cdot))$ is the solution to the next equation in distributions.

$$ds(\cdot)/dt = -s(\cdot)A - \lambda^T(\cdot)N - l^T\delta^{(k)}(\cdot), \quad t \in [t_\alpha, t_\beta + 0]. \tag{60}$$

Lemma 4.1 *With given bounds γ, ν on control $U(\cdot)$ and state constraint (58) the problem of reaching target set $\mathcal{M}$ from given starting position x^α within time $t_\beta - t_\alpha$ is solvable iff inequality (59) is true for all $l \in \mathbb{R}^n$, $\lambda(\cdot) \in D_k^{m*}$.*

To reach the final solution to Problem 4.13 we consider the related *Dual Problem (SCCSM)* of optimization, which is

Problem 4.13 (Dual SCCSM) *Find*

$$\nu^0 = \Phi[l^0, \lambda^0(\cdot)] = \sup_{l,\lambda}\{\Phi(l, \lambda(\cdot)) \mid l, \lambda(\cdot)\}, \tag{61}$$

where

$$\Phi(l, \lambda(\cdot)) = \frac{\Gamma(l, \lambda(\cdot)) - \gamma \|s(\cdot \mid l, \lambda(\cdot))B(\cdot)\|_{D_{k-1}^*}}{\|\lambda(\cdot)\|_{D_k^*}}$$

along solutions $s(\cdot \mid l, \lambda(\cdot))$ to adjoint system (60).

Resolving Problem 4.13 allows to conclude the following

Theorem 4.4 *The minimal norm of the bound ν on state constraint (58) that solves Problem 4.13 is a result of solving the optimization procedure (61) which produces*

$$\nu^0 = \Phi[l^0, \lambda^0(\cdot)] \tag{62}$$

where $\{l^0, \lambda^0\}$ are the maximizers in (61), attained under $\|\lambda^0(\cdot)\|_{D_k^*} \neq 0$.

The optimal control $U^0(\cdot)$ is found from the maximum rule generated due to following reasoning. With $\{l^0, \lambda^0(\cdot)\}$ being the maximizers in (61), we have relations

$$\max_{U(\cdot)}\langle s(\cdot \mid l^0, \lambda^0)B, U(\cdot)\rangle = \langle s(\cdot \mid l^0, \lambda^0(\cdot)B, U^0(\cdot)\rangle =$$

$$= \Gamma(l^0, \lambda^0(\cdot)) - \nu^0\|\lambda^0(\cdot)\|_{D_k^*}, \tag{63}$$

over all bounded $\|U(\cdot)\| \leq \gamma$.

Theorem 4.5 *The optimal control $U^0(\cdot)$ for Problem 4.13 satisfies the Maximum Rule (63) along trajectories of adjoint system(60) governed by optimizers $\{l^0, \lambda^0(\cdot)\}$ in the Dual Problem (61).*

Remark 4.1 With $U^0(\cdot)$,—the output of an integrator being found, one may find the input $\mathsf{U}_0(\cdot)$ of this integrator as $d^{k-1}U^0(\cdot)/dt^{k-1} = \mathsf{U}_0(\cdot)$. A direct minimization of bound ν through finding optimal $\mathsf{U}^0(\cdot)$ instead of $U^0(\cdot)$ is achieved within a similar framework..

4.3.1 Guaranteed State Estimation Under Smooth Inputs

We now return to equation

$$dx/dt = Ax + CV(t), +x^{\beta}\delta(t - t_{\beta}), \quad t \in [t_{\alpha}, t_{\beta}], \tag{64}$$

with constant coefficients, smooth trajectories $x(\cdot) \in D_k^n$, and disturbances : $V(\cdot) \in D_{k-1}^q$, being norm-bounded as $\|V(\cdot)\|_{D_{k-1}} \leq \kappa$. As before symbol *supp* stands for the support of function $x(\cdot)$.

It is also assumed that system (64) is *dissipative* (Willems 2007). Here the smooth disturbances V are in the class D_{k-1} taken as

$$V(t) = V^{(0)}(t) = \int_{t_{\alpha}}^{t} \frac{(t-\xi)^{k-1}}{(k-1)!}\mathbf{V}(s)ds,$$

and $\mathbf{V}(\cdot)$ is an unknown norm bounded input: $\|\mathbf{V}(\cdot)\|_{D_0}^q \leq \kappa$, with given bound κ. Then $V^{(0)}(\cdot) \in D_{k-1}^q[\alpha, \beta]$.

The m-dimensional vector measurement is

$$y(t) = Hx(t), \quad y(t) \in \mathbb{R}^m, \quad y(\cdot) \in D_k^m, \tag{65}$$

where integer $k \geq 0$ and function $y(\cdot) \in D_k^m$ is k-times continuously differentiable. with derivatives arranged as

$$\mathbf{y} = \left\{y, \frac{dy}{dt}, \ldots, \frac{d^{k-1}y}{dt^{k-1}}\right\}.$$

The problem is to identify each coordinate x_j^{β} of $x(t_{\beta}) = x^{\beta}$ from measurement $\mathbf{y}(\cdot)$ by means of a linear operation $< \mathcal{W}^{(j)}(\cdot), \mathbf{y}(\cdot) >$, where $\mathcal{W}^{(j)}(\cdot) \in D_k^{(m\times k)*}$, is a matrix distribution of higher order realized as

$$< \mathcal{W}^{(j)}(\cdot), \mathbf{y}(\cdot) >= \sum_{i=0}^{k-1} \int_{t_{\alpha}}^{t_{\beta}} dw_i^{(j)}(t)\frac{d^i y(t)}{dt^i}, \quad \mathcal{W}^{(j)}(\cdot) = \{w_0^{(j)}(\cdot), \ldots, w_{k-1}^{(j)}(\cdot)\}, \tag{66}$$

with m-vectors $w_i^{(j)}(\cdot) \in \mathbf{BV}^m[t_{\alpha}, t_{\beta} + 0]$, $j = 1, \ldots, n$, and $d^0y(t)/dt^0 = y(t)$.

Under the non-bias condition (28) the mentioned problem is reduced to *the Primal Problem (SESM)* which sounds as follows.

Problem 4.14 (Primal GSESM) *Given system (64), with measurement (65), find the minimal worst-case estimation error for each coordinate* x_j^{β}*, namely,*

$$\Upsilon_j^0[\mathbf{y}(\cdot)] = \|\langle \mathcal{W}(j)(\cdot), \mathbf{y}(\cdot)\rangle - x_j^{\beta}\mathbf{e}_j\| = \min_{\mathcal{W}}\{\Upsilon_j[\mathcal{W}^{(j)}, \mathbf{y}]\},$$

with $\mathcal{W}_i^{(j)}(\cdot) \in (\mathbf{BV})^{m\times k}[t_{\alpha}, t_{\beta} + 0]$*, under condition (28), and* $j = 1, \ldots, n$.

Here each cost function $\Upsilon_j[\mathcal{W}^{(j)}, \mathbf{y}]$ is found trough solving a Dual Problem (SESM), which is,

Problem 4.15 (Dual GSESM) *Find*

$$\Upsilon_j[\mathcal{W}^{(j)}, \mathbf{y}] = \max_{\mathbf{V}}\{\|\langle \mathcal{W}^{(j)}(\cdot), \mathbf{y}(\cdot)\rangle - x^{\beta}\| \mid \|\mathbf{V}(\cdot)\|_{D_0} \leq \kappa\},\ j = 1, \ldots, n. \tag{67}$$

4.3.2 The System Duality Under Smooth Inputs

Considering Problems 4.12, 4.13, and 4.14, 4.15, assume that parameters of the former pair are changed as

$$A_* = -A,\ B_*^T = -H,\ N^T = C,\ U(\cdot) = W(\cdot),\ \Lambda(\cdot) = V(\cdot),\ \ p = q, \tag{68}$$

with related functional spaces also coinciding. Then one may observe that under the new notations Problem 4.12 will coincide with Problem 4.14.

Theorem 4.6 *Assume parameters of Problem 4.12 have been changed as in (68). Then this problem will coincide with Problem 4.14 demonstrating a* ***system duality*** *between problems of state-constrained smooth control and guaranteed state estimation under smooth disturbances.*

Remark The consideration of linear differential equations of controlled systems for the problems of this paper may not be limited to those with constant coefficients as here, provided the matrix system parameters would be differentiable functions. However the solutions will have to be reached through more complicated longer relations as indicated in paper (Kurzhanski and Daryin 2008).

5 Conclusion

This paper gives an analysis of solution schemes for problems of state-constrained control and estimation in linear systems under inputs that range from generalized functions of higher order to highly differentiable smooth functions. Indicated are classes of functional spaces within which the problems may be solved correctly. Emphasized are two types of duality in solving such problems,—the one in *mathematical sense* between Primal and Dual problems of optimization and the one in *system sense* between problems of optimal state-constrained control and optimal guaranteed state estimation.

Note that in problems of *state estimation* with $V(\cdot) \in D_{k-1}$, $x(\cdot) \in D_k$ *the dual variable* $\mathcal{W}(\cdot) \in D_k^*$ is taken as a *generalized function of higher order* that may include *delta*-functions and their derivatives.The same situation occurs in *state constrained control* under smooth inputs, where *the generalized Lagrange-type multiplier* $\lambda(\cdot) \in D_k^*$, attached to the state constraint, is to be chosen among *generalized functions of higher order*. □

References

M.S. Agranovich, Generalized Functions (Obobschenniye Funkcii). Moscow Independent University Publication (2008)

S.M. Aseev, V.M. Veliov, Maximum principle for problems with domination discount. Discrete Impulsive Syst. Ser. B **19**(1–2b), 43–63 (2012)

J.-P. Aubin, *Viability Theory*. SCFA (Birkhäuser, Boston, 1991)

A. Bensoussan, J.-L. Lions, *Contrôle impulsionnel et inéquations quasi-variationnelles* (Dunod, Paris, 1982)

K. Fan, Minimax theorems. Proc. Natl. Acad. Sci. USA **39**(1), 42–47 (1953)

I.M. Gelfand, G.E. Shilov, *Generalized Functions* (Dover Publications, New York, 1991)

R.E. Kalman, A new approach to linear filtering and prediction problems. Trans. ASME (Series D) **82**, 35–45 (1960)

N.N. Krasovski, On the theory of controllability and observability of linear dynamic systems. Appl. Math. Mech. **28**(1), 3–14 (1964)

N.N. Krasovski, A.I. Subbotin, *Game-Theoretical Control Problems*. Springer Series in Soviet Mathematics (Springer, New York, 1988)

A.B. Kurzhanski, Evolution equations for problems of control and estimation of uncertain systems, in *Proceedings of the International Congress of Mathematicians*, pp. 1381–1402 (1983)

A.B. Kurzhanski, Synthesizing impulse controls under state constraints. Problems Control Inf. **N2**, 6–15 (2016)

A.B. Kurzhanski, A.N. Daryin, Dynamic programming for impulse controls. Ann. Rev. Control **32**(2), 213–227 (2008)

A.B. Kurzhanski, Y.S. Osipov, On the control of a linear system by generalized inputs. Differ. Equ. (Differencialnye Uravneniya) **5**(8), 1360–1370 (1969)

A.B. Kurzhanski, P. Varaiya, *Dynamics and Control of Trajectory Tubes* (Birkhauser, Boston, 2014)

A.B. Kurzhanski, V.M. Veliov, *Set-Valued Analysis and Differential Inclusions*. Progress in Systems and Control Theory, vol. 16 (Birkhauser, Boston, 1993)

R.T. Rockafellar, *Convex Analysis*, 2nd edn. (Princeton University Press, Princeton, NJ, 1999)

R.T. Rockafellar, R.J. Wets, *Variational Analysis* (Springer, Berlin, 2005)

L. Schwartz, *Théorie des Distributions*, vols. 1, 2 (Hermann, Paris, 1950–1951)

L. Schwartz, *Mathematics for Physical Sciences* (Addison-Wesley, Reading, 1966)

V. Veliov, Stability-like properties of differential inclusions. Set-Valued Anal. **5**(1), 73–88 (1997)

J.C. Willems, Dissipative dynamical systems. Eur. J. Control **13**(2–3), 134–151 (2007)

Positive Approximations of the Inverse of Fractional Powers of SPD M-Matrices

Stanislav Harizanov and Svetozar Margenov

Abstract This study is motivated by the recent development in the fractional calculus and its applications. During last few years, several different techniques are proposed to localize the nonlocal fractional diffusion operator. They are based on transformation of the original problem to a local elliptic or pseudoparabolic problem, or to an integral representation of the solution, thus increasing the dimension of the computational domain. More recently, an alternative approach aimed at reducing the computational complexity was developed. The linear algebraic system $\mathcal{A}^\alpha \mathbf{u} = \mathbf{f}$, $0 < \alpha < 1$ is considered, where $\mathcal{A}$ is a properly normalized (scalded) symmetric and positive definite matrix obtained from finite element or finite difference approximation of second order elliptic problems in $\Omega \subset \mathbb{R}^d$, $d = 1, 2, 3$. The method is based on best uniform rational approximations (BURA) of the function $t^{\beta-\alpha}$ for $0 < t \leq 1$ and natural β.

The maximum principles are among the major qualitative properties of linear elliptic operators/PDEs. In many studies and applications, it is important that such properties are preserved by the selected numerical solution method. In this paper we present and analyze the properties of positive approximations of $\mathcal{A}^{-\alpha}$ obtained by the BURA technique. Sufficient conditions for positiveness are proven, complemented by sharp error estimates. The theoretical results are supported by representative numerical tests.

1 Introduction

This work is inspired by the recent development in the fractional calculus and its various applications, i.e., to Hamiltonian chaos, Zaslavsky (2002), anomalous diffusion in complex systems, Bakunin (2008), long-range interaction in elastic

S. Harizanov · S. Margenov (✉)
Institute of Information and Communication Technologies, Bulgarian Academy of Sciences, Sofia, Bulgaria
e-mail: margenov@parallel.bas.bg

G. Feichtinger et al. (eds.), *Control Systems and Mathematical Methods in Economics*, Lecture Notes in Economics and Mathematical Systems 687,
https://doi.org/10.1007/978-3-319-75169-6_8

deformations, Silling (2000), nonlocal electromagnetic fluid flows, McCay and Narasimhan (1981), image processing, Gilboa and Osher (2008). A more recent impressive examples of anomalous diffusion models in chemical engineering are provided in Metzler et al. (2014). Such kind of applications lead to fractional order partial differential equations that involve in general non-symmetric elliptic operators see, e.g. Kilbas et al. (2006). An important subclass of this topic are the fractional powers of self-adjoint elliptic operators, which are nonlocal but self-adjoint. In particular, the fractional Laplacian (Pozrikidis 2016) describes an unusual diffusion process associated with random excursions. In general, the parabolic equations with fractional derivatives in time are associated with sub-diffusion, while the fractional elliptic operators are related to super-diffusion.

Let us consider the elliptic boundary value problem in a weak form: find $u \in V$ such that

$$a(u, v) := \int_{\Omega} (\mathbf{a}(x)\nabla u(x) \cdot \nabla v(x) + q(x))\, dx = \int_{\Omega} f(x)v(x)dx, \quad \forall v \in V, \tag{1}$$

where

$$V := \{v \in H^1(\Omega) : \ v(x) = 0 \ \text{on} \ \Gamma_D\},$$

$\Gamma = \partial\Omega$, and $\Gamma = \bar{\Gamma}_D \cup \bar{\Gamma}_N$. We assume that Γ_D has positive measure, $q(x) \geq 0$ in Ω, and $\mathbf{a}(x)$ is an SPD $d \times d$ matrix, uniformly bounded in Ω, i.e.,

$$c\|\mathbf{z}\|^2 \leq \mathbf{z}^T \mathbf{a}(x)\, \mathbf{z} \leq C\|\mathbf{z}\|^2 \quad \forall \mathbf{z} \in \mathbb{R}^d, \forall x \in \Omega, \tag{2}$$

for some positive constants c and C. Also, Ω is a polygonal domain in $\mathbb{R}^d$, $d \in \{1, 2, 3\}$, and $f(x)$ is a given Lebesgue integrable function on Ω that belongs to the space $L_2(\Omega)$. Further, the case when $\mathbf{a}(x)$ does not depend on x is referred to as problem in homogeneous media, while the general case models processes in non homogeneous media. The bilinear form $a(\cdot, \cdot)$ defines a linear operator $\mathcal{L} : V \to V^*$ with V^* being the dual of V. Namely, for all $u, v \in V$ $a(u, v) := \langle \mathcal{L}u, v\rangle$, where $\langle \cdot, \cdot \rangle$ is the pairing between V and V^*.

One possible way to introduce $\mathcal{L}^\alpha$, $0 < \alpha < 1$, is through its spectral decomposition, i.e.

$$\mathcal{L}^\alpha u(x) = \sum_{i=1}^{\infty} \lambda_i^\alpha c_i \psi_i(x), \quad \text{where} \quad u(x) = \sum_{i=1}^{\infty} c_i \psi_i(x). \tag{3}$$

Here $\{\psi_i(x)\}_{i=1}^\infty$ are the eigenfunctions of $\mathcal{L}$, orthonormal in L_2-inner product and $\{\lambda_i\}_{i=1}^\infty$ are the corresponding positive real eigenvalues. This definition generalizes the concept of equally weighted left and right Riemann-Liouville fractional derivative, defined in one space dimension, to the multidimensional case. There is still ongoing research about the relations of the different definitions and their applications, see, e.g. Bates (2006).

The numerical solution of nonlocal problems is rather expensive. The following three approaches (A1–A3) are based on transformation of the original problem

$$\mathcal{L}^\alpha u = f \tag{4}$$

to a local elliptic or pseudo-parabolic problem, or on integral representation of the solution, thus increasing the dimension of the original computational domain.

The Poisson problem is considered in the related papers refereed bellow, i.e.

$$a(u, v) := \int_\Omega \nabla u(x) \cdot \nabla v(x) dx.$$

A1 Extension to a mixed boundary value problem in the semi-infinite cylinder $C = \Omega \times \mathbb{R}_+ \subset \mathbb{R}^{d+1}$

A Neumann to Dirichlet map is used in Chen et al. (2016). Then, the solution of fractional Laplacian problem is obtained by $u(x) = v(x, 0)$ where $v : \Omega \times \mathbb{R}_+ \to \mathbb{R}$ is a solution of the equation

$$-div\left(y^{1-2\alpha}\nabla v(x, y)\right) = 0, \quad (x, y) \in \Omega \times \mathbb{R}_+,$$

where $v(\cdot, y)$ satisfies the boundary conditions of (1) $\forall y \in \mathbb{R}_+$,

$$\lim_{y\to\infty} v(x, y) = 0, \quad x \in \Omega,$$

as well as

$$\lim_{y\to 0^+} \left(-y^{1-2\alpha} v_y(x, y)\right) = f(x), \quad x \in \Omega.$$

It is shown that the variational formulation of this equation is well posed in the related weighted Sobolev space. The finite element approximation uses the rapid decay of the solution $v(x, y)$ in the y direction, thus enabling truncation of the semi-infinite cylinder to a bounded domain of modest size. The proposed multilevel method is based on the Xu-Zikatanov identity (Xu and Zikatanov 2002). The numerical tests for $\Omega = (0, 1)$ and $\Omega = (0, 1)^2$ confirm the theoretical estimates of almost optimal computational complexity.

A2 Transformation to a pseudo-parabolic problem

The problem (1) is considered in Vabishchevich (2014, 2015) assuming the boundary condition

$$a(x)\frac{\partial u}{\partial n} + \mu(x)u = 0, \quad x \in \partial\Omega,$$

which ensures $\mathcal{L} = \mathcal{L}^* \geq \delta\mathcal{I}$, $\delta > 0$. Then the solution of fractional power diffusion problem u can be found as

$$u(x) = w(x, 1), \quad w(x, 0) = \delta^{-\alpha} f,$$

where $w(x, t)$, $0 < t < 1$, is the solution of pseudo-parabolic equation

$$(t\mathcal{D} + \delta\mathcal{I})\frac{dw}{dt} + \alpha\mathcal{D}w = 0,$$

and $\mathcal{D} = \mathcal{L} - \delta\mathcal{I} \geq 0$. Stability conditions are obtained for the fully discrete schemes under consideration. A further development of this approach is presented in Lazarov and Vabishchevich (2017) where the case of fractional order boundary conditions is studied.

A3 Integral representation of the solution

The following representation of the solution of (1) is used in Bonito and Pasciak (2015):

$$\mathcal{L}^{-\alpha} = \frac{2\sin(\pi\alpha)}{\pi}\int_0^\infty t^{2\alpha-1}\left(\mathcal{I} + t^2\mathcal{L}\right)^{-1} dt$$

Among others, the authors introduce an exponentially convergent quadrature scheme. Then, the approximate solution of u only involves evaluations of $(\mathcal{I} + t_i\mathcal{A})^{-1} f$, where $t_i \in (0, \infty)$ is related to the current quadrature node, and where $\mathcal{I}$ and $\mathcal{A}$ stand for the identity and the finite element stiffness matrix corresponding to the Laplacian. The computational complexity of the method depends on the number of quadrature nodes. For instance, the presented analysis shows that approximately 50 auxiliary linear systems have to be solved to get accuracy of the quadrature scheme of order $O(10^{-5})$ for $\alpha \in \{0.25, 0.5, 0.75\}$. A further development of this approach is available in Bonito and Pasciak (2016), where the theoretical analysis is extended to the class of regularly accretive operators.

An alternative approach is applied in Harizanov et al. (2016) where a class of optimal solvers for linear systems with fractional power of symmetric and positive definite (SPD) matrices is proposed. Let $\mathcal{A} \in \mathbb{R}^{N\times N}$ be a normalized SPD matrix generated by a finite element or finite difference approximation of some self-adjoint elliptic problem. An efficient method for solving algebraic systems of linear equations involving fractional powers of the matrix $\mathcal{A}$ is considered, namely for solving the system

$$\mathcal{A}^\alpha \mathbf{u} = \mathbf{f}, \quad \text{where} \quad 0 < \alpha < 1. \tag{5}$$

The fractional power of SPD matrix $\mathcal{A}$, similarly to the infinite dimensional counterpart $\mathcal{L}$, is expressed through the spectral representation of $\mathbf{u}$ through the eigenvalues and eigenvectors $\{(\Lambda_i, \mathbf{\Psi}_i)\}_{i=1}^N$ of $\mathcal{A}$, assuming that the eigenvectors are l_2-orthonormal, i.e. $\mathbf{\Psi}_i^T\mathbf{\Psi}_j = \delta_{ij}$ and $\Lambda_1 \le \Lambda_2 \le \dots \Lambda_N \le 1$. Then $\mathcal{A} = \mathcal{W}\mathcal{D}\mathcal{W}^T$, $\mathcal{A}^\alpha = \mathcal{W}\mathcal{D}^\alpha\mathcal{W}^T$, where the $N \times N$ matrices $\mathcal{W}$ and $\mathcal{D}$ are defined as $\mathcal{W} = (\mathbf{\Psi}_1^T, \mathbf{\Psi}_2^T, \dots, \mathbf{\Psi}_N^T)$ and $\mathcal{D} = diag(\Lambda_1, \dots, \Lambda_N)$, $\mathcal{A}^{-\alpha} = \mathcal{W}\mathcal{D}^{-\alpha}\mathcal{W}^T$, and the solution of $\mathcal{A}^\alpha\mathbf{u} = \mathbf{f}$ can be expressed as

$$\mathbf{u} = \mathcal{A}^{-\alpha}\mathbf{f} = \mathcal{W}\mathcal{D}^{-\alpha}\mathcal{W}^T\mathbf{f}. \tag{6}$$

Instead of the system (5), one can solve the equivalent system $\mathcal{A}^{\alpha-\beta}\mathbf{u} = \mathcal{A}^{-\beta}\mathbf{f} := \mathbf{F}$ with $\beta \ge 1$ an integer. Then the idea is to approximately evaluate $\mathcal{A}^{\beta-\alpha}\mathbf{F}$ using a set of equations involving inversion of $\mathcal{A}$ and $\mathcal{A} - d_jI$, for $j = 1, \dots, k$. The integer parameter $k \ge 1$ is the number of partial fractions of the best uniform rational approximation (BURA) $r_\alpha^\beta(t)$ of $t^{\beta-\alpha}$ on the interval $(0, 1]$. One can observe that the algorithm of Bonito and Pasciak (2015), see A3, can be viewed as a particular rational approximation of $\mathcal{A}^{-\alpha}$. It is also important, that in certain sense the results from Harizanov et al. (2016) are more general, and are applicable to a wider class of sparse SPD matrices.

Assuming that $\mathcal{A}$ is a large-scale matrix, the computational complexity of the discussed methods for numerical solution of fractional diffusion problems is substantially high. Then, the parallel implementation of such methods for real life problems is an unavoidable topic. In this context, there are some serious advantages of the last two approaches, see e.g., in Ciegis et al. (2017).

The maximum principles are among the major qualitative properties of the elliptic or parabolic operators/PDEs. In general, to solve PDEs we use some numerical method, and it is a natural requirement that such qualitative properties are preserved on the discrete level. Most of the studies which deal with such topics give sufficient conditions for the discretization parameters in order to guarantee the certain maximum principle. It is easily see, that under certain such assumptions, the solutions of (A1) and (A3) satisfy certain maximum principle. In this paper, we study positive approximations of $\mathcal{A}^{-\alpha}$, obtained by BURA technique introduced in Harizanov et al. (2016), under rather general assumptions for the normalized SPD matrix $\mathcal{A}$.

The rest of the paper is organized as follows. In Sect. 2 we provide a brief introduction to the topic of monotone matrices including some basic properties of the M-matrices and their relations to FEM discretization of elliptic PDEs. Sufficient conditions for positive approximations of the inverse of a given SPD matrix, based on BURA technique are presented in Sect. 3. The analysis in Sect. 4 is devoted to a class of best rational approximations of $\mathcal{A}^{-\alpha}$, that satisfy such sufficient conditions. Sharp error estimates for a class of BURA approximations are also included in this section. Some numerical tests and short concluding remarks are given at the end.

2 Monotone Matrices and SPD M-Matrices

The maximum principles are some of the most useful properties used to solve a wide range of problems in the PDEs. For instance, their use is often essential to study the uniqueness and necessary conditions of solvability, approximation and boundedness of the solution, as well as, for quantities of physical interest like maximum stress, torsional stiffness, electrostatic capacity, charge density etc. Under certain regularity conditions, a classical maximum principle for elliptic problems reads as follows. Suppose that $\mathcal{L}u \geq 0$ in Ω, then a nonnegative maximum is attained at the boundary $\partial\Omega$. Let us assume additionally that $u \geq 0$ in $\partial\Omega$. Then the positivity preserving property holds, that is, $u(x) > 0$ for $x \in \Omega$ or $u \equiv 0$.

To solve PDEs we use some numerical methods, and it is a natural requirement that such qualitative properties are preserved on the discrete level. Most of the papers which deal with this topic give sufficient conditions for the discretization parameters in order to guarantee the certain maximum principle. For instance, when FEM is applied, the related results are usually described in terms of properties of the related mass and stiffness matrices.

Definition 1 A real square matrix $\mathcal{A}$ is called monotone if for all real vectors v, $\mathcal{A}\mathbf{v} \geq 0$ implies $\mathbf{v} \geq 0$, where $\geq$ is in element-wise sense.

The next property is sometimes used as an alternative definition.

Proposition 1 *Let $\mathcal{A}$ be a real square matrix. $\mathcal{A}$ is monotone if and only if $\mathcal{A}^{-1} \geq 0$.*

Definition 2 The class of Z-matrices are those matrices whose off-diagonal entries are less than or equal to zero. Let $\mathcal{A}$ be a $N \times N$ real Z-matrix, then $\mathcal{A}$ is a non-singular M-matrix if every real eigenvalue of $\mathcal{A}$ is positive. A symmetric M-matrix is sometimes called a Stieltjes matrix.

The M-matrices are among most often used monotone matrices. They arise naturally in some discretizations of elliptic operators.

Let $\mathcal{A}$ be an SPD matrix obtained after FEM approximation of (1) by linear triangle elements. Let us assume also that $\Omega \subset \mathbb{R}^2$ is discretized by a nonobtuse triangle mesh τ_h, and the coefficients $a(x) = a_e$ and $q(x) = q_e$ are piecewise constants on the triangles $e \in \tau_h$. Then $\mathcal{A}$ can be assembled by the element matrices

$$\mathcal{A}_e = \mathcal{K}_e + \mathcal{M}_e,$$

where $\mathcal{K}_e$ is the element stiffness matrix and $\mathcal{M}_e$ is the element mass matrix. Subject to a scaling factor, the off diagonal elements of the symmetric and positive semidefinite matrix $\mathcal{K}_e$ are equal to some of $-\cot(\theta_i) \leq 0$, $i = 1, 2, 3$, where $0 < \theta_i \leq \pi/2$ are the nonobtuse angles of the triangle e. Imposing the boundary conditions we get that the global stiffness matrix is SPD M-matrix. The element mass matrix is positive diagonal matrix if a proper quadrature formula is applied. A similar result can be obtained by the standard diagonalization known as *lumping the mass*. Then, the global mass matrix is positive diagonal matrix and $\mathcal{A}$ is

SPD M-matrix. A more general considerations of this kind are available in Kraus and Margenov (2009) including the case of coefficient anisotropy as well as the nonconforming linear finite elements. Similar representations of the element mass and stiffness matrices are derived if $\Omega \subset \mathbb{R}^3$ is discretized by a nonobtuse tetrahedral mesh τ_h. We get again that $\mathcal{A}$ is again SPD M-matrix, see e.g., Kosturski and Margenov (2009).

Remark 1 Not all monotone matrices are M-matrices, and the sum of two monotone matrices is not always monotone. The next examples prove these statements.

Example 1

E1.1 $\mathcal{A}_1 = \begin{pmatrix} -1 & 3 \\ 2 & -4 \end{pmatrix}$ is not M-matrix, but $\mathcal{A}_1^{-1} = \frac{1}{2}\begin{pmatrix} 4 & 3 \\ 2 & 1 \end{pmatrix}$ and therefore $\mathcal{A}_1$ is monotone.

E1.2 $\mathcal{A}_2 = \mathcal{A}_1 + 6\mathcal{I}$ is a sum of two monotone matrices, but $\mathcal{A}_2^{-1} = \frac{1}{4}\begin{pmatrix} 2 & -3 \\ -2 & 0 \end{pmatrix}$, and therefore $\mathcal{A}_2$ is not monotone.

In what follows, we study positive approximations of $\mathcal{A}^{-\alpha}$ for a given normalized SPD M-matrix $\mathcal{A}$. It follows straightforwardly that the inverse of each such approximation will approximate $\mathcal{A}^{\alpha}$ in the class of monotone functions.

3 Positive Approximations of the Inverse of SPD M-Matrices

Explicit computation and memory storage of the fractional power $\mathcal{A}^{\alpha}$ in (5) for large-scale problems is expensive and impractical. Even when $\mathcal{A}$ is sparse, $\mathcal{A}^{\alpha}$ is typically dense. Therefore, we study possible positive approximations of the action of $\mathcal{A}^{-\alpha}$ based solely on the information of $\mathcal{A}$. For this purpose, we consider the class of rational functions

$$\mathcal{R}(m,k) := \{P_m/Q_k \;:\; P_m \in \mathcal{P}_m,\, Q_k \in \mathcal{P}_k\},$$

fix a positive integer β, and search for an appropriate candidate r in it, that approximates well the univariate function $t^{\beta-\alpha}$ on the unit interval $[0, 1]$. Note that, due to the normalization of $\mathcal{A}$, this interval covers the spectrum of $\mathcal{A}^{\alpha}$,

Definition 3 Let $\alpha \in (0, 1)$, and $\beta, m, k \in \mathbb{N} \setminus \{0\}$. The minimizer $r_{\alpha}^{\beta} \in \mathcal{R}(m,k)$ of the problem

$$\min_{r \in \mathcal{R}(m,k)} \max_{t \in [0,1]} \left| t^{\beta-\alpha} - r(t) \right|, \tag{7}$$

will be called β-Best Uniform Rational Approximation (β-BURA). Its error will be denoted by

$$E_\alpha(m,k;\beta) := \max_{t\in[0,1]} \left| t^{\beta-\alpha} - r_\alpha^\beta(t) \right|.$$

Based on classical Spectral Theory arguments (see Harizanov et al. (2016, Theorem 2.1)), the univariate approximation error $E_\alpha(m,k;\beta)$ is an upper bound for the multivariate relative error $\|\mathcal{A}^{-\beta} r_\alpha^\beta(\mathcal{A})\mathbf{f} - \mathcal{A}^{-\alpha}\mathbf{f}\|_{\mathcal{A}^{\gamma+\beta}} / \|\mathbf{f}\|_{\mathcal{A}^{\gamma-\beta}}$ for the corresponding matrix-valued BURA approximation of the exact solution $\mathbf{u}$ in (6). Here, $\gamma \in \mathbb{R}$ can be arbitrary, and the Krylov norms are defined via standard energy dot product, i.e. $\|\mathbf{f}\|^2_{\mathcal{A}^{\gamma-\beta}} = \langle \mathcal{A}^{\gamma-\beta}\mathbf{f}, \mathbf{f}\rangle$.

Proposition 2 *Let $\mathcal{A} \in \mathbb{R}^{N\times N}$ be an SPD matrix with eigenvalues $0 < \Lambda_1 \le \Lambda_2 \le \dots \le \Lambda_N \le 1$. Let r_α^β be the β-BURA for given α, β, m, k. Then,*

$$\|\mathcal{A}^{-\beta} r_\alpha^\beta(\mathcal{A})\mathbf{f} - \mathcal{A}^{-\alpha}\mathbf{f}\|_{\mathcal{A}^{\gamma+\beta}} \le E_\alpha(m,k;\beta)\|\mathbf{f}\|_{\mathcal{A}^{\gamma-\beta}}, \qquad \forall\gamma \in \mathbb{R},\ \forall \mathbf{f} \in \mathbb{R}^N. \tag{8}$$

For the practical computation of $\mathcal{A}^{-\beta} r_\alpha^\beta(\mathcal{A})\mathbf{f}$ we use the partial fraction decomposition of $t^{-\beta} r_\alpha^\beta(t)$, which is of the form

$$t^{-\beta} r_\alpha^\beta(t) = \sum_{j=0}^{m-k-\beta} b_j t^j + \sum_{j=1}^{\beta} \frac{c_{0,j}}{t^j} + \sum_{j=1}^{k} \frac{c_j}{t-d_j}, \tag{9}$$

provided all r_α^β has no complex poles and the real ones $\{d_j\}_1^k$ are all of multiplicity 1. Later, we will see that for $\beta = 1$ and $m = k$ the above assumption on the poles of r_α^β holds true for any $\alpha \in (0,1)$. Furthermore, in all our numerical experiments with various β, m, k and $\alpha \in \{0.25, 0.5, 0.75\}$ the assumption always remains valid. Hence, it does not seem to restrict the application range of the proposed method. On the other hand, under (9) the approximate solution

$$\mathbf{u}_r := \mathcal{A}^{-\beta} r_\alpha^\beta(\mathcal{A})\mathbf{f} = \sum_{j=0}^{m-k-\beta} b_i \mathcal{A}^j\mathbf{f} + \sum_{j=1}^{\beta} c_{0,j}\mathcal{A}^{-j}\mathbf{f} + \sum_{j=1}^{k} c_j(\mathcal{A} - d_j I)^{-1}\mathbf{f} \tag{10}$$

of $\mathbf{u}$ can be efficiently numerically computed via solving several linear systems, that involve $\mathcal{A}$ and its diagonal variations $\mathcal{A} - d_j I$, for $j = 1, \dots, k$.

Definition 4 A real symmetric matrix $\mathcal{A}^{-1}$ is said to be doubly nonnegative if it is both positive definite, and entrywise nonnegative.

Our first goal is to analyze under what conditions on the coefficients and the poles in (9), the matrix $\mathcal{A}^{-\beta} r_\alpha^\beta(\mathcal{A})$ remains doubly nonnegative. Clearly the matrix is symmetric whenever $\mathcal{A}$ is, so the main investigations are on assuring $\mathcal{A}^{-\beta} r_\alpha^\beta(\mathcal{A}) \ge 0$. The following proposition contains sufficient conditions for positivity.

Proposition 3 *If $\mathcal{A}$ is a normalized SPD M-matrix, $m < k + \beta$, $c_0 \geq 0$, $c \geq 0$, and* $\mathbf{d} < 0$ *(entrywise), then $\mathcal{A}^{-\beta} r_\alpha^\beta(\mathcal{A})$ in* (10) *is doubly nonnegative.*

Proof Since $m < k + \beta$, equation (10) is simplified to

$$\mathcal{A}^{-\beta} r_\alpha^\beta(\mathcal{A}) = \sum_{j=1}^{\beta} c_{0,j} \mathcal{A}^{-j} + \sum_{j=1}^{k} c_j (\mathcal{A} - d_j I)^{-1}.$$

For every $j = 1, \dots, k$, the matrix $\mathcal{A} - d_j I$ is an SPD M-matrix, as $d_j < 0$ and the diagonal elements increase their values, i.e. become stronger dominant. Hence, $(\mathcal{A} - d_j I)^{-1} \geq 0$.We have $\mathcal{A}^{-1} \geq 0$, thus $\mathcal{A}^{-j} = (\mathcal{A}^{-1})^j \geq 0$, $j = 1, \dots, \beta$, as each entry of $\mathcal{A}^{-j}$ is a sum of nonnegative summands. Finally, a linear combination of positively scaled doubly nonnegative matrices is also a doubly nonnegative matrix.

Note that, when applying pure polynomial approximation techniques for $t^{-\alpha}$ on $[\Lambda_1, 1]$ like in Harizanov et al. (2017), there is practically no chance to come up with a positive approximation of $\mathcal{A}^{-\alpha}$. First of all, such an approximant is a linear combination of positive degrees of $\mathcal{A}$ and in particular $\mathcal{A}$ itself appears with a nonzero coefficient. This matrix has non-positive off-diagonal entries. Furthermore, it was numerically observed that the coefficient sequence in the linear combination is sign alternating. Hence, the proposed β-BURA approach seems the right and most natural tool for constructing positive approximations of $\mathcal{A}^{-\alpha}$, or alternatively, monotone approximations of $\mathcal{A}^{\alpha}$. Another disadvantage of the former approach is the restriction on Λ_1 to be well-separated from zero, which is also a restriction on the condition number of $\mathcal{A}$.

4 Analysis of a Class of Best Rational Approximations of Fractional Power of SPD M-Matrices

Among all various classes of best rational approximations, the diagonal sequences $r \in \mathcal{R}(k, k)$ of the *Walsh table* of t^α, $\alpha \in (0, 1)$ are studied in greatest detail (Newman 1964; Ganelius 1979; Stahl 1993; Saff and Stahl 1995). There is an existence and uniqueness of the BURA elements for all k and α. The distribution of poles, zeros, and extreme points of those elements plays a central role in asymptotic convergence analysis, when $k \to \infty$, thus is well known. In this section, we show that the above diagonal class perfectly fits within our positive $\mathcal{A}^{-\alpha}$ approximation framework.

First, we collect some preliminary results that will be later needed for the proof of the main theorem. The following characterization lemma, which we state here without proof, is vital for our further investigations.

Lemma 1 (Saff and Stahl (1995, Lemma 2.1)) *Let $m = k$ and $0 < \alpha < 1$.*

(a) *The best rational approximant r_α^1 is of exact numerator and denominator degree k.*
(b) *All k zeros $\zeta_1, \ldots, \zeta_k$ and poles $d_1, \ldots, d_k$ of r_α^1 lie on the negative half-axis $\mathbb{R}_{<0}$ and are interlacing; i.e., with an appropriate numbering we have*

$$0 > \zeta_1 > d_1 > \zeta_2 > d_2 > \cdots > \zeta_k > d_k > -\infty \tag{11}$$

(c) *The error function $t^{1-\alpha} - r_\alpha^1(t)$ has exactly $2k+2$ extreme points $\eta_1, \ldots, \eta_{2k+2}$ on $[0, 1]$, and with an appropriate numbering we have*

$$0 = \eta_1 < \eta_2 < \cdots < \eta_{2k+2} = 1 \tag{12}$$

$$\eta_j^{1-\alpha} - r_\alpha^1(\eta_j) = (-1)^j E_\alpha(k, k; 1), \qquad j = 1, \ldots, 2k+2. \tag{13}$$

The next lemma builds a bridge between the fractional decompositions of r_α^1 and $t^{-1} r_\alpha^1$.

Lemma 2 *Let $m = k$, $0 < \alpha < 1$, and*

$$r_\alpha^1(t) = b_0^* + \sum_{j=1}^{k} \frac{c_j^*}{t - d_j}, \qquad t^{-1} r_\alpha^1(t) = \frac{c_{0,1}}{t} + \sum_{j=1}^{k} \frac{c_j}{t - d_j}.$$

Then

$$c_{0,1} = E_\alpha(k, k; 1), \qquad c_j = c_j^*/d_j, \quad j = 1, \ldots, k. \tag{14}$$

Proof The second part of (14) follows directly from

$$\frac{1}{t(t - d_j)} = \frac{1}{d_j}\left(\frac{1}{t - d_j} - \frac{1}{t}\right), \quad j = 1, \ldots, k.$$

For the first part, we combine the above identity with (12) and (13)

$$c_{0,1} = b_0^* - \sum_{j=1}^{k} \frac{c_j^*}{d^j} = r_\alpha^1(0) = -\left(\eta_1^{1-\alpha} - r_\alpha^1(\eta_1)\right) = E_\alpha(k, k; 1).$$

The proof of the lemma is completed.

Our last lemma provides an asymptotic bound on $E_\alpha(k, k; 1)$. The proof can be found in Stahl (1993).

Lemma 3 (Stahl (1993, Theorem 1)) *The limit*

$$\lim_{k\to\infty} e^{2\pi\sqrt{\alpha k}} E_{1-\alpha}(k,k;1) = 4^{1+\alpha}|\sin\pi\alpha|$$

holds true for each $\alpha > 0$.

Now, we are ready to formulate and prove our main result.

Theorem 1 *Let* $\beta = 1$ *and* $m = k$. *For every normalized SPD M-matrix* $\mathcal{A}$ *and every* $\alpha \in (0,1)$, *the matrix* $\mathcal{A}^{-1}r_\alpha^1(\mathcal{A})$ *is doubly nonnegative and for all* $\gamma \in \mathbb{R}$

$$\frac{\|\mathcal{A}^{-1}r_\alpha^1(\mathcal{A})\mathbf{f} - \mathcal{A}^{-\alpha}\mathbf{f}\|_{\mathcal{A}^{\gamma+1}}}{\|\mathbf{f}\|_{\mathcal{A}^{\gamma-1}}} \le 4^{2-\alpha}|\sin\pi(1-\alpha)|e^{-2\pi\sqrt{(1-\alpha)k}}(1+\mathrm{o}(1)). \tag{15}$$

Proof Based on the results in Lemma 1, we can quickly derive $\mathcal{A}^{-1}r_\alpha^1(\mathcal{A}) \ge 0$. For this purpose, we study the sign pattern of c and d and assure the applicability of Proposition 3. From (11) we know that all the poles $\{d_j\}$ are real, negative, and of multiplicity 1. The same holds true for the zeros $\{\zeta_j\}$. Since $r_\alpha^1(t)$ is continuous on $\mathbb{R}\setminus\{d_j\}$, the function changes its sign $2k$ times—at each zero ζ_j and at each pole d_j. In Lemma 2, we have already computed that $r_\alpha^1(0) = c_{0,1} = E_\alpha(k,k;1) > 0$, thus, due to interlacing, at each pole d_j we have

$$\begin{matrix} \lim_{t\to d_j^+} r_\alpha^1(t) < 0 \\ \lim_{t\to d_j^-} r_\alpha^1(t) > 0 \end{matrix} \quad\Longrightarrow\quad \begin{matrix} \lim_{t\to d_j^+} r_\alpha^1(t) = -\infty \\ \lim_{t\to d_j^-} r_\alpha^1(t) = +\infty \end{matrix} \quad\Longrightarrow\quad c_j^* < 0.$$

Since $c^* < 0$ and $d < 0$, from Lemma 2 it follows that $c_0 > 0$ and $c > 0$. Hence, Proposition 3 gives rise to $\mathcal{A}^{-1}r_\alpha^1(\mathcal{A}) \ge 0$.

The error estimate (15) is a direct corollary of Proposition 2 and Lemma 3.

Some remarks are in order. For any fixed $\alpha < 1$, the relative error (15) decays exponentially as $k \to +\infty$ with order $\sqrt{(1-\alpha)k}$. When $\alpha \to 1$ the relative error decays linearly independently of k, since $|\sin\pi(1-\alpha)| \to \pi(1-\alpha)$. It is straightforward to extend the coefficient correspondence (14) to $c_j = c_j^*/d_j^\beta$ for any (m,k,β), such that $m < k+\beta$. Therefore, a necessary condition for Proposition 3 to be applicable is the sequence c^* to have constant sign. Due to the proof of Theorem 1, it implies that zeros $\{\zeta_i\}_1^m$ and poles $\{d_j\}_1^k$ of r_α^β should be interlacing, thus $|m-k| \le 1$. Furthermore (see Harizanov et al. (2016, (20))) the following identity always holds true

$$c_{0,1} + \sum_{j=1}^{k} c_j = 0, \qquad m < k+\beta-1.$$

Hence, another necessary condition for applicability of Proposition 3 is $m \geq k + \beta - 1$. Combining all derived constraints, we observe that $\mathcal{A}^{-\beta} r_\alpha^\beta(\mathcal{A})$ could be represented as a sum of doubly nonnegative matrices only if $m = k + \beta - 1$ and $m \leq k + 1$, meaning that we are left with the admissible triples

$$(m, k, \beta) = \{(k, k, 1), (k, k, 2), (k + 1, k, 2)\}.$$

In Saff and Stahl (1995) it is remarked that for the case $(k, k, 2)$ it cannot be theoretically excluded that one root ζ_i and one pole d_i of the 2-BURA $r_\alpha^2(t)$ lie outside of $\mathbb{R}_{<0}$. For the case $(k + 1, k, 2)$ we can show that $c_{0,2} = -E_\alpha(k + 1, k; \beta) < 0$, thus the assumptions of the proposition are again violated. In conclusion, the triple $(k, k, 1)$, investigated in Theorem 1 is the unique choice of parameters for which one can prove positiveness of the approximation $\mathcal{A}^{-\beta} r_\alpha^\beta(\mathcal{A})$ using Proposition 3.

5 Numerical Tests

The main goal of the numerical tests is to illustrate the positive properties of the proposed approximations of inverse of fractional powers of SPD M-matrices. Complementary, we provide a short discussion related to some interpretations of the results from Sect. 4 in the case of fractional diffusion problems.

The following test problems are in 1D. For the purpose of this study, the presented numerical tests are fully representative. Let us remind, that the theoretical error estimates of the BURA based algorithm are robust with respect to the normalized matrix $\mathcal{A}$. We have shown in Harizanov et al. (2016), that in the case $\Omega \subset \mathbb{R}^3$, the accuracy and efficiency of the algorithm are fully preserved, utilizing the BoomerAMG preconditioner in the PCG solver for the arising sparse linear systems.

We consider fractional powers of the Poisson's equation on the unit interval with Dirichlet boundary conditions:

$$\mathcal{L}u := -u''(x) = f(x), \qquad x \in [0, 1], \quad u(0) = u(1) = 0. \tag{16}$$

On a uniform grid with mesh parameter $h = 1/(N + 1)$, using central finite differences, the operator $\mathcal{L}$ is approximated by the $N \times N$ matrix $\mathcal{A}_h := tridiag(-1, 2, -1)/h^2$, which in turn can be rewritten as

$$\mathcal{A}_h = 4h^{-2}\mathcal{A}, \qquad \mathcal{A} := tridiag\left(-\frac{1}{4}, \frac{1}{2}, -\frac{1}{4}\right). \tag{17}$$

The matrix $\mathcal{A}$ is a normalized, SPD M-matrix, which eigenvectors and eigenvalues are explicitly known:

$$\Lambda_i = \sin^2\left(\frac{i\pi}{2(N+1)}\right), \qquad \boldsymbol{\Psi}_i = \left\{\sin\frac{im\pi}{N+1}\right\}_{m=1}^{N}, \qquad i = 1, \ldots, N.$$

We approximate $\mathcal{L}^\alpha$ by $\mathcal{A}_h^\alpha$ and, due to Theorem 1, the ℓ^2 relative error is bounded by an h-dependent constant.

Corollary 1 *Let* $\mathbf{u}_h := \mathcal{A}_h^{-\alpha}\mathbf{f} = \left(\frac{h}{2}\right)^{2\alpha}\mathcal{A}^{-\alpha}\mathbf{f}$ *be the exact solution of the discretized fractional Poisson's equation* $\mathcal{A}_h^\alpha\mathbf{u} = \mathbf{f}$. *Let* $r_\alpha^1 \in \mathcal{R}(k,k)$ *be 1-BURA and denote by* $\mathbf{u}_{h,r} := \left(\frac{h}{2}\right)^{2\alpha}\mathcal{A}^{-1}r_\alpha^1(\mathcal{A})\mathbf{f}$. *Then*

$$\frac{\|\mathbf{u}_{h,r} - \mathbf{u}_h\|_2}{\|\mathbf{f}\|_2} \le \left(\frac{4}{h}\right)^{2(1-\alpha)} |\sin\pi(1-\alpha)| e^{-2\pi\sqrt{(1-\alpha)k}}(1 + o(1)).$$

Proof Indeed, let $\mathbf{u} = \mathcal{A}^{-\alpha}\mathbf{f}$ and $\mathbf{u}_r = \mathcal{A}^{-1}r_\alpha^1(\mathcal{A})\mathbf{f}$. Applying

$$\|\cdot\|_2 = \|\cdot\|_{\mathcal{A}^0} \le \mathrm{k}(\mathcal{A})\|\cdot\|_{\mathcal{A}^2} < h^{-2}\|\cdot\|_{\mathcal{A}^2},$$

where $\mathrm{k}(\mathcal{A})$ is the condition number of $\mathcal{A}$, we derive

$$\frac{\|\mathbf{u}_{h,r} - \mathbf{u}_h\|_2}{\|\mathbf{f}\|_2} \le h^{-2}\frac{\|\mathbf{u}_{h,r} - \mathbf{u}_h\|_{\mathcal{A}^2}}{\|\mathbf{f}\|_{\mathcal{A}^0}} \le h^{-2}\left(\frac{h}{2}\right)^{2\alpha}\frac{\|\mathbf{u}_r - \mathbf{u}\|_{\mathcal{A}^2}}{\|\mathbf{f}\|_{\mathcal{A}^0}}. \tag{18}$$

The result follows from (15) for $\gamma = 1$.

Due to Corollary 1, we can compute the minimal degree k that guarantees $\|\mathbf{u}_{h,r} - \mathbf{u}_h\|_2/\|\mathbf{f}\|_2 < \varepsilon$ for every given pair (ε, h). Such an ℓ^2 error analysis is outside of the scope of this paper, so we will not further elaborate on it.

In our numerical experiments, we choose $\alpha \in \{0.25, 0.5, 0.75\}$ and $k \in \{5, 6, 7, 8, 9\}$. The square root of an M-matrix is again an M-matrix (Alefeld and Schneider 1982) and it is easy to check that $\mathcal{A}_h^3$ is also an M-matrix. Therefore, all considered $\mathcal{A}_h^\alpha$ are M-matrices, their inverse matrices are doubly nonnegative (but dense!), and constructing computationally cheep approximants within the same class is of great practical importance. The univariate error estimates $E_\alpha(k,k;1)$ for the used choice of the above parameters are summarized in Table 1. Note that each

Table 1 Errors $E_\alpha(k,k;1)$ of BURA $r_\alpha^1(t)$ of $t^{1-\alpha}$ on $[0,1]$

α	$E_\alpha(5,5,1)$	$E_\alpha(6,6,1)$	$E_\alpha(7,7,1)$	$E_\alpha(8,8,1)$	$E_\alpha(9,9,1)$
0.25	2.8676e−5	9.2522e−6	3.2566e−6	1.2288e−6	4.9096e−7
0.50	2.6896e−4	1.0747e−4	4.6037e−5	2.0852e−5	
0.75	2.7348e−3	1.4312e−3	7.8269e−4		

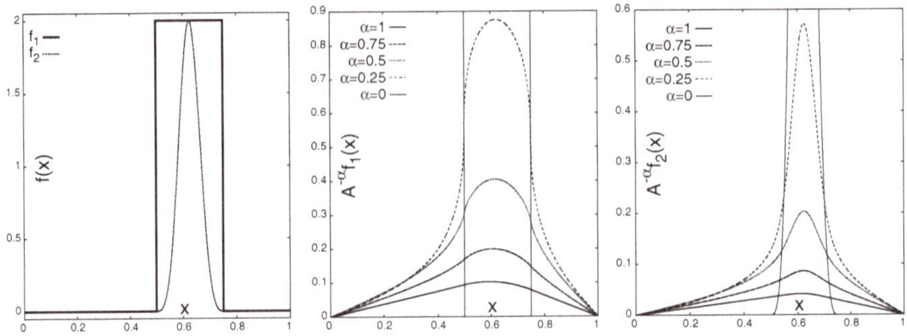

Fig. 1 Test data and their exact fractional diffusions

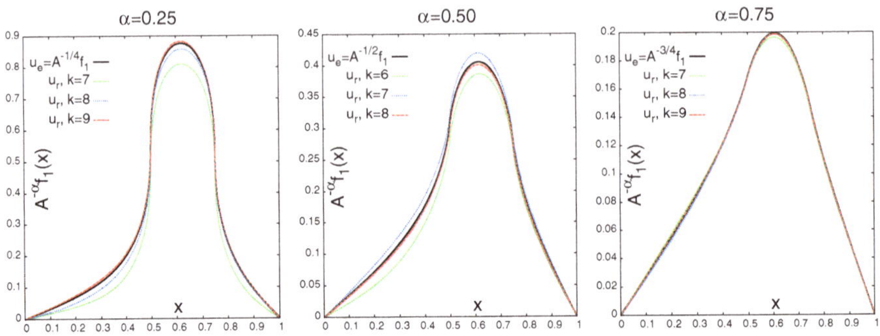

Fig. 2 Positive approximations of $\mathcal{A}_h^{-\alpha}\mathbf{f}_1$ for $h = 2^{-11}$

of them satisfies the inequality (15) even without introducing the low-order term o(1) in the right-hand-side. A modified Remez algorithm is used for the derivation of r_α^1 (Marinov and Andreev 1987; Cheney and Powell 1987).

For f in (4) we take two different positive functions, supported on the interval $[1/2, 3/4]$. The first one f_1 is piecewise constant and discontinuous, while the second one f_2 is a C^2 cubic spline function, corresponding to the Irwin-Hall distribution. Together with the exact discretized solutions $\mathcal{A}_h^{-\alpha}\mathbf{f}$, they are illustrated on Fig. 1.

We consider mesh parameters $h = 2^{-m}$, $m \in \{5, 6, \dots, 11\}$. On Fig. 2 the corresponding approximants $\mathbf{u}_{h,r}$, $h = 2^{-11}$, of $\mathbf{u}_h = \mathcal{A}_h^{-\alpha}\mathbf{f}_1$ are plotted. As suggested by Corollary 1, $\mathbf{u}_{h,r}$ fails to approximate well $\mathbf{u}_h$ on a fine grid for smaller k and α, thus we use $k = 9$ for $\alpha = 0.25$ and $k = 8$ for $\alpha = 0.5$ to get a balanced relative accuracy of order $O(10^{-3})$ (see the last row of Table 2). For larger α, the exponential growth of the ℓ^2 relative error with $h \to 0$ is less significant, as it is of order $2(1-\alpha)$, thus for $\alpha = 0.75$, even $k = 5$ is enough to get the same order of accuracy. On Fig. 3 and in Table 2 we numerically confirm the asymptotic behavior of the relative ℓ^2 error from Corollary 1.

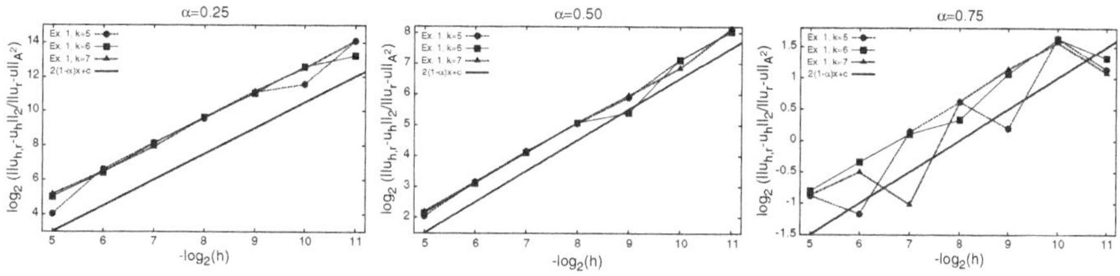

Fig. 3 Numerical confirmation for $\frac{\|\mathbf{u}_{h,r}-\mathbf{u}_h\|_2}{\|\mathbf{f}_1\|_2} / \frac{\|\mathbf{u}_r-\mathbf{u}\|_{\mathcal{A}^2}}{\|\mathbf{f}_1\|_{\mathcal{A}^0}} = \mathrm{O}(h^{-2(1-\alpha)})$ in (18)

Table 2 ℓ^2 relative error $\frac{\|\mathbf{u}_{h,r}-\mathbf{u}_h\|_2}{\|\mathbf{f}_2\|_2}$

	$\alpha = 0.25$			$\alpha = 0.5$			$\alpha = 0.75$		
h	$k = 7$	$k = 8$	$k = 9$	$k = 6$	$k = 7$	$k = 8$	$k = 5$	$k = 6$	$k = 7$
2^{-5}	7.5e−5	2.6e−5	2.6e−6	3.1e−4	1.0e−4	3.1e−5	8.5e−4	3.5e−4	2.0e−4
2^{-6}	1.2e−4	7.1e−5	3.4e−5	6.2e−4	2.6e−4	3.7e−5	2.9e−4	7.3e−4	2.0e−4
2^{-7}	2.2e−4	8.5e−5	4.9e−5	1.3e−3	5.5e−4	1.7e−4	1.6e−3	1.0e−3	6.7e−5
2^{-8}	1.0e−3	1.1e−4	3.9e−5	2.4e−3	1.0e−3	3.8e−4	2.7e−3	6.7e−4	4.3e−4
2^{-9}	4.9e−3	7.9e−4	1.7e−4	1.3e−3	1.2e−3	6.4e−4	8.7e−4	1.3e−3	1.1e−3
2^{-10}	9.5e−3	4.9e−3	2.9e−4	8.8e−3	1.4e−3	4.7e−4	5.9e−3	3.0e−3	1.5e−3
2^{-11}	3.6e−2	1.1e−2	4.7e−3	1.4e−2	8.9e−3	2.3e−3	1.7e−3	9.9e−4	4.2e−4

The sufficient conditions from Proposition 3 hold true for the cases under consideration. This means that positivity of all considered approximations is guaranteed. Therefore, the discrete maximum principle is always inherited. The presented numerical results are fully aligned with the theory. What is very important is the numerical robustness of positivity with respect to both accuracy parameters h and k which is confirmed for all $\alpha \in \{0.25, 0.5, 0.75\}$. Even in the case of lower accuracy, we do not observe any oscillations. The monotonicity preservation of the data is clearly expressed, capturing their geometrical shape.

6 Concluding Remarks

This study is inspired by some quite recent results in the numerical methods for fractional diffusion problems. In the Introduction, we discussed three methods based on reformulation of the original nonlocal problem into local (elliptic, pseudo parabolic, and integral) problems. In all cases, the cost is in the increased dimension of computational domain from d to $d + 1$.

Our approach is based on best uniform rational approximations of $t^{\beta-\alpha}$, $0 \le t \le 1$. The primal motivation is to reduce the computational complexity. A next important step is made in this paper. Here, we provide sufficient conditions to guarantee positive approximation of the inverse of fractional powers of normalized SPD M-matrices. Therefore, we get a numerical method which preserves the maximum principle. The presented numerical results clearly confirm the monotone behaviour of the solution, without any observed oscillations. Further research has to be devoted to the topic of accuracy of mass conservation.

The currently available methods and algorithms for numerical solution of boundary value problems with fractional power of elliptic operators have a quite different nature. A serious theoretical and experimental study is required to get a comparative analysis of their advantages and disadvantages for particular classes of problems. For instance, the error analysis is in different functional spaces assuming different conditions for smoothness. The comparison of the computational complexity is also an open question.

As a part of our analysis, Theorem 1 provides a sharp error estimates for the 1-BURA based approximations $E(k, k, 1)$. Then, at the beginning of Sect. 5, we showed how this result can be used to derive relative error estimates of the numerical solution of fractional order elliptic problems in ℓ^2. The numerical tests are well aligned with this theoretical estimates. The presented approach has a strong potential for further development addressing different pairs of functional spaces in the relative error estimates, varying the smoothness assumptions, for $d = 1, 2, 3$.

In addition, a lot of new numerical tests are needed to evaluate/confirm/compare the computational efficiency for more realistic towards real-life large-scale super diffusion problems.

Acknowledgements This research has been partially supported by the Bulgarian National Science Fund under Grant BNSF-DN12/1. The work of S. Harizanov has been also partially supported by the Bulgarian National Science Fund under Grant BNSF-DM02/2.

References

G. Alefeld, N. Schneider, On square roots of M-matrices. Linear Algebra Appl. **42**, 119–132 (1982)

O.G. Bakunin, *Turbulence and Diffusion: Scaling Versus Equations* (Springer Science & Business Media, Berlin, 2008)

P.W. Bates, On some nonlocal evolution equations arising in materials science. Nonlinear Dyn. Evol. Equ. **48**, 13–52 (2006)

A. Bonito, J. Pasciak, Numerical approximation of fractional powers of elliptic operators. Math. Comput. **84**(295), 2083–2110 (2015)

A. Bonito, J. Pasciak, Numerical approximation of fractional powers of regularly accretive operators. IMA J. Numer. Anal. **37**, drw042v1–drw042 (2016)

L. Chen, R. Nochetto, O. Enrique, A.J. Salgado, Multilevel methods for nonuniformly elliptic operators and fractional diffusion. Math. Comput. **85**, 2583–2607 (2016)

E.W. Cheney, M.J.D. Powell, The differential correction algorithm for generalized rational functions. Constr. Approx. **3**(1), 249–256 (1987)

R. Ciegis, V. Starikovicius, S. Margenov, R. Kriauziene, Parallel solvers for fractional power diffusion problems. Concurrency Comput. Practice Exp. (2017). https://doi.org/10.1002/cpe.4216

T. Ganelius, Rational approximation of x^α on $[0, 1]$. Anal. Math. **5**, 19–33 (1979)

G. Gilboa, S. Osher, Nonlocal operators with applications to image processing. Multiscale Model. Simul. **7**(3), 1005–1028 (2008)

S. Harizanov, R. Lazarov, S. Margenov, P. Marinov, Y. Vutov, Optimal solvers for linear systems with fractional powers of sparse SPD matrices. Numer Linear Algebra Appl. e2167, (2018). https://doi.org/10.1002/nla.2167

S. Harizanov, S. Margenov, P. Marinov, Y. Vutov, Volume constrained 2-phase segmentation method utilizing a linear system solver based on the best uniform polynomial approximation of $x^{-1/2}$. J. Comput. Appl. Math. **310**, 115–128 (2017)

A. Kilbas, H. Srivastava, J. Trujillo, *Theory and Applications of Fractional Differential Equations* (Elsevier, Amsterdam, 2006)

N. Kosturski, S. Margenov, MIC(0) preconditioning of 3D FEM problems on unstructured grids: conforming and non-conforming elements. J. Comput. Appl. Math. **226**(2), 288–297 (2009)

J. Kraus, S. Margenov, *Robust Algebraic Multilevel Methods and Algorithms*, vol. 5. (de Gruyter, Berlin, 2009)

R. Lazarov, P. Vabishchevich, A numerical study of the homogeneous elliptic equation with fractional order boundary conditions. Fract. Calc. Appl. Anal. **20**(2), 337–351 (2017)

P.G. Marinov, A.S. Andreev, A modified Remez algorithm for approximate determination of the rational function of the best approximation in Hausdorff metric. Comptes rendus de l'Academie bulgare des Scieces **40**(3), 13–16 (1987)

B. McCay, M. Narasimhan, Theory of nonlocal electromagnetic fluids. Archives Mech. **33**(3), 365–384 (1981)

R. Metzler, J.H. Jeon, A.G. Cherstvy, E. Barkai, Anomalous diffusion models and their properties: non-stationarity, non-ergodicity, and ageing at the centenary of single particle tracking. Phys. Chem. Chem. Phys. **16**(44), 24128–24164 (2014)

D.J. Newman, Rational approximation to $|x|$. Mich. Math. J. **11**(1), 11–14 (1964)

C. Pozrikidis, *The Fractional Laplacian* (Chapman and Hall/CRC, Boca Raton, 2016)

E.B. Saff, H. Stahl, Asymptotic distribution of poles and zeros of best rational approximants to x^{α} on $[0, 1]$, in *Topics in Complex Analysis*. Banach Center Publications. vol. 31 (Institute of Mathematics, Polish Academy of Sciences, Warsaw, 1995)

S.A. Silling, Reformulation of elasticity theory for discontinuities and long-range forces. J. Mech. Phys. Solids **48**(1), 175–209 (2000)

H. Stahl, Best uniform rational approximation of x^{α} on $[0, 1]$. Bull. Am. Math. Soc. **28**(1), 116–122 (1993)

P.N. Vabishchevich, Numerical solving the boundary value problem for fractional powers of elliptic operators. CoRR abs/1402.1636 (2014). http://arxiv.org/abs/1402.1636

P.N. Vabishchevich, Numerically solving an equation for fractional powers of elliptic operators. J. Comput. Phys. **282**, 289–302 (2015)

J. Xu, L. Zikatanov, The method of alternating projections and the method of subspace corrections in Hilbert space. J. Am. Math. Soc. **15**(3), 573–597 (2002)

G.M. Zaslavsky, Chaos, fractional kinetics, and anomalous transport. Phys. Rep. **371**(6), 461–580 (2002)

A General Lagrange Multipliers Theorem and Related Questions

Andrei V. Dmitruk and Nikolai P. Osmolovskii

Abstract The paper deals with a general optimization problem with equality and inequality constraints in a Banach space, the latter being given by closed convex cones with nonempty interiors. A necessary optimality condition in the form of Lagrange multipliers rule is presented.

1 Introduction

The aim of this paper is to present a self-contained theory of a general Lagrange multipliers rule (LMR) for an abstract optimization problem in a Banach space with both equality and inequality constraints, that covers most of theoretical and applied problems, including in particular optimal control problems with purely state and mixed state-control constraints.

To the theory of LMR as the most important necessary optimality condition, a highly vast literature is devoted concerning a variety of classes of problems, finite and infinite dimensional, convex and nonconvex, smooth and nonsmooth, etc. (see e.g. Hurwicz (1958); Dubovitskii and Milyutin (1965); Pshenichnyi (1982); Gamkrelidze and Kharatishvili (1967); Varaiya (1967); Nagahisa and Sakawa (1969); Kurcyusz (1976); Maurer and Zowe (1979); Norris (1973); Pourciau (1980); Rockafellar (1993); Tamminen (1994); Jahn (1994) and the literature therein).

A. V. Dmitruk
Russian Academy of Sciences, Central Economics and Mathematics Institute, Moscow, Russia

Lomonosov Moscow State University, Moscow, Russia
e-mail: dmitruk@member.ams.org

N. P. Osmolovskii (✉)
Department of Informatics and Mathematics, Kazimierz Pulaski University of Technology and Humanities in Radom, Radom, Poland

Department of Applied Mathematics, Moscow State University of Civil Engineering, Moscow, Russia
e-mail: osmolovski@uph.edu.pl

G. Feichtinger et al. (eds.), *Control Systems and Mathematical Methods in Economics*, Lecture Notes in Economics and Mathematical Systems 687,
https://doi.org/10.1007/978-3-319-75169-6_9

We do not aim to survey all the corresponding results, nor cover totally all the known problems. Instead, we propose a formulation (Theorem 2.1), which, not pretending to be new, is, in our opinion, reasonably general and, at the same time, most simple and convenient for the practical usage in many situations, such as optimal control problems with state and mixed control-state constraints Milyutin et al. (2004); Dmitruk and Osmolovskii (2014), with age structured systems Osmolovskii and Veliov (2017), etc. Note only that probably the closest to our result is paper (Norris 1973) with slightly different assumptions and proof.

Any proof of LMR is essentially based on some results from convex analysis and functional analysis, which therefore will be given first. The material presented here constitutes a self-contained piece of theory, involving only standard notions and facts, and not involving any difficult concepts from nonsmooth analysis, which are highly specific and thus are not available for the nonspecialist. So, this theory is entirely available for students of mathematical specialties and can be used for teaching purposes.

The paper is organized as follows. In Sect. 2 we formulate the main result: the Lagrange multipliers rule for an abstract optimization problem (Theorem 2.1). Section 3 is devoted to separation theorems and related results used in the extremum theory. Some properties of sublinear functionals are presented in Sect. 4. In Sect. 5 we consider questions related to the classical Lyusternik theorem on the tangent manifold. Note again that all concepts and facts in Sects. 3–5 are to some extent well-known, so we sometimes do not give references to their initial sources. On the other hand, for the reader's convenience, we give proofs for the most of these facts. The reader who is familiar with this preliminary material can skip it and go directly to the proof of the main result which is given in Sect. 6.

2 Main Result

Let X, Y, and $Z_i, i = 1, \ldots, \nu$ be Banach spaces, $\mathcal{D} \subset X$ an open set, and $K_i \subset Z_i$, $i = 1, \ldots, \nu$ closed convex cones with nonempty interiors. Let $F_0 : \mathcal{D} \to \mathbb{R}$, $g : \mathcal{D} \to Y$, and $f_i : \mathcal{D} \to Z_i$, $i = 1, \ldots, \nu$, be given mappings. Consider the following optimization problem:

$$F_0(x) \to \min, \qquad f_i(x) \in K_i\,, \quad i = 1, \ldots, \nu, \qquad g(x) = 0. \tag{1}$$

Let $K_i^0 := \{z_i^* \in Z_i^* : \langle z_i^*, z_i \rangle \leqslant 0 \text{ for every } z_i \in K_i\}$ be the polar cone to K_i, $i = 1, \ldots, \nu$. Here $\langle z_i^*, z_i \rangle$ is the duality pairing between Z_i and its dual space Z_i^*. We study the local minimality of an admissible point $x^0 \in \mathcal{D}$.

It is worth noting that the inequality constraints $f_i(x) \leqslant 0$, where $f_i : \mathcal{D} \to \mathbb{R}$ are given functionals, may also be presented in the form $f_i(x) \in K_i$ if we put $K_i = \mathbb{R}_- := (-\infty, 0]$. Then $K_i^0 = \mathbb{R}_+ := [0, \infty)$. On the other hand, all the inequality constraints $f_i(x) \in K_i$ can be written as one constraint $f(x) \in K$ if we define a mapping $f : X \to Z = Z_1 \times \ldots \times Z_\nu$ by $f(x) = (f_1(x), \ldots, f_\nu(x))$, and a cone $K = K_1 \times \ldots \times K_\nu$ in Z. However, we choose the form (1) to keep a visual relation with convenient statements.

We impose the following

Assumptions (1) The objective function F_0 and the mappings f_i are Fréchet differentiable at x_0; the operator g is strictly differentiable at x_0 (smoothness of the data functions), (2) the image of the derivative $g'(x_0)$ is closed in Y (weak regularity of equality constraint).

(The definition of strictly differentiable operator will be recalled in Sect. 5.3.)

The following theorem gives necessary conditions for a point $x_0 \in \mathcal{D}$ to be a local minimizer for problem (1).

Theorem 2.1 *Let x_0 provide a local minimum in problem (1). Then there exist Lagrange multipliers $\alpha_0 \geqslant 0$, $z_i^* \in K_i^0$, $i = 1, \ldots, \nu$, and $y^* \in Y^*$, satisfying the nontriviality condition*

$$\alpha_0 + \sum_{i=1}^{\nu} \|z_i^*\| + \|y^*\| > 0, \tag{2}$$

the complementary slackness conditions

$$\langle z_i^*, f_i(x_0)\rangle = 0, \qquad i = 1, \ldots, \nu, \tag{3}$$

and such that the Lagrange function

$$\mathcal{L}(x) = \alpha_0 F_0(x) + \sum_{i=1}^{\nu} \langle z_i^*, f_i(x)\rangle + \langle y^*, g(x)\rangle$$

is stationary at x_0: $\mathcal{L}'(x_0) = 0$, i.e.,

$$\alpha_0 F_0'(x_0) + \sum_{i=1}^{\nu} z_i^* f_i'(x_0) + y^* g'(x_0) = 0. \tag{4}$$

We prove this theorem in Sect. 6, but first we need a number of auxiliary notions and assertions. We start with some facts of linear functional analysis, most of which are, of course, well known, while others are specific for the extremum theory.

3 Some Facts of Linear Functional Analysis

Each of these facts holds in a proper vector space—linear topological space, locally convex space, normed space, Banach space, and sometimes even in a vector space without any topology. The first time reader can harmlessly assume that all happens in a Banach space.

3.1 Separation of Convex Sets

Let X be a topological vector space, $x_1, x_2 \in X$. By $[x_1, x_2]$ we denote the interval in X with the ends x_1, x_2, i.e., $[x_1, x_2] = \text{co}\{x_1, x_2\}$. By X^* we denote the dual space to X, consisting of all linear continuous functionals $x^* : X \to \mathbb{R}$.

Definition Let A and B be two sets in X. A nonzero functional $x^* \in X^*$ *separates* these sets if

$$\sup_{x \in A} \langle x^*, x \rangle \leqslant \inf_{x \in B} \langle x^*, x \rangle. \tag{5}$$

Obviously, this is equivalent to the existence of a number c such that $\langle x^*, x \rangle \leqslant c$ on A, and $\langle x^*, x \rangle \geqslant c$ on B. It is said that the hyperplane $\langle x^*, x \rangle = c$ *separates* A and B.

The following theorem is a key abstract assertion, on which a comprehensive theory of necessary extremum conditions is based.

Theorem 3.1 (Hahn–Banach) *Let A be an open convex set, B a convex set. Suppose that $A \cap B = \emptyset$. Then there exists a nonzero linear continuous functional separating A and B.*

This is the Hahn–Banach theorem in the "geometric form", or the *separation theorem.* Its proof can be found in any textbook on functional analysis (e.g. in Kolmogorov and Fomin (1968); Dunford and Schwartz (1968)).

Corollary 3.1 *Let X be a locally convex space, A a convex closed set, and B a convex compact set. Suppose that $A \cap B = \emptyset$. Then there exists a linear continuous functional $x^* \in X^*$ strictly separating A and B, i.e.,*

$$\sup_{x \in A} \langle x^*, x \rangle < \inf_{x \in B} \langle x^*, x \rangle. \tag{6}$$

(Any such x^* is necessarily nonzero.)

The proof follows from the fact that the compact set B can be placed into an open convex set $\widetilde{B}$ that still does not intersect A. Then any x^* separating A and $\widetilde{B}$ satisfies (6).

Along with Theorem 3.1, the following "dual" fact holds as well (Dunford and Schwartz 1968).

Theorem 3.2 (Hahn–Banach) *Let A be a convex set in X^*, open in the weak-* topology, and B be a convex set in X^*. Suppose that $A \cap B = \emptyset$. Then A and B can be separated by a nonzero element $x \in X$.*

Now, let be given a set $M \subset X$.

Definition 3.1 An element $x^* \in X^*$ is a support functional to the set M if

$$\inf x^*(M) := \inf_{x \in M} \langle x^*, x \rangle > -\infty.$$

This obviously means that $\langle x^*, x \rangle \geqslant a$ on M for some real a. The set of all support functionals to M we denote by M^*. Clearly, M^* is a convex and closed cone. (The set $-M^*$ is called *barrier cone* to M.) Note that if K is a cone, then $x^* \in K^*$ iff $\langle x^*, x \rangle \geqslant 0$ for all $x \in K$, and in this case, K^* is said to be *dual* (or *conjugate*) to the cone K. An easy property is that, if K_1 and K_2 are two nonempty cones, then $(K_1 + K_2)^* = K_1^* \cap K_2^*$ (while the question about $(K_1 \cap K_2)^*$ is not that simple). Also note that, if one of the two sets is a cone, then the constant c in the separation theorem can be taken zero, and then x^* or $-x^*$ is an element of the dual cone.

Let $L \subset X$ be a subspace. Then L^* consists of all functionals vanishing on L. The set of such functionals is denoted by $L^{\perp}$ and called *annihilator* of the subspace L. So, the dual cone for a subspace coincides with its annihilator.

Another basic example is given by the following

Lemma 3.1 *Let $l : X \to \mathbb{R}$ be a nonzero linear functional on a vector space X. Define an open half-space $K = \{x \in X : \langle l, x \rangle > 0\}$. Let $m \in K^*$, i.e., $\langle m, x \rangle \geqslant 0$ for all $x \in K$. Then there is $\alpha \geqslant 0$ such that $m = \alpha l$. Thus, the dual cone to a half-space is the ray spanned by the functional defining this half-space.*

Proof Consider the mapping $A : X \to \mathbb{R}^2$, $Ax = (\langle l, x \rangle, \langle m, x \rangle)$. Obviously, it is not onto: $AX \neq \mathbb{R}^2$ (since it does not include the point (1,-1)), and so, there is a nonzero pair $(\alpha, \beta) \in \mathbb{R}^2$ such that $\alpha \langle l, x \rangle + \beta \langle m, x \rangle = 0$ for all $x \in X$, i.e. $\alpha\, l + \beta\, m = 0$. If $\beta = 0$, then $\alpha \neq 0$ and $l = 0$, a contradiction with the assumption. Therefore, $\beta \neq 0$ and we can set $\beta = -1$. Thus, $m = \alpha l$. Obviously, $\alpha \geqslant 0$, since $\langle l, x \rangle > 0$ implies $\langle m, x \rangle \geqslant 0$. □

Finally, one can easily show that two sets A and B in X can be separated if and only if there exist nonzero functionals x^* and y^* such that

$$x^* \in A^*, \quad y^* \in B^*, \quad x^* + y^* = 0, \quad \inf x^*(A) + \inf y^*(B) \geqslant 0\,.$$

This equivalent formulation of separation of two sets makes it possible to generalize the separation theorem for the case of a finite number of sets. Let us pass to this generalization.

3.2 *The Dubovitskii–Milyutin Theorem on the Nonintersection of Cones*

Let, as before, X be a linear topological space.

Theorem 3.3 (Dubovitskii–Milyutin (1965)) *Let $\Omega_1, \ \ldots, \ \Omega_s, \Omega$ be nonempty convex cones in X, among which the cones $\Omega_1, \ \ldots, \Omega_s$ are open. Then*

$\Omega_1 \cap \ldots \cap \Omega_s \cap \Omega = \emptyset$ *iff there exist linear functionals*

$$x_1^* \in \Omega_1^*, \quad \ldots, \quad x_s^* \in \Omega_s^*, \quad x^* \in \Omega^*, \tag{7}$$

not all equal zero, such that

$$x_1^* + \ldots + x_s^* + x^* = 0. \tag{8}$$

Dubovitskii and Milyutin called relation (8) the *Euler (Euler–Lagrange) equation* for the given system of cones. Although the name may seem strange, the fact is that all necessary conditions of the first order for a local minimum in various problems on a conditional extremum, including the Euler-Lagrange equation in the calculus of variations, and even Maximum principle in optimal control, can be obtained by using this (at first glance, primitive) equality. The corresponding procedure is called Dubovitskii–Milyutin's scheme (or approach) and will be presented below.

To prove Theorem 3.3, we need the following simple fact. Let $X_1, \ldots, X_s$ be Banach spaces, and $W = X_1 \times \ldots \times X_s$ their product. Then $w^* \in W^*$ iff there exist $x_i^* \in X_i^*$, $i = 1, \ldots, s$ such that, for any $w = (x_1, \ldots, x_s) \in W$ we have $\langle w^*, w \rangle = \langle x_1^*, x_1 \rangle + \ldots + \langle x_s^*, x_s \rangle$.

Proof of Theorem 3.3. ($\Longrightarrow$) Suppose that $\Omega_1 \cap \ldots \cap \Omega_s \cap \Omega = \emptyset$. In the product W of s copies $X \times \ldots \times X = X^s$ of the space X consider two cones: $K = \Omega_1 \times \ldots \times \Omega_s$ and $D = \{w = (x_1, \ldots, x_s) \mid x_1 = \ldots = x_s = x \in \Omega\}$. Thus, D is the "diagonal" in the product $\Omega \times \ldots \times \Omega$ of s copies of Ω. We claim that $K \cap D = \emptyset$.

Suppose not, and let $w \in K \cap D$. Since $w \in K$, then $w = (x_1, \ldots, x_s)$, where $x_1 \in \Omega_1, \ldots, x_s \in \Omega_s$. Since $w \in D$, we have $x_1 = \ldots = x_s = x \in \Omega$. Thus, $x \in \Omega_1 \cap \ldots \cap \Omega_s \cap \Omega$, which contradicts the nonintersection of the cones.

Since the cones K and D are convex, and K is open, the Hahn–Banach separation theorem says that condition $K \cap D = \emptyset$ implies the existence of a nonzero $w^* \in W^*$ such that

$$\langle w^*, K \rangle \geqslant 0 \quad \text{and} \quad \langle w^*, D \rangle \leqslant 0.$$

But $w^* = (x_1^*, \ldots, x_s^*)$, where all $x_i^* \in X^*$, and the condition $\langle w^*, K \rangle \geqslant 0$ means that $\langle x_1^*, x_1 \rangle + \ldots + \langle x_s^*, x_s \rangle \geqslant 0$ for any $x_1 \in \Omega_1, \ldots, x_s \in \Omega_s$. From the positive homogeneity of all Ω_i it easily follows that all $x_i^* \in \Omega_i^*$, $i = 1, \ldots, s$. Moreover, not all $x_1^*, \ldots, x_s^*$ equal zero, since $w^* \neq 0$.

Further, condition $\langle w^*, D \rangle \leqslant 0$ means that $\langle x_1^* + \ldots + x_s^*, x \rangle \leqslant 0$ for all $x \in \Omega$. Set $x^* = -(x_1^* + \ldots + x_s^*)$. Then $x^* \in \Omega^*$ and $x_1^* + \ldots + x_s^* + x^* = 0$.

($\Longleftarrow$) Suppose there is a nonzero collection of functionals $x_1^*, \ldots, x_s^*, x^*$ satisfying conditions (7) and (8), but the cones $\Omega_1, \ldots, \Omega_s, \Omega$ intersect. Let $\hat{x}$ be their common element. Among the functionals $x_1^*, \ldots, x_s^*$, there is at least one nonzero: $x_{i_0}^* \neq 0$ (otherwise, the last functional $x^* = 0$ too, so the whole collection is trivial, a contradiction). Obviously, $\langle x_{i_0}^*, \hat{x} \rangle > 0$ since $\hat{x} \in \operatorname{int} \Omega_{i_0}$.

For the rest of x_i^* we have $\langle x_i^*, \hat{x}\rangle \geqslant 0$, and $\langle x^*, \hat{x}\rangle \geqslant 0$ as well. Consequently, $\left\langle\sum_{i=1}^{s} x_i^* + x^*,\ \hat{x}\right\rangle > 0$, which contradicts the equality (8). □

We will also give an analog (generalization) of this theorem for a finite number of convex sets, not necessarily cones.

Theorem 3.4 (Dubovitskii–Milyutin (1965)) *Let $M_1, \ldots, M_s$, M be nonempty convex sets in a linear topological space X, among which the sets $M_1, \ldots, M_s$ are open. Then the condition*

$$M_1 \cap \ldots \cap M_s \cap M = \emptyset$$

is equivalent to the existence of a nontrivial collection of functionals $(x_1^*, \ldots, x_s^*, x^*)$ from X^* such that

$$x_1^* + \ldots + x_s^* + x^* = 0, \tag{9}$$

$$\inf\langle x_1^*, M_1\rangle + \ldots + \inf\langle x_s^*, M_s\rangle + \inf\langle x^*, M\rangle \geqslant 0. \tag{10}$$

The proof can be reduced to the case of cones, where one can apply Theorem 3.3. However, we will give another, maybe even more simple proof by a direct use of the separation theorem.

Proof ($\Longrightarrow$) In the space $W = X^s$, consider the set A of elements of the form $w = (x_1 - x, \ldots, x_s - x)$, where $x_1 \in M_1, \ldots, x_s \in M_s$, $x \in M$. Obviously, A is open and convex. The nonintersection of $M_1, \ldots, M_s$ and M yields that A does not contain zero element $(0, \ldots 0) \in X^s$. Consequently, there exists a functional $w^* = (x_1^*, \ldots, x_s^*)$ that separates A from zero: $\langle w^*, w\rangle \geqslant 0 \forall\, w \in A$, that is

$$\langle x_1^*, x_1\rangle + \ldots + \langle x_s^*, x_s\rangle - \langle x_1^* + \ldots + x_s^*, x\rangle \geqslant 0$$

for all $x_1 \in M_1, \ldots, x_s \in M_s$, $x \in M$. Set $x^* = -(x_1^* + \ldots + x_k^*)$. Obviously, the collection $(x_1^*, \ldots, x_k^*, x^*)$ is nonzero and satisfies conditions (9), (10).

($\Longleftarrow$) Suppose there is a nonzero collection of functionals $x_1^*, \ldots, x_s^*, x^*$ satisfying conditions (9), (10). Denote for brevity $m_i = \inf\langle x_i^*, M_i\rangle$ and $m_i = \inf\langle x^*, M\rangle$. Obviously, at least one $x_{i_0}^* \neq 0$, whence $\langle x_{i_0}^*, \hat{x}\rangle > m_{i_0}$ because $\hat{x} \in \operatorname{int} M_{i_0}$, while all other $\langle x_i^*, \hat{x}\rangle \geqslant m_i$ and $\langle x^*, \hat{x}\rangle \geqslant m$. Then, in view of (10),

$$\left\langle\sum_{i=1}^{s} x_i^* + x^*,\ \hat{x}\right\rangle > \sum_{i=1}^{s} m_i + m \geqslant 0,$$

which contradicts the equality (9). □

Note that Theorem 3.3 directly follows from the last one, because if M is a cone, then $\inf\langle x^*, M\rangle = c > -\infty$ is equivalent to that $c = 0$ and $x^* \in M^*$.

Also note that both Theorems 3.3 and 3.4 are in fact equivalent to Theorem 3.1, but are more prepared for application in the theory of optimization.

3.3 The Dual Cone to the Intersection of Cones

An important corollary of Theorem 3.3 is the following formula (Dubovitskii and Milyutin 1965).

Lemma 3.2 *Let K_1 and K_2 be convex cones, one of which intersects with the interior of other:* $(\operatorname{int} K_1) \cap K_2 \neq \emptyset$. *Then*

$$(K_1 \cap K_2)^* = K_1^* + K_2^* . \tag{11}$$

Proof The inclusion $\supset$ is trivial, so we just have to prove $\subset$. Take any $p \in (K_1 \cap K_2)^*$, i.e., $(p,\, K_1 \cap K_2) \geqslant 0$. We need to represent it as $p = p_1 + p_2$, where $p_1 \in K_1^*$ $p_2 \in K_2^*$. Assume that $p \neq 0$ (otherwise there is a trivial representation $0 = 0 + 0$).

Introduce one more cone, an open half-space $K_0 = \{x \mid (p, x) < 0\}$. It is nonempty, since $p \neq 0$. Obviously, $K_0 \cap \operatorname{int} K_1 \cap K_2 = \emptyset$ (otherwise $\exists\, x \in \operatorname{int} K_1 \cap K_2$ such that $x \in K_0$, i.e., $(p, x) < 0$, which contradicts the choice of p.)

Since the cones K_0 and $\operatorname{int} K_1$ are open, by the Dubovitskii–Milyutin Theorem 3.3 there exist $p_i \in K_i^*$, $i = 0, 1, 2$, not all zero, such that $p_0 + p_1 + p_2 = 0$ (we took into account that $(\operatorname{int} K_1)^* = K_1^*$). But $p_0 = -\alpha p$ with some $\alpha \geqslant 0$, i.e., we have $\alpha p = p_1 + p_2$. If $\alpha = 0$, we get $p_1 + p_2 = 0$, and both these functionals are not zero. Then, again by Theorem 3.3, we get $\operatorname{int} K_1 \cap K_2 = \emptyset$, which contradicts the assumption. Therefore, $\alpha > 0$, and hence $p = \frac{1}{\alpha}(p_1 + p_2) \in K_1^* + K_2^*$. □

3.4 The Support Cone to a Convex Set at a Point

Let X be a linear topological space, $M \subset X$ a convex set.

Definition An element $l \in X^*$ is called *support functional* or *inner normal* to the set M *at a point* $x_0 \in \overline{M}$ if $\langle l, M\rangle \geqslant \langle l, x_0\rangle$. The set of all such functionals will be denoted by $M^*(x_0)$. Obviously, it is a closed convex cone. In particular case when M itself is a cone and $x_0 = 0$, we obtain $M^*(x_0) = M^*$, i.e., the support cone to a cone M at the origin is exactly the above dual (conjugate) cone M^*.

Together with the cone $M^*(x_0)$, introduce the set $\Omega(x_0) = \{h \in X : \exists\, \alpha > 0$ such that $x_0 + \alpha h \in \operatorname{int} M\}$. Obviously, it is an open convex cone (may be empty), which is called *the cone of interior directions* to the set M at the point x_0.

Lemma 3.3 *If M has a nonempty interior, then $\Omega^*(x_0) = M^*(x_0)$.*

Proof With no loss of generality we take $x_0 = 0$. Then $\Omega(0) = \bigcup_{\alpha>0} \alpha(\operatorname{int} M)$, so $\langle l, \Omega(0)\rangle \geqslant 0 \iff \langle l, \operatorname{int} M\rangle \geqslant 0 \iff \langle l, M\rangle \geqslant 0$, q.e.d. □

Corollary 3.2 *If K is a convex cone, and $x_0 \in \overline{K}$, then*

$$K^*(x_0) = \{l \in K^* : \langle l, x_0\rangle = 0\}.$$

Proof We have to show that the inequality $\langle l, K\rangle \geqslant \langle l, x_0\rangle$ holds $\iff \langle l, K\rangle \geqslant 0$ and $\langle l, x_0\rangle = 0$. Let us prove the implication $\Rightarrow$.

Since $x_0 + K \subset K$, we obtain $\langle l, x_0 + K\rangle \geqslant \langle l, x_0\rangle$, hence $\langle l, K\rangle \geqslant 0$. In particular, $\langle l, x_0\rangle \geqslant 0$. But since $0 \in \overline{K}$ and $l \in K^*(x_0)$, we have $0 = \langle l, 0\rangle \geqslant \langle l, x_0\rangle$, whence $\langle l, x_0\rangle \leqslant 0$, and so $\langle l, x_0\rangle = 0$. The reverse implication is trivial. □

The Tangential Cone Note by the way that there is yet another cone related to a convex set M at a point x_0, called the *tangential cone*

$$T(x_0) = \overline{\bigcup_{\alpha\geqslant 0} \alpha(M - x_0)} = \overline{\mathbb{R}_+(M - x_0)}.$$

Obviously, it is the minimal closed cone containing the set $M - x_0$. An easy fact is that $T(x_0) = \overline{\Omega(x_0)}$ provided that M has nonempty interior. (Indeed, the inclusion $T(x_0) \supset \overline{\Omega(x_0)}$ is trivial. On the other hand, for any $\alpha \geqslant 0$ we have $\alpha(M - x_0) \subset \overline{\alpha(M - x_0)}$, hence $T(x_0) \subset \overline{\Omega(x_0)}$.) However, we will not use this cone in what follows.

3.5 Lemma on the Nontriviality of Annihilator

Let $L \subset X$ be a linear manifold. The set $L^\perp = \{x^* \in X^* \mid \langle x^*, x\rangle = 0\ \forall\, x \in L\}$ is called *annihilator* of L. As it was noted above, any $x^* \in L^\perp$ is a support functional to L, and the converse is also true. Thus, $L^\perp = L^*$. Obviously, $L^\perp \subset X^*$ is a subspace, i.e., a closed linear manifold.

A simple corollary of the Hahn–Banach separation Theorem 3.1 is the following

Lemma 3.4 *Let X be a locally convex space and $L \subset X$ a subspace which does not coincide with X. Then $L^\perp$ contains a nonzero element.*

(Recall that a topological vector space X is locally convex if any nonempty open set in X contains a nonempty open convex subset.)

Proof Take any $x \in X \setminus L$. Since L is closed and the space X is locally convex, there exists an open convex set $U \ni x$ which does not intersect with L. Then U and L can be separated by a nonzero functional x^*, and therefore, x^* is bounded on L either from above or from below, which in turn implies that $\langle x^*, L\rangle = 0$, q.e.d. □

Note that here the assumption of the closedness of L is essential.

3.6 The Banach Open Mapping Theorem

The following theorem is one of the basic principles of functional analysis.

Theorem 3.5 (S. Banach) *Let X and Y be Banach spaces and $A : X \to Y$ a linear surjective operator. Then the image of the unit ball contains a ball of some radius $r > 0$:*

$$AB_1(0) \supset B_r(0).$$

This assertion can be equivalently formulated as follows: *for any $y \in Y$ there is an $x \in X$ such that $Ax = y$ and $||x|| \leqslant C\,||y||$, where the constant C does not depend on x or y.* (One can take $C = 1/r$.)

The latter formulation is more convenient for application in optimization theory than the standard formulation saying that the image of any open set is open.

3.7 Lemma on the Closed Image of a Combined Operator

An important role in optimization theory is played by the following fact (Alekseev et al. 1987; Levitin et al. 1974).[1]

Lemma 3.5 *Let X, Y, Z be Banach spaces, $A : X \to Y$ and $B : X \to Z$ continuous linear operators. Let $AX = Y$ and the set $B(\mathrm{Ker}\, A)$ be closed in Z. Then the "combined" operator $T : X \to Y \times Z$, such that $x \mapsto (Ax, Bx)$, has a closed image.*

Proof Let a sequence $x_n \in X$ be such that $A\,x_n = y_n$, $B\,x_n = z_n$, and $(y_n, z_n) \to (y_0, z_0)$. We must show that there exists $x \in X$ such that $Ax = y_0$ and $Bx = z_0$. Since $AX = Y$, we can assume that $y_0 = 0$. (Otherwise, taking $\hat{x}$ such that $A\hat{x} = y_0$, we obtain $\delta y_n = A\delta x_n \to 0$.)

Thus, $Ax_n \to 0$. By the Banach open mapping theorem, there exists a sequence $x'_n \to 0$ such that $Ax'_n = Ax_n$. Define $\bar{x}_n = x_n - x'_n$. Obviously, $\bar{x}_n \in Ker\,A$ and $B\bar{x}_n = Bx_n - Bx'_n \to z_0$ (tends to the same point z_0), so, z_0 is the limit point of the sequence $B\bar{x}_n \in B(Ker\,A)$. Since $B(Ker\,A)$ is closed by the assumption, $z_0 \in B(Ker\,A)$, which means that $\exists\, x_0$ such that $Ax_0 = 0$ and $Bx_0 = z_0$. □

Clearly, this lemma remains valid if the image $Y_1 := AX$ is just a closed subspace in Y, not necessarily the whole space Y. Then Y_1 itself is a Banach space, and the operator $A : X \to Y_1$ maps onto, hence the image of $T = (A, B)$ is closed in $Y_1 \times Z$, and therefore, in $Y \times Z$.

[1] May be this fact was also noted in some earlier works.

The following particular case of Lemma 3.5 is most usable in optimal control theory for systems of ODEs. (There, the operator A is the linearized control system, and B is the derivative of endpoint constraints.)

Corollary 3.3 *Let $A : X \to Y$ be a linear surjective operator, and $B : X \to Z$ a finite dimensional linear operator ($dim\ Z < \infty$). Then the combined operator $Tx = (Ax, Bx)$ has a closed image in $Y \times Z$.*

The proof follows from the fact that the subspace $B(\text{Ker } A)$ is always finite dimensional, and therefore closed.

3.8 Annihilator of the Kernel of a Surjective Operator

Let X and Y be Banach spaces and $A : X \to Y$ a linear operator. Recall that its adjoint operator $A^* : Y^* \to X^*$ maps any $y^* \in Y^*$ to a functional $x^* \in X^*$ defined by the relation $\langle x^*, x\rangle = \langle y^*, Ax\rangle$. Thus, $x^* = A^* y^*$. Sometimes we will also use another, "physical" notation: $x^* = y^* A$, which is more convenient in formulas.

The following fact is well-known and widely used.

Lemma 3.6 *Let the operator $A : X \to Y$ be surjective. Then*

$$(Ker\ A)^* = A^* Y^*.$$

In other words, any linear functional $x^* \in X^*$ vanishing on Ker A *has the form $\langle x^*, x\rangle = \langle y^*, Ax\rangle$ with some $y^* \in Y^*$. And vice versa, any functional of this form vanishes on* Ker A. *(The last statement does not require the surjectivity of A).*

Proof

a) Show that $A^*Y^* \subset (\text{Ker } A)^*$. Indeed, let $x^* = A^*y^*$, i.e., $\langle x^*, x\rangle = \langle y^*, Ax\rangle\ \forall x \in X$. Then, obviously, for any $x \in \text{Ker } A$ we have $\langle x^*, x\rangle = 0$. (Here the condition $AX = Y$ is not required.)
b) Show that $(\text{Ker } A)^* \subset A^*Y^*$. Let a linear functional x^* vanish on $\text{Ker } A$. Consider the operator $T : X \to Y \times \mathbb{R},\ x \to (Ax, \langle x^*, x\rangle)$.

   By Corollary 3.3, its image TX is closed in $Y \times \mathbb{R}$. This image does not coincide with the whole space $Y \times R$, because it does not contain the point $(0, 1)$, since $Ax = 0$ implies $\langle x^*, x\rangle = 0$. Then, by Lemma 3.4 the annihilator of TX contains a nonzero element $(y^*, c) \in Y^* \times \mathbb{R}$, i.e.,

$$\langle y^*, Ax\rangle + c\langle x^*, x\rangle = 0 \qquad \text{for all } x \in X.$$

   We claim that $c \neq 0$. Indeed, if $c = 0$, then $\langle y^*, Ax\rangle = 0$ on the whole space X, which means that $\langle y^*, y\rangle = 0$ on the whole Y (since $AX = Y$), whence $y^* = 0$, so $(y^*, c) = (0, 0)$, a contradiction.

Thus, $c \neq 0$. Then $\langle x^*, x\rangle = -\langle \frac{1}{c}\, y^*, Ax\rangle$ for all $x \in X$, i.e., $\langle x^*, x\rangle = \langle y_1^*, Ax\rangle$, where $y_1^* = -\frac{1}{c}\, y^* \in Y^*$, as required. □

Remark Note an important particular case of this lemma. If $Y = \mathbb{R}^m$, then $Ax = (\langle a_1, x\rangle \ldots, \langle a_m, x\rangle)$, where all $a_i \in X^*$, and $y^* = (\beta_1 \ldots, \beta_m) \in \mathbb{R}^m$. Lemma 3.6 says that, if a linear functional x^* is *subjected* to functionals $a_1 \ldots, a_m$ in the sense that $\langle a_1, x\rangle = 0, \ldots, \langle a_m, x\rangle = 0$ implies $\langle x^*, x\rangle = 0$, then x^* is a *linear combination* of these a_i, i.e. $x^* = \beta_1 a_1 + \ldots + \beta_m a_m$ for some $\beta \in \mathbb{R}^m$. Lemma 3.6 is nothing more than the straightforward generalization of this well known fact of finite-dimensional linear algebra to the case of infinite-dimensional spaces, and its proof is in fact the same.

3.9 The Farkas–Minkowski Theorem

Lemma 3.6 has an analogue in the case when Ker A is replaced by $A^{-1}K$, where K is a cone in the image space Y (see e.g. Hurwicz (1958)).

Theorem 3.6 *Let X and Y be linear topological spaces, $A : X \to Y$ a linear continuous operator, $K \subset Y$ a convex cone with a nonempty interior, $\Omega = A^{-1}K = \{x \in X : \; Ax \in K\}$. Suppose that $\exists\, x_0$ such that $Ax_0 \in \operatorname{int} K$.*

Then $x^* \in \Omega^*$ if and only if there exists an $y^* \in K^*$ such that $x^* = A^* y^*$. That is, $\Omega^* = A^* K^*$. (In "physical" notation, $x^* = y^* A$ and $\Omega^* = K^* A$, respectively.)

Proof If $y^* \in K^*$ and $x^* = y^* A$, then obviously $x^* \in \Omega^*$. So, we have to prove only the reverse implication.

Take any $x^* \in \Omega^*$. In the space $X \times Y$, consider a subspace $\Gamma = \{(x, y) : \; Ax = y\}$ (the graph of operator A), and a convex cone $C = X \times K$. Since A is continuous, the subspace Γ is closed. Define a linear functional $p : X \times Y \to \mathbb{R}$ by setting $p(x, y) = \langle x^*, x\rangle$. Obviously, $p \geqslant 0$ on $\Gamma \cap C = \{(x, y) : \; Ax = y \in K\}$, i.e., $p \in (\Gamma \cap C)^*$. By the assumption, the pair $(x_0,\; Ax_0) \in \Gamma \cap (\operatorname{int} C)$, whence, by Lemma 3.2, $p = h + q$, where $h \in \Gamma^*$ and $q \in C^*$. Since Γ is the kernel of linear operator $X \times Y \to Y$, $(x, y) \mapsto Ax - y$, which is obviously surjective, $h \in \Gamma^*$ means that $h(x, y) = \langle y^*, \; Ax - y\rangle$ for all x, y with some $y^* \in Y^*$.

Now, let $q(x, y) = \langle \lambda, x\rangle + \langle \mu, y\rangle$ with some $\lambda \in X^*$, $\mu \in Y^*$. The condition $q \in C^* = X^* \times K^*$ means that $\lambda = 0$ and $\mu \in K^*$, so $q(x, y) = \langle \mu, y\rangle$, and then $p = h + q$ means that $\langle x^*, x\rangle = \langle y^*, \; Ax - y\rangle + \langle \mu, y\rangle$ for all $x \in X$, $y \in Y$. This implies that $y^* = \mu \in K^*$ and $\langle x^*, x\rangle = \langle y^*, Ax\rangle$, q.e.d. □

4 Sublinear Functionals

Let X be a normed space. A functional $\varphi : X \to \mathbb{R}$ is called *sublinear* if it is positively homogeneous and convex:

(a) $\varphi(\lambda x) = \lambda\varphi(x) \quad \forall x \in X,\ \lambda > 0,$
(b) $\varphi(x + y) \leqslant \varphi(x) + \varphi(y) \quad \forall x, y \in X.$

The first condition is the *positive homogeneity*. Condition (b) is called *subadditivity*; for positively homogeneous functionals it is equivalent to the convexity.

Similar to linear functionals, a sublinear functional φ is called *bounded* if there exists a constant C such that

(c) $\quad |\varphi(x)| \leqslant C\|x\| \quad \forall x \in X.$

By virtue of homogeneity of φ, this condition is equivalent to the boundedness of φ on the unit ball in X, and in view of convexity it is also equivalent to the boundedness *from above* on the unit ball.

In what follows, we consider only bounded sublinear functionals.

4.1 Support Functionals of Sublinear Functionals

Definition A linear functional $l \in X^*$ is called *support* to a sublinear functional φ if $l(x) \leqslant \varphi(x)$ for all $x \in X$. The set of all support functionals to φ is denoted by $\partial\varphi$ and called *subdifferential* of φ, while the elements $l \in \partial\varphi$ are called the *subgradients* of φ.

If C is such that $\varphi(x) \leqslant C\|x\|$ for all $x \in X$, then $\partial\varphi \subset B_C(0)$. Indeed, if $l \in \partial\varphi$, then $\langle l, x\rangle \leqslant \varphi(x) \leqslant C\|x\| \ \ \forall x$. Hence $|\langle l, x\rangle| \leqslant C\|x\| \ \ \forall x$, and therefore $\|l\| \leqslant C$.

So, any bounded sublinear functional has a bounded set $\partial\varphi$. It is easy to verify that $\partial\varphi$ is convex and closed. Moreover, the separation theorem implies that it is weakly* closed. By the Banach–Alaoglu theorem (Kolmogorov and Fomin 1968; Dunford and Schwartz 1968), any bounded weakly* closed set in the dual space is compact in the weak* topology of this space. Thus, $\partial\varphi$ is convex and weakly* compact in X^*.

We now show that $\partial\varphi$ is not empty. Moreover, even a stronger fact holds true.

Theorem 4.1 (G. Minkowski) *Let $\varphi : X \to \mathbb{R}$ be a bounded sublinear functional on a normed space X. Then the set $\partial\varphi$ of its support functionals is nonempty, and the following representation holds:*

$$\varphi(x) = \max_{l \in \partial\varphi} \langle l, x\rangle \qquad \forall x \in X. \tag{12}$$

Note that the maximum in the right-hand side of (12) is always attained because of the weak* compactness of $\partial\varphi$.

Proof If $l \in \partial\varphi$, then $\langle l, x\rangle \leqslant \varphi(x) \;\; \forall\, x$, from which we obtain the inequality

$$\sup_{l\in\partial\varphi} \langle l, x\rangle \leqslant \varphi(x) \qquad \forall\, x.$$

Now, take any $x_0 \in X$ and show that there exists $l \in \partial\varphi$ such that $\langle l, x_0\rangle = \varphi(x_0)$. Then equality (12) will be established.

In the product $X \times \mathbb{R}$, consider the set $K = \{(x,t) \,:\, \varphi(x) < t\}$. Obviously, it is a nonempty convex cone. Moreover, it is open since φ is continuous. Set $t_0 = \varphi(x_0)$. Then the point (x_0, t_0) does not belong to K. Therefore, by the Hahn–Banach Theorem 3.1, there is a nonzero functional $(l, \alpha) \in X^* \times \mathbb{R}$ separating it from K:

$$\langle l, x\rangle + \alpha t \;\leqslant\; \langle l, x_0\rangle + \alpha t_0 \qquad \forall\,(x,t) \in K\,. \tag{13}$$

Let us analyze this condition. Since, for a fixed x, it holds for all $t > \varphi(x)$, then clearly $\alpha \leqslant 0$. If $\alpha = 0$, then $\langle l, x - x_0\rangle \leqslant 0$ for all $x \in X$, whence $l = 0$, which contradicts the nontriviality of the pair (l, α). Therefore, $\alpha < 0$, and we can put $\alpha = -1$. Then (13) becomes:

$$\langle l, x\rangle - \langle l, x_0\rangle \;\leqslant\; t - t_0 \qquad \forall\, t > \varphi(x),$$

whence this is true also for $t = \varphi(x)$, and so, we have

$$\langle l, x\rangle - \langle l, x_0\rangle \;\leqslant\; \varphi(x) - \varphi(x_0) \qquad \forall\, x \in X. \tag{14}$$

Now, take any $\bar{x} \in X$, and set $x = N\bar{x}$, where N is a large number. Then

$$\langle l, \bar{x}\rangle - \frac{1}{N}\langle l, x_0\rangle \;\leqslant\; \varphi(\bar{x}) - \frac{1}{N}\varphi(x_0),$$

whence, letting $N \to \infty$, we obtain $\langle l, \bar{x}\rangle \leqslant \varphi(\bar{x})$. Since $\bar{x}$ is arbitrary, this means that $l \in \partial\varphi$, and therefore, $\partial\varphi$ is nonempty. Next, setting $x = 0$ in (14), we get $\langle l, x_0\rangle \geqslant \varphi(x_0)$. Since $l \in \partial\varphi$, the opposite inequality is also true. Therefore, we obtain equality $\langle l, x_0\rangle = \varphi(x_0)$, which proves the theorem. □

An easy observation is that, taking an arbitrary nonempty bounded set $Q \subset X^*$, we obtain a bounded sublinear functional

$$\varphi(x) = \sup_{x^*\in Q} \langle x^*, x\rangle \qquad \forall\, x \in X, \tag{15}$$

where Q can be harmlessly replaced by the weak-* closure of its convex hull. Due to Theorem 4.1, this formula describes all possible sublinear bounded functionals on X, while the next lemma ensures the uniqueness of representation (15) if Q is also convex and closed.

Lemma 4.1 *Let Q be a bounded, convex, and closed set in X^*. Then the functional (15) is sublinear, bounded, and has $\partial\varphi = Q$.*

Proof The sublinearity and boundedness are obvious. It remains to prove the last equality. The inclusion $Q \subset \partial\varphi$ is obvious. To prove the reverse inclusion, take any $x^* \notin Q$ and show that $x^* \notin \partial\varphi$. Indeed, since Q is weakly-* closed and $x^* \notin Q$, the separation Theorem 3.2 says that there is $x_0 \in X$ such that $\langle x^*, x_0\rangle > \sup\langle Q, x_0\rangle = \varphi(x_0)$, hence $x^* \notin \partial\varphi$. □

(If Q is unbounded, the functional (15) is still sublinear, but not bounded. Such functionals are not of much interest for our purposes.)

4.2 Subdifferential of the Maximum of Sublinear Functionals

Theorem 4.2 (Dubovitskii–Milyutin (1965)) *Let $\varphi_i : X \to \mathbb{R}$, $i = 1, \ldots, m$, be bounded sublinear functionals on a normed space X, and*

$$\Phi(x) := \max_{1 \leqslant i \leqslant m} \varphi_i(x).$$

Then

$$\partial\Phi = co\Big(\bigcup_{1 \leqslant i \leqslant m} \partial\varphi_i\Big). \tag{16}$$

Proof Define the sets $A_i = \partial\varphi_i$. Since all of them are weakly-* compact, their union $Q = \bigcup_{1 \leqslant i \leqslant m} A_i$ is weakly-* compact too, and then its convex hull as well.

For any $x \in X$, we have

$$\Phi(x) = \max_{1 \leqslant i \leqslant m} \varphi_i(x) = \max_{1 \leqslant i \leqslant m} \max\langle A_i, x\rangle =$$

$$= \max\langle \bigcup_{1 \leqslant i \leqslant m} A_i, x\rangle = \max\langle co \bigcup_{1 \leqslant i \leqslant m} A_i, x\rangle = \max\langle Q, x\rangle.$$

By Lemma 4.1 this yields $Q = \partial\Phi$. □

4.3 Subdifferential of a Composite Sublinear Functional

Recall the Hahn–Banach theorem in the "algebraic form", called *extension theorem* (also proved in standard courses of functional analysis, see e.g. Kolmogorov and Fomin (1968); Dunford and Schwartz (1968)).

Theorem 4.3 (Hahn–Banach) *Let X be a arbitrary vector space, $\Gamma \subset X$ a subspace, $\varphi : X \to \mathbb{R}$ a sublinear functional, and $\varphi_\Gamma : \Gamma \mapsto \mathbb{R}$ its restriction to Γ. Then for any $l : \Gamma \to \mathbb{R}$ satisfying $l \in \partial\varphi_\Gamma$ there exists $\widetilde{l} : X \to \mathbb{R}$ satisfying $\widetilde{l} \in \partial\varphi$ such that $\widetilde{l}(x) = l(x)$ on Γ, i.e., $\widetilde{l}$ is an extension of l from the subspace Γ to the entire space X.*

The next theorem, proposed in Dubovitskii and Milyutin (1965), is an analog of Theorem 3.6.

Theorem 4.4 *Let $A : X \to Y$ be a linear operator, $\varphi : Y \to \mathbb{R}$ a sublinear functional. Then the functional $f(x) = \varphi(Ax)$ is sublinear, and its subdifferential*

$$\partial f \;=\; A^* \, \partial\varphi, \tag{17}$$

i.e., any $p \in \partial f$ has the form $\langle p, x\rangle = \langle \mu, Ax\rangle$ with some $\mu \in \partial\varphi$. (Or, in "physical" notation, $\partial f = \partial\varphi \cdot A$ and $p = \mu A$, respectively.) The reverse inclusion is trivial.

Proof Let $p \in \partial f$ i.e., $\langle p, x\rangle \leqslant \varphi(Ax) \;\; \forall x \in X$. In the product $X \times Y$ define a sublinear functional $\widehat{\varphi}(x, y) = \varphi(y)$, and on the subspace $\Gamma = \{(x, y) \mid y = Ax\}$ define a linear functional $l(x, y) = p(x)$. Then $l(x, y) \leqslant \widehat{\varphi}(x, y)$ on Γ (since $\langle p, x\rangle \leqslant \varphi(Ax) \;\; \forall x \in X$.) By Theorem 4.3, the functional l can be extended to $\widehat{l}$ defined on the whole space $X \times Y$ preserving the property $\widehat{l} \leqslant \widehat{\varphi}$. But every linear functional $\widehat{l}$ on $X \times Y$ has the form $\widehat{l}(x, y) = \langle \lambda, x\rangle + \langle \mu, y\rangle$ with some $\lambda \in X^*$ and $\mu \in Y^*$. So, we have

$$\langle \lambda, x\rangle + \langle \mu, y\rangle \;\leqslant\; \widehat{\varphi}(x, y) \;=\; \varphi(y), \qquad \forall\, (x, y) \in X \times Y,$$

$$\langle \lambda, x\rangle + \langle \mu, y\rangle \;=\; l\langle x, y\rangle \;=\; \langle p, x\rangle \qquad \forall\, (x, y) \in \Gamma.$$

The first relation implies that $\lambda = 0$ and $\mu \in \partial\varphi$, and the second one that $\langle p, x\rangle = \langle \mu, Ax\rangle$, since $y = Ax$ on Γ. Thus, $p = \mu A$, q.e.d. □

4.4 Subdifferential of the Derivative of a Sublinear Functional

Let $\varphi : X \to \mathbb{R}$ be a bounded sublinear functional. Then its directional derivative

$$\psi(\bar{x}) = \varphi'(x_0, \bar{x}) := \lim_{\varepsilon \to 0+} \frac{\varphi(x_0 + \varepsilon\bar{x}) - \varphi(x_0)}{\varepsilon}$$

at any point $x_0 \in X$ is also a bounded sublinear functional. (This simple fact holds for any convex function φ Lipschitz continuous around x_0). Note that

$$\varphi'(x_0, \bar{x}) \leqslant \frac{\varphi(x_0 + \varepsilon\bar{x}) - \varphi(x_0)}{\varepsilon} \qquad \text{for all} \quad \varepsilon > 0, \tag{18}$$

because, by the convexity of φ, the right-hand side monotonically decreases as ε decreases to $0+$.

The next lemma gives a relation between the subdifferentials of functionals φ and ψ (see e.g. Dubovitskii and Milyutin (1965); Pshenichnyi (1982)).

Lemma 4.2 $\partial\psi = \{x^* \in \partial\varphi \; : \; \langle x^*, x_0\rangle = \varphi(x_0)\}$.
In particular case when $x_0 = 0$, *we have* $\partial\psi = \partial\varphi$.

Proof ($\supset$) Let $\langle x^*, x\rangle \leqslant \varphi(x)$ for all x, and $\langle x^*, x_0\rangle = \varphi(x_0)$. Then $\forall\, \bar{x} \in X$ and $\varepsilon > 0$ we have $\langle x^*, \; x_0 + \varepsilon\bar{x}\rangle \leqslant \varphi(x_0 + \varepsilon\bar{x})$, and since $\varepsilon\bar{x} = (x_0 + \varepsilon\bar{x}) - x_0$, we have $\langle x^*, \varepsilon\bar{x}\rangle \leqslant \varphi(x_0 + \varepsilon\bar{x}) - \varphi(x_0)$, and therefore,

$$\langle x^*, \bar{x}\rangle \leqslant \lim_{\varepsilon\to 0+} \frac{\varphi(x_0 + \varepsilon\bar{x}) - \varphi(x_0)}{\varepsilon} = \varphi'(x_0, \bar{x}), \qquad \text{i.e.} \quad x^* \in \partial\psi.$$

($\subset$). Let $\langle x^*, \bar{x}\rangle \leqslant \varphi'(x_0, \bar{x})$ for all $\bar{x} \in X$. Setting $\varepsilon = 1$ in (18), we have $\langle x^*, \bar{x}\rangle \leqslant \varphi(x_0 + \bar{x}) - \varphi(x_0) \leqslant \varphi(\bar{x})$ (the last inequality holds by the subadditivity). Taking $\bar{x} = -x_0$, we get $-\langle x^*, x_0\rangle \leqslant -\varphi(x_0)$, while the reverse inequality $\langle x^*, x_0\rangle \leqslant \varphi(x_0)$ always holds since $x^* \in \partial\varphi$. Thus, $\langle x^*, x_0\rangle = \varphi(x_0)$, q.e.d. □

4.5 Cones Defined by Sublinear Functionals and Their Dual Cones

Let X be a normed space and $\varphi : X \to \mathbb{R}$ a bounded sublinear functional.

Consider the convex closed cone $K = \{x \in X : \; \varphi(x) \leqslant 0\}$ and the convex open cone $\Omega = \{x \in X : \; \varphi(x) < 0\}$.

Theorem 4.5 (Dubovitskii and Milyutin (1965)) *Suppose the cone* Ω *is nonempty. Then 1)* $\operatorname{int} K = \Omega$, $\; K = \overline{\Omega}$, *and 2)* $K^* = -\mathbb{R}_+\partial\varphi$, *i.e., for every linear functional* $\mu \in K^*$ *there exist* $\alpha \geqslant 0$ *and* $l \in \partial\varphi$ *such that* $\mu = -\alpha l$. *(The converse is obviously true.)*

Proof The first assertion is an easy consequence of the convexity and continuity of φ. To prove the second one, take any $\mu \in K^*$. If $\mu = 0$, we can take $\alpha = 0$ and any $l \in \partial\varphi$, thus obtaining the required relation $\mu = -\alpha l$.

Further, consider the case $\mu \neq 0$. By the assumption, we have:

$$\varphi(x) < 0 \implies \mu(x) \geqslant 0. \tag{19}$$

In the product $X \times \mathbb{R}$, define the cones:

$$C_1 = \{(x,t) : \ \varphi(x) < t\} \quad \text{and} \quad C_2 = \{(x,t) : \ \mu(x) < 0, \ \ t = 0\}.$$

Clearly, they are nonempty and convex, the cone C_1 being open. Furthermore, $C_1 \bigcap C_2 = \emptyset$. Indeed, if they have a common element $(x, 0)$, then $\varphi(x) < 0$ and $\mu(x) < 0$, which contradicts (19).

By the separation theorem there exists a functional $(\lambda, \beta) \in X^* \times \mathbb{R}$ which is strictly negative on C_1 and nonnegative on C_2, that is,

$$\varphi(x) < t \implies l(x) + \beta t < 0\,, \tag{20}$$

$$\mu(x) < 0 \implies l(x) \geqslant 0\,. \tag{21}$$

Taking in (20) $x = 0$ and $t = 1$, we obtain $\beta < 0$, so we can set $\beta = -1$. Then (20) becomes:

$$\varphi(x) < t \implies l(x) < t \qquad \forall\, x, t,$$

whence $l(x) \leqslant \varphi(x) \ \ \forall\, x$, i.e. $l \in \partial\varphi$. By Lemma 3.1 condition (21) implies $l = -\alpha\mu$, where $\alpha \geqslant 0$. If $\alpha = 0$, then $l = 0$, hence, $\varphi(x) \geqslant 0 \ \ \forall\, x$, which contradicts the assumption that K is nonempty. Therefore, $\alpha > 0$, and then $\mu = -\frac{1}{\alpha}\, l$, q.e.d. □

As was already noted, any sublinear functional φ obviously generates a closed convex cone $K = \{x \in X : \ \varphi(x) \leqslant 0\}$. The next theorem shows that, under some "regularity" assumption, the reverse relation also holds.

Theorem 4.6 *Let $K \subset X$ be a closed convex cone with a nonempty interior. Then there exists a bounded sublinear functional φ that generates this cone, i.e., $K = \{x \in X : \ \varphi(x) \leqslant 0\}$.*

Proof Consider the polar cone K^0 and define a set $Q = \{x^* \in K^0 : \ ||x^*|| \leqslant 1\}$. Define a bounded sublinear functional $\varphi(x) = \sup\, \langle Q, x\rangle$. By the Alaoglu theorem, Q is compact in the weak-* topology, and obviously convex, hence by Lemma 4.1 $\partial\varphi = Q$.

Now, define a set $C = \{x \mid \ \varphi(x) \leqslant 0\} = \{x \mid \ \langle Q, x\rangle \leqslant 0\}$. Obviously, it is a closed convex cone. We claim that $C = K$. Indeed, if $x \in K$, then by the definition of K^0 we have $\langle Q, x\rangle \leqslant 0$, that is, $\varphi(x) \leqslant 0$, so $x \in C$. On the other hand, if $x \in C$, i.e., $\langle Q, x\rangle \leqslant 0$, then also $\langle K^0, x\rangle \leqslant 0$, which implies that $x \in K$ (otherwise, if $x \notin K$, then by the separation theorem $\exists\, x^* \in K^0$ such that $\langle x^*, x\rangle > 0$, a contradiction). □

In a standard case, when there is a finite number of smooth scalar inequalities $f_i(x) \leqslant 0$, $i = 1, \ldots, m$, they can be replaced by one vector-valued inequality

$$f(x) := (f_1(x), \ \ldots, \ f_m(x)) \in K = \mathbb{R}^m_-\,,$$

or by one nonsmooth inequality $\Phi(x) := \max\limits_{1 \leqslant i \leqslant m} f_i(x) \leqslant 0$.

Here, the cone $\mathbb{R}^m_-$ is given by the sublinear inequality $\varphi(x) := \max_{1 \leqslant i \leqslant m} x_i \leqslant 0$.

All the above are facts of linear and convex functional analysis. However, to handle the nonlinear equality constraint $g(x) = 0$ in optimization problem (1), we also need some facts of nonlinear functional analysis.

5 Covering and Metric Regularity

Our main goal in this section will be to obtain a "correction theorem" for the operator g in a neighborhood of a reference point x_0, which is now often called *generalized Lyusternik theorem*. It is essentially related with (and, in fact, based on) the following concept of *covering mapping* and on a theorem on the stability of covering property under small perturbations.

5.1 Milyutin's Covering Theorem

Recall the following important notion proposed by Milyutin (see Levitin et al. (1974, 1978); Dmitruk et al. (1980)).

Let X be a complete metric space, Y a vector space with a translation-invariant metric (e.g. a normed space), G a set in X, and $T : X \to Y$ a mapping. We denote the metrics in X and Y by the same letter d, and the ball $B_r(x)$ is sometimes denoted by $B(x, r)$.

Definition 5.1 The mapping T covers on G with a rate $a > 0$ if

$$\forall\, B_r(x) \subset G \qquad T(B_r(x)) \supset B_{ar}(T(x)). \tag{22}$$

(Obviously, this property makes sense only if the set G has a nonempty interior; moreover, one can assume that $G \subset \overline{\operatorname{int} G}$.)[2]

Note that this property remains valid if the mapping T is replaced by $T + y$ for any fixed $y \in Y$. Note also, that a covering mapping can be not continuous on G. (A simple example is presented by the mapping $T : \mathbb{R}^2 \to \mathbb{R}$, $(x, y) \mapsto x + f(y)$, where f is an arbitrary function.)

The most simple and, at the same time fundamental, example of covering mapping is a linear surjective operator $A : X \to Y$ between Banach spaces. By Theorem 3.5, it covers with some rate $r > 0$ on the whole space X.

Now, let another mapping $S : X \to Y$ be also given.

---

[2] Actually, the original Milyutin's definition is slightly more general, but for our purposes the given one would be completely enough.

Definition 5.2 The mapping S contracts on G with a rate $b \geqslant 0$ if

$$\forall\, B_r(x) \subset G \qquad S(B_r(x)) \subset B_{br}(S(x)). \tag{23}$$

Obviously, any such mapping is continuous on G. On the other hand, any mapping b–Lipschitz continuous on G contracts on G with rate b.

The following theorem presents the stability of covering property under small contracting additive perturbations. Its formulation was proposed by A.A. Milyutin in Levitin et al. (1974, Sec.2) and the proof was published in Levitin et al. (1978, Sec.2), Dmitruk et al. (1980).

Theorem 5.1 (A.A. Milyutin) *Let T cover on G with a rate $a > 0$ and have a closed graph (e.g., be continuous on G), and let S contract on G with a rate $b < a$. Then their sum $F = T + S$ covers on G with the rate $a - b > 0$.*

The proof is so important and at the same time transparent, that it worth to be given here completely. Take any ball $B(x_0, \rho) \subset G$. We must show that

$$F(B(x_0, \rho)) \supset B(F(x_0), (a - b)\rho).$$

Without loss of generality, assume that $a = 1$ and hence $b < 1$. Denote for brevity $y_0 = F(x_0), r = (1 - b)\rho$. Take any $\hat{y} \in B(y_0, r)$. We have to show that $\exists\, \hat{x} \in B(x_0, \rho)$ such that $F(\hat{x}) = \hat{y}$. The point $\hat{x}$ will be obtained as the limit of a sequence $\{x_n\}$, which will be generated by a special iteration process.

At the beginning, we have the following relation:

$$T(x_0) + S(x_0) = y_0\,, \tag{24}$$

and we need to obtain $T(\hat{x})+S(\hat{x}) = \hat{y}$. Since the mapping T is 1-covering, so does $T + S(x_0)$, and since $\hat{y} \in B_r(y_0)$ and $B(x_0, r) \subset G$, there exists $x_1 \in B(x_0, r)$ such that

$$T(x_1) + S(x_0) = \hat{y}\,. \tag{25}$$

Now, replace here $S(x_0)$ by $S(x_1)$. Since the mapping S is b–contracting on the ball $B(x_0, r)$, we have $d(S(x_1), S(x_0)) \leqslant br$, and so

$$T(x_1) + S(x_1) = y_1\,, \tag{26}$$

where $d(\hat{y}, y_1) \leqslant br$. So, we moved from Eq. (24) for a "base" point x_0 to Eq. (26) for a new "base" point x_1, where

$$d(x_0, x_1) \leqslant r, \qquad y_1 \in B(y_0, br),$$

and we still need to obtain $T(\hat{x}) + S(\hat{x}) = \hat{y}$. Since

$$r + br < r(1 + b + b^2 + \ldots) = r\frac{1}{1-b} = \rho,$$

the ball $B(x_1, br)$ is contained in the initial ball $B(x_0, \rho)$; hence the 1–covering of T and $b-$contracting of S hold on $B(x_1, br)$. Then, by analogy with the preceding step, there exists $x_2 \in B(x_1, br)$, such that

$$T(x_2) + S(x_2) = y_2 \qquad \text{with} \qquad d(\hat{y}, y_2) \leqslant b^2 r,$$

and so on. Continuing this process infinitely, we obtain a sequence of points x_n, y_n such that

$$F(x_n) = T(x_n) + S(x_n) = y_n\,, \tag{27}$$

$$d(x_{n-1}, x_n) \leqslant b^{n-1} r, \qquad d(\hat{y}, y_n) \leqslant b^n r. \tag{28}$$

Moreover, we have

$$d(x_0, x_n) + b^n r \leqslant d(x_0, x_1) + d(x_1, x_2) + \ldots + d(x_{n-1}, x_n) + b^n r \leqslant$$

$$\leqslant r + br + \ldots + b^{n-1} r + b^n r < r\frac{1}{1-b} = \rho, \tag{29}$$

whence the ball $B(x_n, b^n r)$ is contained in the initial ball $B(x_0, \rho)$, which makes the next step possible.

Consider the obtained sequence $\{x_n\}$. The first inequality in (28) implies that it is a Cauchy sequence, and since X is complete, this sequence has a limit $\hat{x}$. By (29) we get $d(x_0, \hat{x}) \leqslant \rho$, i.e., $\hat{x} \in B(x_0, \rho)$. The second inequality in (28) implies that $y_n \to \hat{y}$, and then, from (27) and continuity of F on the initial ball (or from the closedness of its graph) we get $F(\hat{x}) = \hat{y}$, which is exactly what was required. □

Remark The iteration process in this proof is much similar to that in the Newton method: the role of derivative F', involved in the Newton method, is played in our case by the mapping T, and the role of the small nonlinear residual is played by the mapping S. The covering property of mapping T allows us to "solve" Eq. (25) with respect to x_1, while the perturbational term S applied to x_1 pushes us aside the desired goal $\hat{y}$. Repeating iteratively this procedure, we obtain a sequence which gives in the limit a desired solution: $T(\hat{x}) + S(\hat{x}) = \hat{y}$. This abstract Newton-like method can be called *Lyusternik iteration process* (Lyusternik 1934). Note however, that this process does not completely coincide with the Newton method, because the mapping T is not one-to-one, in general, and therefore, Eq. (25) is not solved uniquely, in contrast to the Newton method. So, the Lyusternik process is more general than the latter. (For example, the Lyusternik iteration process is actually used in the standard proof of the Banach open mapping Theorem 3.5, whereas the Newton method cannot be used there).

5.2 Local Covering and Metric Regularity

In what follows, we will actually need *a local version* of the covering property, when a mapping g covers *in a neighborhood* of a given point x_0, i.e., when inclusion (5.3) holds for any ball containing in this neighborhood.[3] A closely related to this property is the following central (and the most simple) notion of nonsmooth analysis.

Set $y_0 = g(x_0)$. The mapping g is said to be *metric regular with constant* C in neighborhoods of x_0 and y_0 if there exist neighborhoods $\mathcal{U}(x_0)$ and $\mathcal{V}(y_0)$ of x_0 and y_0, respectively, such that for every $x \in \mathcal{U}(x_0)$ and every $y \in \mathcal{V}(y_0)$ the following estimate holds:

$$dist\,(x, g^{-1}(y)) \leqslant C\,d(g(x), y). \tag{30}$$

(Here, *dist* denotes the distance from a point to a set.)

Between these two notions the following simple relation holds (see e.g. Dmitruk et al. (1980)).

Theorem 5.2 *Suppose that a mapping $g : X \to Y$ covers with a rate $a > 0$ in a neighborhood of x_0. Then g is metric regular with constant $C = 1/a$ in neighborhoods of x_0 and y_0.*

If g is continuous at x_0, the reverse is also true: if a mapping g is metric regular with a constant C in neighborhoods of x_0 and y_0, then it covers in some neighborhood of x_0 with any rate $a < 1/C$.

Proof ($\Longrightarrow$) Again, we set $a = 1$, so g is 1-covering on $B_\varepsilon(x_0)$ for some $\varepsilon > 0$. Set $\delta = \varepsilon/3$, $y_0 = g(x_0)$ and show that g is metric 1-regular on $B_\varepsilon(x_0) \times B_\varepsilon(y_0)$.

Take any $x \in B(x_0, \delta)$, $y' \in B(y_0, \delta)$ and denote $g(x) = y$, $d(y', y) = r$. Thus, $y' \in B_r(y)$, and we have to show that $d(x, g^{-1}(y')) \leqslant r$. We will actually show a bit more, that

$$\exists\, x' \in B_r(x) \quad \text{such that} \quad g(x') = y'. \tag{31}$$

The following two cases are possible:

(a) $r \geqslant 2\delta$ (y and y' are "far" from each other),
(b) $r < 2\delta$ (y and y' are "close").

In case (a), since g is 1-covering, the image of $B(x_0, \delta)$ contains the ball $B(y_0, \delta)$, hence $\exists\, x' \in B(x_0, \delta)$ such that $g(x') = y'$. Then

$$d(x', x) \leqslant d(x', x_0) + d(x_0, x) \leqslant 2\delta \leqslant r,$$

and so, (31) is proved.

[3] As to connection between the local covering in a neighborhood of a point and "nonlocal" covering on an open set, see Dmitruk (2005).

In case (b), we have $\delta + r < \delta + 2\delta = \varepsilon$, therefore, by the triangle inequality, $B_r(x) \subset B_\varepsilon(x_0)$, and since g covers with 1 on $B_\varepsilon(x_0)$, we have $g(B_r(x)) \supset B_r(y)$. Taking into account that $y' \in B_r(y)$, we get (31). □

($\Longleftarrow$) Now, we assume that $C = 1$. So, let g be continuous at x_0, and for some $\delta > 0$, $\varepsilon > 0$ metric regular with constant 1 on $B_\delta(x_0) \times B_\varepsilon(y_0)$. Reduce, if necessary, δ so that

$$0 < \delta < \varepsilon/3 \qquad \text{and} \qquad g(B_\delta(x_0)) \subset B_{\varepsilon/3}(y_0) \tag{32}$$

(the last is possible since g is continuous), and show that g covers on $B_\delta(x_0)$ with any rate $a < 1$.

Consider any ball $B(x, r) \subset B_\delta(x_0)$. Obviously, its radius $r \leqslant 2\delta$ (because, its any point x' satisfies $d(x', x) \leqslant d(x', x_0) + d(x_0, x) \leqslant 2\delta$).

Denote $F(x) = y$ and take any $r' < r$. It suffices to show that

$$g(B_r(x)) \supset B_{r'}(y), \tag{33}$$

which readily would imply the a–covering of g for any $a < 1$.

Take any point $y' \in B_{r'}(y)$. In view of (32),

$$d(y_0, y') \leqslant d(y_0, y) + d(y, y') \leqslant \frac{\varepsilon}{3} + r' < \frac{\varepsilon}{3} + r \leqslant \frac{\varepsilon}{3} + 2\delta < \varepsilon,$$

hence $y' \in B_\varepsilon(y_0)$. Since $x \in B_\delta(x_0)$, then, by the metric 1-regularity, the points x, y' satisfy the estimate:

$$d(x,\, g^{-1}(y')) \leqslant d(y, y') \leqslant r' < r.$$

This implies that $\exists\, x' \in B_r(x)$ such that $g(x') = y'$, which proves (33), and hence, the theorem is completely proved. □

This theorem says that the local metric regularity and local covering are, in fact, one and the same property. Our experience shows that, *to obtain* this property is more convenient in the "geometrical" form of covering, while *to use* it is more convenient in the "analytical" form of metric regularity.

5.3 Covering and Metric Regularity for Strictly Differentiable Operators

Let now X and Y be Banach spaces, $G \subset X$ an open set, $x_0 \in G$ a given point. Recall the following

Definition An operator $g : G \to Y$ is said to be *strictly differentiable* at x_0 if there exists a linear operator $T : X \to Y$ such that, for any $\varepsilon > 0$ there exists a neighborhood $\mathcal{O}(x_0)$ such that, for all $x_1, x_2 \in \mathcal{O}(x_0)$ the following inequality holds:

$$\|g(x_2) - g(x_1) - T(x_2 - x_1)\| \leqslant \varepsilon \|x_2 - x_1\|. \tag{34}$$

The latter means that the operator $g(x) - Tx$ is Lipschitz continuous in $\mathcal{O}(x_0)$ with constant ε. Also, it means that

$$g(x_2) - g(x_1) - T(x_2 - x_1) = \zeta(x_1, x_2)\|x_2 - x_1\|, \tag{35}$$

where $\zeta(x_1, x_2) \to 0$ as $\|x_1 - x_0\| + \|x_2 - x_0\| \to 0$.

The operator T is called *strict derivative* of g at x_0. Fixing $x_1 = x_0$, we obtain that g is Frechet differentiable at x_0 and its Frechet derivative $g'(x_0) = T$. It can be easily shown (by using the mean value theorem) that any operator continuously Frechet differentiable at x_0 is strictly differentiable at this point, but the reverse is not true even in the 1-dimensional case.

An important corollary of Theorem 5.1 for strictly differentiable operators is the following

Theorem 5.3 *Let X and Y be Banach spaces, $G \subset X$ an open set, $x_0 \in G$ a given point, and $g : G \to Y$ an operator strictly differentiable at x_0. Suppose that $g'(x_0)X = Y$ (the so-called Lyusternik condition). Then the operator g covers with some rate $a > 0$ in a neighborhood of x_0.*

Proof Since $g'(x_0)X = Y$, the Banach open mapping theorem guarantees that the linear operator $g'(x_0)$ covers with some rate $a > 0$ on the whole space X.

Further, let us represent g in the form

$$g(x) = g(x_0 + (x - x_0)) = g(x_0) + g'(x_0)(x - x_0) + S(x), \tag{36}$$

where $S(x) = g(x) - g(x_0) - g'(x_0)(x - x_0)$.

The operator $g(x_0) + g'(x_0)(x - x_0)$ covers on X with the same rate a, while, according to (34), the operator S can be made Lipschitz continuous with any constant $\varepsilon > 0$ in a properly chosen neighborhood $\mathcal{O}(x_0)$. Taking any $\varepsilon < a$, we obtain by Theorem 5.1 that g covers on $\mathcal{O}(x_0)$ with the rate $a - \varepsilon > 0$. □

Theorems 5.2 and 5.3 directly imply the following theorem on a correction to the level set of a nonlinear operator, which was in fact first proposed by L.A. Lyusternik in his seminal paper (Lyusternik 1934).

Theorem 5.4 *Let X and Y be Banach spaces, $G \subset X$ an open set, $x_0 \in G$ a fixed point, and $g : G \to Y$ a strictly differentiable at x_0 operator. Suppose that $y_0 = g(x_0) = 0$ and $g'(x_0)X = Y$. Then g is metric regular in some neighborhoods of x_0 and y_0, and hence, there exist a neighborhood $\mathcal{O}(x_0)$ and a constant C such that, for every $x \in \mathcal{O}(x_0)$ one can find an $h = h(x) \in X$ such that*

$$g(x + h) = 0 \quad \textit{and} \quad \|h\| \leqslant C\|g(x)\|. \tag{37}$$

This theorem is a highly convenient and efficient tool in handling a wide range of nonlinear equations, and because of this, plays a key role in studying optimization problems with equality constraints.

6 Proof of Theorem 2.1

Our proof follows the Dubovitskii–Milyutin scheme (Dubovitskii and Milyutin 1965). It consists of two steps. At the first step, we pass from the local minimality to the incompatibility of (sub)linear approximations of all the constraints and the cost of the problem, and at the second step, this incompatibility of approximations is rewritten, by the Dubovitskii–Milyutin separation theorem, as the corresponding Euler–Lagrange equation.

First of all, following (Levitin et al. 1974, 1978), it is convenient to introduce the next

Definition We say that x_0 is a point of *s-necessity* (or *strongest necessity*) in Problem (1) if there is no sequence $x_n \to x_0$ such that

$$F_0(x_n) < F_0(x_0), \quad f_i(x_n) \in \operatorname{int} K_i, \quad i = 1, \dots \nu, \qquad g(x_n) = 0. \tag{38}$$

Obviously, the local minimum at the point x_0 implies the s-necessity at this point.[4]

Like in all problems with inequality constraints, it is convenient to distinguish between active and inactive indices. If i is such that $f_i(x_0) \in \operatorname{int} K_i$, this index is called *inactive*. Otherwise, if $f_i(x_0) \in \partial K_i$, the index i is *active*. The index $i = 0$ is always active by definition. The set of all active indices is denoted by $I(x_0)$.

6.1 Lemma on Incompatibility of a System of Approximations

For any $i = 1, \dots, \nu$, define

$$C_i = \{\bar{z} \in Z_i : \ \exists \alpha > 0 \ \text{ such that } \ f_i(x_0) + \alpha z \in \operatorname{int} K_i \}$$

- the cone of interior directions to the cone K_i at the point $z_i = f_i(x_0)$. Clearly, $C_i = \operatorname{int} K_i - \mathbb{R}_+ f_i(x_0)$, it is nonempty, convex and open. Along with this, define an open convex cone

$$\Omega_i = \{\bar{x} \in X : \ \bar{z} = f_i'(x_0)\,\bar{x} \in C_i \}$$

[4]The notion of s-necessity exactly corresponds to the notion of *Pareto weak efficiency* in the theory of vector optimization.

– the preimage of cone C_i under the linear mapping $f_i'(x_0) : X \to Z_i$. Note that, if i is inactive, i.e. $f_i(x_0) \in \operatorname{int} K_i$, then $C_i = Z_i$ and $\Omega_i = X_i$.

Now, consider the following system of approximations for Problem (1) at the point x_0:

$$F_0'(x_0)\,\bar{x} < 0, \qquad \bar{x} \in \Omega_i\,, \quad i = 1, \ldots, \nu, \qquad g'(x_0)\,\bar{x} = 0. \tag{39}$$

Lemma 6.1 *If x_0 is a point of s-necessity in Problem (1) and $g'(x_0)X = Y$ (the Lyusternik regularity condition), then there is no $\bar{x} \in X$ satisfying system (39).*

Proof Suppose, on the contrary, that there exists $\bar{x}$ satisfying system (39). Consider the "sequence" $x_0 + \varepsilon\bar{x}$ with $\varepsilon \to 0+$. For this sequence, we have

$$g(x_0 + \varepsilon\bar{x}) \;=\; g(x_0) + g'(x_0)\varepsilon\bar{x} + o(\varepsilon) \;=\; o(\varepsilon),$$

hence, by Theorem 5.4, there is a correction $r_\varepsilon \in X$ such that $g(x_0 + \varepsilon\bar{x} + r_\varepsilon) = 0$ and $\|r_\varepsilon\| = o(\varepsilon)$.

Further, take any $i \in \{1, \ldots, \nu\}$. If $i \notin I(x_0)$, then obviously $f_i(x_0 + \varepsilon\bar{x} + r_\varepsilon) \in \operatorname{int} K_i$. If $i \in I(x_0)$, we have $f_i(x_0) \in K_i$ and $f_i(x_0) + \alpha f_i'(x_0)\bar{x} \in \operatorname{int} K_i$ for some $\alpha > 0$, whence the whole half-interval $(f_i(x_0),\ f_i(x_0) + \alpha f_i'(x_0)\bar{x}]$ lies in $\operatorname{int} K_i$. Then

$$f_i(x_0 + \varepsilon\bar{x} + r_\varepsilon) = \; f_i(x_0) + \varepsilon f_i'(x_0)\bar{x} \; + o(\varepsilon) \;\in \operatorname{int} K_i$$

for all small enough $\varepsilon > 0$. Similarly, $F_0(x_0 + \varepsilon\bar{x} + r_\varepsilon) < F_0(x_0)$ for small enough $\varepsilon > 0$, since $F_0'(x_0)\bar{x} < 0$ and $\|r_\varepsilon\| = o(\varepsilon)$. Thus, the sequence $x_0 + \varepsilon\bar{x} + r_\varepsilon$ satisfies system (38), which contradicts the s-necessity at x_0 . □

6.2 *Passage to the Euler–Lagrange Equation*

Let x_0 be a point of s-necessity in problem (1). Without loss of generality assume that all indices $i \geqslant 1$ are active. Consider the regular case, when $g'(x_0)X = Y$. Then by Lemma 6.1 the system of approximations (39) is incompatible. We aim to apply the Dubovitskii–Milyutin "separation" Theorem 3.3.

Assume first that $F_0'(x_0) \neq 0$ and $\operatorname{Im} f_i'(x_0) \cap C_i \neq \emptyset$ for all $i = 1, \ldots, \nu$ (the main, nondegenerate case). Then all the cones in system (39) are nonempty. By Theorem 3.3, there exist a support functional x_0^* to the half-space $\Omega_0 = \{\bar{x} : F_0'(x_0)\bar{x} < 0\}$, support functionals x_i^* to the cones Ω_i , $i = 1, \ldots, \nu$, and a support functional x^* to the subspace $\Omega = \{\bar{x} : g'(x_0)\bar{x} = 0\}$, not all of which are equal to zero, such that $x_0^* + x_1^* + \ldots + x_\nu^* + x^* = 0$. By Lemma 3.1 $x_0^* = -\alpha_0 f_0'(x_0)$ with some $\alpha_0 \geqslant 0$, by the Farkas–Minkowski Theorem 3.6 each $x_i^* = z_i^* f_i'(x_0)$ with some $z_i^* \in K_i^*$ such that $\langle z_i^*, f_i(x_0)\rangle = 0$, and by Lemma 3.6 $x^* = y^* g'(x_0)$ with some $y^* \in Y^*$. Thus, we get

$$-\alpha_0 F_0'(x_0) + \sum_{i=1}^{\nu} z_i^* f_i'(x_0) + y^* g'(x_0) = 0.$$

Obviously, $\alpha_0 + \sum_{i=1}^{\nu} \|z_i^*\| + \|y^*\| > 0$ (otherwise all the support functionals equal zero). Finally note that $-z_i^* \in K_i^0$ for all $i = 1, \ldots, \nu$, and changing y^* to $-y^*$, we obtain the Euler–Lagrange equation in the required form (4).

Now, consider the degenerate cases. If $F_0'(x_0) = 0$, we take $\alpha_0 = 1$ and all other functionals equal to zero, thus obtaining (4). If $\exists\, i$ such that $\operatorname{Im} f_i'(x_0) \cap C_i = \emptyset$, we can separate the subspace $\operatorname{Im} f_i'(x_0)$ and the open cone C_i by a nonzero z_i^*. By Lemma 3.3 $\langle z_i^* K_i \rangle \geqslant 0$ and $\langle z_i^*, f_i(x_0) \rangle = 0$. Setting $\alpha_0 = 0$, all $z_j^* = 0$ for $j \neq i$, and $y^* = 0$, we obtain (4). Finally, if $g'(x_0)X \neq Y$, then by the assumption, the subspace $g'(x_0)X$ is closed in Y, and hence, by Lemma 3.4, there is a nonzero functional $y^* \in Y^*$ vanishing on $g'(x_0)X$, which means that $y^* g'(x_0) = 0$. Taking all other functionals equal zero, we again obtain Eq. (4). □

Remarks

1. As one could see, the incompatibility of system (39) follows not only from the local minimum at x_0 in problem (1), but even from the s−necessity in this problem. In the conditions of s−necessity, the cost functional plays the same role as any active inequality constraint (the inactive constraints can be removed from the study altogether). Therefore, the inequality $F_0(x) - F_0(x_0) < 0$ can be replaced by a more general condition $f_0(x) \in \operatorname{int} K_0$, where K_0 is a convex cone with a nonempty interior in a Banach space Z_0. Thus, instead of study the local minimum at x_0 in problem (1), one can study the presence of s−necessity at x_0 in a general system of inequalities and equalities:

$$f_i(x) \in K_i\,, \qquad i = 0, 1, \ldots \nu, \qquad g(x) = 0, \tag{40}$$

in which all f_i come symmetrically. In this case, the following analog of Theorem 2.1 holds:

Theorem 6.1 *Let x_0 be a point of s−necessity in system (40). Then there exist Lagrange multipliers $z_i^* \in K_i^0$, $i = 0, 1, \ldots, \nu$, and $y^* \in Y^*$, not all of which equal zero, satisfying the complementary slackness conditions $\langle z_i^*, f_i(x_0) \rangle = 0$, $i = 0, 1, \ldots, \nu$, such that the Lagrange function*

$$\mathcal{L}(x) = \sum_{i=0}^{\nu} \langle z_i^*, f_i(x) \rangle + \langle y^*, g(x) \rangle$$

is stationary at x_0: $$\mathcal{L}'(x_0) = \sum_{i=0}^{\nu} z_i^* f_i'(x_0) + y^* g'(x_0) = 0.$$

2. The functional F_0 in problem (1) can be taken not necessarily smooth; one can take it in the form $F_0(x) = \varphi_0(f_0(x))$, where $f_0 : X \to Z_0$, similarly to other f_i, is a smooth mapping, and φ_0 is a sublinear functional on Z_0 such that $\varphi_0(f_0(x_0)) = 0$ and the cone $K_0 = \{x : \varphi_0(x) \leqslant 0\}$ has a nonempty interior. Then the inequality $F_0(x) < 0$ is equivalent to that $f_0(x) \in \operatorname{int} K_0$, and we again come to the $s-$necessity in system (40), symmetric with respect to all $i = 0, 1, \ldots, \nu$.
3. On the other hand, by virtue of Theorem 4.6, any "vector-valued" inequality $f_i(x) \in K_i$ can be represented as a scalar inequality $\varphi_i(f_i(x)) \leqslant 0$, where $\varphi_i : Z_i \to \mathbb{R}$ is a sublinear functional such that $K_i = \{x : \varphi_i(x) \leqslant 0\}$. In this case, we come to a system of a finite number of nonsmooth scalar inequalities and one (generally, infinite-dimensional) equality:

$$\varphi_i(f_i(x)) \leqslant 0, \qquad i = 0, 1, \ldots, \nu, \qquad g(x) = 0. \tag{41}$$

The conditions of $s-$necessity for this system are still given by Theorem 6.1 with $z_i^* = \alpha_i \hat{z}_i^*$, where $\alpha_i \geqslant 0$ and $\hat{z}_i^* \in \partial\varphi_i(f_i(x_0))$. In its turn, system (41) can be reduced to a system with only one nonsmooth scalar inequality:

$$\Phi(x) := \max_{0 \leqslant i \leqslant \nu} \varphi_i(f_i(x)) \leqslant 0, \qquad g(x) = 0. \tag{42}$$

All these representations are equivalent and give the same conditions of $s-$necessity.
4. Finally note that Theorem 6.1 (and equivalent Theorem 2.1) can be proved in another way, if one starts from the nonsmooth representation (42) instead of the smooth "infinite-dimensional" system (40). On this way, instead of the Dubovitskii–Milyutin Theorem 3.3 on the nonintersection of cones one should use the Dubovitskii–Milyutin Theorem 4.2 on the subdifferential of maximum of sublinear functionals. Let us show this, keeping the relations between the functionals φ_i and the cones K_i as in Remark 3.

Let a functional $\Phi : X \to \mathbb{R}$ be Lipschitz continuous around x_0, whose directional derivative $\Psi(\bar{x}) = \Phi'(x_0, \bar{x})$ is convex in $\bar{x}$, hence is a bounded sublinear functional. Consider the system

$$\Phi(x) \leqslant 0, \qquad g(x) = 0.$$

Suppose that $g'(x_0)X = Y$ and the $s-$necessity for this system at x_0 holds. Then it easily follows that there is no $\bar{x}$ such that

$$\Phi'(x_0, \bar{x}) < 0, \qquad g'(x_0)\bar{x} = 0.$$

This means that $\Psi(\bar{x}) \geqslant 0$ for all $\bar{x} \in L = Ker\, g'(x_0)$, i.e. the functional $\Psi|_L : L \to \mathbb{R}$ is nonnegative on the subspace L. Therefore, $\partial\Psi|_L$ contains the functional $l^* = 0$. Then, by the Hahn–Banach Theorem 4.3, there exists its extension $\widetilde{l^*} \in X^*$

such that $\widetilde{l}^* = l^* = 0$ on L, and $\widetilde{l}^* \in \partial\Psi$, i.e., $\Psi(\bar{x}) - \langle \widetilde{l}^*, \bar{x}\rangle \geqslant 0$ on the whole space X. Since $\widetilde{l}^* = -y^* g'(x_0)$ for some $y^* \in Y^*$, we obtain the inclusion

$$\partial\Psi + y^* g'(x_0) \ni 0. \tag{43}$$

Now, let Φ have the form as in (42), where all indices $i = 1, \dots, \nu$ are active, i.e. all $f_i(x_0) = 0$. It easily follows that

$$\Phi'(x_0, \bar{x}) = \max_{1 \leqslant i \leqslant \nu} \varphi_i'(f_i(x_0), f_i'(x_0)\bar{x}) \qquad \forall\, \bar{x} \in X.$$

Define sublinear functionals $\psi_i(\bar{z}) = \varphi_i'(f_i(x_0), \bar{z})$ on Z_i, linear operators $A_i = f_i'(x_0) : X \to Z_i$, and sublinear functionals $P_i(\bar{x}) = \psi_i(A_i\bar{x})$. Then

$$\Phi'(x_0, \bar{x}) = \max_{1 \leqslant i \leqslant \nu} P_i(\bar{x}).$$

By Theorem 4.4, $\partial P_i = \partial\psi_i \cdot A_i$, that is $x_i^* \in \partial P_i$ iff $x_i^* = z_i^* f_i'(x_0)$ for some $z_i^* \in \partial\psi_i$. The latter means that $z_i^* \in \partial\varphi_i$ and $\langle z_i^*, f_i(x_0)\rangle = \varphi_i(f_i(x_0)) = 0$.

Since $\Psi(\bar{x}) = \max_i P_i(\bar{x})$, by the Dubovitskii–Milyutin Theorem 4.2 $x^* \in \partial\Psi$ iff $x^* = \sum_i \alpha_i x_i^*$, where all $x_i^* \in \partial P_i$, all $\alpha_i \geqslant 0$ and $\sum \alpha_i = 1$. Therefore, $x^* = \sum \alpha_i z_i^* f_i'(x_0)$, where $z_i^* \in \partial\varphi_i$ (hence $z_i^* \in K_i^0$) and $\langle z_i^*, f_i(x_0)\rangle = 0$.

In view of (43), we have

$$\sum_i \alpha_i z_i^* f_i'(x_0) + y^* g'(x_0) = 0.$$

Note that $\hat{z}_i^* = \alpha_i z_i^*$ still satisfy the conditions $\hat{z}_i^* \in K_i^0$ and $\langle \hat{z}_i^*, f_i(x_0)\rangle = 0$, so we come to the Euler–Lagrange equation

$$\sum_i \hat{z}_i^* f_i'(x_0) + y^* g'(x_0) = 0,$$

which proves Theorem 6.1.

Acknowledgements This research was partially supported by the Russian Foundation for Basic Research under grants 16-01-00585 and 17-01-00805.

References

V.M. Alekseev, V.M. Tikhomirov, S.V. Fomin, *Optimal control*. (Transl. from the Russian) (Consultants Bureau, New York, 1987) 309 p.

A.V. Dmitruk, On a nonlocal metric regularity of nonlinear operators. Control Cybern. **34**(3), 723–746 (2005)

A.V. Dmitruk, N. P. Osmolovskii, Necessary conditions for a weak minimum in optimal control problems with integral equations subject to state and mixed constraints. SIAM J. Control Optim. **52**, 3437–3462 (2014)

A.V. Dmitruk, A.A. Milyutin, N.P. Osmolovsky, Lyusternik's theorem and the theory of extrema. Russ. Math. Surv. **35**, 11–51(1980)

A.Ya. Dubovitskii, A.A. Milyutin, Extremum problems in the presence of restrictions. Zh. Vychisl. Mat. Mat. Fiz. (USSR Comput. Math. Math. Phys.) **5**(3), 1–80, (1965)

N. Dunford, J. Schwartz, *Linear Operators, Part 1: General Theory* (Wiley-Interscience, London, 1968)

R.V. Gamkrelidze, G.L. Kharatishvili, *Extremal Problems in Linear Topological Spaces, I.* Math. Syst. Theory **1**(3), 229–256 (1967)

L. Hurwicz, Programming in linear spaces, in *Studies in Linear and Nonlinear Programming*, ed. by K.J. Arrow, L. Hurwicz, H. Uzawa (Stanford University Press, Stanford, 1958)

J. Jahn, *Introduction to the Theory of Nonlinear Optimization* (Springer, Berlin, 1994)

A.N. Kolmogorov, S.V. Fomin, *Elements of Function Theory and Functional Analysis* (Nauka, Moscow, 1968, in Russian)

S. Kurcyusz, On the existence and nonexistence of lagrange multipliers in Banach Spaces. J. Optim. Theory Appl. **20**(1), 81–110 (1976)

E.S. Levitin, A.A. Milyutin, N.P. Osmolovskii, Conditions for a local minimum in a problem with constraints, in *Mathematical Economics and Functional Analysis* (Nauka, Moscow, 1974, in Russian), pp. 139–202.

E.S. Levitin, A.A. Milyutin, N.P. Osmolovskii, *Conditions of high order for a local minimum in problems with constraints*. Russ. Math. Surv. **33**(6), 97–168 (1978)

L.A. Lyusternik, On the conditional extrema of functionals. Matem. Sbornik, **41**, 390–401 (1934, in Russian)

H. Maurer, J. Zowe, First and second order necessary and sufficient optimality conditions for infinite-dimensional programming problems. Math. Program. **16**, 98–110 (1979)

A.A. Milyutin, A.V. Dmitruk, N.P. Osmolovskii, *Maximum principle in optimal control* (Moscow State University, Faculty of Mechanics and Mathematics, Moscow, 2004, in Russian), 168 p.

Y. Nagahisa, Y. Sakawa, Nonlinear programming in Banach Spaces. J. Optim. Theory Appl. **4**(3), 182–190 (1969)

D.O. Norris, Nonlinear programming applied to state-constrained optimization problems. J. Math. Anal. Appl. **43**, 261–272 (1973)

N.P. Osmolovskii, V.M. Veliov, Optimal control of age-structured systems with mixed state-control constraints. J. Math. Anal. Appl. **455**(1), 396–421 (2017)

B.H. Pourciau, Modern multiplier rules. Am. Math. Mon. **87**, 433–452 (1980)

B.N. Pshenichnyi, *Necessary Conditions of Extremum* (Nauka, Moscow, 1969, 1982, in Russian)

R.T. Rockafellar, Lagrange multipliers and optimality. SIAM Rev. **35**(2), 183–238 (1993)

E.V. Tamminen, Sufficient conditions for the existence of multipliers and Lagrangian duality in abstract optimization problems. J. Optim. Theory Appl. **82**(1), 93–104 (1994)

P.P. Varaiya, Nonlinear programming in Banach Space. SIAM J. Appl. Math. **15**(2), 147–152 (1967)

Strict Dissipativity Implies Turnpike Behavior for Time-Varying Discrete Time Optimal Control Problems

Lars Grüne, Simon Pirkelmann, and Marleen Stieler

Abstract We consider the turnpike property for infinite horizon undiscounted optimal control problems in discrete time and with time-varying data. We show that, under suitable conditions, a time-varying strict dissipativity notion implies the turnpike property and a continuity property of the optimal value function. We also discuss the relation of strict dissipativity to necessary optimality conditions and illustrate our results by an example.

1 Introduction

Infinite horizon optimal control problems are notoriously difficult to solve if the problem data is time-varying. Unlike the time invariant case, global approaches like dynamic programming do not lead to a stationary Bellman equation but—in the discrete time setting considered in this paper—rather to an infinite sequence of such equations. Since we consider undiscounted problems in this paper, the dynamic programming approach has the additional difficulty that the Bellman equation is not a contraction. Pontryagin-type necessary optimality conditions (see, e.g., Aseev et al. (2016); Blot and Hayek (2014)) appear somewhat more suitable for this class of problems, however, they still lead to an infinite dimensional system of coupled difference equations for which no general solution method exists.

It has been observed in various papers (e.g., in Anderson and Kokotović (1987); Porretta and Zuazua (2013); Trélat and Zuazua (2015)), that the turnpike property facilitates the computation of optimal trajectories on long finite time horizons. In the time-invariant setting of these papers, the turnpike property, which has its origins in mathematical economy (Dorfman et al. 1987; McKenzie 1986), describes the

L. Grüne (✉) · S. Pirkelmann · M. Stieler
Chair of Applied Mathematics, Mathematical Institute, University of Bayreuth, Bayreuth, Germany
e-mail: lars.gruene@uni-bayreuth.de; simon.pirkelmann@uni-bayreuth.de; marleen.stieler@uni-bayreuth.de

G. Feichtinger et al. (eds.), *Control Systems and Mathematical Methods in Economics*, Lecture Notes in Economics and Mathematical Systems 687,
https://doi.org/10.1007/978-3-319-75169-6_10

fact that an optimal trajectory on a finite time horizon stays close to an optimal equilibrium most of the time. In order to compute an (approximately) optimal trajectory, it thus suffices to compute the optimal equilibrium as well as optimal paths to and from the optimal equilibrium. For the infinite horizon problem, the turnpike property demands that the optimal trajectory converges to the optimal equilibrium. Under suitable conditions, the finite horizon turnpike property holds if and only if the infinite horizon turnpike property holds (Grüne et al. 2017).

In the time-varying setting of this paper, the optimal equilibrium is replaced by a time-varying infinitely long trajectory, at which the system is operated optimally in an overtaking sense. Since this trajectory is very difficult to compute compared to the time-invariant setting the situation reverses: instead of using the turnpike property and the knowledge about the optimal equilibrium for the approximation of finite horizon optimal trajectories, now we may use finite horizon optimal trajectories (which can be efficiently computed numerically if the horizon is not too long) and the turnpike property in order to approximate the infinite-horizon optimal trajectory. This can be done directly by numerically computing optimal trajectories on finite horizons with increasing length, or indirectly via a receding horizon or model predictive control (MPC) approach, see Remark 1 and Grüne and Pirkelmann (2017). However, in order to decide whether these methods can be employed, we need to find ways to check whether the given optimal control problem exhibits the turnpike property.

In the time-invariant case it is known that there is a strong relation between strict dissipativity in the sense of Willems (1972) and the turnpike property, see Grüne and Müller (2016). The main result in this paper shows that under suitable conditions a time-varying version of strict dissipativity implies the time-varying turnpike property. Moreover, we show that together with a local controllability assumption this property also implies a continuity property for the optimal value function which is useful for the analysis of MPC schemes. We finally discuss the relation between strict dissipativity and necessary optimality conditions for uniformly convex problems and illustrate our results by a simple yet nontrivial example.

2 Problem Statement and Definitions

2.1 Setting

Consider the following time-varying control system

$$x(k+1) = f(k, x(k), u(k)), \quad x(k_0) = x_0, \tag{1}$$

with $f : \mathbb{N}_0 \times X \times U \to X$ and normed spaces X and U. In this setting $k \in \mathbb{N}_0$ represents a time instant, $x(k) \in X$ is the state of the system at that time and $u(k) \in$

U is the control applied to the system during the next sampling interval. For a given initial state $x_0 \in X$ at initial time k_0 and a control sequence $u \in U^N$ of length $N \in \mathbb{N}$ we denote the state trajectory which results from iteratively applying (1) by $x_u(\cdot; k_0, x_0)$. To shorten the notation we may omit the initial time when it is clear from the context and write $x_u(\cdot, x_0)$ instead.

We define $\mathbb{X}(k) \subseteq X$ to be the sets of admissible states at time k and $\mathbb{U}(k, x) \subseteq U$ as the set of admissible control values for $x \in \mathbb{X}(k)$. We denote by $\mathbb{U}^N(k, x)$ the set of admissible control sequences for initial state $x \in \mathbb{X}(k)$, i.e. control sequences $u \in U^N$ that satisfy

$$u(j) \in \mathbb{U}(k + j, x_u(j; k, x)) \quad \text{and} \quad x_u(j + 1; k, x) \in \mathbb{X}(k + j + 1)$$

for all $j \in \{0, \dots, N - 1\}$ and similarly $\mathbb{U}^\infty(k, x)$ as the set of control sequences $u \in U^\infty$ satisfying

$$u(j) \in \mathbb{U}(k + j, x_u(j; k, x)) \quad \text{and} \quad x_u(j + 1; k, x) \in \mathbb{X}(k + j + 1)$$

for all $j \in \mathbb{N}_0$.

The goal in our setting is to investigate the structure and properties of solutions to the infinite-horizon optimal control problem

$$\underset{u \in \mathbb{U}^\infty(k_0, x_0)}{\text{minimize}} \underbrace{\sum_{j=0}^{\infty} \ell(k_0 + j, x_u(j; k_0, x_0), u(j))}_{=: J_\infty(k_0, x_0, u)}, \tag{2}$$

where $\ell : \mathbb{N}_0 \times X \times U \to \mathbb{R}$ is the stage cost function.

2.2 *Overtaking Optimality*

The objective function in (2) will not necessarily assume a finite value for all control sequences $u \in U^\infty$. In particular, it may happen that $J_\infty(k, x, u) = -\infty$ for several control sequences $u \in U^\infty$, i.e. we do not get a unique minimal value which means it is not obvious how to decide which control sequence performs best. Similarly, it may happen that $J_\infty(k, x, u) = \infty$ for all control sequences in which case the usual definition of optimality also breaks down. To deal with this issue we use the concept of *overtaking optimality*[1] which was first introduced by Gale (1967).

Definition 1 (Overtaking Optimality) Let $x \in \mathbb{X}(k)$ and consider a control sequence $u^* \in \mathbb{U}^\infty(k, x)$ with corresponding trajectory $x_{u^*}(\cdot; k, x)$. The pair

[1]In particular in the economic literature, this property is also referred to as *catching up optimality*, see e.g. Bewley (2009).

(x_{u^*}, u^*) is called overtaking optimal if

$$\liminf_{K\to\infty} \sum_{j=0}^{K-1} \ell(k+j, x_u(j;k,x), u(j)) - \ell(k+j, x_{u^*}(j), u^*(j)) \geq 0 \tag{3}$$

for all $u \in \mathbb{U}^\infty(k,x)$.

Using the above definition we can handle the case of infinite values of $J_\infty(k,x,u)$. The definition considers a trajectory pair (x_{u^*}, u^*) as optimal if in the limit inferior its cost is overtaken by the cost of any other trajectory. The following definition characterizes for which trajectory the system yields optimal performance, where optimality is now thought of in the sense of Definition 1. Note that both definitions just differ in the fact, that in the second one the initial value is no longer fixed.

Definition 2 (Optimal Operation) Let $x \in \mathbb{X}(k)$ and consider a control sequence $u^* \in \mathbb{U}^\infty(k,x)$ with corresponding state trajectory $x^* = x_{u^*}(\cdot;k,x)$. We say the system (1) is optimally operated at (x^*, u^*) if

$$\liminf_{K\to\infty} \sum_{j=0}^{K-1} \ell(k+j, x_u(j;k,x'), u(j)) - \ell(k+j, x^*(j), u^*(j)) \geq 0 \tag{4}$$

for all $x' \in \mathbb{X}(k)$ and $u \in \mathbb{U}^\infty(k,x')$.

To better understand the difference between both definitions it is insightful to consider the second definition from a viewpoint of a time-invariant setting where there exists an optimal equilibrium at which the system performs best. In our setting the optimal equilibrium corresponds to a more general time-varying pair (x^*, u^*) that is defined in Definition 2, whereas the first definition formally introduces the optimality notion we are using.

In the subsequent sections we will always assume that a trajectory pair (x^*, u^*) at which the system is optimally operated exists.

We also consider the finite horizon optimal control problem

$$\underset{u\in\mathbb{U}^N(k_0,x_0)}{\text{minimize}} \underbrace{\sum_{j=0}^{N-1} \ell(k_0+j, x_u(j;k_0,x_0), u(j))}_{=:J_N(k_0,x_0,u)}, \tag{5}$$

where again $\ell : \mathbb{N}_0 \times X \times U \to \mathbb{R}$ is the stage cost function. For this problem the optimal value is always finite and we can use the 'usual' definition of optimality. In this case we denote the optimal control sequence by $u_N^* \in U^N$ and the corresponding state trajectory by $x_{u_N^*}(\cdot;\cdot,x)$ for initial value $x \in X$, or $u_{N,x}^* \in U^N$ and $x_{u_{N,x}^*}(\cdot;\cdot,x)$ if we want to emphasize the dependence on the initial value.

3 Definitions of Turnpike and Continuity Property

We will consider two different versions of the turnpike property, one for the finite and one for the infinite-horizon optimal control problem. In order to be able to treat both in a unified way without having to distinguish between the optimality notions on finite or infinite horizon we introduce a shifted cost functional, which always has a finite value along the optimal trajectory.

Definition 3 (Shifted Stage Cost) We define the shifted stage cost $\hat{\ell} : \mathbb{N}_0 \times X \times U \to \mathbb{R}$ as

$$\hat{\ell}(k, x, u) := \ell(k, x, u) - \ell(k, x^*(k), u^*(k))$$

and the shifted cost functional as

$$\hat{J}_N(k, x, u) := \sum_{j=0}^{N-1} \hat{\ell}(k + j, x_u(j; k, x), u(j))$$

for $N \in \mathbb{N} \cup \{\infty\}$. The corresponding optimal value function is given by

$$\begin{aligned} \hat{V}_N(k, x) &:= \inf_{u \in \mathbb{U}^N(k,x)} \hat{J}_N(k, x, u) = \inf_{u \in \mathbb{U}^N(k,x)} J_N(k, x, u) - J_N^*(k) \\ &= V_N(k, x) - J_N^*(k), \end{aligned}$$

with $J_N^*(k) := \sum_{j=k}^{k+N-1} \ell(j, x^*(j), u^*(j))$.

In the following we will write

$$|(x, u)|_{(\bar{x},\bar{u})} := \|x - \bar{x}\| + \|u - \bar{u}\|$$

to shorten the notation, using the norms on the spaces X and U. We use the following comparison functions commonly encountered in control literature. A good overview of properties of comparison functions can be found in Kellett (2014).

Definition 4 (Comparison Functions)

$$\begin{aligned} \mathcal{K}_\infty := \{\alpha : \mathbb{R}_0^+ \to \mathbb{R}_0^+ \mid \alpha \text{ is continuous, strictly increasing} \\ \text{and unbounded with } \alpha(0) = 0\} \end{aligned}$$

$$\mathcal{L} := \{\sigma : \mathbb{R}_0^+ \to \mathbb{R}_0^+ | \sigma \text{ is continuous and strictly decreasing with } \lim_{s\to\infty} \sigma(s) = 0\}$$

Furthermore, let $\#D$ denote the cardinality of a set D, i.e. the number of elements contained in the set. With these definitions we are now able to define the turnpike property on finite and infinite time horizons.

Definition 5 (Time-Varying Turnpike Property) Consider a pair (x^*, u^*) at which the system (1) is optimally operated.

(a) The optimal control problem on infinite horizon with shifted stage cost $\hat{\ell}$ has the time-varying turnpike property at (x^*, u^*) if the following holds: There exists $\rho \in \mathcal{L}$ such that for each $k \in \mathbb{N}_0$, each optimal trajectory $x_{u^*_\infty}(\cdot, x)$, $x \in \mathbb{X}(k)$ and all $P \in \mathbb{N}$ there is a set $\mathcal{Q}(k, x, P, \infty) \subseteq \mathbb{N}_0$ with $\#\mathcal{Q}(k, x, P, \infty) \leq P$ and

$$|(x_{u^*_\infty}(j; k, x), u^*_\infty(j))|_{(x^*(k+j), u^*(k+j))} \leq \rho(P)$$

for all $j \in \mathbb{N}_0$ with $j \notin \mathcal{Q}(k, x, P, \infty)$.

(b) The optimal control problem on finite horizon has the time-varying turnpike property at (x^*, u^*) if the following holds: There exists $\sigma \in \mathcal{L}$ such that for each $k \in \mathbb{N}_0$, each optimal trajectory $x_{u^*_N}(\cdot, x)$, $x \in \mathbb{X}(k)$ and all $N, P \in \mathbb{N}$ there is a set $\mathcal{Q}(k, x, P, N) \subseteq \{0, \ldots, N\}$ with $\#\mathcal{Q}(k, x, P, N) \leq P$ and

$$|(x_{u^*_N}(j; k, x), u^*_N(j))|_{(x^*(k+j), u^*(k+j))} \leq \sigma(P)$$

for all $j \in \{0, \ldots, N\}$ with $j \notin \mathcal{Q}(k, x, P, N)$.

The turnpike property describes the fact that optimal solutions on the infinite and finite horizon are close to the optimal trajectory of the system most of the time. This is illustrated in Fig. 1 for the finite-horizon case.

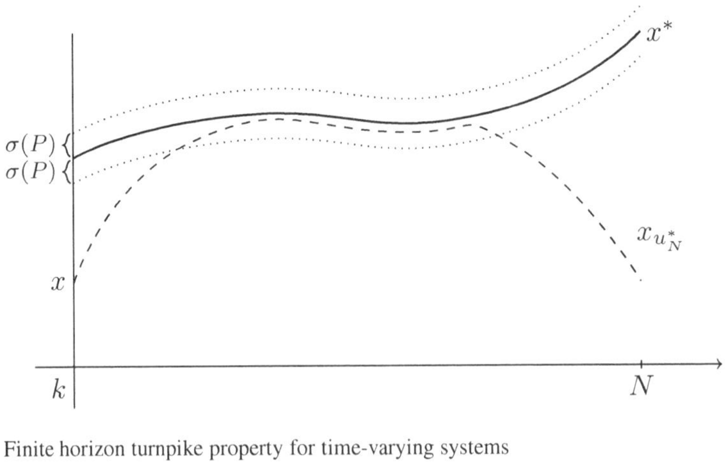

Fig. 1 Finite horizon turnpike property for time-varying systems

Definition 6 (Continuity Property of $\hat{V}_N$ and $\hat{V}_\infty$) The optimal value functions $\hat{V}_N$ and $\hat{V}_\infty$ are (approximately) continuous at x^* if for each $k \in \mathbb{N}_0$ there is an open ball $\mathcal{B}_\varepsilon(x^*(k))$, $\varepsilon > 0$, around $x^*(k)$ and a function $\gamma_V : \mathbb{R}_0^+ \times \mathbb{R}_0^+ \to \mathbb{R}_0^+$ with $\gamma_V(N, r) \to 0$ if $N \to \infty$ and $r \to 0$, and $\gamma_V(\cdot, r)$, $\gamma_V(N, \cdot)$ monotonous for fixed r and N, such that for all $x \in \mathcal{B}_\varepsilon(x^*(k)) \cap \mathbb{X}(k)$ and all $N \in \mathbb{N} \cup \{\infty\}$ the inequality

$$|\hat{V}_N(k, x) - \hat{V}_N(k, x^*(k))| \leq \gamma_V(N, \|x - x^*(k)\|)$$

holds, where we make the assumption that $\gamma_V(\infty, r) =: \omega_V(r)$ with $\omega_V \in \mathcal{K}_\infty$.

Remark 1 As mentioned in the introduction, the turnpike property is not only an interesting phenomenon in general system theory and allows to relate finite and infinite horizon optimal trajectories, but also plays an important role in the context of model predictive control (MPC). In this control method, a control input is synthesized by iteratively solving finite horizon optimal control problems and concatenating the initial pieces of the resulting optimal trajectories. In particular, the turnpike property guarantees that the optimal open-loop trajectories, which are calculated in the MPC iterations, are close to the infinite horizon optimal trajectory for a certain number of time steps. Together with continuity of the optimal value function, this allows for the construction of a Lyapunov function as well as convergence and performance estimates for time-invariant MPC, see Grüne (2013) and Grüne and Stieler (2014), and for performance estimates of the MPC closed-loop solution in the time-varying setting, see Grüne and Pirkelmann (2017).

4 From Dissipativity to Turnpike

While the turnpike and continuity properties are handy tools to use in the construction of approximately optimal trajectories and for the analysis of MPC schemes, they are in general difficult to verify directly. As an alternative we consider the concept of dissipativity,[2] which is a property of the system that can be checked more easily. Our goal in this section is to prove that the turnpike and continuity properties are satisfied if we assume that the system is (strictly) dissipative as follows.

Definition 7 (Dissipativity) The system (1) is dissipative with respect to the supply rate $s : \mathbb{N}_0 \times X \times U$ if there exists a storage function $\lambda : \mathbb{N}_0 \times X \to \mathbb{R}$ bounded from below on X such that for all $k \in \mathbb{N}_0$ and all $(x, u) \in \mathbb{X}(k) \times \mathbb{U}(k, x)$ the following holds:

$$\lambda(k+1, f(k, x, u)) - \lambda(k, x) \leq s(k, x, u). \tag{6}$$

[2] Introduced in the context of control systems by Jan Willems in 1972, see Willems (1972).

The system (1) is strictly dissipative with respect to the supply rate $s : \mathbb{N}_0 \times X \times U$ and the optimal trajectory (x^*, u^*), if there exists $\alpha \in \mathcal{K}_\infty$ such that

$$\lambda(k+1, f(k, x, u)) - \lambda(k, x) \leq s(k, x, u) - \alpha(|(x, u)|_{(x^*(k), u^*(k))}) \tag{7}$$

holds for all $k \in \mathbb{N}_0$ and all $(x, u) \in \mathbb{X}(k) \times \mathbb{U}(k, x)$.

In the sequel we will assume that the system (1) is strictly dissipative with respect to the supply rate $s(k, x, u) = \hat{\ell}(k, x, u) = \ell(k, x, u) - \ell(k, x^*(k), u^*(k))$. We further assume that the optimal trajectory x^* from Definition 2 is cheaply reachable, which expresses that it can be reached from any initial state with bounded cost. Since the shifted cost along x^* is zero, this can be expressed via a bound on the shifted optimal value functions.

Assumption 1 (Cheap Reachability) The trajectory pair (x^*, u^*) is called *cheaply reachable* if there exists $E \in \mathbb{R}$ such that for each $k \in \mathbb{N}_0$ and for all $x \in \mathbb{X}(k)$, $N \in \mathbb{N} \cup \{\infty\}$ the inequality

$$\hat{V}_N(k, x) \leq E \tag{8}$$

holds.

Using dissipativity and cheap reachability it can be shown that both the finite and infinite optimal control problems have the turnpike property from Sect. 3.

Theorem 1 (Strict Dissipativity and Cheap Reachability Imply Turnpike) *Let (x^*, u^*) be an optimal pair. If the optimal control problem is strictly dissipative wrt the supply rate $s(k, x, u) = \hat{\ell}(k, x, u) = \ell(k, x, u) - \ell(k, x^*(k), u^*(k))$ with bounded storage function λ for the trajectory pair (x^*, u^*) and (x^*, u^*) is cheaply reachable, then the turnpike property from Definition 5 holds.*

Proof We first prove the finite-horizon turnpike property from Definition 5(b). Let $k \in \mathbb{N}_0$, $x \in \mathbb{X}(k)$ and consider a control sequence $u \in \mathbb{U}(k, x)$ with corresponding state trajectory $x_u(\cdot; k, x)$. From strict dissipativity we have

$$\begin{aligned}\hat{\ell}(k+j, x_u(j; k, x), u(j)) \geq{}& \lambda(k+j+1, f(k+j, x_u(j; k, x), u(j))) \\ &- \lambda(k+j, x_u(j)) + \alpha(|(x_u(j; k, x), u(j))|_{(x^*(j), u^*(j))})\end{aligned}$$

for all $j \in \mathbb{N}_0$. This yields

$$\begin{aligned}\hat{J}_N(k, x, u) &= \sum_{j=0}^{N-1} \hat{\ell}(k+j, x_u(j; k, x), u(j)) \\ &\geq \lambda(k+N, f(k+N-1, x_u(N-1; k, x), u(N-1))) - \lambda(k, x_u(k; k, x)) \\ &+ \sum_{j=0}^{N-1} \alpha(|(x_u(j; k, x), u(j))|_{(x^*(j), u^*(j))}).\end{aligned} \tag{9}$$

We prove the finite-horizon turnpike property by contradiction. Suppose the turnpike property does not hold for

$$\sigma(P) := \alpha^{-1}\left(\frac{2M_\lambda + E}{P}\right),$$

in which $M_\lambda > 0$ is a bound on $|\lambda|$ and with E from Assumption 1. This means that there are $N \in \mathbb{N}$, $x \in \mathbb{X}(k)$ and $P \in \mathbb{N}$ such that the number of elements $j \in \mathcal{Q}(k, x, P, N)$, i.e. those elements for which $|(x_{u_N^*}(j; k, x), u_N^*(j))|_{(x^*(j), u^*(j))} > \sigma(P)$ is larger than P. Using (9) with the optimal control sequence $u = u_N^*$ and taking only those elements in the sum into account for which $|(x_{u_N^*}(j; k, x), u_N^*(j))|_{(x^*(j), u^*(j))} > \sigma(P)$ holds (the other summands are lower-bounded by zero), this implies

$$\hat{V}_N(k, x) = \hat{J}_N(k, x, u_N^*) > -2M_\lambda + P\alpha(\sigma(P)) = -2M_\lambda + 2M_\lambda + E = E.$$

However, this contradicts Assumption 1.

The proof for the infinite horizon follows analogously with

$$\rho(P) := \alpha^{-1}\left(\frac{2M_\lambda + E}{P}\right).$$

□

To show that not only the turnpike property but also continuity of the optimal value function holds, we need some additional assumptions, first of all local controllability near the optimal trajectory of the system.

Assumption 2 (Local Controllability) Assume that the system is locally controllable along the trajectory pair (x^*, u^*), i.e. there exists a time $d \in \mathbb{N}$, $\delta_c > 0$, and $\gamma_x, \gamma_u, \gamma_c \in \mathcal{K}_\infty$ such that for each $k \in \mathbb{N}_0$ and for any two points $x \in \mathcal{B}_{\delta_c}(x^*(k))$, $y \in \mathcal{B}_{\delta_c}(x^*(k + d))$ there exists $u \in \mathbb{U}^d(x)$ satisfying $x_u(d, x) = y$ and the estimates $\|x_u(j; k, x) - x^*(k + j)\| \leq \gamma_x(\delta)$, $\|u(j) - u^*(k + j)\| \leq \gamma_u(\delta)$ and $|\hat{\ell}(j + k, x_u(j; k, x), u(j))| \leq \gamma_c(\delta)$ for all $j = 0, \ldots, d - 1$, where $\delta := \max\{\|x - x^*(k)\|, \|y - x^*(k + d)\|\}$.

Clearly, local controllability means that any two points within a tube along the optimal trajectory can be connected in forward time as illustrated by Fig. 2. The following definition is closely related to strict dissipativity. The cost function $\tilde{\ell}$ defined therein is sometimes also called *rotated* stage cost.

Definition 8 (Modified Stage Cost) Consider the modified stage cost $\tilde{\ell} : \mathbb{N}_0 \times X \times U \to \mathbb{R}_{\geq 0}$ defined by:

$$\tilde{\ell}(k, x, u) := \hat{\ell}(k, x, u) + \lambda(k, x) - \lambda(k + 1, f(k, x, u))$$

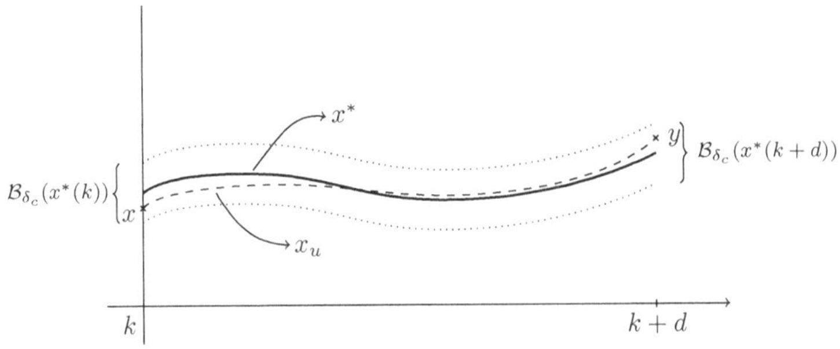

Fig. 2 Local controllability along the optimal trajectory

using the storage function λ from the assumed strict dissipativity of the system. We also define the modified cost functional by

$$\tilde{J}_N(k, x, u) := \sum_{j=0}^{N-1} \tilde{\ell}(k + j, x_u(j; k, x), u(j)). \tag{10}$$

In the previous definition we could also have written the supply rate s instead of the shifted stage cost function $\hat{\ell}$ since by our assumptions the two functions coincide.

Note that the modified stage cost is bounded from below by a function $\alpha_l \in \mathcal{K}_\infty$, i.e.

$$\tilde{\ell}(k, x, u) \geq \alpha_l(|(x, u)|_{(x^*(k), u^*(k))}) \tag{11}$$

holds for all $(x, u) \in \mathbb{X}(k) \times \mathbb{U}(k, x)$. This is immediately concluded from strict dissipativity of the system, with $\alpha_l := \alpha$. One easily sees that for the modified cost functional the following identity holds:

$$\tilde{J}_N(k, x, u) = \hat{J}_N(k, x, u) + \lambda(k, x) - \lambda(k + N, x_u(N; k, x)). \tag{12}$$

Assumption 3 There exists an upper bound $\alpha_u \in \mathcal{K}_\infty$ such that the modified stage cost from Definition 8 satisfies the inequality

$$\tilde{\ell}(k, x, u) \leq \alpha_u(\|(x, u)\|_{(x^*(k), u^*(k))}) \tag{13}$$

for all $(x, u) \in \mathbb{X}(k) \times \mathbb{U}(k, x)$.

We point out that the inequalities (11) and (13) imply that $\tilde{\ell}(k, x^*(k), u^*(k)) = 0$ for each $k \in \mathbb{N}_0$. The following preliminary result shows that an optimal trajectory starting in a neighbourhood of the optimal pair (x^*, u^*) will stay near the optimal pair for some time.

Lemma 1 *Suppose that the system* (1) *is strictly dissipative and that Assumptions 1, 2 and 3 hold. Then there exist* $N_1 > 0$, $R \geq N/2$ *and* $\eta : \mathbb{N} \times \mathbb{R}_0^+ \to \mathbb{R}_0^+$ *with* $\eta(N, r) \to 0$ *if* $N \to \infty$ *and* $r \to 0$, *such that for each* $k > 0$ *the open loop optimal trajectories with horizon* $N \geq N_1$ *starting in* $x_1 \in \mathcal{B}_{\delta_c}(x^*(k))$ *satisfy*

$$|(x_{u^*_{N,x_1}}(j; k, x_1), u^*_{N,x_1}(j))|_{(x^*(k+j), u^*(k+j))} \leq \eta(N, \|x_1 - x^*(k)\|)$$

for all $j \in \{0, \ldots, R\}$ *and* δ_c *from Assumption 2.*

Proof [3] Let $k \in \mathbb{N}_0$. Choose an arbitrary $x_1 \in \mathcal{B}_{\delta_c}(x^*(k))$. By Theorem 1 we know that for the optimal trajectory $x_{u^*_{N,x_1}}(\cdot; k, x_1)$ the finite horizon turnpike property holds. This means we can choose $0 < \varepsilon \leq \delta_c$ and N, $P \leq N - 2d$ (d from Assumption 2), such that there are at least $N - P \geq 2d$ time instants $j \in \{0, \ldots, N\}$ at which

$$|(x_{u^*_{N,x_1}}(j; k, x_1), u^*_{N,x_1}(j))|_{(x^*(k+j), u^*(k+j))} \leq \sigma(P) \leq \varepsilon$$

holds. In particular, for those time instants we also have

$$\|x_{u^*_{N,x_1}}(j; k, x_1) - x^*(k + j)\| \leq \varepsilon \leq \delta_c.$$

Let R denote the largest such time index and note that $R \geq N - P \geq 2d$. We now construct a control sequence $\bar{u} \in \mathbb{U}^N$ as follows: By applying Assumption 2 with $x = x_1$, $y = x^*(k + d)$ we know that there exists a control sequence $u_1 \in \mathbb{U}^d$ with $x_{u_1}(d; k, x_1) = x^*(k + d)$. We define $\bar{u}(j) = u_1(j)$ for $j \in \{0, \ldots, d - 1\}$. For $j \in \{d, \ldots, R - d - 1\}$ we choose $\bar{u}(j) = u^*(k + j)$, and thus get $x_{\bar{u}}(R - d) = x^*(k + R - d)$. Using Assumption 2 again, this time with $x = x^*(k + R - d) \in \mathcal{B}_{\delta_c}(x^*(k + R - d))$ and $y = x_{u^*_{N,x_1}}(R, x_1) \in \mathcal{B}_{\delta_c}(x^*(k + R))$, we obtain the control sequence $u_2 \in \mathbb{U}^d$. We finish by defining $\bar{u}(j) = u_2(j - R + d)$ for $j \in \{R - d, \ldots, R - 1\}$ and $\bar{u}(j) = u^*_{N,x_1}(j)$ for $j \in \{R, \ldots, N - 1\}$.

Observe that by construction the trajectories $x_{\bar{u}}(j; k, x_1)$ and $x_{u^*_{N,x_1}}(j; k, x_1)$ coincide for $j \in \{R, \ldots, N\}$. Due to the optimality principle, and because $x_{u^*_{N,x_1}}(j; k, x_1)$ is the tail of an optimal trajectory for $j \in \{R, \ldots, N\}$, the initial pieces of the control sequences u^*_{N,x_1} and $\bar{u}$ up to time $R - 1$ satisfy

$$J_R(k, x_1, u^*_{N,x_1}) \leq J_R(k, x_1, \bar{u})$$

as well as

$$\hat{J}_R(k, x_1, u^*_{N,x_1}) \leq \hat{J}_R(k, x_1, \bar{u}). \tag{14}$$

[3]The proof uses a construction similar to the one of Lemma 6.3 in Grüne (2013).

Now consider the modified cost functionals $\tilde{J}_R$. From (12) with $N = R$ and the fact that $x_{\bar{u}}(R, x_1) = x_{u^*_{N,x_1}}(R, x_1)$ it follows that

$$\begin{aligned}\tilde{J}_R(k, x_1, u^*_{N,x_1}) &= \hat{J}_R(k, x_1, u^*_{N,x_1}) + \lambda(k, x_1) - \lambda(k + R, x_{u^*_{N,x_1}}(R; k, x_1)) \\ &\overset{(14)}{\leq} \hat{J}_R(k, x_1, \bar{u}) + \lambda(k, x_1) - \lambda(k + R, x_{u^*_{N,x_1}}(R; k, x_1)) \\ &= \hat{J}_R(k, x_1, \bar{u}) + \lambda(k, x_1) - \lambda(k + R, x_{\bar{u}}(R; k, x_1)) = \tilde{J}_R(k, x_1, \bar{u}).\end{aligned} \tag{15}$$

We abbreviate $r := \|x_1 - x^*(k)\|$. From the construction of $\bar{u}$ we know that

$$\|x_{\bar{u}}(j; k, x_1) - x^*(k + j)\| \leq \gamma_x(r) \text{ and } \|\bar{u}(j) - u^*(k + j)\| \leq \gamma_u(r)$$

for $j = \{0, \ldots, d - 1\}$, and similarly $\|x_{\bar{u}}(j; k, x_1) - x^*(k + j)\| \leq \gamma_x(\varepsilon)$ as well as $\|\bar{u}(j) - u^*(k + j)\| \leq \gamma_u(\varepsilon)$ for $j \in \{R - d, \ldots, R - 1\}$. In addition, we have $x_{\bar{u}}(j; k, x_1) = x^*(k + j)$ and $\bar{u}(j) = u^*(k + j)$ for $j \in \{d, \ldots, R - d - 1\}$. Recalling that the modified stage cost satisfies $\tilde{\ell}(k, x^*(k), u^*(k)) = 0$ and using Assumption 3, we thus get the following estimate for the modified cost functional with the control sequence $\bar{u}$:

$$\begin{aligned}\tilde{J}_R(k, x_1, \bar{u}) &= \sum_{j=0}^{R-1} \tilde{\ell}(k + j, x_{\bar{u}}(j; k, x_1), \bar{u}(j)) \\ &= \sum_{j=0}^{d-1} \underbrace{\tilde{\ell}(k+j, x_{\bar{u}}(j; k, x_1), \bar{u}(j))}_{\leq \alpha_u(|(x_{\bar{u}}(j;k,x_1),\bar{u}(j))|_{(x^*(k+j),u^*(k+j))})} + \underbrace{\sum_{j=d}^{R-d-1} \tilde{\ell}(k+j, x_{\bar{u}}(j; k, x_1), \bar{u}(j))}_{=0} \\ &\quad + \sum_{j=R-d}^{R-1} \underbrace{\tilde{\ell}(k + j, x_{\bar{u}}(j; k, x_1), \bar{u}(j))}_{\leq \alpha_u(|(x_{\bar{u}}(j;k,x_1),\bar{u}(j))|_{(x^*(k+j),u^*(k+j))})} \\ &\leq \sum_{j=0}^{d-1} \alpha_u(\underbrace{|(x_{\bar{u}}(j; k, x_1), \bar{u}(j))|_{(x^*(k+j),u^*(k+j))}}_{\leq \gamma_x(r)+\gamma_u(r)}) \\ &\quad + \sum_{j=R-d}^{R-1} \alpha_u(\underbrace{|(x_{\bar{u}}(j; k, x_1), \bar{u}(j))|_{(x^*(k+j),u^*(k+j))}}_{\leq \gamma_x(\varepsilon)+\gamma_u(\varepsilon)}) \\ &\leq d\alpha_u(\gamma_x(r) + \gamma_u(r)) + d\alpha_u(\gamma_x(\varepsilon) + \gamma_u(\varepsilon))\end{aligned} \tag{16}$$

Now assume that $|(x_{u^*_{N,x_1}}(\tilde{j}; k, x_1), u^*_{N,x_1}(\tilde{j}))|_{(x^*(k+\tilde{j}),u^*(k+\tilde{j}))} \geq \Delta$ holds for some $\tilde{j} \in \{0, \ldots, R - 1\}$ and $\Delta > \alpha_l^{-1}(d\alpha_u(\gamma_x(r) + \gamma_u(r)) + d\alpha_u(\gamma_x(\varepsilon) + \gamma_u(\varepsilon)))$. By summing up to time R the modified stage cost for the control sequence u^*_{N,x_1} and using (11) and (16) we get the estimate

$$\begin{aligned}
\tilde{J}_R(k, x_1, u^*_{N,x_1}) &= \sum_{j=0}^{R-1} \tilde{\ell}(k+j, x_{u^*_{N,x_1}}(j; k, x_1), u^*_{N,x_1}(j)) \\
&\overset{(11)}{\geq} \sum_{j=0}^{R-1} \alpha_l(|(x_{u^*_{N,x_1}}(j; k, x_1), u^*_{N,x_1}(j))|_{(x^*(k+j), u^*(k+j))}) \\
&\geq \underbrace{\alpha_l(|(x_{u^*_{N,x_1}}(\tilde{j}; k, x_1), u^*_{N,x_1}(\tilde{j}))|_{(x^*(k+\tilde{j}), u^*(k+\tilde{j}))})}_{>\Delta} \\
&> d\alpha_u(\gamma_x(r) + \gamma_u(r)) + d\alpha_u(\gamma_x(\varepsilon) + \gamma_u(\varepsilon)) \overset{(16)}{\geq} \tilde{J}_R(k, x_1, \bar{u}).
\end{aligned}$$

But this contradicts (15) and thus we get $\Delta \leq \alpha_l^{-1}(d\alpha_u(\gamma_x(r) + \gamma_u(r)) + d\alpha_u(\gamma_x(\varepsilon) + \gamma_u(\varepsilon)))$. Finally, choose $\varepsilon = \sigma(\frac{N}{2})$, which satisfies $\varepsilon \to 0$ for $N \to \infty$, and define $\eta(N, r) := \alpha_l^{-1}(d\alpha_u(\gamma_x(r) + \gamma_u(r)) + d\alpha_u(\gamma_x(\varepsilon) + \gamma_u(\varepsilon)))$. By choice of R we know that $R \geq N - P$, which for $P = \frac{N}{2}$ yields the assertion, i.e. $R \geq \frac{N}{2}$. It remains to ensure that $N - P = \frac{N}{2} \geq 2d$ as well as $\varepsilon \leq \delta_c$, which can be achieved by setting $N_1 \geq \max\{4d, 2\sigma^{-1}(\delta_c)\}$. □

As a final assumption in order to prove continuity of the optimal value function we require the stage cost to be continuous.

Assumption 4 (Continuity of the Stage Cost) We assume that the function ℓ is continuous in the sense that there exists $\eta_\ell \in \mathscr{K}_\infty$ such that for each $k \in \mathbb{N}_0$ and each compact set $\mathbb{Y} \subseteq \mathbb{X}(k) \times \mathbb{U}(k)$ the inequality

$$|\ell(k, x, u) - \ell(k, x', u')| \leq \eta_\ell(|(x, u)|_{(x', u')}) \tag{17}$$

holds for all $(x, u), (x', u') \in \mathbb{Y}$.

Theorem 2 (Continuity Property of the Optimal Value Function) *If the optimal control problem (2) is strictly dissipative and Assumptions 1–4 are satisfied, then the optimal value function is (approximately) continuous in the sense of Definition 6.*

Proof[4] Let $k \geq 0$ and pick $\delta \in (0, \delta_c]$ with δ_c from Assumption 2. To shorten the notation we write $x_1 = x^*(k)$ and choose $x_2 \in \mathscr{B}_\delta(x_1) \cap \mathbb{X}(k)$. We denote the optimal control sequence for N steps starting in x_1 by u^*_{N,x_1}, and the one starting in

[4]The idea is similar to the proof of Theorem 16 in Müller and Grüne (2016).

x_2 by u^*_{N,x_2}. According to Lemma 1 we can choose $N \geq N_1$ sufficiently large such that both

$$|(x_{u^*_{N,x_1}}(j;k,x_1), u^*_{N,x_1}(j))|_{(x^*(k+j),u^*(k+j))} \leq \eta(N, \|x_1 - x^*(k)\|) \leq \eta(N,\delta) \leq \delta_c$$

and

$$|(x_{u^*_{N,x_2}}(j;k,x_2), u^*_{N,x_2}(j))|_{(x^*(k+j),u^*(k+j))} \leq \eta(N, \|x_2 - x^*(k)\|) \leq \eta(N,\delta) \leq \delta_c$$

hold for all $j \in \{0, \ldots, R\}$. From the proof of Lemma 1 we also know that $R \geq 2d > d$.

Define $\varepsilon := \eta(N,\delta)$, $\hat{\delta} := \max\{\delta, \varepsilon\}$ and let $x_3 := x_{u^*_{N,x_1}}(d;k,x_1)$. Because of Assumption 4 we know that

$$\begin{aligned}|\ell(k+j, x_{u^*_{N,x_1}}(j;k,x_1), u^*_{N,x_1}(j)) - \ell(k+j, x^*(k+j), u^*(k+j))| \\ \leq \eta_\ell(|(x_{u^*_{N,x_1}}(j;k,x_1), u^*_{N,x_1}(j))|_{(x^*(k+j),u^*(k+j))}) \leq \eta_\ell(\varepsilon).\end{aligned}$$

This leads to the estimate

$$\sum_{j=0}^{d-1} \underbrace{\ell(k+j, x_{u^*_{N,x_1}}(j;k,x_1), u^*_{N,x_1}(j))}_{\geq \ell(k+j,x^*(k+j),u^*(k+j))-\eta_\ell(\varepsilon)} \geq J_d^*(k) - d\eta_\ell(\varepsilon).$$

Furthermore, we can apply Assumption 2 with $x = x_2$, $y = x_3$ to conclude that there exists a control sequence $u_1 \in \mathbb{U}^d$ such that $x_{u_1}(d, x_2) = x_3$ and the estimate

$$\begin{aligned}|\ell(k+j, x_{u_1}(j,x_2), u_1(j)) - \ell(k+j, x^*(k+j), u^*(k+j))| \\ \leq \gamma_c(\max\{\|x_2 - x^*(k)\|, \|x_3 - x^*(k+d)\|\}) \leq \gamma_c(\hat{\delta})\end{aligned}$$

holds for all $j \in \{0, \ldots, d-1\}$. This yields

$$\sum_{j=0}^{d-1} \underbrace{\ell(k+j, x_{u_1}(j;k,x_2), u_1(j))}_{\leq \ell(k+j,x^*(k+j),u^*(k+j))+\gamma_c(\hat{\delta})} \leq J_d^*(k) + d\gamma_c(\hat{\delta}).$$

Now define a control sequence $\bar{u} \in \mathbb{U}^N$ by $\bar{u}(j) = u_1(j)$ for $j \in \{0, \ldots, d-1\}$ and $\bar{u}(j) = u^*_{N,x_1}(j)$ for $j \in \{d, \ldots, N-1\}$ and note that by construction of $\bar{u}$ the trajectories $x_{\bar{u}}(j;k,x_2)$ and $x_{u^*_{N,x_1}}(j;k,x_1)$ coincide for $j \in \{d, \ldots, N\}$. Thus we get

$$
\begin{aligned}
V_N(k, x_2) &\leq J_N(k, x_1, \bar{u}) \\
&= \sum_{j=0}^{d-1} \ell(k+j, x_{\bar{u}}(j; k, x_2), \bar{u}(j)) + \sum_{j=d}^{N-1} \ell(k+j, x_{\bar{u}}(j; k, x_2), \bar{u}(j)) \\
&= \underbrace{\sum_{j=0}^{d-1} \ell(k+j, x_{u_1}(j; k, x_2), u_1(j))}_{\leq J_d^*(k)+d\gamma_c(\hat{\delta})} - \underbrace{\sum_{j=0}^{d-1} \ell(k+j, x_{u^*_{N,x_1}}(j; k, x_1), u^*_{N,x_1}(j))}_{\geq J_d^*(k)-d\eta_\ell(\varepsilon)} \\
&\quad + \sum_{j=0}^{N-1} \ell(k+j, x_{u^*_{N,x_1}}(j; k, x_1), u^*_{N,x_1}(j)) \\
&\leq V_N(k, x_1) + d(\gamma_c(\hat{\delta}) + \eta_\ell(\varepsilon)).
\end{aligned}
$$

Setting $\tilde{\gamma}_V(N, \delta) = d(\gamma_c(\hat{\delta}) + \eta_\ell(\varepsilon)))$ and using the definition of $\hat{V}_N$ then yields

$$
\hat{V}_N(k, x_2) \leq \hat{V}_N(k, x_1) + \tilde{\gamma}_V(N, \delta). \tag{18}
$$

Observe that $\tilde{\gamma}_V \to 0$ if both $N \to \infty$ and $\delta \to 0$. Finally, to get the required monotonicity we define

$$
\gamma_V(N, r) := \sup_{\tilde{N} \geq N, \tilde{\delta} \leq r} \tilde{\gamma}_V(\tilde{N}, \tilde{\delta}),
$$

for which (18) remains true. The converse inequality follows by exchanging the roles of x_1 and x_2 which concludes the proof. □

5 Optimality Conditions Imply Dissipativity

In this section we show how strict dissipativity can be established if optimality conditions for the infinite horizon optimal control problem (2) are satisfied. The proof extends those for discounted and non-discounted time-invariant optimal control problems, see Grüne et al. (2016) and Damm et al. (2014). The optimality conditions in the literature which most easily lead to the desired result are those derived in Blot and Hayek (2014, Theorem 2.2), which we will hence use in the sequel. However, we believe that using other optimality conditions strict dissipativity can be proved, too. We will elaborate more on this with respect to the results stated in Aseev et al. (2016) at the end of the section.

To be consistent with (Blot and Hayek 2014, Theorem 2.2), let us assume that $X = \mathbb{R}^n$ and $U = \mathbb{R}^m$ and that no constraints are imposed on the state and control variables. We first define the Hamiltonian which is the key ingredient for deriving optimality conditions.

Definition 9 (Hamiltonian) For all times $k \in \mathbb{N}_0$ the *Hamiltonian* $H_k : X \times U \times \mathbb{R}^n \times \mathbb{R} \to \mathbb{R}$ of problem (2) is defined by

$$H_k(x, u, p, \eta) := -\eta \ell(k, x, u) + p^T f(k, x, u).$$

For the readers' convenience we state (Blot and Hayek 2014, Theorem 2.2) in our notation. Note that the sign of ℓ has been changed in the definition above and theorem below because we are considering minimization problems here.

Theorem 3 *Let (x^*, u^*) be an overtaking optimal pair for (2). If it holds:*

1. *For all $k \in \mathbb{N}_0$ the functions $\ell(k, \cdot, \cdot)$ and $f(k, \cdot, \cdot)$ are continuous on a neighborhood of (x^*, u^*) and differentiable at (x^*, u^*).*
2. *For all $k \in \mathbb{N}_0$ the partial differential $\frac{\partial f}{\partial x}(k, x^*(k), u^*(k))$ is invertible.*

Then, there are $\eta_0 \in \mathbb{R}$, and $p_{k+1} \in \mathbb{R}^n$ for all $k \in \mathbb{N}_0$ satisfying the following conditions:

(i) $(\eta_0, p_1) \neq (0, 0)$.
(ii) $\eta_0 \geq 0$.
(iii) *For all $k \in \mathbb{N}_0$ it holds*

$$p_k = p_{k+1}^T \frac{\partial f}{\partial x}(k, x^*(k), u^*(k)) - \eta_0 \frac{\partial \ell}{\partial x}(k, x^*(k), u^*(k)).$$

(iv) *For all $k \in \mathbb{N}_0$ it holds* $\dfrac{\partial H_k}{\partial u}(x^*(k), u^*(k), p_{k+1}, \eta_0) = 0.$

In what follows, structural assumptions on the optimal control problems are imposed.

Assumption 5 We assume that the dynamics $f(k, \cdot, \cdot)$ are affine for each $k \in \mathbb{N}_0$. We also assume that there is $\kappa \in \mathbb{R}_{>0}$ and $F \in \mathcal{K}_\infty$ such that for all $k \in \mathbb{N}_0$ it holds

$$\ell(k, t(x_1, u_1) + (1-t)(x_2, u_2)) \leq t\ell(k, x_1, u_1) + (1-t)\ell(k, x_2, u_2) - \frac{\kappa}{2} t(1-t) F(\|(x_1, u_1) - (x_2, u_2)\|) \tag{19}$$

for all (x_1, u_1), $(x_2, u_2) \in X \times U$ and $t \in [0, 1]$.

Remark 1

1. We call the property introduced in Assumption 5 *uniform strict convexity of ℓ wrt κ and F*. The word uniform refers to the fact that κ and F do not depend on the time k.
2. It follows from the definitions, that *strong convexity* (see e.g. Nesterov (2004) for a definition) implies (19) and this property itself implies strict convexity.

Theorem 4 (Optimality Conditions Imply Strict Dissipativity) *Let Assumption 5 and those of Theorem 3 hold. If $\eta_0 \neq 0$ and $\sup_{k\in\mathbb{N}_0} \|p_k\| < \infty$, then the optimal control problem (2) is strictly dissipative on every bounded set[5] $\mathbb{X}_0$ wrt supply rate $s(k, x, u) = \hat{\ell}(k, x, u)$ and optimal pair (x^*, u^*).*

Proof In order to prove strict dissipativity we have to verify that there is $\alpha \in \mathcal{K}_\infty$ and a storage function λ such that (7) holds. We claim that making the ansatz $\lambda(k, x) = \frac{1}{\eta_0} p_k^T (x - x^*(k))$ yields the desired property. Note that the restriction to bounded sets $\mathbb{X}_0$ is needed here in order to ensure that λ is bounded from below as required in Definition 7.

Let $\mathbb{X}_0$ be an arbitrary bounded set in $\mathbb{R}^n$. This yields boundedness of λ. Conditions (iii) and (iv) in Theorem 3 read

(iii) $\forall k \in \mathbb{N}_0 : \ p_k = -\eta_0 \frac{\partial \ell}{\partial x}(k, x^*(k), u^*(k)) + p_{k+1}^T \frac{\partial f}{\partial x}(k, x^*(k), u^*(k))$ and
(iva) $\forall k \in \mathbb{N}_0 : -\eta_0 \frac{\partial \ell}{\partial u}(k, x^*(k), u^*(k)) + p_{k+1}^T \frac{\partial f}{\partial u}(k, x^*(k), u^*(k)) = 0.$

Let us consider the modified stage cost $\tilde{\ell}$ (cf. Definition 8) using our ansatz for the storage function:

$$\begin{aligned}\tilde{\ell}(k, x, u) &= \hat{\ell}(k, x, u) + \frac{1}{\eta_0} p_k^T (x - x^*(k)) - \frac{1}{\eta_0} p_{k+1}^T (f(k, x, u) - x^*(k+1)) \\ &= \ell(k, x, u) - \ell(k, x^*(k), u^*(k)) \\ &\qquad + \frac{1}{\eta_0} p_k^T (x - x^*(k)) - \frac{1}{\eta_0} p_{k+1}^T (f(k, x, u) - x^*(k+1))\end{aligned}$$

Since ℓ is uniformly strictly convex wrt κ and F, p_k linear and f affine for each k, the modified cost $\tilde{\ell}$ is uniformly strictly convex wrt κ and F (and in particular strictly convex for all $k \in \mathbb{N}_0$). This means that a point $(\bar{x}(k), \bar{u}(k))$ satisfying $\frac{\partial \tilde{\ell}}{\partial x}(k, \bar{x}(k), \bar{u}(k)) = \frac{\partial \tilde{\ell}}{\partial u}(k, \bar{x}(k), \bar{u}(k)) = 0$ is a unique strict minimizer of $\tilde{\ell}(k, \cdot, \cdot)$. Let us therefore consider the partial derivatives of $\tilde{\ell}$. For all $k \in \mathbb{N}_0$ we have

$$\begin{aligned}\frac{\partial \tilde{\ell}}{\partial x}(k, x, u) &= \frac{\partial \ell}{\partial x}(k, x, u) + \frac{1}{\eta_0} p_k - \frac{1}{\eta_0} p_{k+1}^T \frac{\partial f}{\partial x}(k, x, u) \text{ and} \\ \frac{\partial \tilde{\ell}}{\partial u}(k, x, u) &= \frac{\partial \ell}{\partial u}(k, x, u) - \frac{1}{\eta_0} p_{k+1}^T \frac{\partial f}{\partial u}(k, x, u).\end{aligned}$$

Now plugging in $(x^*(k), u^*(k))$ and conditions (iii) and (iva) for the first and second equation, respectively, we obtain

$$\frac{\partial \tilde{\ell}}{\partial x}(k, x^*(k), u^*(k)) = 0 \text{ and } \frac{\partial \tilde{\ell}}{\partial u}(k, x^*(k), u^*(k)) = 0.$$

[5] This means that dissipativity holds for all $x \in \mathbb{X}_0$.

For each $k \in \mathbb{N}_0$ the point $(x^*(k), u^*(k))$ is thus the unique strict minimizer of $\tilde{\ell}$ at time k. By definition of the modified stage cost $\tilde{\ell}$ we have

$$\begin{aligned}\tilde{\ell}(k, x^*(k), u^*(k)) &= \hat{\ell}(k, x^*(k), u^*(k)) + \lambda(k, x^*(k)) - \lambda(k+1, f(k, x^*(k), u^*(k))) \\ &= \frac{1}{\eta_0} p_k^T (x^*(k) - x^*(k)) - \frac{1}{\eta_0} p_{k+1}^T (f(k, x^*(k), u^*(k)) - x^*(k+1)) \\ &= 0.\end{aligned}$$

Fix an arbitrary $t \in (0, 1)$. For $k \in \mathbb{N}_0$ consider an arbitrary point $(x, u) \in X \times U$. We define $(\bar{x}, \bar{u}) := t(x, u) + (1-t)(x^*(k), u^*(k)) \in X \times U$. Assumption 5 implies

$$\begin{aligned}\tilde{\ell}(k, \bar{x}, \bar{u}) &+ \frac{\kappa}{2} t(1-t) F(\|(x, u) - (x^*(k), u^*(k))\|) \\ &\leq t\tilde{\ell}(k, x, u) + (1-t)\tilde{\ell}(k, x^*(k), u^*(k)) = t\tilde{\ell}(k, x, u) \\ \Rightarrow \tilde{\ell}(k, x, u) &> \frac{1}{t}\tilde{\ell}(k, x^*(k), u^*(k)) + \frac{\kappa}{2}(1-t) F(\|(x, u) - (x^*(k), u^*(k))\|) \\ &= \frac{\kappa}{2}(1-t) F(\|(x, u) - (x^*(k), u^*(k))\|).\end{aligned}$$

This implies (7) if we set $\alpha(r) := \frac{\kappa}{2}(1-t)F(r)$, which is of class $\mathcal{K}_\infty$ because $F \in \mathcal{K}_\infty$ and $\frac{\kappa}{2}(1-t) \in \mathbb{R}_{>0}$.

Remark 2 The assumption of ℓ being uniformly strictly convex is needed in order to establish that $\alpha \in \mathcal{K}_\infty$ in (7) does not depend on the time k.

As indicated at the beginning of the section the optimality conditions of the reference (Blot and Hayek 2014, Theorem 2.2) fit our purpose very well but are just exemplary and we conjecture that alternative conditions can also be taken to establish strict dissipativity and thus the turnpike property. We will point out similarities and differences of the conditions above with those in Aseev et al. (2016). Firstly, let us mention that an important part of Aseev et al. (2016) is that the authors are able to establish a *transversality condition*. Such conditions are a valuable tool to restrict the set of candidates of optimal solutions to the infinite-horizon optimal control problem and, moreover, can be used in order to ensure $\sup_{k \in \mathbb{N}_0} \|p_k\| < \infty$ in Theorem 4. A comparable result does not exist in Blot and Hayek (2014, Section 2.2) (but in other results in that reference).

The assumptions that are imposed in Aseev et al. (2016), Blot and Hayek (2014) are in general difficult to compare. However, the main assumption (Assumption A) in Aseev et al. (2016) can be simplified if Condition 2 in Theorem 3 holds. Moreover, reference Aseev et al. (2016) assumes weakly overtaking optimality whereas the theorem we used from Blot and Hayek (2014) assumes overtaking optimality. The statements in the theorems are strongly related: Condition (iii) in Theorem 3 is the same as (Aseev et al. 2016, Corollary 2.3), and Condition (iv) is similar to the maximum condition in Aseev et al. (2016, Theorem 2.2), that reads

(adapted to our notation)

$$\forall\, k \in \mathbb{N}_0 : \left(-\frac{\partial \ell}{\partial u}(k, x^*(k), u^*(k)) + p_{k+1}^T \frac{\partial f}{\partial u}(k, x^*(k), u^*(k)) \right) v \le 0 \qquad (20)$$

$\forall v \in T_{U_k}(u^*(k))$. The set $T_{U_k}(u^*(k))$ denotes the Bouligand tangent cone of U_k (the constraint set for u at time k in Aseev et al. (2016)) at point $u^*(k)$. Certainly, (20) is obtained under weaker assumptions than (Blot and Hayek 2014, Theorem 2.2), yet it also yields a weaker statement and it is currently an open question whether it is still sufficient to prove strict dissipativity.

6 Example

In this section we provide an example of a time-varying optimal control problem, that was introduced in Grüne and Pirkelmann (2017). It can be interpreted as a very simple room heating/cooling model that has to react to external influences (the weather). We will verify that the example meets the assumptions needed for strict dissipativity and for the turnpike property. The latter will also be illustrated by means of numerical simulations.

The system dynamics is given by

$$f : \mathbb{N}_0 \times \mathbb{R} \times \mathbb{R} \to \mathbb{R}, \;\; f(k, x, u) = x + u + w_k,$$

with $w_k = -2\sin\left(\frac{k\pi}{12}\right) + a_k$ and in which the a_k are iid random numbers on the interval $[-\frac{1}{4}, \frac{1}{4}]$. In a physical interpretation of the example the state x corresponds to the temperature within a room, the control u to the heating/cooling and the time-varying data w_k to the changes of the external temperature over time that also influence the inside temperature. The stage cost of the system is

$$\ell(k, x, u) = u^2 + \varepsilon x^2,$$

for $0 < \varepsilon \ll 1$. Note that the term εx^2 is a regularization term that renders the original cost u^2, that was used in Grüne and Pirkelmann (2017), strictly convex wrt x and u. However, numerical experiments show, that the optimal trajectories for both versions of ℓ do not differ for sufficiently small ε. The system has to be operated subject to the control constraints $\mathbb{U}(k) = [-3, 3]$ and the state constraints $\mathbb{X}(k) = [-1/2, 1/2]$ if $k \in [24j + 12, 24(j + 1)$, $j \in \mathbb{N}_0$ and $\mathbb{X}(k) = [-2, 2]$ if $k \in [24j, 24j + 12)$. We assume that we have a perfect prediction of the external influence w_k, which means that its values are known whenever we optimize. Since a correct weather forecast is hardly possible for a few days, let alone on an infinite horizon, this may not be realistic. However, a verification of the turnpike property would allow us to apply model predictive control, in which only finite horizon problems of moderate horizon length have to be solved.

In what follows, we aim to verify the assumptions of Theorem 4. Since this result was stated for unconstrained problems, we first rewrite the example above using penalty functions $b_1 : \mathbb{N}_0 \times \mathbb{R} \to \mathbb{R}_{\geq 0}$ and $b_2 : \mathbb{N}_0 \times \mathbb{R} \to \mathbb{R}_{\geq 0}$. Then, the reformulated stage cost is given as follows (the dynamics remain unchanged):

$$L(k, x, u) := l(k, x, u) + b_1(k, x) + b_2(k, u),$$

$$b_1(k, x) = \begin{cases} c_x(|x| - 2)^4 & , x \notin [-2, 2] \\ 0 & , x \in [-2, 2] \end{cases}, \quad k \in [24j, 24j + 12), \ j \in \mathbb{N}_0,$$

$$b_1(k, x) = \begin{cases} c_x(|x| - 1/2)^4 & , x \notin [-1/2, 1/2] \\ 0 & , x \in [-1/2, 1/2] \end{cases}, \quad k \in [24j + 12, 24(j + 1)), \ j \in \mathbb{N}_0,$$

$$b_2(k, u) = \begin{cases} c_u(|u| - 3)^4 & , u \notin [-3, 3] \\ 0 & , u \in [-3, 3] \end{cases}, \quad k \in \mathbb{N}_0,$$

with c_x and $c_u \in \mathbb{R}_{>0}$.

We claim, that the reformulated optimal control problem satisfies Assumption 5. It is clear that for predictable a_k the dynamics are affine for each $k \in \mathbb{N}_0$. The Hessian of the stage cost reads

$$H_{(x,u)}L(k, x, u) = \begin{pmatrix} 2\varepsilon + \frac{d^2 b_1}{dx^2}(k, x) & 0 \\ 0 & 2 + \frac{d^2 b_2}{du^2}(k, u) \end{pmatrix}.$$

It is easily seen, that $\frac{d^2 b_1}{dx^2}(k, x) \geq 0$ and $\frac{d^2 b_2}{du^2}(k, u) \geq 0$ for all $k \in \mathbb{N}_0$, $x \in \mathbb{R}$ and $u \in \mathbb{R}$ such that we can conclude positive semidefiniteness of the matrix $H_{(x,u)}L(k, x, u) - 2\varepsilon I$, in which I is the identity matrix of dimension 2. For two times continuously differentiable functions this property is equivalent to L being strongly convex wrt 2ε (see e.g. Nesterov (2004)) for all $k \in \mathbb{N}_0$ and this implies uniform strict convexity of L wrt $\kappa = 2\varepsilon$ and $F(r) = r^2$.

Let us now check the assumptions of Theorem 3. Clearly, the continuity and differentiability requirements are met. The second condition also holds because $\frac{\partial f}{\partial x}(k, x, u) = 1$. For this example it moreover holds, that $\eta_0 \neq 0$: If $\eta_0 = 0$ then Theorem 3 yields that $p_1 \neq 0$. From condition (iii) applied to this example we get $p_k = p_{k+1}$ for all $k \in \mathbb{N}_0$. This contradicts (iva), which in case $\eta_0 = 0$ implies $p_{k+1} = 0$. It is left to show that the adjoints p_k are bounded. A formal proof appears technically involved, however, we can give evidence why it is reasonable to expect bounded p_k. The adjoint p_k is a measurement of how much the value of the trajectory differs from the optimal value if the trajectory value at time k differs (slightly) from $x^*(k)$. In our example the absence of constraints allows to steer the trajectory to $x^*(k + 1)$ in one step after having been disturbed at time k. Thus, the value of the disturbed trajectory and the optimal trajectory only differ in the first term and this difference can be estimated on bounded sets by a bound which is independent of k. This implies boundedness of the p_k and thus by Theorem 3 strict dissipativity for our example.

In what follows we will investigate Assumption 1 to conclude by Theorem 1 that the example exhibits the turnpike property on any compact set $\mathbb{X}_0 \subset \mathbb{R}^n$. For the cheap reachability in Assumption 1 one first shows that the optimal pair (x^*, u^*) satisfies the (uniform) estimates

$$|x^*(k)| \leq \sqrt[4]{\frac{81-4\varepsilon}{16c_x}} + 2 \tag{21}$$

and

$$|u^*(k)| \leq \sqrt[4]{\frac{81-4\varepsilon}{16c_u}} + 3. \tag{22}$$

The idea of the proof is as follows: We compare the cost of an admissible trajectory that is constructed such that it is constantly zero after the first time step, to the cost of the optimal pair. If the estimates above are violated this contradicts the fact that (x^*, u^*) is overtaking optimal. For cheap reachability we need to show that there exists $E \in \mathbb{R}$ such that for all $k \in \mathbb{N}_0$, $x \in \mathbb{X}_0$ and $N \in \mathbb{N} \cup \{\infty\}$ it holds $\hat{V}_N(k, x) \leq E$. To see this we consider a control sequence $\tilde{u}(\cdot)$ of length N given by $\tilde{u}(0) = -x + x^*(k+1) - w_k$, $\tilde{u}(j) = u^\star_{N-1,x^*(k+1)}(j-1),\ j \in \{1, \dots, N-1\}$. This yields

$$\begin{aligned}\hat{V}_N(k,x) &\leq \hat{\ell}(k,x,\tilde{u}(0)) + \underbrace{\hat{V}_{N-1}(k+1,x^*(k+1))}_{\leq 0} \leq \ell(k,x,\tilde{u}(0)) - \underbrace{\ell(k,x^*(k),u^*(k))}_{\geq 0} \\ &\leq \varepsilon x^2 + (-x + x^*(k+1) - w_k)^2 + b_1(k,x) + b_2(k, -x + x^*(k+1) - w_k).\end{aligned}$$

Using compactness of $\mathbb{X}_0$, boundedness of $(w_k)_{k\in\mathbb{N}_0}$, $(x^*(k))_{k\in\mathbb{N}_0}$ and $(u^*(k))_{k\in\mathbb{N}_0}$, the fact that the b_i can be bounded uniformly in k using (21), (22) we obtain a bound E that does not depend on k, x and N and conclude the assertion.

We performed several numerical simulations that illustrate that the system in the example has the turnpike property. For the purpose of the simulations the trajectory of optimal operation on an infinite horizon has been approximated by computing an optimal trajectory on a large finite horizon of $N = 100$ and leaving the initial value free. In the figures this trajectory is depicted in black. The regularization factor was chosen as $\varepsilon = 10^{-10}$ and the penalty parameters as $c_x = c_u = 10^{10}$. Figure 3 depicts open-loop trajectories of the state for different horizon lengths. As one can see the trajectories are close to the trajectory of optimal operation most of the time. It is also visible that the finite horizon trajectories will at some point turn away from

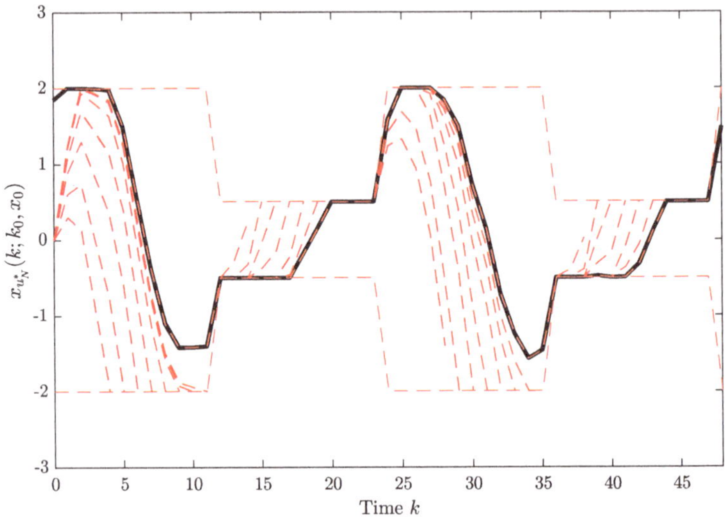

Fig. 3 Numerical simulations of the trajectory of optimal operation (black line) and open-loop trajectories of the state (dashed red lines) with different fixed initial value $x_0 = 0$ and different horizon lengths of N

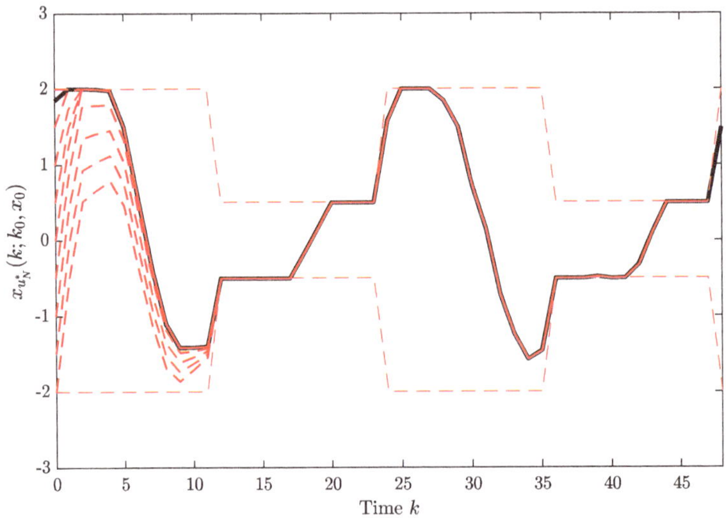

Fig. 4 Numerical simulations of the trajectory of optimal operation (black line) and open-loop trajectories of the state (dashed red lines) with different initial values x_0 and fixed horizon length of $N = 48$

the optimal trajectory and hit the constraints. This is due to the fact that it is cheaper to deviate from the infinite horizon optimal trajectory than it would be to stay close to it. Such a behavior is characteristic for the turnpike property.

In Fig. 4 open-loop trajectories for different initial values and fixed horizon length of $N = 48$ are shown. One observes that the open-loop solutions quickly converge to the trajectory of optimal operation.

Acknowledgements The research was supported by the DFG Grants GR1569/13-1 and 16-1.

References

B.D.O. Anderson, P.V. Kokotović, Optimal control problems over large time intervals. Automatica **23**(3), 355–363 (1987)

S.M. Aseev, M.I. Krastanov, V.M. Veliov, Optimality conditions for discrete-time optimal control on infinite horizon. Research Report 2016-09 (2016). Available online

T.F. Bewley, *General Equilibrium, Overlapping Generations Models, and Optimal Growth Theory* (Harvard University Press, Cambridge, 2009)

J. Blot, N. Hayek, *Infinite-Horizon Optimal Control in the Discrete-Time Framework* (Springer, New York, 2014)

T. Damm, L. Grüne, M. Stieler, K. Worthmann, An exponential turnpike theorem for dissipative discrete time optimal control problems. SIAM J. Control Optim. **52**(3), 1935–1957 (2014)

R. Dorfman, P.A. Samuelson, R.M. Solow, *Linear Programming and Economic Analysis* (Dover Publications, New York, 1987). Reprint of the 1958 original

D. Gale, On optimal development in a multi-sector economy. Rev. Econ. Stud. **34**(1), 1–18, (1967)

L. Grüne, Economic receding horizon control without terminal constraints. Automatica **49**(3), 725–734 (2013)

L. Grüne, C.M. Kellett, S.R. Weller, On a discounted notion of strict dissipativity. IFAC-PapersOnLine **49**(18), 247–252 (2016)

L. Grüne, C.M. Kellett, S.R. Weller, On the relation between turnpike properties for finite and infinite horizon optimal control problems. J. Optim. Theory Appl. **173**, 727 (2017). Online version available via https://doi.org/10.1007/s10957-017-1103-6

L. Grüne, M.A. Müller, On the relation between strict dissipativity and the turnpike property. Syst. Control Lett. **90**, 45–53 (2016)

L. Grüne, S. Pirkelmann, Closed-loop performance analysis for economic model predictive control of time-varying systems, in *Proceedings of the 56th IEEE Annual Conference on Decision and Control (CDC 2017)*, Melbourne, pp. 5563–5569 (2017)

L. Grüne, M. Stieler, Asymptotic stability and transient optimality of economic MPC without terminal conditions. J. Proc. Control **24**(8), 1187–1196 (2014)

C.M. Kellett, A compendium of comparison function results. Math. Control Sign. Syst. **26**(3), 339–374 (2014)

L.W. McKenzie, Optimal economic growth, turnpike theorems and comparative dynamics, in *Handbook of Mathematical Economics*, vol. III (North-Holland, Amsterdam, 1986), pp. 1281–1355

M.A. Müller, L. Grüne, Economic model predictive control without terminal constraints for optimal periodic behavior. Automatica **70**, 128–139 (2016)

Y. Nesterov, *Introductory Lectures on Convex Optimization: A Basic Course*. Applied Optimization, vol. 87, 1st edn. (Springer, Berlin, 2004)

A. Porretta, E. Zuazua, Long time versus steady state optimal control. SIAM J. Control Optim. **51**(6), 4242–4273 (2013)

E. Trélat, E. Zuazua, The turnpike property in finite-dimensional nonlinear optimal control. J. Differ. Equ. **258**(1), 81–114 (2015)

J.C. Willems, Dissipative dynamical systems part I: general theory. Arch. Ration. Mech. Anal. **45**(5), 321–351 (1972)

Part II
Economics and Environmental Models

Optimal Exploitation of Renewable Resources: Lessons in Sustainability from an Optimal Growth Model of Natural Resource Consumption

Sergey Aseev and Talha Manzoor

Abstract We study an optimal growth model for a single resource based economy. The resource is governed by the standard model of logistic growth, and is related to the output of the economy through a Cobb-Douglas type production function with exogenously driven knowledge stock. The model is formulated as an infinite-horizon optimal control problem with unbounded set of control constraints and non-concave Hamiltonian. We transform the original problem to an equivalent one with simplified dynamics and prove the existence of an optimal admissible control. Then we characterize the optimal paths for all possible parameter values and initial states by applying the appropriate version of the Pontryagin maximum principle. Our main finding is that only two qualitatively different types of behavior of sustainable optimal paths are possible depending on whether the resource growth rate is higher than the social discount rate or not. An analysis of these behaviors yields general criterions for sustainable and strongly sustainable optimal growth (w.r.t. the corresponding notions of sustainability defined herein).

S. Aseev (✉)
Steklov Mathematical Institute of Russian Academy of Sciences, Moscow, Russia

International Institute for Applied Systems Analysis, Laxenburg, Austria

Krasovskii Institute of Mathematics and Mechanics, Ural Branch of Russian Academy of Sciences, Yekaterinburg, Russia
e-mail: aseev@mi.ras.ru

T. Manzoor
Department of Electrical Engineering, Center for Water Informatics & Technology, Lahore University of Management Sciences, Lahore, Pakistan
e-mail: talha.manzoor@lums.edu.pk

G. Feichtinger et al. (eds.), *Control Systems and Mathematical Methods in Economics*, Lecture Notes in Economics and Mathematical Systems 687,
https://doi.org/10.1007/978-3-319-75169-6_11

1 Introduction

Following the first analysis conducted by Ramsey (1928), the mathematical problem of inter-temporal resource allocation has attracted a significant amount of attention over the past decades, and has driven the evolution of first exogenous, and then endogenous growth theory (see Acemoglu 2009; Barro and Sala-i-Martin 1995). Employed growth models are typically identified by the production of economic output, the dynamics of the inputs of production, and the comparative mechanism of alternate consumption paths. Our framework considers a renewable resource, whose reproduction is logistic in nature, as the only input to production. The relationship of the resource with the output of the economy is explained through a Cobb-Douglas type production function with an exogenously driven knowledge stock. Alternate consumption paths are compared via a discounted utilitarian approach. The question that we concern ourselves with for our chosen framework, is the following: what are the conditions of sustainability for optimal development?

In the context of sustainability, the discounted utilitarian approach may propose undesirable solutions in certain scenarios. For instance, discounted utilitarianism has been reported to force consumption asymptotically to zero even when sustainable paths with non-decreasing consumption are feasible (Asheim and Mitra 2010). The Brundtland Commission defines sustainable development as development that meets the needs of the present, without compromising the ability of future generations to meet their own needs (Brundtland Commission 1987). In this spirit, we employ the notion of sustainable development, as a consumption path ensuring a non-decreasing welfare for all future generations. This notion of sustainability is natural, and has also been used by various authors in their work. For instance, Valente (2005) evaluates this notion of sustainability for an exponentially growing natural resource, and derives a condition necessary for sustainable consumption. We extend this model by allowing the resource to grow at a declining rate (the logistic growth model). We build on the work presented previously in Manzoor et al. (2014) which proves the existence of an optimal path only in the case when the resource growth rate is higher than the social discount rate and admissible controls are uniformly bounded.

Our model is formulated as an infinite-horizon optimal control problem with logarithmic instantaneous utility. The problem involves unbounded controls and the non-concave Hamiltonian. These preclude direct application of the standard existence results and Arrow's sufficient conditions for optimality. We transform the original problem to an equivalent one with simplified dynamics and prove the general existence result. Then we apply a recently developed version of the maximum principle (Aseev and Veliov 2012, 2014, 2015) to our problem and describe the optimal paths for all possible parameter values and initial states in the problem. Our analysis of the Hamiltonian phase space reveals that there are only two qualitatively different types of behavior of the sustainable optimal paths in the model. In the first case the instantaneous utility is a non-decreasing function in the long run along the optimal path (we call such paths *sustainable*).

The second case corresponds to the situation when the optimal path is sustainable and in addition the resource stock is asymptotically nonvanishing (we call such paths *strongly sustainable*). We show that a strongly sustainable equilibrium is attainable only when the resource growth rate is higher than the social discount rate. When this condition is violated, we see that the optimal resource exploitation rate asymptotically follows the Hotelling rule of optimal depletion of an exhaustible resource (Hotelling 1974). In this case optimal consumption is sustainable only if the depletion of the resource is compensated by appropriate growth of the knowledge stock and/or decrease of the output elasticity of the resource.

The paper is organized as follows. Section 2 sets up the problem. Section 3 establishes the equivalence of the problem with a simpler version, and applies the maximum principle after proving the existence of an optimal control. Section 4 presents an analysis of the associated Hamiltonian system and formulates the optimal feedback law. We conclude in Sect. 5 where we develop conditions for sustainability and strong sustainability of the optimal paths in our model.

The paper draws on a companion working paper (Aseev and Manzoor 2016), which contains proofs for several auxiliary results related to our model.

2 Problem Formulation

Consider a society consuming a single renewable resource. The resource, whose quantity is given by $S(t) > 0$ at each instant of time $t \geq 0$, is governed by the standard model of logistic growth. In the absence of consumption, it regenerates at rate $r > 0$ and saturates at carrying capacity $K > 0$. The society consumes the resource by exerting effort (exploitation rate) $u(t) > 0$ resulting in a total consumption velocity of $u(t)S(t) > 0$ at time $t \geq 0$ respectively. The dynamics of the resource stock are then given by the following equation:

$$\dot{S}(t) = r\, S(t) \left(1 - \frac{S(t)}{K}\right) - u(t)S(t), \qquad u(t) \in (0, \infty).$$

The initial stock of the resource is $S(0) = S_0 > 0$.

We assume a single resource economy whose output $Y(t) > 0$ at instant $t \geq 0$ is related to the resource by the Cobb-Douglas type production function

$$Y(t) = A(t)\big(u(t)S(t)\big)^{\alpha}, \qquad \alpha \in (0, 1]. \tag{1}$$

Here $A(t) > 0$ represents an exogenously driven knowledge stock at time $t \geq 0$. We assume $\dot{A}(t) \leq \mu A(t)$, where $\mu \geq 0$ is a constant, and $A(0) = A_0 > 0$.

The whole output $Y(t)$ produced at each instant $t \geq 0$ is consumed and the corresponding instantaneous utility is measured by the logarithmic function $t \mapsto \ln Y(t) = \ln A(t) + \alpha\,[\ln S(t) + \ln u(t)]$, $t \geq 0$.

This leads to the following optimal control problem $(P1)$:

$$J(S(\cdot), u(\cdot)) = \int_0^\infty e^{-\rho t} [\ln S(t) + \ln u(t)] \, dt \to \max, \tag{2}$$

$$\dot{S}(t) = r S(t) \left(1 - \frac{S(t)}{K}\right) - u(t) S(t), \qquad S(0) = S_0, \tag{3}$$

$$u(t) \in (0, \infty), \tag{4}$$

where $\rho > 0$ is the subjective discount rate.

By an *admissible control* in problem $(P1)$ we mean a Lebesgue measurable locally bounded function $u\colon [0, \infty) \mapsto \mathbb{R}^1$ which satisfies the control constraint (4) for all $t \geq 0$. By definition, the corresponding to $u(\cdot)$ *admissible trajectory* is a (locally) absolutely continuous function $S(\cdot) : [0, \infty) \mapsto \mathbb{R}^1$ which is a Caratheodory solution (see Filippov 1988) to the Cauchy problem (3) on the whole infinite time interval $[0, \infty)$. Due to the local boundedness of the admissible control $u(\cdot)$ such admissible trajectory $S(\cdot)$ always exists and is unique (see Filippov 1988, Section 7). A pair $(S(\cdot), u(\cdot))$ where $S(\cdot)$ is an admissible control and $S(\cdot)$ is the corresponding admissible trajectory is called an *admissible pair* in problem $(P1)$.

Due to (3) for any admissible trajectory $S(\cdot)$ the following estimate holds:

$$S(t) \leq S_{\max} = \max\{S_0, K\}, \qquad t \geq 0. \tag{5}$$

The integral in (2) is understood in improper sense, i.e.

$$J(S(\cdot), u(\cdot)) = \lim_{T \to \infty} \int_0^T \mathrm{e}^{-\rho t} [\ln S(t) + \ln u(t)] \, dt \tag{6}$$

if the limit exists.

Using estimate (5) and control system (3) it can be easily shown that there is a decreasing function $\omega : [0, \infty) \mapsto (0, \infty)$ such that $\omega(t) \to +0$ as $t \to \infty$ and for any admissible pair $(S(\cdot), u(\cdot))$ the following inequality holds:

$$\int_T^{T'} e^{-\rho t} [\ln S(t) + \ln u(t)] \, \mathrm{d}t < \omega(T), \qquad 0 \leq T < T'. \tag{7}$$

This fact immediately implies that for any admissible pair $(S(\cdot), u(\cdot))$ the limit in (6) always exists and is either finite or equals $-\infty$ (see Aseev and Manzoor 2016 for details).

Due to (7) for any admissible pair $(S(\cdot), u(\cdot))$ the value $\sup_{(S(\cdot),u(\cdot))} J(S(\cdot), u(\cdot))$ is finite. This allows us to understand the optimality of an admissible pair $(S_*(\cdot), u_*(\cdot))$ in the strong sense (Carlson et al. 1991). By definition, an admissible pair $(S_*(\cdot), u_*(\cdot))$ is *strongly optimal* (or, for brevity, simply *optimal*) in the problem $(P1)$ if the functional (2) takes the maximal possible value on this pair, i.e.

$$J(S_*(\cdot), u_*(\cdot)) = \sup_{(S(\cdot),u(\cdot))} J(S(\cdot), u(\cdot)) < \infty.$$

Notice, that the set of control constraints in problem $(P1)$ (see (4)) is nonclosed and unbounded. Due to this circumstance the standard existence theorems (see e.g. Balder 1983; Cesari 1983) are not applicable to problem $(P1)$ directly. Moreover, the situation is complicated here by the fact that the Hamiltonian of problem $(P1)$ is non-concave in the state variable S. These preclude the usage of Arrow's sufficient conditions for optimality (see Carlson et al. 1991).

Our analysis below is based on application of the recently developed normal form version of the Pontryagin maximum principle (Pontryagin et al. 1964) for infinite-horizon optimal control problems with adjoint variable specified explicitly via the Cauchy type formula (see Aseev and Veliov 2012, 2014, 2015). However, such approach assumes that the optimal control exists. So, the proof of the existence of an optimal admissible pair $(S_*(\cdot), u_*(\cdot))$ in problem $(P1)$ and establishing of the corresponding version of the maximum principle will be our primary goal in the next section.

3 Existence of an Optimal Control and the Maximum Principle

Let us transform problem $(P1)$ into a more appropriate equivalent form.

Due to (3) along any admissible pair $(S(\cdot), u(\cdot))$ we have

$$\frac{d}{dt}\left[e^{-\rho t}\ln S(t)\right] \overset{\text{a.e.}}{=} -\rho e^{-\rho t}\ln S(t) + re^{-\rho t} - e^{-\rho t}\left(\frac{r}{K}S(t) + u(t)\right), \quad t > 0.$$

Integrating this equality on arbitrary time interval $[0, T]$, $T > 0$, we obtain

$$\int_0^T e^{-\rho t}\ln S(t)\,dt = \frac{\ln S_0 - e^{-\rho T}\ln S(T)}{\rho} + \frac{r}{\rho^2}\left(1 - e^{-\rho T}\right) - \int_0^T e^{-\rho t}\left(\frac{r}{\rho K}S(t) + \frac{u(t)}{\rho}\right)dt.$$

Hence, for any admissible pair $(S(\cdot), u(\cdot))$ and arbitrary $T > 0$ we have

$$\int_0^T e^{-\rho t}\left[\ln S(t) + \ln u(t)\right]dt = \frac{\ln S_0 - e^{-\rho T}\ln S(T)}{\rho} + \frac{r}{\rho^2}\left(1 - e^{-\rho T}\right) - \frac{r}{\rho K}\int_0^T e^{-\rho t}S(t)\,dt + \int_0^T e^{-\rho t}\left(\ln u(t) - \frac{u(t)}{\rho}\right)dt. \quad (8)$$

Here due to estimate (7) limits of the both sides in (8) as $T \to \infty$ exist and equal either a finite number or $-\infty$ simultaneously, and due to (5) either (i) $\lim_{T\to\infty} e^{-\rho T} \ln S(T) = 0$ or (ii) $\liminf_{T\to\infty} e^{-\rho T} \ln S(T) < 0$.

In the case (i) passing to the limit in (8) as $T \to \infty$ we get

$$\int_0^\infty e^{-\rho t} [\ln S(t) + \ln u(t)]\, dt = \frac{\ln S_0}{\rho} + \frac{r}{\rho^2}$$
$$- \frac{r}{\rho K} \int_0^\infty e^{-\rho t} S(t)\, dt + \int_0^\infty e^{-\rho t} \left(\ln u(t) - \frac{u(t)}{\rho} \right) dt, \qquad (9)$$

where both sides in (9) are equal to a finite number or $-\infty$ simultaneously.

In the case (ii) condition $\liminf_{T\to\infty} e^{-\rho T} \ln S(T) < 0$ implies

$$\int_0^\infty e^{-\rho t} [\ln S(t) + \ln u(t)]\, dt = \lim_{T\to\infty} \int_0^T e^{-\rho t} [\ln S(t) + \ln u(t)]\, dt = -\infty$$

(see Aseev and Manzoor 2016 for details). Hence, in the case (ii) (9) also holds as $-\infty = -\infty$.

Neglecting now the constant terms in the right-hand side of (9) we obtain the following optimal control problem $(\tilde{P}1)$ which is equivalent to $(P1)$:

$$\tilde{J}(S(\cdot), u(\cdot)) = \int_0^\infty e^{-\rho t} \left[\ln u(t) - \frac{u(t)}{\rho} - \frac{r}{\rho K} S(t) \right] dt \to \max,$$

$$\dot{S}(t) = r S(t) \left(1 - \frac{S(t)}{K} \right) - u(t) S(t), \qquad S(0) = S_0, \qquad (10)$$

$$u(t) \in (0, \infty). \qquad (11)$$

Further, the function $u \mapsto \ln u - u/\rho$ is increasing on $(0, \rho]$ and it reaches the global maximum at point $u_* = \rho$. Hence, any optimal control $u_*(\cdot)$ in $(\tilde{P}1)$ (if such exists) must satisfy to inequality $u_*(t) \geq \rho$ for almost all $t \geq 0$. Hence, without loss of generality the control constraint (11) in $(\tilde{P}1)$ (and hence the control constraint (4) in $(P1)$) can be replaced by the control constraint $u(t) \in [\rho, \infty)$. Thus we arrive to the following (equivalent) problem $(P2)$:

$$J(S(\cdot), u(\cdot)) = \int_0^\infty e^{-\rho t} [\ln u(t) + \ln S(t)]\, dt \to \max,$$

$$\dot{S}(t) = r S(t) \left(1 - \frac{S(t)}{K} \right) - u(t) S(t), \qquad S(0) = S_0,$$

$$u(t) \in [\rho, \infty). \qquad (12)$$

Here the class of admissible controls in problem $(P2)$ consists of all locally bounded functions $u(\cdot)$ satisfying the control constraint (12) for all $t \geq 0$.

To simplify dynamics in $(P2)$ let us introduce the new state variable $x(\cdot)$: $x(t) = 1/S(t)$, $t \geq 0$. As it can be verified directly, in terms of the state variable $x(\cdot)$ problem $(P2)$ can be rewritten as the following (equivalent) problem $(P3)$:

$$J(x(\cdot), u(\cdot)) = \int_0^{\infty} e^{-\rho t} [\ln u(t) - \ln x(t)] \, dt \to \max, \tag{13}$$

$$\dot{x}(t) = [u(t) - r] x(t) + a, \qquad x(0) = x_0 = \frac{1}{S_0}, \tag{14}$$

$$u(t) \in [\rho, \infty). \tag{15}$$

Here $a = r/K$. The class of admissible controls $u(\cdot)$ in $(P3)$ consists of all measurable locally bounded functions $u \colon [0, \infty) \mapsto [\rho, \infty)$.

Notice, that due to linearity of (14) for arbitrary admissible control $u(\cdot)$ the corresponding trajectory $x(\cdot)$ can be expressed via the Cauchy formula (see Hartman 1964):

$$x(t) = x_0 e^{\int_0^t u(\xi)\, d\xi - rt} + a e^{\int_0^t u(\xi)\, d\xi - rt} \int_0^t e^{-\int_0^s u(\xi)\, d\xi + rs} \, ds, \qquad t \geq 0. \tag{16}$$

Since the problems $(P1)$, $(P2)$ and $(P3)$ are equivalent we will focus our analysis below on problem $(P3)$ with simplified dynamics (see (14)) and the closed set of control constraints (see (15)).

The constructed problem $(P3)$ is a particular case of the following autonomous infinite-horizon optimal control problem $(P4)$ with exponential discounting:

$$J(x(\cdot), u(\cdot)) = \int_0^{\infty} e^{-\rho t} g(x(t), u(t)) \, dt \to \max,$$

$$\dot{x}(t) = f(x(t), u(t)), \qquad x(0) = x_0, \tag{17}$$

$$u(t) \in U.$$

Here U is a nonempty closed subset of $\mathbb{R}^m$, $x_0 \in G$ is an initial state, G is an open convex subset of $\mathbb{R}^n$, $\rho > 0$ is the discount rate, and $f : G \times U \mapsto \mathbb{R}^n$ and $g : G \times U \mapsto \mathbb{R}^m$ are given functions. The class of admissible controls in $(P4)$ consists of all measurable locally bounded functions $u \colon [0, \infty) \mapsto U$. The optimality of admissible pair $(x_*(\cdot), u_*(\cdot))$ is understood in the strong sense (Carlson et al. 1991).

Problems of type $(P4)$ were intensively studied in last decades (see Aseev 2015a,b, 2016; Aseev et al. 2012; Aseev and Kryazhimskiy 2004; Aseev and Kryazhimskii 2007; Aseev and Veliov 2012, 2014, 2015). Here we will employ the existence result and the variant of the Pontryagin maximum principle for problem

$(P4)$ developed in Aseev (2015b, 2016) and Aseev and Veliov (2012, 2014, 2015) respectively.

We will need to verify validity of the following conditions (see Aseev 2015b, 2016; Aseev et al. 2012; Aseev and Veliov 2012, 2014, 2015).

(A1) *The functions $f(\cdot,\cdot)$ and $g(\cdot,\cdot)$ together with their partial derivatives $f_x(\cdot,\cdot)$ and $g_x(\cdot,\cdot)$ are continuous and locally bounded on $G \times U$.*

(A2) *There exists a number $\beta > 0$ and a nonnegative integrable function $\lambda : [0,\infty) \mapsto \mathbb{R}^1$ such that for every $\zeta \in G$ with $\|\zeta - x_0\| < \beta$ Eq. (17) with $u(\cdot) = u_*(\cdot)$ and initial condition $x(0) = \zeta$ (instead of $x(0) = x_0$) has a solution $x(\zeta;\cdot)$ on $[0,\infty)$ in G, and*

$$\max_{\theta \in [x(\zeta;t), x_*(t)]} \left| e^{-\rho t} \langle g_x(\theta, u_*(t)), x(\zeta;t) - x_*(t) \rangle \right| \overset{a.e.}{\leq} \|\zeta - x_0\| \lambda(t).$$

Here $[x(\zeta;t), x_*(t)]$ denotes the line segment with vertices $x(\zeta;t)$ and *$x_*(t)$.*

(A3) *There is a positive function $\omega(\cdot)$ decreasing on $[0,\infty)$ such that $\omega(t) \to +0$ as $t \to \infty$ and for any admissible pair $(x(\cdot), u(\cdot))$ the following estimate holds:*

$$\int_T^{T'} e^{-\rho t} g(x(t), u(t))\, dt \leq \omega(T), \qquad 0 \leq T \leq T'.$$

Obviously, condition $(A1)$ is satisfied because $f(x,u) = [u - r]x + a$, $g(x,u) = \ln u - \ln x$, $f_x(x,u) = u - r$ and $g_x(x,u) = -1/x$, $x > 0$, $u \in [\rho,\infty)$, in $(P3)$.

Let us show that $(A2)$ also holds for any admissible pair $(x_*(\cdot), u_*(\cdot))$ in $(P3)$. Set $\beta = x_0/2$ and define the nonnegative integrable function $\lambda : [0,\infty) \mapsto \mathbb{R}^1$ as follows: $\lambda(t) = 2e^{-\rho t}/x_0$, $t \geq 0$. Then, as it can be seen directly, for any real ζ: $|\zeta - x_0| < \beta$, the Cauchy problem (14) with $u(\cdot) = u_*(\cdot)$ and the initial condition $x(0) = \zeta$ (instead of $x(0) = x_0$) has a solution $x(\zeta;\cdot)$ on $[0,\infty)$ and

$$\max_{\theta \in [x(\zeta;t), x_*(t)]} \left| e^{-\rho t} g_x(\theta, u_*(t))\, (x(\zeta;t) - x_*(t)) \right| \overset{\text{a.e.}}{\leq} |\zeta - x_0| \lambda(t).$$

Hence, for any admissible pair $(x_*(\cdot), x_*(\cdot))$ condition $(A2)$ is also satisfied.

Validity of $(A3)$ for any admissible pair $(x_*(\cdot), u_*(\cdot))$ follows from (7) directly.

For an admissible pair $(x(\cdot), u(\cdot))$ consider the following linear system:

$$\dot{z}(t) = -\left[f_x(x(t), u(t))\right]^* z(t) = [-u(t) + r]\, z(t). \tag{18}$$

The normalized fundamental solution $Z(\cdot)$ to Eq. (18) is defined as follows:

$$Z(t) = e^{-\int_0^t u(\xi)\, d\xi + rt}, \qquad t \geq 0. \tag{19}$$

Due to (16) and (19) for any admissible pair $(x(\cdot), u(\cdot))$ we have

$$\left|e^{-\rho t} Z^{-1}(t) g_x(x(t), u(t))\right|$$
$$= \left|\frac{e^{-\rho t} e^{\int_0^t u(\xi)\,d\xi - rt}}{x_0 e^{\int_0^t u(\xi)\,d\xi - rt} + a e^{\int_0^t u(\xi)\,d\xi - rt} \int_0^t e^{-\int_0^s u(\xi)\,d\xi + rs}\,ds}\right| \le \frac{e^{-\rho t}}{x_0}, \qquad t \ge 0.$$

Hence, for any $T > 0$ the function $\psi_T : [0, T] \mapsto \mathbb{R}^1$ defined as

$$\psi_T(t) = Z(t) \int_t^T e^{-\rho s} Z^{-1}(s) g_x(x(s), u(s))\,ds$$
$$= -e^{-\int_0^t u(\xi)\,d\xi + rt} \int_t^T \frac{e^{\int_0^s u(\xi)\,d\xi - rs} e^{-\rho s}}{x(s)}\,ds, \qquad t \in [0, T], \tag{20}$$

is absolutely continuous, and the function $\psi : [0, \infty) \mapsto \mathbb{R}^1$ defined as

$$\psi(t) = Z(t) \int_t^\infty e^{-\rho s} Z^{-1}(s) g_x(x(s), u(s))\,ds$$
$$= -e^{-\int_0^t u(\xi)\,d\xi + rt} \int_t^\infty \frac{e^{\int_0^s u(\xi)\,d\xi - rs} e^{-\rho s}}{x(s)}\,ds, \qquad t \ge 0, \tag{21}$$

is locally absolutely continuous.

Define the normal form Hamilton-Pontryagin function $\mathcal{H} : [0, \infty) \times (0, \infty) \times [\rho, \infty) \times \mathbb{R}^1 \mapsto \mathbb{R}^1$ and the normal-form Hamiltonian $H : [0, \infty) \times (0, \infty) \times \mathbb{R}^1 \mapsto \mathbb{R}^1$ for problem $(P3)$ in the standard way:

$$\mathcal{H}(t, x, u, \psi) = \psi f(x, u) + e^{-\rho t} g(x, u) = \psi[(u - r)x + a] + e^{-\rho t}[\ln u - \ln x],$$

$$H(t, x, \psi) = \sup_{u \ge \rho} \mathcal{H}(t, x, u, \psi),$$

$$t \in [0, \infty), \quad x \in (0, \infty), \quad u \in [\rho, \infty), \quad \psi \in \mathbb{R}^1.$$

Theorem 1 *There is an optimal admissible control $u_*(\cdot)$ in problem $(P3)$. Moreover, for any optimal admissible pair $(x_*(\cdot), u_*(\cdot))$ we have*

$$u_*(t) \overset{a.e.}{\le} \left(1 + \frac{1}{K x_*(t)}\right)(r + \rho), \qquad t \ge 0. \tag{22}$$

Proof Let us show that there is a continuous function $M : [0, \infty) \mapsto \mathbb{R}^1$, $M(t) \ge 0$, $t \ge 0$, and a function $\delta : [0, \infty) \mapsto \mathbb{R}^1$, $\delta(t) > 0$, $t \ge 0$, $\lim_{t\to\infty} (\delta(t)/t) = 0$, such that for any admissible pair $(x(\cdot), u(\cdot))$, satisfying on a set $\mathfrak{M} \subset [0, \infty)$, meas $\mathfrak{M} > 0$, to inequality $u(t) > M(t)$, for all $t \in \mathfrak{M}$ we have

$$\inf_{T: T-\delta(T)\geq t} \left\{ \sup_{u\in[\rho, M(t)]} \mathcal{H}(t, x(t), u, \psi_T(t)) - \mathcal{H}(t, x(t), u(t), \psi_T(t)) \right\} > 0, \tag{23}$$

where the function $\psi_T(\cdot)$ is defined on $[0, T]$, $T > 0$, by equality (20).

Let $(x(\cdot), u(\cdot))$ be an arbitrary admissible pair in $(P3)$. Then due to (16) and (19), for any $T > 0$ and arbitrary $t \in [0, T]$ we get (see (20))

$$\begin{aligned} -x(t)\psi_T(t) &= \left[x_0 + a \int_0^t e^{-\int_0^s u(\xi)\,d\xi + rs}\, ds \right] \int_t^T \frac{e^{-\rho s}}{x_0 + a \int_0^s e^{-\int_0^\tau u(\xi)\,d\xi + r\tau}\, d\tau}\, ds \\ &\geq x_0 \int_t^T \frac{e^{-\rho s}}{x_0 + a \int_0^s e^{r\tau}\, d\tau}\, ds \geq \frac{r x_0 e^{-(r+\rho)t}}{(r x_0 + a)(r+\rho)} \left[1 - e^{-(r+\rho)(T-t)} \right]. \end{aligned} \tag{24}$$

For a $\delta > 0$ define the function $M_\delta \colon [0, \infty) \mapsto \mathbb{R}^1$ by equality

$$M_\delta(t) = \frac{(r x_0 + a)(r+\rho)}{r x_0 \left[1 - e^{-(r+\rho)\delta} \right]} e^{rt} + \frac{1}{\delta}, \qquad t \geq 0. \tag{25}$$

Then for any $T\colon T - \delta > t$ and arbitrary $(x(\cdot), u(\cdot))$ the function $u \mapsto \mathcal{H}(t, x(t), u, \psi_T(t))$ reaches its maximal value on $[\rho, \infty)$ at the point (see (24))

$$u_T(t) = -\frac{e^{-\rho t}}{x(t)\psi_T(t)} \leq \frac{(r x_0 + a)(r+\rho)}{r x_0 \left[1 - e^{-(r+\rho)(T-t)} \right]} e^{rt} \leq M_\delta(t) - \frac{1}{\delta}. \tag{26}$$

Now, set $\delta(t) \equiv \delta$ and $M(t) \equiv M_\delta(t)$, $t \geq 0$. Let $(x(\cdot), u(\cdot))$ be an admissible pair such that inequality $u(t) > M_\delta(t)$ holds on a set $\mathfrak{M} \subset [0, \infty)$, meas $\mathfrak{M} > 0$. For arbitrary $t \in \mathfrak{M}$ define the function $\Phi \colon [t+\delta, \infty) \mapsto \mathbb{R}^1$ as follows

$$\begin{aligned} \Phi(T) &= \sup_{u\in[\rho, M(t)]} \mathcal{H}(t, x(t), u, \psi_T(t)) - \mathcal{H}(t, x(t), u(t), \psi_T(t)) \\ &= \psi_T(t) u_T(t) x(t) + e^{-\rho t} \ln u_T(t) - \left[\psi_T(t) u(t) x(t) + e^{-\rho t} \ln u(t) \right], \qquad T \geq t + \delta. \end{aligned}$$

Due to (26) we have

$$\begin{aligned} \Phi(T) = -e^{-\rho t} + e^{-\rho t} \left[-\rho t - \ln(-\psi_T(t)) - \ln x(t) \right] \\ - \left[\psi_T(t) u(t) x(t) + e^{-\rho t} \ln u(t) \right], \qquad T \geq t + \delta. \end{aligned}$$

Hence, due to (20) and (26) for a.e. $T \geq t + \delta$ we get

$$\frac{d}{dT}\Phi(T) = -\frac{e^{-\rho t}}{\psi_T(t)}\frac{d}{dT}\left[\psi_T(t)\right] - u(t)x(t)\frac{d}{dT}\left[\psi_T(t)\right]$$

$$= x(t)\frac{d}{dT}\left[\psi_T(t)\right]\left[\frac{e^{-\rho t}}{-\psi_T(t)x(t)} - u(t)\right] = x(t)\frac{d}{dT}\left[\psi_T(t)\right](u_T(t) - u(t)) > 0.$$

Hence,

$$\inf_{T>0:\, t\leq T-\delta}\left\{\sup_{u\in[\rho,M(t)]} \mathcal{H}(t, x(t), u, \psi_T(t)) - \mathcal{H}(t, x(t), u(t), \psi_T(t))\right\}$$

$$= \inf_{T>0:\, t\leq T-\delta} \Phi(T) = \Phi(t+\delta) > 0.$$

Thus, for any $t \in \mathfrak{M}$ inequality (23) is proved.

Since the instantaneous utility in (13) is concave in u, the system (14) is affine in u, the set U is closed (see (15)), conditions $(A1)$ and $(A3)$ are satisfied, and since $(A2)$ also holds for any admissible pair $(x_*(\cdot), u_*(\cdot))$ in $(P3)$, all conditions of the existence result in Aseev (2016) are fulfilled (see also Aseev 2015b, Theorem 1). Hence, there is an optimal admissible control $u_*(\cdot)$ in $(P3)$ and, moreover, $u_*(t) \overset{a.e.}{\leq} M_\delta(t)$, $t \geq 0$. Passing to a limit in this inequality as $\delta \to \infty$ we get (see (25))

$$u_*(t) \overset{\text{a.e.}}{\leq} \left(1 + \frac{1}{Kx_0}\right)(r+\rho)e^{rt}, \qquad t \geq 0. \tag{27}$$

Further, it is easy to see that for any $\tau > 0$ the pair $(\tilde{x}_*(\cdot), \tilde{u}_*(\cdot))$ defined as $\tilde{x}_*(t) = x_*(t+\tau)$, $\tilde{u}_*(t) = u_*(t+\tau)$, $t \geq 0$, is an optimal admissible pair in the problem $(P3)$ taken with initial condition $x(0) = x_*(\tau)$. Hence, using the same arguments as above we get the following inequality for $(\tilde{x}_*(\cdot), \tilde{u}_*(\cdot))$ (see (27)):

$$\tilde{u}_*(t) \overset{\text{a.e.}}{\leq} \left(1 + \frac{1}{K\tilde{x}_*(0)}\right)(r+\rho)e^{rt}, \qquad t \geq 0.$$

Hence, for arbitrary fixed $\tau > 0$ we have

$$u_*(t) = \tilde{u}_*(t-\tau) \overset{\text{a.e.}}{\leq} \left(1 + \frac{1}{Kx_*(\tau)}\right)(r+\rho)e^{r(t-\tau)}, \qquad t \geq \tau.$$

Due to arbitrariness of $\tau > 0$ this implies (22). □

Theorem 2 *Let $(x_*(\cdot), u_*(\cdot))$ be an optimal admissible pair in problem $(P3)$. Then the function $\psi : [0, \infty) \mapsto \mathbb{R}^1$ defined for pair $(x_*(\cdot), u_*(\cdot))$ by formula* (21) *is (locally) absolutely continuous and satisfies the conditions of the normal form maximum principle, i.e.* $\psi(\cdot)$ *is a solution of the adjoint system*

$$\dot{\psi}(t) = -\mathcal{H}_x\left(x_*(t), u_*(t), \psi(t)\right), \tag{28}$$

and the maximum condition holds:

$$\mathcal{H}(x_*(t), u_*(t), \psi(t)) \stackrel{a.e.}{=} H(x_*(t), \psi(t)). \tag{29}$$

Proof As it already have been shown above condition $(A1)$ is satisfied and $(A2)$ holds for any admissible pair $(x_*(\cdot), u_*(\cdot))$ in $(P3)$. Hence, due to the variant of the maximum principle developed in Aseev and Veliov (2012, 2014, 2015) the function $\psi : [0, \infty) \mapsto \mathbb{R}^1$ defined for pair $(x_*(\cdot), u_*(\cdot))$ by formula (21) satisfies the conditions (28) and (29). □

Notice, that the Cauchy type formula (21) implies (see (16) and (19))

$$\psi(t) = -e^{-\int_0^t u_*(\xi)\,d\xi + rt} \int_t^\infty \frac{e^{-\rho\tau} e^{\int_0^\tau u_*(\xi)\,d\xi - r\tau}}{e^{\int_0^\tau u_*(\xi)\,d\xi - r\tau}\left[x_0 + a\int_0^\tau e^{-\int_0^\theta u_*(\xi)\,d\xi + r\theta}\,d\theta\right]}\,d\tau$$

$$> -\frac{e^{-\int_0^t u_*(\xi)\,d\xi + rt}}{x_0 + a\int_0^t e^{-\int_0^\theta u_*(\xi)\,d\xi + r\theta}\,d\theta} \int_t^\infty e^{-\rho\tau}\,d\tau = -\frac{e^{-\rho t}}{\rho x_*(t)}, \qquad t \geq 0. \tag{30}$$

Thus, due to (21) the following condition holds:

$$0 < -\psi(t)x_*(t) < \frac{e^{-\rho t}}{\rho}, \qquad t \geq 0. \tag{31}$$

Note also, that due to (Aseev 2015a, Corollary to Theorem 3) formula (21) implies the following stationarity condition for the Hamiltonian (see Aseev and Kryazhimskii 2007; Michel 1982):

$$H(t, x_*(t), \psi(t)) = \rho \int_t^\infty e^{-\rho s} g(x_*(s), u_*(s))\,ds, \qquad t \geq 0. \tag{32}$$

It can be shown directly that if an admissible pair (not necessary optimal) $(x(\cdot), u(\cdot))$ together with an adjoint variable $\psi(\cdot)$ satisfies the core conditions (28) and (29) of the maximum principle and $\lim_{t\to\infty} H(t, x(t), \psi(t)) = 0$ then condition (32) holds for the triple $(x(\cdot), u(\cdot), \psi(\cdot))$ as well (see Aseev and Kryazhimskii 2007, Section 3).

Further, due to the maximum condition (29) for a.e. $t \geq 0$ we have

$$u_*(t) = \arg\max_{u\in[\rho,\infty)} \left[\psi(t)x_*(t)u + e^{-\rho t}\ln u\right].$$

This implies (see (31))

$$u_*(t) \stackrel{a.e.}{=} -\frac{e^{-\rho t}}{\psi(t)x_*(t)} > \rho, \qquad t \in [0, \infty). \tag{33}$$

Substituting this formula for $u_*(\cdot)$ in (14) and in (28) due to Theorem 2 we get that any optimal trajectory $x_*(\cdot)$ together with the corresponding adjoint variable $\psi(\cdot)$ must satisfy to the Hamiltonian system of the maximum principle:

$$\begin{aligned} \dot{x}(t) &= -rx(t) - \frac{e^{-\rho t}}{\psi(t)} + a, \\ \dot{\psi}(t) &= r\psi(t) + \frac{2e^{-\rho t}}{x(t)}. \end{aligned} \tag{34}$$

Moreover, estimate (31) and condition (32) must hold as well.

In the terms of the current value adjoint variable $\lambda(\cdot)$, $\lambda(t) = e^{\rho t}\psi(t)$, $t \geq 0$, one can rewrite system (34) as follows:

$$\begin{aligned} \dot{x}(t) &= -rx(t) - \frac{1}{\lambda(t)} + a, \\ \dot{\lambda}(t) &= (\rho + r)\lambda(t) + \frac{2}{x(t)}. \end{aligned} \tag{35}$$

In terms of variable $\lambda(\cdot)$ estimate (31) takes the following form:

$$0 < -\lambda(t)x_*(t) < \frac{1}{\rho}, \qquad t \geq 0. \tag{36}$$

Accordingly, the optimal control $u_*(\cdot)$ can be expressed as follows (see (33)):

$$u_*(t) \stackrel{a.e.}{=} -\frac{1}{\lambda(t)x_*(t)}, \qquad t \geq 0. \tag{37}$$

Define the normal form current value Hamiltonian $M : (0, \infty) \times \mathbb{R}^1 \mapsto \mathbb{R}^1$ for problem $(P3)$ in the standard way (see Aseev and Kryazhimskii 2007, Section 3):

$$M(x, \lambda) = e^{\rho t} H(t, x, \psi), \quad x \in (0, \infty), \quad \lambda \in \mathbb{R}^1. \tag{38}$$

Then in the current value terms the stationarity condition (32) takes the form

$$M(x_*(t), \lambda(t)) = \rho e^{\rho t} \int_t^{\infty} e^{-\rho s} g(x_*(s), u_*(s))\, ds, \qquad t \geq 0. \tag{39}$$

In the next section we will analyze the system (35) coupled with the estimate (36) and the stationarity condition (39). We will show that there are only two qualitatively different types of behavior of the optimal paths that are possible. If $r > \rho$ then the optimal path asymptotically approaches an optimal nonvanishing steady state while the corresponding optimal control tends to $(r + \rho)/2$ as $t \to \infty$. If $r \leq \rho$ then the optimal path $x_*(\cdot)$ goes to infinity, while the corresponding optimal control $u_*(\cdot)$ tends to ρ as $t \to \infty$, i.e. asymptotically it follows the Hotelling rule of optimal depletion of an exhaustible resource (Hotelling 1974).

4 Analysis of the Hamiltonian System

Due to Theorem 2 it is sufficient to analyze the behavior of system (35) only in the open set $\Gamma = \{ (x, \lambda) \colon x > 0, \lambda < 0\}$ in the phase plane $\mathbb{R}^2$.

Let us introduce functions $y_1 \colon (1/K, \infty) \mapsto (-\infty, 0)$ and $y_2 \colon (0, \infty) \mapsto (-\infty, 0)$ as follows (recall that $a = r/K$):

$$y_1(x) = \frac{1}{a - rx}, \qquad x \in \left(\frac{1}{K}, \infty\right), \qquad y_2(x) = -\frac{2}{(\rho + r)x}, \qquad x \in (0, \infty).$$

Due to (35) the curves $\gamma_1 = \{(x, \lambda) \colon \lambda = y_1(x), x \in (1/K, \infty)\}$ and $\gamma_2 = \{(x, \lambda) \colon \lambda = y_2(x), x \in (0, \infty)\}$ are the nullclines at which the derivatives of variables $x(\cdot)$ and $\lambda(\cdot)$ vanish respectively.

Two qualitatively different cases are possible: (i) $r > \rho$ and (ii) $r \leq \rho$.

Consider case (i). In this case the nullclines γ_1 and γ_2 have a unique intersection point $(\hat{x}, \hat{\lambda})$ which is a unique equilibrium of system (35) in Γ:

$$\hat{x} = \frac{2r}{(r - \rho)K}, \qquad \hat{\lambda} = \frac{(\rho - r)K}{(\rho + r)r}. \tag{40}$$

The corresponding equilibrium control $\hat{u}(\cdot)$ is

$$\hat{u}(t) \equiv \hat{u} = \frac{\rho + r}{2}, \qquad t \geq 0. \tag{41}$$

The eigenvalues of the system linearized around the equilibrium are given by

$$\sigma_{1,2} = \frac{\rho}{2} \pm \frac{1}{2}\sqrt{2r^2 - \rho^2},$$

which are real and distinct with opposite signs when $r > \rho$. Hence, by the Grobman-Hartman theorem in a neighborhood Ω of the equilibrium state $(\hat{x}, \hat{\lambda})$ the system (35) is of saddle type (see Hartman 1964, Chapter 9).

The nullclines γ_1 and γ_2 divide the set Γ in four open regions:

$$\Gamma_{-,-} = \left\{(x,\lambda) \in \Gamma \colon \lambda < y_1(x), \frac{1}{K} < x \leq \hat{x}\right\} \bigcup \left\{(x,\lambda) \in \Gamma \colon \lambda < y_2(x), \hat{x} < x < \infty\right\},$$

$$\Gamma_{+,-} = \left\{(x,\lambda) \in \Gamma \colon \lambda < y_2(x), 0 < x \leq \frac{1}{K}\right\} \bigcup \left\{(x,\lambda) \in \Gamma \colon y_1(x) < \lambda < y_2(x), \frac{1}{K} < x < \hat{x}\right\},$$

$$\Gamma_{+,+} = \left\{(x,\lambda) \in \Gamma \colon y_2(x) < \lambda < 0, 0 < x \leq \hat{x}\right\} \bigcup \left\{(x,\lambda) \in \Gamma \colon y_1(x) < \lambda < 0, \hat{x} < x < \infty\right\},$$

$$\Gamma_{-,+} = \left\{(x,\lambda) \in \Gamma \colon y_2(x) < \lambda < y_1(x), x > \hat{x}\right\}.$$

Any solution $(x(\cdot), \lambda(\cdot))$ of (35) in Γ has definite signs of derivatives of its (x, λ)-coordinates in the sets $\Gamma_{-,-}$, $\Gamma_{-,+}$, $\Gamma_{+,+}$, and $\Gamma_{-,+}$. These signs are indicated by the corresponding subscripts.

The behavior of the flows is shown in Fig. 1 through the phase portrait.

For any initial state $(\xi, \beta) \in \Gamma$ there is a unique solution $(x_{\xi,\beta}(\cdot), \lambda_{\xi,\beta}(\cdot))$ of the system (35) satisfying initial conditions $x(0) = \xi$, $\lambda(0) = \beta$, and due to the standard extension result this solution is defined on some maximal time interval $[0, T_{\xi,\beta})$ in Γ where $0 < T_{\xi,\beta} \leq \infty$ (see Hartman 1964, Chapter 2).

Let us consider behaviors of solutions $(x_{\xi,\beta}(\cdot), \lambda_{\xi,\beta}(\cdot))$ of system (35) in Γ for all possible initial states $(\xi, \beta) \in \Gamma$ as $t \to T_{\xi,\beta}$.

The standard analysis of system (35) shows that only three types of behavior of solutions $(x_{\xi,\beta}(\cdot), \lambda_{\xi,\beta}(\cdot))$ of (35) in Γ as $t \to T_{\xi,\beta}$ are possible:

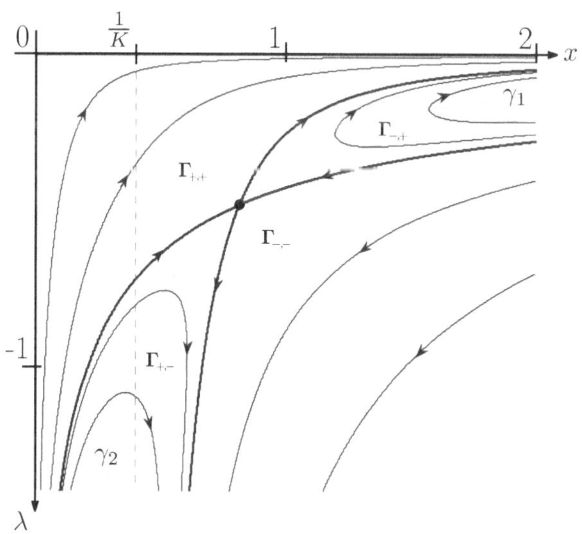

Fig. 1 Phase portrait of (35) around $(\hat{x}, \hat{\lambda})$. Here $r = 5$, $\rho = 0.1$, and $K = 2.5$

1. $(x_{\xi,\beta}(t), \lambda_{\xi,\beta}(t)) \in \Gamma_{-,-}$ or $(x_{\xi,\beta}(t), \lambda_{\xi,\beta}(t)) \in \Gamma_{+,-}$ for all sufficiently large times t. In this case $T_{\xi,\beta} = \infty$ and $\lim_{t\to\infty} \lambda_{\xi,\beta}(t) = -\infty$ while $\lim_{t\to\infty} x_{\xi,\beta}(t) = 1/K$. Due to Theorem 2 such asymptotic behavior does not correspond to an optimal path because it contradicts the necessary condition (36).
2. $(x_{\xi,\beta}(t), \lambda_{\xi,\beta}(t)) \in \Gamma_{+,+}$ for all sufficiently large times t. In this case $\lim_{t\to T_{\xi,\beta}} x_{\xi,\beta}(t) = \infty$ and $\lim_{t\to T_{\xi,\beta}} \lambda_{\xi,\beta}(t) = 0$. If $(x_{\xi,\beta}(\cdot), \lambda_{\xi,\beta}(\cdot))$ corresponds to an optimal pair $(x_*(\cdot), u_*(\cdot))$ in $(P3)$ then due to Theorem 2 $x_*(\cdot) \equiv x_{\xi,\beta}(\cdot)$, $T_{\xi,\beta} = \infty$, $\lim_{t\to\infty} x_*(t) = \infty$, and $\lim_{t\to\infty} \lambda_{\xi,\beta}(t) = 0$. Set $\lambda_*(\cdot) \equiv \lambda_{\xi,\beta}(\cdot)$ in this case and define the function $\phi_*(\cdot)$ by equality $\phi_*(t) = \lambda_*(t)x_*(t)$, $t \in [0, \infty)$.

By direct differentiation for a.e. $t \in [0, \infty)$ we get (see (35))

$$\dot{\phi}_*(t) \overset{\text{a.e.}}{=} (\rho + r)\lambda_*(t)x_*(t) + 2 - r\lambda(t)x_*(t) - 1 + a\lambda_*(t) = \rho\phi_*(t) + 1 + a\lambda_*(t).$$

Hence,

$$\phi_*(t) = e^{\rho t}\left[\phi_*(0) + \int_0^t e^{-\rho s}\,(1 + a\lambda_*(s))\,ds\right], \qquad t \in [0, \infty). \tag{42}$$

Since $\lim_{t\to\infty} \lambda_*(t) = 0$ the improper integral $\int_0^\infty e^{-\rho s}\,(1 + a\lambda_*(s))\,ds$ converges, and due to (36) we have $0 > \phi_*(t) = \lambda_*(t)x_*(t) > -1/\rho$ for all $t > 0$. Due to (42) this implies

$$\phi_*(0) = -\int_0^\infty e^{-\rho s}\,(1 + a\lambda_*(s))\,ds = -\frac{1}{\rho} - a\int_0^\infty e^{-\rho s}\lambda_*(s)\,ds.$$

Substituting this expression for $\phi_*(0)$ in (42) we get

$$\phi_*(t) = -\frac{1}{\rho} - ae^{\rho t}\int_t^\infty e^{-\rho s}\lambda_*(s)\,ds, \qquad t \in [0, \infty).$$

Due to the L'Hospital rule we have

$$\lim_{t\to\infty} e^{\rho t}\int_t^\infty e^{-\rho s}\lambda_*(s)\,ds = \lim_{t\to\infty} \frac{\int_t^\infty e^{-\rho s}\lambda_*(s)\,ds}{e^{-\rho t}} = \lim_{t\to\infty} \frac{\lambda_*(t)}{\rho} = 0.$$

Hence,

$$\lim_{t\to\infty} u_*(t) = \lim_{t\to\infty} \frac{-1}{\lambda_*(t)x_*(t)} = \lim_{t\to\infty} \frac{-1}{\phi_*(t)} = \rho.$$

But due to the system (35) and the inequality $r > \rho$ this implies $\lim_{t\to\infty} x_*(t) \le a < \infty$ that contradicts the equality $\lim_{t\to\infty} x_*(t) = \infty$. So, all these trajectories of (35) are the blow up ones. Thus, there are not any trajectories of (35) that correspond to optimal admissible pairs due to Theorem 2 in the case 2).

3. $\lim_{t\to\infty}(x(t), \lambda(t)) = (\hat{x}, \hat{\lambda})$ as $t \to \infty$. In this case, since the equilibrium $(\hat{x}, \hat{\lambda})$ is of saddle type, there are only two trajectories of (35) which tend to the equilibrium point $(\hat{x}, \hat{\lambda})$ asymptotically as $t \to \infty$ and lying on the stable manifold of $(\hat{x}, \hat{\lambda})$. One such trajectory $(x_1(\cdot), \lambda_1(\cdot))$ approaches the point $(\hat{x}, \hat{\lambda})$ from the left from the set $\Gamma_{+,+}$ (we call this trajectory *the left equilibrium trajectory*), while the second trajectory $(x_2(\cdot), \lambda_2(\cdot))$ approaches the point $(\hat{x}, \hat{\lambda})$ from the right from the set $\Gamma_{-,-}$ (we call this trajectory *the right equilibrium trajectory*). It is easy to see that both these trajectories are fit to estimate (36) and stationarity condition (39). Hence, $(x_1(\cdot), \lambda_1(\cdot))$, $(x_2(\cdot), \lambda_2(\cdot))$ and the stationary trajectory $(\hat{x}(\cdot), \hat{\lambda}(\cdot))$, $\hat{x}(\cdot) \equiv \hat{x}$, $\hat{\lambda}(\cdot) \equiv \hat{\lambda}$, $t \geq 0$, are unique trajectories of (35) which can correspond to the optimal pairs in problem $(P3)$ due to Theorem 2.

Due to Theorem 1 for any initial state $x_0 > 0$ an optimal control $u_*(\cdot)$ in problem $(P3)$ exists. Hence, for any initial state $\xi \in (0, \hat{x})$ there is a unique $\beta < 0$ such that the corresponding trajectory $(x_{\xi,\beta}(\cdot), \lambda_{\xi,\beta}(\cdot))$ coincides (up to a shift in time) with the left equilibrium trajectory $(x_1(\cdot), \lambda_1(\cdot))$ on time interval $[0, \infty)$. Analogously, for any initial state $\xi > \hat{x}$ there is a unique $\beta < 0$ such that the corresponding trajectory $(x_{\xi,\beta}(\cdot), , \lambda_{\xi,\beta}(\cdot))$ coincides (up to a shift in time) with the right equilibrium trajectory $(x_2(\cdot), \lambda_2(\cdot))$ on $[0, \infty)$. The corresponding optimal control is defined uniquely by (37). Thus, for any initial state $x_0 > 0$ the corresponding optimal pair $(x_*(\cdot), u_*(\cdot))$ in $(P3)$ is unique, and due to Theorem 2 the corresponding current value adjoint variable $\lambda_*(\cdot)$ is also unique.

Further, to the left of the point $(\hat{x}, \hat{\lambda})$ in the set $\Gamma_{+,+}$, the function $x_1(\cdot)$ increases. Therefore, while $(x_1(\cdot), \lambda_1(\cdot))$ lies in $\Gamma_{+,+}$, the time can be uniquely expressed in terms of the first coordinate of the trajectory $(x_1(\cdot), \lambda_1(\cdot))$ as a smooth function $t = t_1(x)$, $x \in (0, \hat{x})$. Changing the time variable $t = t_1(x)$ on interval $(0, \hat{x})$, we find that the function $\lambda_-(x) = \lambda_1(t_1(x))$, $x \in (0, \hat{x})$, is a solution to the following differential equation on $(0, \hat{x})$:

$$\frac{d\lambda(x)}{dx} = \frac{d\lambda(t_1(x))}{dt} \times \frac{dt_1(x)}{dx} = \frac{\lambda(x)\,((\rho + r)\lambda(x)x + 2)}{x(-r\lambda(x)x - 1 + a\lambda(x))} \tag{43}$$

with the boundary condition

$$\lim_{x\to\hat{x}-0} \lambda(x) = \hat{\lambda}. \tag{44}$$

Obviously, the curve $\lambda_- = \{(x, \lambda) : \lambda = \lambda_-(x), x \in (0, \hat{x})\}$ corresponds to the region of the stable manifold of $(\hat{x}, \hat{\lambda})$ where $x < \hat{x}$.

Analogously, to the right of the point $(\hat{x}, \hat{\lambda})$ in the set $\Gamma_{-,-}$, while $(x_1(\cdot), \lambda_1(\cdot))$ lies in $\Gamma_{-,-}$,the function $x_1(\cdot)$ decreases. Hence, the time can be uniquely expressed in terms of the first coordinate of the trajectory $(x_1(\cdot), \lambda_1(\cdot))$ as a smooth function $t = t_2(x)$, $x \in (\hat{x}, \infty)$. Changing the time variable $t = t_2(x)$ on interval $(\hat{x}, \infty)$, we find that the function $\lambda_+(x) = \lambda_2(t_2(x))$, $x > \hat{x}$, is a solution to the differential equation (43) on $(\hat{x}, \infty)$ with the boundary condition

$$\lim_{x \to \hat{x}+0} \lambda(x) = \hat{\lambda}. \tag{45}$$

As above, the curve $\lambda_+ = \{(x, \lambda) \colon \lambda = \lambda_+(x), x \in (\hat{x}, \infty)\}$ corresponds to the region of the stable manifold of $(\hat{x}, \hat{\lambda})$ where $x > \hat{x}$.

Using solutions $\lambda_-(\cdot)$ and $\lambda_+(\cdot)$ of differential equation (43) along with (37) we can get an expression for the optimal feedback law as follows

$$u_*(x) = \begin{cases} -\frac{1}{\lambda_-(x)x}, & \text{if } x < \hat{x}, \\ \frac{\rho+r}{2}, & \text{if } x = \hat{x}, \\ -\frac{1}{\lambda_+(x)x}, & \text{if } x > \hat{x}. \end{cases}$$

Now, consider the case (ii) when $r \leq \rho$. In this case $y_2(x) > y_1(x)$ for all $x > 1/K$ and hence the nullclines γ_1 and γ_2 do not intersect in Γ. Accordingly, the system (35) does not have an equilibrium point in Γ.

The nullclines γ_1 and γ_2 divide the set Γ in three open regions:

$$\hat{\Gamma}_{-,-} = \left\{(x, \lambda) \in \Gamma \colon \lambda < y_1(x), x > \frac{1}{K}\right\},$$

$$\hat{\Gamma}_{+,-} = \left\{(x, \lambda) \in \Gamma \colon \lambda < y_2(x), 0 < x \leq \frac{1}{K}\right\} \bigcup \left\{(x, \lambda) \in \Gamma \colon y_1(x) < \lambda < y_2(x), x > \frac{1}{K}\right\},$$

$$\hat{\Gamma}_{+,+} = \left\{(x, \lambda) \in \Gamma \colon y_2(x) < \lambda < 0, 0 < x \leq \hat{x}\right\} \bigcup \left\{(x, \lambda) \in \Gamma \colon y_1(x) < \lambda < 0, \hat{x} < x < \infty\right\},$$

The behavior of the flows is shown in Fig. 2 through the phase portrait.

Any solution $(x(\cdot), \lambda(\cdot))$ of (35) in Γ has the definite signs of derivatives of its (x, λ) coordinates in each set $\hat{\Gamma}_{-,-}$, $\hat{\Gamma}_{+,+}$, and $\hat{\Gamma}_{-,+}$ as indicated by the subscripts.

The standard analysis of the behaviors of solutions $(x(\cdot), \lambda(\cdot))$ of system (35) in each of sets $\hat{\Gamma}_{-,-}$, $\hat{\Gamma}_{+,-}$ and $\Gamma_{+,+}$ shows that there are only two types of asymptotic behavior of solutions $(x(\cdot), \lambda(\cdot))$ of (35) that are possible:

1. $\lim_{t\to\infty} x(t) = 1/K$, $\lim_{t\to\infty} \lambda(t) = -\infty$. In this case $(x(t), \lambda(t)) \in \hat{\Gamma}_{-,-}$ for all sufficiently large times $t \geq 0$. Due to Theorem 2 such asymptotic behavior does not correspond to an optimal admissible pair because in this case $\lim_{t\to\infty} \lambda(t)x(t) = -\infty$ that contradicts condition (36). Thus this case can be eliminated from the consideration.
2. $\lim_{t\to\infty} x(t) = \infty$, $\lim_{t\to\infty} \lambda(t) = 0$. In this case $(x(t), \lambda(t)) \in \hat{\Gamma}_{+,+}$ for all $t \geq 0$. Since the case (1) can be eliminated from the consideration, we conclude that the case (2) is the only one that can be realized for an optimal admissible pair $(x_*(\cdot), u_*(\cdot))$ (which exists) in $(P3)$ due to the maximum principle (Theorem 2).

Fig. 2 Phase portrait of (35) in the case $r < \rho$. Here $r = 0.1$, $\rho = 0.5$, and $K = 2.5$

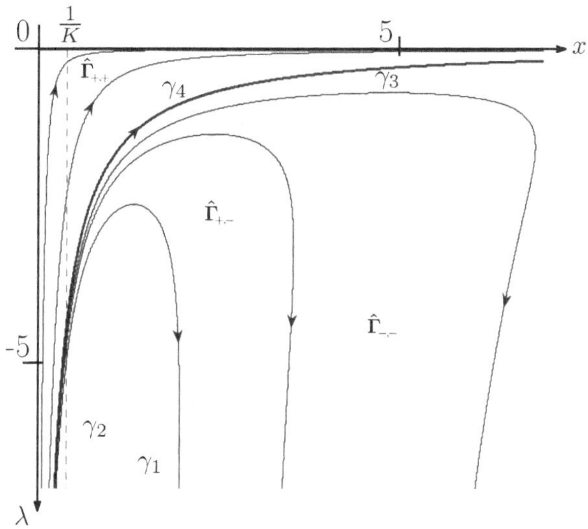

Let us consider behavior of trajectory $(x_*(\cdot), \lambda_*(\cdot))$ of system (35) that corresponds to the optimal pair $(x_*(\cdot), u_*(\cdot))$ in the set $\hat{\Gamma}_{+,+}$ in more details.

As in the subcase (b) of case (i) above, define the function $\phi_*(\cdot)$ as follows:

$$\phi_*(t) = \lambda_*(t)x_*(t), \qquad t \in [0, \infty).$$

Repeating the calculations presented in the subcase (b) of case (i) we get

$$\phi_*(t) = -\frac{1}{\rho} - ae^{\rho t}\int_t^{\infty} e^{-\rho s}\lambda_*(s)\,ds, \qquad t \in [0, \infty).$$

As in the subcase (b) of case (i) above, due to the L'Hospital rule this implies

$$\lim_{t\to\infty} e^{\rho t}\int_t^{\infty} e^{-\rho s}\lambda_*(s)\,ds = \lim_{t\to\infty}\frac{\int_t^{\infty} e^{-\rho s}\lambda_*(s)\,ds}{e^{-\rho t}} = \lim_{t\to\infty}\frac{\lambda_*(t)}{\rho} = 0.$$

Hence,

$$\lim_{t\to\infty} u_*(t) = \lim_{t\to\infty}\frac{-1}{\lambda_*(t)x_*(t)} = \lim_{t\to\infty}\frac{-1}{\phi_*(t)} = \rho.$$

Thus, asymptotically, any optimal admissible control $u_*(\cdot)$ satisfies the Hotelling rule (Hotelling 1974) of optimal depletion of an exhaustible resource in the case (ii).

Now let us show that the optimal control $u_*(\cdot)$ is defined uniquely by Theorem 2 in the case (ii).

Define the function $y_3\colon (0,\infty) \mapsto \mathbb{R}^1$ and the curve $\gamma_3 \subset \Gamma$ as follows:

$$y_3(x) = -\frac{1}{\rho x}, \quad x \in (0,\infty), \qquad \gamma_3 = \{(x,\lambda)\colon \lambda = y_3(x), x \in (0,\infty)\}.$$

It is easy to see that $y_3(x) \geq y_2(x)$ for all $x > 0$ and $y_3(x) > y_1(x)$ for all $x > 1/K$ in the case (ii). Hence, the curve γ_3 is located not below γ_2 and strictly above γ_1 in $\hat{\Gamma}_{+,+}$ (see Fig. 2). Notice that if $r = \rho$ then γ_3 coincide with γ_2 while if $r < \rho$ then γ_3 lies strictly above γ_2 in $\hat{\Gamma}_{+,+}$. It can be demonstrated directly that any trajectory $(x(\cdot), \lambda(\cdot))$ of system (35) can intersect curve γ_3 only one time and only in the upward direction.

Due to (36) a trajectory $(x_*(\cdot), \lambda_*(\cdot))$ of system (35) that corresponds to the optimal pair $(x_*(\cdot), u_*(\cdot))$ lies strictly above γ_3. Since the system (35) is autonomous by virtue of the theorem on uniqueness of a solution of first-order ordinary differential equation (see Hartman 1964, Chapter 3) trajectories of system (35) that lie above γ_3 do not intersect the curve $\gamma_4 = \{(x,\lambda)\colon x = x_*(t), \lambda = \lambda_*(t), t \geq 0\}$ which is the graph of the trajectory $(x_*(\cdot), \lambda_*(\cdot))$.

Further, trajectory $(x_*(\cdot), \lambda_*(\cdot))$ is defined on infinite time interval $[0,\infty)$. This implies that all trajectories $\big(x_{x_0,\beta}(\cdot), \lambda_{x_0,\beta}(\cdot)\big)$, $\beta \in (-1/(\rho x_0), \lambda_*(0))$, are also defined on the whole infinite time interval $[0,\infty)$, i.e. $T_{x_0,\beta} = \infty$ for all $\beta \in (-1/(\rho x_0), \lambda_*(0))$. Thus, we have proved that there is a nonempty set (a continuum) of trajectories $\big\{(x_{x_0,\beta}(\cdot), \lambda_{x_0,\beta}(\cdot))\big\}$, $\beta \in (-1/(\rho x_0), \lambda_*(0))$, $t \in [0,\infty)$, of system (35) lying strictly between the curves γ_3 and γ_4. All these trajectories are defined on the whole infinite time interval $[0,\infty)$ and, hence, all of them correspond to some admissible pairs $\big\{(x_{x_0,\beta}(\cdot), u_{x_0,\beta}(\cdot))\big\}$. Since these trajectories are located above γ_3 they satisfy also the estimate (36).

Consider the current value Hamiltonian $M(\cdot,\cdot)$ for (x,λ) lying above γ_3 in $\hat{\Gamma}_{+,+}$ (see (38)):

$$\begin{aligned} M(x,\lambda) &= \sup_{u \geq \rho}\{u\lambda x + \ln u\} + (a - rx)\lambda - \ln x \\ &= -1 - \ln(-\lambda x) + (a - rx)\lambda - \ln x, \qquad -\frac{1}{\rho x} < \lambda < 0. \end{aligned} \tag{46}$$

For any trajectory $(x_{x_0,\beta}(\cdot), \lambda_{x_0,\beta}(\cdot))$ of system (35) lying above γ_3 in $\hat{\Gamma}_{+,+}$ we have

$$x_{x_0,\beta}(t) \geq e^{(\rho - r)t} x_0, \quad t \geq 0.$$

On the other hand for any trajectory $(x_{x_0,\beta}(\cdot), \lambda_{x_0,\beta}(\cdot))$ of system (35) lying between γ_3 and γ_4 in $\hat{\Gamma}_{+,+}$ we have

$$\frac{1}{2(r+\rho)} < -\lambda_{x_0,\beta}(t) x_{x_0,\beta}(t) < \frac{1}{\rho} \quad \text{if} \quad x_{x_0,\beta}(t) > \frac{1}{K}.$$

These imply that for any trajectory $(x_{x_0,\beta}(\cdot), \lambda_{x_0,\beta}(\cdot))$ of system (35) lying between γ_3 and γ_4 in $\hat{\Gamma}_{+,+}$ and for corresponding adjoint variable $\psi_{x_0,\beta}(\cdot)$, $\psi_{x_0,\beta}(t) = e^{-\rho t}\lambda_{x_0,\beta}(t)$, $t \geq 0$, we have

$$\lim_{t\to\infty} H(t, x_{x_0,\beta}(t), \psi_{x_0,\beta}(t)) = \lim_{t\to\infty} \left\{e^{-\rho t} M(x_{x_0,\beta}(t), \lambda_{x_0,\beta}(t))\right\} = 0.$$

Hence, for any such trajectory $(x_{x_0,\beta}(\cdot), \lambda_{x_0,\beta}(\cdot))$ of system (35) we have (see (39))

$$M(x_{x_0,\beta}(t), \lambda_{x_0,\beta}(t)) = \rho e^{\rho t} \int_t^{\infty} e^{-\rho s} g(x_{x_0,\beta}(t), \lambda_{x_0,\beta}(t))\, ds, \qquad t \geq 0.$$

Let $u_{x_0,\beta}(\cdot)$ be the control corresponding to $x_{x_0,\beta}(\cdot)$, i.e. $u_{x_0,\beta}(t) = -1/(x_{x_0,\beta}(t)\lambda_{x_0,\beta}(t))$. Then taking in the last equality $t = 0$ we get

$$J(x_{x_0,\beta}(\cdot), u_{x_0,\beta}(\cdot)) = \int_0^{\infty} e^{-\rho s} g(x_{x_0,\beta}(t), \lambda_{x_0,\beta}(t))\, ds = \frac{1}{\rho} M(x_{x_0,\beta}(0), \lambda_{x_0,\beta}(0)).$$

For any $t \geq 0$ function $M(x_*(t), \cdot)$ (see (46)) increases on $\{\lambda\colon -1/(\rho x_*(t)) < \lambda < 0\}$. Hence, $M(x_*(t), \cdot)$ reaches its maximal value in λ on the set $\{\lambda\colon -1/(\rho x) < \lambda \leq \lambda_*(t)\}$ at the point $\lambda_*(t)$ that correspond to the optimal path $x_*(\cdot)$. Thus, all trajectories $(x_{x_0,\beta}(\cdot), \lambda_{x_0,\beta}(\cdot))$ of system (35) lying between γ_3 and γ_4 in $\hat{\Gamma}_{+,+}$ do not correspond to optimal admissible pairs in $(P3)$.

From this we can also conclude that all trajectories $(x(\cdot), \lambda(\cdot))$ of system (35) lying above γ_4 also do not correspond to optimal admissible pars in $(P3)$. Indeed, if such trajectory $(x(\cdot), \lambda(\cdot))$ corresponds to an optimal pair $(x(\cdot), u(\cdot))$ in $(P3)$ then it must satisfy to condition (39). But in this case we have $\lambda(0) > \lambda_*(0)$ and

$$J(x(\cdot), u(\cdot)) = \frac{1}{\rho} M(x_0, \lambda(0)) = \frac{1}{\rho} M(x_0, \lambda_*(0)) = J(x_*(\cdot), \lambda_*(\cdot)),$$

that contradicts the fact that function $M(x_0, \cdot)$ increases on $\{\lambda\colon -1/(\rho x) < \lambda < 0\}$.

Thus, for any initial state x_0 there is a unique optimal pair $(x_*(\cdot), u_*(\cdot))$ in $(P3)$ in the case (ii). The corresponding current value adjoint variable $\lambda_*(\cdot)$ is also defined uniquely as the maximal negative solution to equation (see (35))

$$\dot{\lambda}(t) = (\rho + r)\lambda(t) + \frac{2}{x_*(t)} \tag{47}$$

on the whole infinite time interval $[0, \infty)$.

The function $x_*(\cdot)$ increases on $[0, \infty)$. Therefore, the time can be uniquely expressed as a smooth function $t = t_*(x)$, $x \in (0, \infty)$. Changing the time variable $t = t_*(x)$, we find that the function $\lambda_0(x) = \lambda_*(t_*(x))$ is solution to the differential equation (43) on the infinite interval $(0, \infty)$.

Using solution $\lambda_0(\cdot)$ of differential equation (43) along with (37) we can get an expression for the optimal feedback law as follows

$$u_*(x) = -\frac{1}{\lambda_0(x)x}, \qquad x > 0.$$

Thus, to find the optimal feedback, we must determine for an initial state $x_0 > 0$ the corresponding initial state $\lambda_0 < 0$ such that solution $(x_*(\cdot), \lambda_*(\cdot))$ of system (35) with initial conditions $x(0) = x_0$ and $\lambda(0) = \lambda_0$ exists on $[0, \infty)$ and $\lambda_*(\cdot)$ is the maximal negative function among all such solutions.

Let us summarize the results obtained in this section in the following theorem.

Theorem 3 *For any initial state $x_0 > 0$ there is a unique optimal admissible pair $(x_*(\cdot), u_*(\cdot))$ in problem $(P3)$, and there is a unique adjoint variable $\psi(\cdot)$ that corresponds $(x_*(\cdot), u_*(\cdot))$ due to the maximum principle (Theorem 2).*

If $r > \rho$ then there is a unique equilibrium $(\hat{x}, \hat{\lambda})$ (see (40)*) in the corresponding current value Hamiltonian system* (35) *and the optimal synthesis is defined as follows*

$$u_*(x) = \begin{cases} -\frac{1}{\lambda_-(x)x}, & \text{if } x < \hat{x}, \\ \frac{r+\rho}{2}, & \text{if } x = \hat{x}, \\ -\frac{1}{\lambda_+(x)x}, & \text{if } x > \hat{x}, \end{cases}$$

where $\lambda_-(\cdot)$ and $\lambda_+(\cdot)$ are the unique solutions of (43) *that satisfy the boundary conditions* (44) *and* (45) *respectively. In this case optimal path $x_*(\cdot)$ is either decreasing, or increasing on $[0, \infty)$, or $x_*(t) \equiv \hat{x}$, $t \geq 0$, depending on the initial state x_0. For any optimal admissible pair $(x_*(\cdot), u_*(\cdot))$ we have $\lim_{t\to\infty} x_*(t) = \hat{x}$ and $\lim_{t\to\infty} u_*(t) = \hat{u}$ (see* (41)*).*

If $r \leq \rho$ then for any initial state x_0 the corresponding optimal path $x_*(\cdot)$ in problem $(P3)$ is an increasing function, $\lim_{t\to\infty} x_*(t) = \infty$, and the corresponding optimal control $u_*(\cdot)$ satisfies asymptotically to the Hotelling rule of optimal depletion of an exhaustible resource, i.e. $\lim_{t\to\infty} u_*(t) = \rho$. The corresponding current value adjoint variable $\lambda_*(\cdot)$ is defined uniquely as the maximal negative solution to Eq. (47) *on $[0, \infty)$. The optimal synthesis is defined as*

$$u_*(x) = -\frac{1}{\lambda_0(x)x}, \qquad x > 0,$$

where $\lambda_0(x) = \lambda_*(t_*(x))$ is the corresponding solution of (43).

In the next section we discuss the issue of sustainability of optimal paths for different values of the parameters in the model.

5 Conclusion

Following Solow (1956) we assume that the knowledge stock $A(\cdot)$ grows exponentially, i.e. $A(t) = A_0 e^{\mu t}$, $t \geq 0$, where $\mu \geq 0$ and $A_0 > 0$ are constants.

Similar to Valente (2005) we say that an admissible pair $(S(\cdot), u(\cdot))$ is *sustainable* in our model if the corresponding instantaneous utility function $t \mapsto \ln Y(t)$, $t \geq 0$, non-decreases in the long run, i.e.

$$\lim_{T\to\infty} \inf_{t\geq T} \frac{d}{dt} \ln Y(t) = \lim_{T\to\infty} \inf_{t\geq T} \frac{\dot{Y}(t)}{Y(t)} \geq 0.$$

Substituting $Y(t) = A(t)\,(u(t)S(t))^{\alpha}$, $A(t) = A_0 e^{\mu t}$, $t \geq 0$, (see (1)) we get the following characterization of sustainability of an admissible pair $(S(\cdot), u(\cdot))$:

$$\frac{\mu}{\alpha} + \lim_{T\to\infty} \inf_{t\geq T} \left[\frac{\dot{u}(t)}{u(t)} + \frac{\dot{S}(t)}{S(t)} \right] \geq 0. \tag{48}$$

We call an admissible pair $(S(\cdot), u(\cdot))$ *strongly sustainable* if it is sustainable and, moreover, the resource stock $S(\cdot)$ is non-vanishing in the long run, i.e.

$$\lim_{T\to\infty} \inf_{t\geq T} S(t) = S_\infty > 0. \tag{49}$$

Consider case (i) when $r > \rho$. In this case due to Theorem 3 there is a unique optimal equilibrium pair in the problem (see (40) and (41)): $\hat{u}(t) \equiv \hat{u} = (r+\rho)/2$, $\hat{S}(t) \equiv \hat{S} = (r-\rho)K/(2r) > 0$, $t \geq 0$, and for any initial state S_0 the corresponding optimal path $S_*(\cdot)$ approaches asymptotically to the optimal equilibrium state $\hat{S}$ while the corresponding optimal exploitation rate $u_*(\cdot)$ approaches asymptotically to the optimal equilibrium value $\hat{u}$. Hence, both conditions (48) and (49) are satisfied. Thus the optimal admissible pair $(S_*(\cdot), u_*(\cdot))$ is strongly sustainable in this case.

Consider case (ii) when $r \leq \rho$. In this case due to Theorem 3 for any initial state S_0 the corresponding optimal control $u_*(\cdot)$ asymptotically satisfies the Hotelling rule of optimal depletion of an exhaustible resource (Hotelling 1974), i.e. $\lim_{t\to\infty} u_*(t) = \rho$, and $\lim_{t\to\infty} \dot{u}_*(t)/u_*(t) = 0$. The corresponding optimal path $S_*(\cdot)$ is asymptotically vanishing, and

$$\lim_{t\to\infty} \dot{S}_*(t)/S_*(t) = \lim_{t\to\infty} (r - u_*(t) - rS_*(t)/K) = r - \rho.$$

Hence, in the case (ii) the sustainability condition (48) takes the following form:

$$\frac{\mu}{\alpha} + r \geq \rho. \tag{50}$$

Notice, that in the case $\alpha = 1$ condition (50) coincides with Valente's necessary condition for sustainability in his capital-resource model with a renewable resource growing exponentially (see Valente 2005).

Since in the case (i) the inequality (50) holds automatically we conclude that (50) is a necessary and sufficient condition (a criterion) for sustainability of the optimal pair $(S_*(\cdot), u_*(\cdot))$ in our model while the stronger inequality

$$r > \rho$$

gives a criterion of its strong sustainability.

The criterion (50) gives the following guidelines for sustainable optimal growth: (1) Take measures to increase growth rate r; (2) Increase ratio of growth rate of knowledge stock μ to output elasticity α; and (3) Decrease social discount ρ i.e., plan long term. The sustainability criterion (50) gives a relationship between the state of technology (depicted by α), the environment (depicted by r), accumulation of knowledge (depicted by μ) and foresight of the social planner (depicted by ρ). According to the guideline 2 above, it is the *ratio* between μ and α that matters but not the individual quantities.

Acknowledgements This work was initiated when Talha Manzoor participated in the 2013 Young Scientists Summer Program (YSSP) at IIASA, Laxenburg, Austria. T. Manzoor is grateful to Pakistan National Member Organization for financial support during the YSSP. Sergey Aseev was supported by the Russian Science Foundation under grant 15-11-10018 in developing of methodology of application of the maximum principle to the problem.

References

D. Acemoglu, *Introduction to Modern Economic Growth* (Princeton University Press, Princeton NJ, 2009)

S.M. Aseev, Adjoint variables and intertemporal prices in infinite-horizon optimal control problems. Proc. Steklov Inst. Math. **290**, 223–237 (2015a)

S.M. Aseev, On the boundedness of optimal controls in infinite-horizon problems. Proc. Steklov Inst. Math. **291**, 38–48 (2015b)

S.M. Aseev, Existence of an optimal control in infinite-horizon problems with unbounded set of control constraints. Trudy Inst. Mat. i Mekh. UrO RAN **22**(2), 18–27 (2016) (in Russian)

S.M. Aseev, A.V. Kryazhimskiy, The Pontryagin maximum principle and transversality conditions for a class of optimal control problems with infinite time horizons. SIAM J. Control Optim. **43**, 1094–1119 (2004)

S.M. Aseev, A.V. Kryazhimskii, The Pontryagin maximum principle and optimal economic growth problems. Proc. Steklov Inst. Math. **257**, 1–255 (2007)

S. Aseev, T. Manzoor, Optimal growth, renewable resources and sustainability, International Institute for Applied Systems Analysis (IIASA), Laxenburg, Austria, WP-16-017, 29 pp., 2016

S.M. Aseev, V.M. Veliov, Maximum principle for infinite-horizon optimal control problems with dominating discount. Dyn. Contin. Discrete Impuls. Syst. Ser. B Appl. Algorithms **19**, 43–63 (2012)

S.M. Aseev, V.M. Veliov, Needle variations in infinite-horizon optimal control, in *Variational and Optimal Control Problems on Unbounded Domains*, ed. by G. Wolansky, A.J. Zaslavski. Contemporary Mathematics, vol. 619 (American Mathematical Society, Providence, 2014), pp. 1–17

S.M. Aseev, V.M. Veliov, Maximum principle for infinite-horizon optimal control problems under weak regularity assumptions. Proc. Steklov Inst. Math. **291**(supplement 1), 22–39 (2015)

S.M. Aseev, K.O. Besov, A.V. Kryazhimskii, Infinite-horizon optimal control problems in economics. Russ. Math. Surv. **67**(2), 195–253 (2012)

G.B. Asheim, T. Mitra, Sustainability and discounted utilitarianism in models of economic growth. Math. Soc. Sci. **59**(2), 148–169 (2010)

E.J. Balder, An existence result for optimal economic growth problems. J. Math. Anal. Appl. **95**, 195–213 (1983)

R.J. Barro, X. Sala-i-Martin, *Economic Growth* (McGraw Hill, New York, 1995)

Brundtland Commission, Our common future: report of the world commission on evironment and development, United Nations, 1987

D.A. Carlson, A.B. Haurie, A. Leizarowitz, *Infinite Horizon Optimal Control. Deterministic and Stochastic Systems* (Springer, Berlin, 1991)

L. Cesari, *Optimization – Theory and Applications. Problems with Ordinary Differential Equations* (Springer, New York, 1983)

A.F. Filippov, *Differential Equations with Discontinuous Right-Hand Sides* (Kluwer, Dordrecht, 1988)

P. Hartman, *Ordinary Differential Equations* (J. Wiley & Sons, New York, 1964)

H. Hotelling, The economics of exhaustible resources. J. Polit. Econ. **39**, 137–175 (1974)

T. Manzoor, S. Aseev, E. Rovenskaya, A. Muhammad, Optimal control for sustainable consumption of natural resources, in *Proceedings, 19th IFAC World Congress, vol.19, part 1 (Capetown, South Africa, 24–29 August, 2014)*, ed. by E. Boje, X. Xia, pp. 10725–10730 (2014)

P. Michel, On the transversality conditions in infinite horizon optimal problems. Econometrica **50**,975–985 (1982)

L.S. Pontryagin, V.G. Boltyanskii, R.V. Gamkrelidze, E.F. Mishchenko, *The Mathematical Theory of Optimal Processes* (Pergamon, Oxford, 1964)

F.P. Ramsey, A mathematical theory of saving. Econ. J. **38**, 543–559 (1928)

R.M. Solow, A contribution to the theory of economic growth. Q. J. Econ. **70** (1), 65–94 (1956)

S. Valente, Sustainable development, renewable resources and technological progress. Environ. Resour. Econ. **30**(1), 115–125 (2005)

(LNG) Arbitrage, Intertemporal Market Equilibrium and (Political) Uncertainty

Franz Wirl

Abstract Since 2009, natural gas markets are characterized by large spreads in liquefied natural gas (LNG) prices between the United States (Henry hub) and Europe and Japan. Moreover these differences are forecasted to persist (at a lower level but still above transport costs) against the law of one price. This paper explores this persistence of apparent arbitrage by investigating an intertemporal competitive equilibrium under uncertainty. Investments of an individual arbitrageur must account (at least) for: (1) rational expectation that this arbitrage will be eroded over time by competitive agents' similar investments; (2) risk of export regulations because governments may intervene and destroy this opportunity in order to protect the interest of local (i.e. U.S.) firms and consumers. The paper analyzes a corresponding stochastic and dynamic (partial) equilibrium that leads to a reduction in investments and implies persistence of apparent arbitrage. This is in line with forecasted but unexplained price differences.

1 Prologue

I first met Vladimir in Gustav's (famous) seminar, where he made a lasting impression in the way he got to the point and this by taking much less time to explain than I did. I well recall one of the many discussions about what characterizes a 'Skiba-point', as we called it at that time. I was also dissatisfied with this literature that was played up and down in hundreds of convex-concave problems. While we were discussing minor issues, Vladimir almost immediately captured the core characteristic and proposed a precise, mathematical definition. Similarly, I enjoyed (but did not follow) his extension of control problems for distributions leading to fine papers.

F. Wirl (✉)
University of Vienna, Faculty of Business, Economics and Statistics, Vienna, Austria
e-mail: Franz.wirl@univie.ac.at

G. Feichtinger et al. (eds.), *Control Systems and Mathematical Methods in Economics*, Lecture Notes in Economics and Mathematical Systems 687,
https://doi.org/10.1007/978-3-319-75169-6_12

Given my appreciation for Vladimir, I feel honored to have been invited to contribute to his Festschrift and thank the editors for this invitation. Since I cannot match Vladimir's mathematical skills but look for a unique selling proposition for my contribution, I opt for a mathematically rather simple analysis (an application of a trick developed by Kamien and Schwartz long time ago) but one that addresses an in my opinion important international economic puzzle: How can the price of a good as homogenous as natural gas differ by factor of five violating the law of one price (of course after accounting for transport costs, which are crucial in this context). Given my (partial) explanation based on the impossibility of governments to commit and therefore nothing is 'sacred' in politics, some readers may link the suggested resolution to other current political events.

2 Introduction

The recent dynamics of natural gas markets are characterized by large spreads in liquefied natural gas (LNG) prices across locations in particular the low prices in the United States (Henry hub) compared with Europe or Japan. Figure 1 shows how the international evolution of gas prices has de-coupled from the US evolution. Although the suspicion about arbitrage in natural gas markets is not new, compare Kleit (1998) about the effect of US natural gas deregulation, the current international price differences are of a much larger magnitude (absolute and relative). Maxwell and Zhu (2011) provide Granger causality tests between natural gas and LNG prices

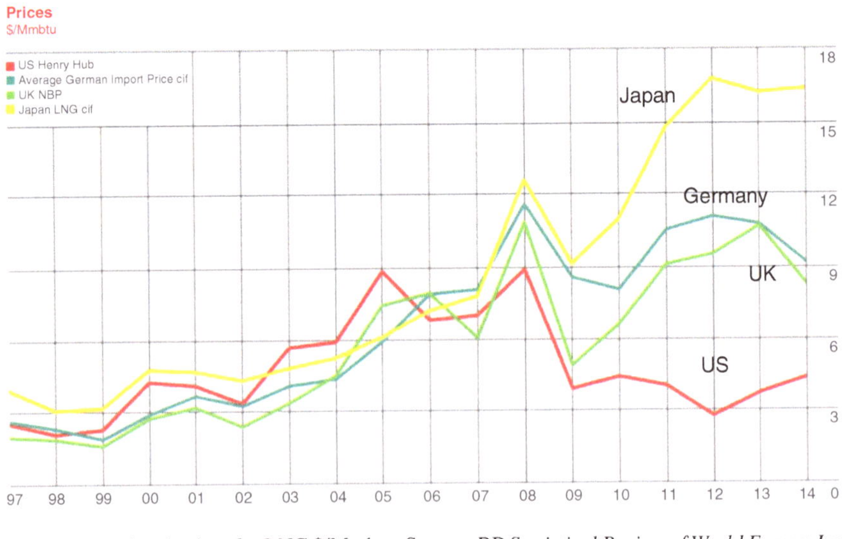

Fig. 1 International prices for LNG $/Mmbtu. Source: BP *Statistical Review of World Energy June (2015)*, page 27

up to 2007. Irrespective of the underlying reasons—a competitive gas market in North America and oil price linked contracts in the rest of the world—such price differences are puzzling for a good as homogenous as natural gas (essentially methane); these different contractual relationships over space and time and the changing degree of vertical integration are not subject of this paper and von Hirschhausen and Neumann (2008) is an empirical analysis of these issues. Given these large price differences shown in Fig. 1, US gas producers should have a strong interest to ship their gas from the low cost US region to high cost regions even after accounting for substantial costs for liquefaction and shipping (re-gasification plants exist and have surplus capacity, see Dehnavi et al. 2015).

The International Energy Agency predicts in its recent World Energy Outlooks 2013 and 2014, IEA (2013, 2014) that substantial differences beyond transport costs will persist during the next 20 years. The persistence of this difference for a homogeneous good requires however an explanation. In order to explain (at least partially) low investment in exploiting this arbitrage a political economic argument is used: US consumers as well as US industry have an interest to keep this comparative advantage, which is threatened if US gas producers start exporting on a large scale. Therefore, the US government may intervene and limit LNG trade, directly or indirectly. However, is this likely or even possible? Is the US not a usual (and often the only) defender of free trade? The haven of property rights (see La Porta et al. 1997)? At a normative level, the US government considers national interests, at a positive level, politicians have an interest to satisfy populist demands, which seem to rise on the left and on the right. LNG exports have positive and negative effects which offer different options: The US may use these exports to reduce its balance of payments deficit or it may want to keep this addition to its gas reserves (mostly non-conventional) for future generations. The latter is however not a good explanation of actual policies given the rush with which the US economy makes use of this bonanza unhindered if not even encouraged by its policy makers. Another option is that the US wants to use it to obtain energy autarchy (it will also turn almost self sufficient in oil, partially due to the expansion of natural gas in almost all applications, IEA 2013) in the short and medium term, which is more compatible with politicians' planning horizons. Aside from questions about energy security, the US (politicians, consumers and industry) may want to keep natural gas within the US due to the implicit subsidies (including environmental risks that the US population faces) for shale gas. The shale gas revolution in 2006 has decreased US natural gas prices remarkably (see Fig. 1). This has created a competitive advantage for US industries even stimulating re-shoring of formerly off-shored industries (see e.g., *The Economist* January 19th 2013, special report on outsourcing and offshoring, p 6–8), and avoiding according to Weber (2014) the often observed resource curse (for a recent investigation of resource curses in the broader context of 35 resource rich countries see Crivelli and Gupta 2014). The benefit from the gas boom may vanish if the US starts to export LNG on a large scale.

US governments can and do intervene in particular in energy markets. Of the many examples consider the US oil price regulation (outside of Alaska) from

1973 to 1981 which was enacted by a Republican President (Nixon). At the same time, the rest of the Western World, which included countries with comparatively little respect for property rights paid their domestic producers the much higher world market price. The new US president simply strengthens this observation. Therefore, past experience and the fact that governments cannot commit, suggest that politically motivated intervention cannot be ruled out and may be even likely under certain circumstances. And indeed, lobbying took already place in order to keep the advantage of relatively low energy costs within the United States and this from both sides, from industry (Andrew Liveris, the CEO of Dow Chemical) and labor (e.g., Representative Edward Mankey warns that exports will harm American workers, both quotes are from FAZ, December 11th, 2012, p 12). This adds a substantial uncertainty to the already complex investment decisions. Levi (2013) highlights that the US geopolitical gain is due to the shift from "natural gas importer to a neutral player" by lessening the dependence on Qatar or Russia. Moreover, any export outside of the Free Trade Agreement countries has to be approved by the US Department of Energy, which may account for the public interest of low domestic energy prices in order to provide the US-economy with a competitive advantage.

This paper addresses the issue of price differentials persisting in the long term in order to resolve the puzzle of current and past large price differences. These high natural gas price differences indicate huge arbitrage but one that is surprisingly hardly exploited although it lasts now for a few years. Indeed, the usual explanations of violations of the law of one price fail (Sect. 2). This paper tries to explain this puzzle at least partially. More precisely, it investigates how an emerging arbitrage induces rational and competitive agents to invest accounting for: (1) the rational expectation that this arbitrage will be eroded over time by the investments of competitors; (2) these investments face the risk that governments intervene and destroy this opportunity in order to grant local (i.e. U.S.) firms and consumers a crucial cost advantage. The combination of these two factors in particular the second one explains—as assumed/predicted in IEA (2014)—the persistence of substantial price differences violating the 'law of one price'.

3 Rounding Up the Usual Suspects

The substantial differences (well above transport costs) in international gas prices are caused by many factors. Drawing on Dehnavi et al. (2015) the usual suspects are:

1. Costs for transport and for sunk investments. However, only (US) liquefaction capacities are needed, because of excess capacity of re-gasification plants, globally and regionally.
2. Constraints of capacities along the supply chain (ships, liquefaction and regasification plants). However, only the shortage of US liquefaction capacities is binding.

3. Long term contracts can bind the direction of flows but again this constraint is hardly binding given the increased flexibility of contracts.
4. Uncertainty and risk aversion of investors. Risk averse competitive ('small') producers have to repay their debt to banks and other investors. This is also not a very good explanation as financial intermediaries and the large energy companies have deep pockets and are able to shoulder this risk.
5. The complexity of LNG trade. In particular, expectations, and the limited time window for recovering the high up front and sunk investments (any arbitrage opportunity must end in markets with free entry) can deter otherwise profitable investments.
6. Unconventional gas reserves (in particular, tight gas and shale gas deposits) elsewhere in the world might be developed. Among others, China has the largest global reserves of non-conventional gas. If it would be able to develop them fast, repeating US shale gas revolution, this can help China to meet its growing demand for natural gas and China may even become a net gas exporter. However, there are several obstacles—geological (deeper location of shale gas formations) and geographical (water scarcity) that prevent China from becoming net gas exporter; for further details see Yegorov and Dehnavi (2014).

Still, all these points cannot explain the observed regional price differences between the US and Europe and Japan. A particular point is that even if all current US gas trades were blocked by the above first five points, US gas producers can speculate on their own. More precisely, they can delay their extraction to future periods with higher prices after all or some of the above mentioned trade hurdles are removed since a price difference of a factor of 4 cannot persist. Therefore, all these obstacles can only delay and slow down investments but cannot hinder them such that prices will come closer until all arbitrage opportunities are eliminated.

However, these economic obstacles are aggravated by political uncertainty whether the US government will allow these exports, or whether it reverses its initial permission if domestic gas prices get too high (e.g. due to an initiative by populist politicians). And if indeed enacted, prices need not converge with any arbitrage then being blocked by political interventions. Therefore, given the high investment costs, the fact that any profit will be transient, possibly only short lived, and the uncertainty including the possibility or even (tacit) threat of an 'export tax' if natural gas exports become large and raise US prices to world market prices, can lead many investors to the belief that the current profit opportunity may quickly turn into a fata morgana. Therefore, they continue to sell to the domestic market in spite of the apparent arbitrage from going abroad.

4 Model

Consider two isolated markets, a and b, for a homogenous good where initially different prices prevail, say, $P^b - P^a > 0$. This creates an opportunity for arbitrage if this difference is larger than the associated costs of delivery. Costs arise from transport (c) and from investments into capacities necessary for shipping, in the case of LNG for gas liquefaction capacities and maybe for vessels (there is sufficient regasification surplus, see Dehnavi et al. (2015)). Let X denote the aggregate flow from the low to the high price region; capitalized variables refer to aggregates, small letters refer to variables that are controlled by competitive agents. The local prices depend on the total volume available. They are determined by indigenous supply including already contracted imports, M^j, and the flow X induced by the price difference between $P^a = P^a(M^a - X)$ and $P^b = P^b\left(M^b + X\right)$. Therefore the gross arbitrage per unit

$$A(X) := P^b\left(M^b + X\right) - P^a\left(M^a - X\right) - c,\ A' < 0 \tag{1}$$

depends only on the aggregate trade X from a to b.

I assume infinitesimally small competitive firms with aggregate mass normalized to 1 and rational expectations, i.e., perfect foresight in a deterministic model. Since all competitors face an identical decision, an individual firm can predict the consequences on market outcomes from its own decision. This allows to treat firm individual (small letters) and market aggregate variables (capital letters) as identical. However, there is a crucial difference: each firm can control only its own variables but must take market aggregates as given, because a small competitive firm has no influence on the aggregate. Let x denote the shipping capacity of a small competitive firm located in region a. This firm will choose its exports y to b by $\max_{y \le x} Ay$ and thus equal to its capacity, $y = x$, whenever $A > 0$. The currently binding constraint in the case of US LNG exports is the lack of gas liquefaction facilities as the US east coast was dotted with plans for LNG importing facilities and re-gasification plants have been constructed from which investors incur now huge losses. Investments in future US liquefaction capacities are sunk (as those in the past in regasification) and the profitability of such investments requires a dynamic and forward looking approach. In particular, a potential investor must account for (1) the high fixed and sunk cost of liquefaction plants in the USA and (2) for how long this arbitrage opportunity exists. Assuming that investment u provides everlasting additional capacity at the costs $k(u)$ and free trade (i.e., no government intervention), a potential arbitrageur chooses an investment path that maximizes the expected net present value of profits,

$$\max_{u(t) \ge 0,\ y(t) \le x} E \int_0^\infty e^{-rt}\left[A(t)\,y(t) - k(u(t))\right]dt, \tag{2}$$

$$\text{s.t. } \dot{x}(t) = u(t),\ x(0) = x_0 = 0; \tag{3}$$

E denotes the expectation with respect to the future arbitrage which is affected by aggregate investment X according to (1) and political uncertainty. Under rational expectations, each firm realizes that it is not the only one trying to seize this opportunity and that their collective investments will reduce the arbitrage in the future. As a consequence, no excess investments will be built such that $y = x$ and the investor will substitute his own plan for the expectation of what the competitors will do, i.e.,

$$X(t) = x(t) \ \forall t \geq 0. \tag{4}$$

Ignoring further, here political uncertainty, the corresponding (saddle point stable) steady state is given by the economically intuitive condition that the net present value of marginal profit is equal to the costs for an incremental (infinitesimal) unit of capacity,

$$\frac{A(X_\infty)}{r} = k'(0). \tag{5}$$

Myopic agents who extrapolate the current arbitrage into the indefinite future yet revise this forecast continuously end up at the same long run capacity characterized above but faster (due to excessive investments that ignore the future declines of the arbitrage).

Investors realize that their arbitrage can and will be contested in the political arena in particular in the light of the fact that the domestic price P^a will increase as more and more is shipped to other locations (here to b). Public intervention can take many forms: It can block, hinder, or increase the costs of export facilities citing environmental or protectionist reasons, it may regulate the local price and/or charge an export tax, etc. And that even free market economies can do so is well demonstrated and not only by the already mentioned US oil price regulation but also by the current US government's talk about trade restrictions. In order to simplify, it is assumed that the government can eliminate this arbitrage opportunity from one day to the next. Whether this is done in a command and control way by forbidding natural gas exports or by adding an export tax that renders exports unprofitable is not crucial, neither for the following analysis nor presumably in practice. Indeed, exports outside the FTA requires regulatory approval. For reasons of simplicity, we consider a complete elimination of this, if still existing, arbitrage opportunity by the government, although one can easily extend the model to a partial elimination.

The approach pioneered in Kamien and Schwartz (1971) allows transferring the stochastic into a deterministic optimization problem. This approach is extended to a competitive intertemporal market equilibrium in which the individual optimization depends in addition on market data that is beyond the agent's control and thus exogenous for the agent's optimization problem (the decision maker controls all process relevant states in Kamien and Schwartz (1971) and its many follow ups). Let $F(t)$ denote the probability that the public intervention—elimination of all arbitrage opportunities—has occurred prior to date t. Then $(1 - F(t))$ denotes the probability that the arbitrage opportunity is still open at time t and the density $\dot{F}(t)$ describes the probability of this event at date t. The objective of the arbitrageur, maximizing the expected net present value of profits, can be expressed then in the following way,

$$\max_{u(t)\geq 0} \int_0^\infty e^{-rt} \left[A(X(t))\, x(t)\, (1 - F(t)) - k(u(t))\right] dt, \tag{6}$$

subject to (3) and (4). With $F(t)$ exogenously specified, the model would be complete. However, we extend the framework and assume that the hazard rate, $h = \dot{F}/(1-F)$, is independent of calendar time t but increases with the aggregate volume shipped (for the reasons given above), $h = h(X)$, $h' > 0$, and $h(0) = 0$, i.e., no exports at all avoid the intervention for sure,

$$\dot{F} = h(X)(1-F), \; F(0) = 0. \tag{7}$$

Remark 1 The assumption $h' > 0$ is not crucial as similar (even numerically similar) results hold for declining hazard rates, $h' < 0$, which means that the hazard of an intervention is high if the arbitrage is large (because the arbitrage declines with respect to X) but becomes smaller as this arbitrage diminishes; see Appendix.

Remark 2 Although the above model is motivated by the current apparent arbitrage from shipping LNG from the US to Japan (or Europe) it applies to other examples in energy and regulated industries, e.g. for nationalization threats with recent examples in Bolivia (power industry) and Russia (natural gas involving Shell and BP).

Example 1 The special case of linear arbitrage (normalized and accounting for delivery costs, i.e., including or normalizing, $c = 0$), linear-quadratic investment costs, and a hazard rate linear in the shipping volume,

$$A(X) = 1 - X, \; k = \kappa u + \frac{\gamma}{2}u^2, \; h(X) = \alpha X, \tag{8}$$

is used for the numerical examples below. It implies for the deterministic case the steady state (= long run capacity):

$$x_\infty = 1 - r\kappa. \tag{9}$$

5 Market Equilibrium

The Hamiltonian for the optimization problem of an individual firm is

$$H = A(X)x(1-F) - k(u) + \lambda u, \tag{10}$$

which does not include a costate for the state F, because the competitive firm has no control over it. The first order optimality conditions are:

$$H_u = -k' + \lambda \Longrightarrow u^* = \begin{matrix} (k')^{-1}(\lambda) \\ 0 \end{matrix} \text{ if } \lambda \begin{matrix} \geq \\ < \end{matrix} k'(0), \tag{11}$$

$$\dot{\lambda} = r\lambda - A(1-F). \tag{12}$$

Therefore the following equations system describes a competitive, rational expectation and symmetric (i.e., substituting (4)) equilibrium (at least for an interior, $u > 0$, solution):

$$\dot{x} = \left(k'\right)^{-1}(\lambda) = \frac{\lambda - \kappa}{\gamma}, \tag{13}$$

$$\dot{\lambda} = r\lambda - A(x)(1 - F) = r\lambda - (1 - x)(1 - F), \tag{14}$$

$$\dot{F} = h(x)(1 - F) = \alpha x(1 - F). \tag{15}$$

The right hand side expressions result for the functional forms chosen in (8).

The stationary solutions of the last differential equation (15) are either $x = 0$ or $F = 1$. Focusing on the latter (the other is a trivial outcome of no interest), Eq. (14) implies that $\lambda \to 0$. Hence, investment must stop in finite time T when $\lambda(T) = k'(0) > 0$.

Therefore the infinite horizon problem is replaced by one that includes the optimal stopping of investment at T such that $u = 0$ for all $t \geq T$.

$$\max_{u(t),T} \int_0^T e^{-rt}\left[A(X(t))\,x(t)(1 - F(t)) - k(u(t))\right]dt + e^{-rT} S(x(T), T). \tag{16}$$

The salvage value, S, is the expected net present value of profits that the firm earns from the remaining and constant capacities $x(T)$ after stopping all investments at time T and accounting for the probability that the intervention has not taken place yet:

$$S(x(T), T) = A(X(T))\,x(T)\int_0^\infty e^{-rt}(1 - F(t))\,dt. \tag{17}$$

In spite of unchanged states, $x(t) = X(t) = x(T)$ for $t \geq T$ the probability of an intervention is still evolving according (15). Since the hazard rate is constant once investments have stopped, the evolution of the cumulative distribution function can be determined in this domain of no further investments,

$$\dot{F} = \alpha X(T)(1 - F) \Longrightarrow F(t) = 1 - (1 - F(T))\,e^{-h(X(T))(t-T)}. \tag{18}$$

Substituting this solution of F, using the linear hazard rate from (8), and dropping the argument T yields

$$S(x, T) = A(X)\,x(1 - F)\int_0^\infty e^{-(r+\alpha X)t}dt = \frac{A(X)\,x(1 - F)}{r + \alpha X}. \tag{19}$$

This salvage value (19) imposes a transversality condition for the costate,

$$\lambda(T) = S_x = \frac{A(X)(1-F)}{r+\alpha X}. \tag{20}$$

The crucial feature is that the market aggregate $X(T)$ cannot be chosen along the optimization of an individual agent (of measure 0) but must be accepted as given although it is identical to the choice variable $x(T)$. While $X(T)$ is constant and known to be equal to the own state for all $t \geq T$, $F(T)$ remains an exogenous time trend for the firm's optimal choice when to stop investing. Therefore, the salvage value S is time dependent from an individual firm's perspective such that the following optimal stopping rule applies (see any of the excellent books on optimal control, e.g., Grass et al. (2008) for a more recent text book),

$$H(T) = rS - \frac{\partial S}{\partial T}. \tag{21}$$

Therefore, we have to compute the derivative on the right hand side (by the chain rule),

$$\frac{\partial S}{\partial T} = -\frac{A(X)\,x}{r+\alpha X}\frac{\partial F}{\partial T},$$

which in turn requires the computation of $\frac{\partial F}{\partial T}$. Differentiating (18) with respect to time and evaluation at $t = T$ yields,

$$-(1-F)\frac{\partial}{\partial t}e^{-\alpha X(t-T)} = \alpha X(1-F)\,e^{-\alpha X(t-T)}\,|_{t=T} = \alpha X(1-F). \tag{22}$$

Considering the left and right hand sides of the optimal terminal time condition (21) in isolation yields,

$$H = A(X)\,x\,(1-F) - k(u) + \frac{A(X)(1-F)}{r+\alpha X}u \text{ at } t = T, \tag{23}$$

and

$$\begin{aligned} rS - \frac{\partial S}{\partial T} &= \frac{rA(X)(1-F)\,x}{r+\alpha X} + \frac{A(X)\,x\alpha X(1-F)}{r+\alpha X} \\ &= Ax(1-F) \text{ at } t = T. \end{aligned} \tag{24}$$

The last expression follows after substituting the property (4) for a symmetric equilibrium.

An economically intuitive conjecture is that of a continuous termination of investments, i.e., $u(T) = 0$, which indeed solves the optimal stopping condition (21). This conjecture simplifies the left hand side of (21) considerably (also accounting for the symmetric equilibrium, $X = x$), $H(T) = Ax(1-F)$, and that equals the right hand side as computed in (24). Therefore, the choice of $u(T) = 0 \Leftrightarrow \lambda(T) = k$ satisfies the optimal stopping condition (21), determines the costate value and implies the additional condition needed,

$$\lambda(T) = S_x = \frac{A(1-F)}{r+\alpha x} = k \Longrightarrow A(1-F) = (r+\alpha x)k, \tag{25}$$

for the determination of the optimal stopping time. This last condition has an intuitive economic interpretation: the expected (accounting for the hazard) net present value of marginal revenue from an infinitesimal increase of capacity equals the costs for an infinitesimal investment,

$$\frac{A}{r} > \frac{A(1-F)}{r+\alpha x} = k'(0), \tag{26}$$

where the left hand side of the inequality is the deterministic counterpart; the inequality holds strictly for $0 < x < \bar{X}$. And of course the larger the government's threat is perceived, i.e., a higher value of α, the more conservative investment will proceed. The opposite holds for first order stochastic dominance of F over G (i.e., concerning the probabilities of a political intervention such that $F < G$ for all x, of course a higher α leads directly a stochastically dominated distribution). Figure 2 shows how this intervention risk reduces the ultimate capacity $x(T)$ even if the intervention does not take place (prior to T) compared with the deterministic counterpart. Two factors lead to this reduction. The term $(1-F)$ in the numerator lowers the current arbitrage (i.e., a lower intercept for fixed $F(T)$ as assumed in Fig. 2), while the denominator increases with respect to x depending on the hazard rate parameter α.

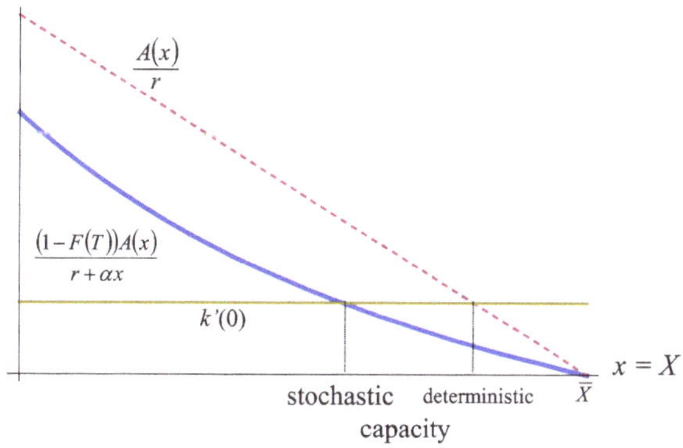

Fig. 2 Determination of capacity: deterministic (steady state) vs stochastic (maximum, i.e. in case of no prior intervention) for F fixed at terminal value

Applying the implicit function to (26), which links the terminal supply capacity $x(T)$ to the stopping time T, yields after differentiating

$$\frac{\partial}{\partial T} = -\frac{A(x(T))\dot{F}(T)}{r+\alpha x(T)} = -\frac{A(x(T))\alpha x(T)(1-F(T))}{r+\alpha x(T)} < 0,$$

$$\frac{\partial}{\partial x(T)} = \frac{(1-F(T))\left((r+\alpha x(T))A'(x(T)) - \alpha A(x(T))\right)}{(r+\alpha x(T))^2}$$

$$= -\frac{(1-F)((r+\alpha x)+\alpha A)}{(r+\alpha x(T))^2} < 0,$$

that a longer duration increases (not surprisingly) the ultimate capacity. However, this holds only ceteris paribus, i.e., without changing model parameters (see the following examples).

Proposition 1 *The competitive rational expectation equilibrium leads to a stopping of investments at $x(T)$ implicitly determined by (26), and explicitly for the example (8),*

$$x(T) = \frac{1-F(T)-r\kappa}{\alpha\kappa+1-F(T)}.$$

This stopping level is below the deterministic counterpart and thus creates the false impression of still existing arbitrage (if the intervention has not occurred already prior to T) with the price difference

$$P^b - P^a = c + (\alpha + r)\kappa$$

for the specification (8). The paths satisfy the following monotonicity properties: $\dot{x} > 0$, $\dot{u} < 0$, and $\dot{F} > 0$.

The explicit claims about the stopping level, and thus for the persisting arbitrage and price difference (assuming no intervention) follow after substituting the specifications of the example (8) into (26) and (1). The monotonicity characterizations follow from (13) to (15). By construction, capacities and cumulative distribution must increase, i.e., $\dot{x} > 0$ and $\dot{F} > 0$. Investment must decline at least close to T since $u \to 0$ and $\lambda \to k'(0)$, thus $\dot{u} < 0 \Longleftrightarrow \dot{\lambda} < 0$ at least for $t \to T$. Moreover, this property must hold globally for the following reason: Non-monotonic behavior of u requires $\dot{u} > 0 \Longleftrightarrow \lambda > k'(0)$ and $\dot{\lambda} > 0$. However, once λ is in the domain where $\dot{\lambda} > 0 \Longleftrightarrow \lambda > \frac{A(x)(1-F)}{r}$ it can never reach the required domain of decline, $\dot{\lambda} < 0$, because the threshold $\frac{A(x)(1-F)}{r}$ moves simultaneously to the left due to $\dot{F} > 0$ (see the phase diagram in Fig. 3).

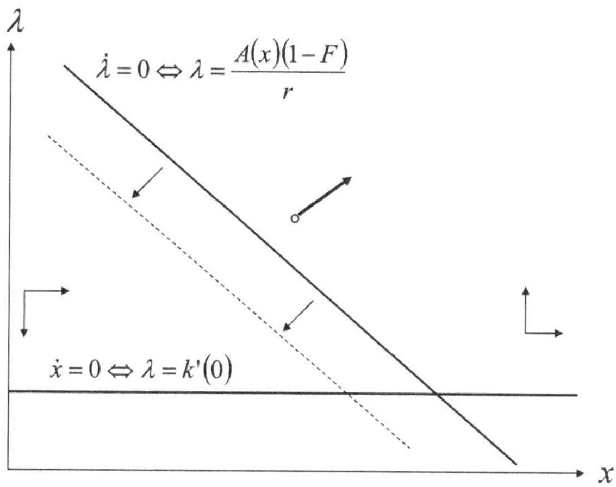

Fig. 3 Phase diagram (x, λ) accounting for shifts due to increasing F. As a consequence, it is impossible to reach the declining domain from points above the bold line $= \dot{\lambda} = 0$

6 Examples

The above analysis and in particular the chart in Fig. 2 depends on the stopping level $F(T)$ and in order to account for that one has to solve the entire (nonlinear) system, unfortunately, numerically:

$$\dot{x} = \frac{\lambda - \kappa}{\gamma}, \; x(0) = 0, \tag{27}$$

$$\dot{\lambda} = r\lambda - (1-x)(1-F), \; \lambda(T) = \kappa, \tag{28}$$

$$\dot{F} = \alpha x(1-F), \; F(0) = 0, \tag{29}$$

$$A(x(T))(1-F(T)) = (r + \alpha x(T))k. \tag{30}$$

The last condition is necessary to determine the stopping time T. A solution can be obtained in the following way: the two-point boundary problem (27)–(29) is solved for a guess $\hat{T}$, which is then varied (e.g., by using a simple binary search or a Newton algorithm) until the last condition (30) is met.

The purpose of the following examples is to get an idea how far this political threat affects investment and how the model parameters affect that. Starting from the Example (8) the maximum surplus is normalized to 1, which may correspond to the current surplus of 4\$/MBtu = a price difference of up to 10\$/MBtu minus transportation costs, which are between 1–2\$ per MBtu (see Dehnavi et al. (2015), further discounted by 50% for taxes, royalties etc. Costs per unit capacity[1] are assumed to be $k'(0) = \kappa = 2$ which corresponds 4\$/MBtu capacity and thus to an annuity $r\kappa = 0.2$ for $r = 0.10$; i.e., investment costs equal to 20% of the profits

[1] Liquefaction costs have substantially decreased over the past 10 years to below US\$200 per ton of annual liquefaction capacity, US Energy Information Administration (2003).

at the maximal = initial level of the arbitrage (i.e., at $X = x = 0$). These costs apply only to investments that proceed in very small steps while γ accounts for the cost escalation due to larger investments and thus faster capacity expansion. It seems reasonable to assume that closing the entire arbitrage within one year would be unprofitable. Thus $rk(1) = r(\kappa + \gamma/2) > 1$ provides a lower bound (of 6) and hence a higher value, $\gamma = 10$, is used. The hazard rate parameter is the last and presumably the most difficult one to choose and the assumption $\alpha = 0.10$ puts this effect at maximum (since $x < 1$) on par with discounting.

Figure 4 shows the time paths for this reference case for investment, capacity and the cumulative distribution function. The terminal value $x(T)$ characterizes the maximum capacity attained in the absence of an intervention prior to T (below 8 years) and $F(T)$ characterizes probability of an intervention prior to this date (at a rather low level of 27%). Given the assumed linear relation for A, the difference between the deterministic (at 0.8 in Fig. 6) and the stochastic level $x(T) = 0.57$ determines the 'arbitrage' as perceived by agents ignoring the political risk. Considering the above calibration, this implies that an 'arbitrage' of around1$/MBtu persists, which corresponds to a price difference of around 4$/MBtu. This terminal value for the 'arbitrage' is comparatively robust across the scenarios but larger for higher adjustment costs and hazards and below IEA's (2013, Fig. 1.3, p 46) prediction for 2035 of around $6/MBtu for Europe and 8$/MBtu for Japan. Figures 5 and 6 investigate the consequences of parameters variations. Figure 5 shows the stopping time T, i.e. the date at which investors stop investing if politicians did not intervene already at a date $t < T$. It shows that higher adjustment costs ($\gamma \uparrow$) and a higher discount rate ($r \uparrow$) expand the (maximal) duration of investing. Increasing costs ($\kappa \uparrow$) and hazard ($\alpha \uparrow$) lower T. Overall, the stopping time is very insensitive with respect to κ and r (the charts at the bottom of Fig. 5) and moderately with respect to γ and α. Figure 6 compares, again within the above proviso of no earlier intervention, the long run outcomes for the cumulative distribution $F(T)$ and capacity $x(T)$. The qualitative consequences of

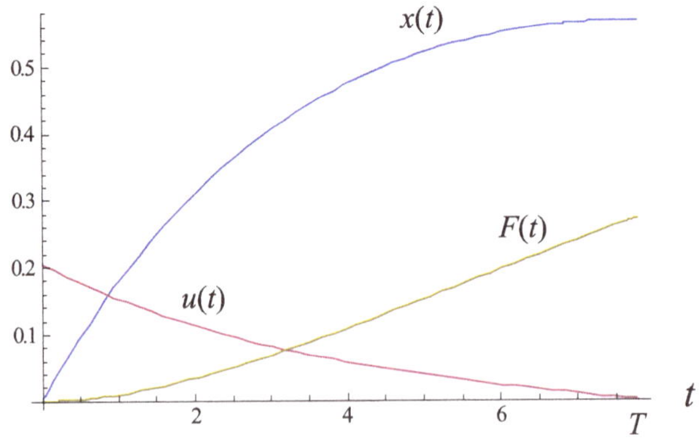

Fig. 4 Time paths (assuming no intervention prior to T), reference example: $\alpha = 0.1, \kappa = 2, \gamma = 10, r = 0.1$

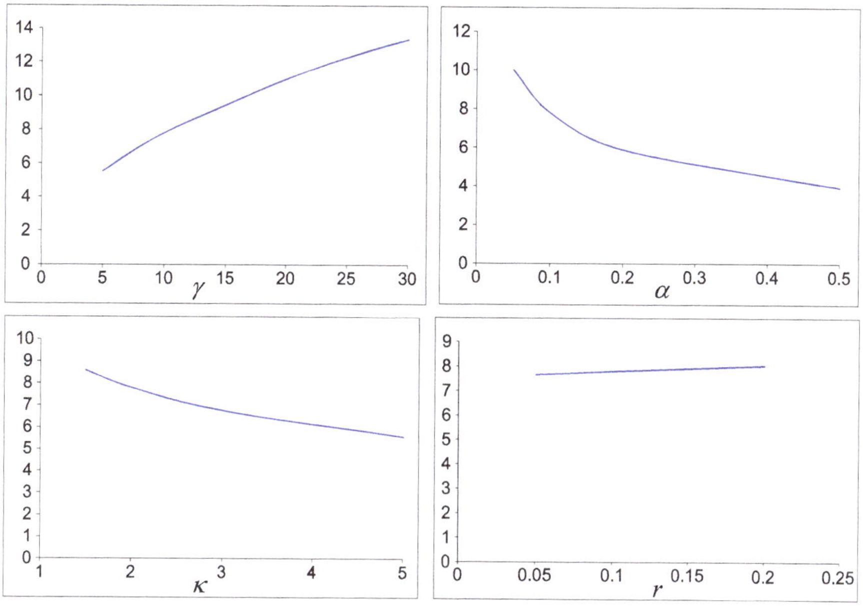

Fig. 5 Stopping time T for parameter variations around reference example: $\alpha = 0.1, \kappa = 2, \gamma = 10, r = 0.1$

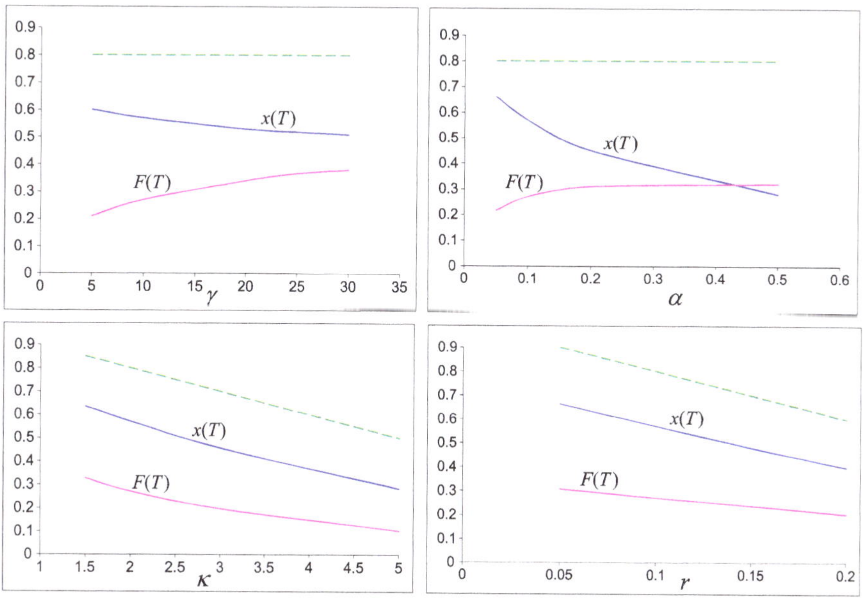

Fig. 6 Capacity (x) and cumulative distribution function (F) at T for parameter variations around the reference example: $\alpha = 0.1, \kappa = 2, \gamma = 10, r = 0.1$; the dashed line refer to the deterministic stationary capacity solution according to (4)

increases of α and γ are similar with countervailing effects between $x\ (T)$, which is declining, while $F\ (T)$ is increasing; quantitatively, the hazard rate parameter has a much larger impact as it can lower terminal capacity to below 1/2 of the reference case. In contrast, increasing the other two parameters (κ, r) lowers both states and also the deterministic solution for long run capacity. The capacity outcome is most sensitive (measured as arc elasticity around the reference case) with respect to the cost parameter κ (after all it affects also the deterministic outcome) and least with respect to the adjustment cost parameter γ. However in all cases and even for small hazard rate parameters α long run capacities are substantially below their deterministic counterpart and create, as argued above, the false impression of an arbitrage. This apparent arbitrage is most sensitive (again in terms of arc elasticities) with respect to the hazard rate (α) followed by the adjustment cost parameter (γ) but fairly insensitive to discounting (r) and the cost parameter κ. Finally, the implicit relation between stopping time and long run capacity as addressed above need not hold if model parameters are changed simultaneously. Using the results from Figs. 5 and 6 imply the expected positive relation between $x\ (T)$ and T for α and κ but a negative relation for r and γ.

7 Conclusion and Policy Implications

This paper investigated how intertemporal investments into an existing arbitrage opportunity are affected if accounting for: (1) the decisions of other market participants and (2) the threat of a public intervention, which increases as the local advantage is eroded due to arbitrageurs exploiting this opportunity. This question is motivated by the current situation in the international natural gas markets where the US shale gas bonanza has led to large differences in international natural gas prices. The major conclusion of the investigated framework is that the threat of such an intervention deters investment substantially. This could be one explanation why investments remain conservative in spite of this huge arbitrage existing now already for some years. It also suggests that a significant price difference above transport costs (including liquefaction) can persist (as predicted e.g., by the IEA) in spite of the law of one price for homogeneous goods.

This prediction is confirmed by the US reluctance to grant permits to natural gas exports to Non-FTA countries, see Table 1.

Table 1 Application received by DOE to export US LNG, as of May, 2016 (Numbers: Bcf/day)

	Total of all applications	Approved	Pending
FTA application	53.43	46.13	7.3
Non-FTA application	50.61	15.8	35.53

Source: DOE, Summary of Export Application August 2013, 2013. Available online at: http://energy.gov/sites/prod/files/2013/08/f2/

Admittedly, the oil price collapse at the end of 2015 reduced the arbitrage opportunities significantly only the price ratio between LNG prices in Japan (and Europe) and the US remains high. However, for the past three decades, LNG trade in Asia has been based on the JCC (Japan Crude Cocktail) pricing formula, which gives Asian LNG importers a limited space for manoeuvre: existing long term contracts rarely include provisions for price review and, driven by real or perceived fears about security of supply, buyers signed new contracts in the 2010s which accepted ever-closer indexation of LNG to crude oil prices (Rogers and Stern 2014). Yet with higher future oil prices one could expect that huge price differences return to the market.

The assumption that it helps US politicians on the domestic political front to restrict LNG exports holds with the usual proviso: ceteris paribus. Other political events, like the ongoing conflict between Ukraine and Russia, can provide a ‘game changer’, which only highlights the political uncertainty affecting international natural gas markets. This relates to another politico-economic question whether it makes sense for Russia (Gazprom) to enter the LNG market in Japan. Given the (currently) high prices in Japan, this makes economic sense. Indeed, Gazprom is just considering to build an LNG export facility at Vladivostok (*The Economist*, March 23rd, 2013, Gazprom. Russia's wounded giant, p 59–60). Aside from the political front, there is substantial uncertainty about Australia's future LNG exports potential.

There are many directions to extend this analysis. For example, one may consider returns as stochastic process accounting for both Brownian motion and a jump process but otherwise following this paper. Complementary one may seek other explanations for the existence of this (apparent?) arbitrage in international LNG markets. This is not an easy task as gas producers can engage in intertemporal speculation by just leaving the gas in the ground as long as US gas prices are low and it is hard to think of physical, economical or legal constraints that rule out this kind of behavior (compare Dehnavi et al. 2015).

Acknowledgements I acknowledge helpful discussions about the LNG market with Yuri Yegorov and Jalal Dehnavi and comments from an anonymous referee. A very first draft of the paper was presented at World Congress of the International Association of Energy Economists, June 16th–18th 2014, New York and I thank the participants for their helpful comments.

Appendix: Declining Hazard, $h' < 0$

No particular property of h was used up to Eq. (18). Specifics of h enter only with the determination of $S(T)$ in (17) and then for the linear choice as in the example (8). Thus using the same example (8) with the only modification of a linearly declining hazard rate,

$$h = \alpha (1 - X)$$

some of the explicit calculations change,

$$\begin{aligned}\dot{F} &= \alpha\left(1-x_T\right)\left(1-F\right)\\ &\Longrightarrow F\left(t\right)=1-\left(1-F_T\right)e^{-h\left(x_T\right)\left(t-T\right)}=1-\left(1-F_T\right)e^{-\alpha\left(1-x_T\right)\left(t-T\right)}\\ S\left(x\left(T\right),T\right) &= A\left(X\left(T\right)\right)x\left(T\right)\int_0^\infty e^{-rt}\left(1-F\left(t\right)\right)dt\\ &= A\left(x_T\right)x_T\left(1-F_T\right)\int_0^\infty e^{-\left(r+\alpha\left(1-x_T\right)\right)t}dt=\frac{A\left(x_T\right)x_T\left(1-F_T\right)}{r+\alpha\left(1-x_T\right)}.\end{aligned}$$

Reformulated as a stopping problem, the optimal stopping conditions are:

$$\lambda\left(T\right)=S_x=\frac{A\left(x_T\right)\left(1-F\right)}{r+\alpha\left(1-x_T\right)}$$

$$H\left(T\right)=rS-\frac{\partial S}{\partial T}$$

$$\frac{\partial S}{\partial T}=-\frac{A\left(x_T\right)x_T}{r+\alpha\left(1-x_T\right)}\frac{\partial F}{\partial T}$$

$$\frac{\partial F}{\partial t}=-\left(1-F\right)\frac{\partial}{\partial t}e^{-\alpha\left(1-x\left(T\right)\right)\left(t-T\right)}=\alpha\left(1-X\right)\left(1-F\right)e^{-\alpha X\left(t-T\right)}\left.\right|_{t=T}=\alpha\left(1-X\right)\left(1-F\right)$$

$$H=A\left(X\right)x\left(1-F\right)-k\left(u\right)+\frac{A\left(X\right)\left(1-F\right)}{r+\alpha\left(1-X\right)}u\text{ at }t=T,$$

$$\begin{aligned}rS-\frac{\partial S}{\partial T} &= \frac{A\left(x_T\right)x_T\left(1-F_T\right)r}{r+\alpha\left(1-x_T\right)}+\frac{A\left(x_T\right)x_T\alpha\left(1-X\right)\left(1-F\right)}{r+\alpha\left(1-x_T\right)}\\ &= \frac{A\left(x_T\right)x_T\left(1-F_T\right)\left(r+\alpha\left(1-X\right)\right)}{r+\alpha\left(1-x_T\right)}=Ax\left(1-F\right)\text{ at }t=T.\end{aligned}$$

Therefore,

$$Ax_T\left(1-F_T\right)-k\left(u\right)+\frac{A\left(1-F_T\right)}{r+\alpha\left(1-X\right)}u=Ax_T\left(1-F_T\right)$$

which holds for $u\left(T\right)=0$. As a consequence, the corresponding stopping condition,

$$\lambda\left(T\right)=\frac{A\left(x_T\right)\left(1-F\right)}{r+\alpha\left(1-x_T\right)}=k'\left(0\right),$$

is similar

$$A\left(1-F\right)=\left(r+\alpha\left(1-x_T\right)\right)k.$$

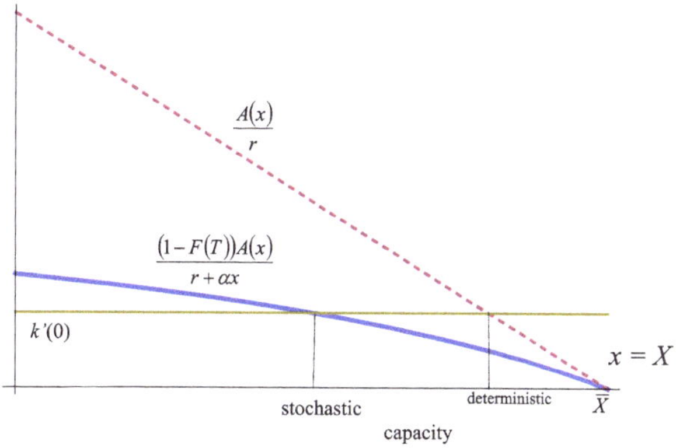

Fig. 7 Counterpart of Fig. 2 for declining hazard rate, $h = \alpha(1 - x)$: Determination of capacity: deterministic (steady state) vs stochastic (maximum, i.e. in case of no prior intervention) for F fixed at terminal value

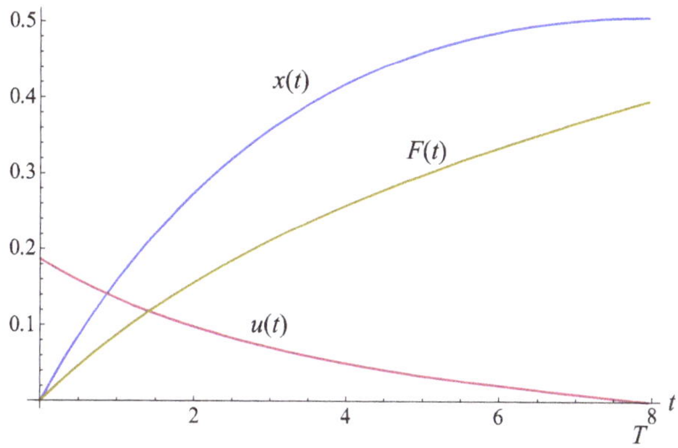

Fig. 8 Counterpart of Fig. 4 for declining hazard rate, $h = \alpha(1 - x)$ Time paths (assuming no intervention prior to T), reference example: $\alpha = 0.1, \kappa = 2, \gamma = 10, r = 0.1$

Figures 7 and 8 are the counterparts of Figs. 2 and 4 highlighting the similarity in spite of the opposite assumptions of either $h' > 0$ or $h' < 0$.

References

BP, BP Statistical Review of World Energy June 2015 (2015). bp.com/statisticalreview

E. Crivelli, S. Gupta, Resource blessing, revenue curse? Domestic revenue effort in resource-rich countries. Eur. J. Polit. Econ. **35**, 88–101 (2014)

J. Dehnavi, F. Wirl , Y. Yuri, Arbitrage in natural gas markets. Int. J. Energy Stat. **3**(4) (2015). https://doi.org/10.1142/S2335680415500180

D. Grass, J.P. Caulkins, G. Feichtinger, G. Tragler, D.A. Behrens, *Optimal Control of Nonlinear Processes: With Applications in Drugs, Corruption and Terror* (Springer, Heidelberg, 2008)

International Energy Agency, *World Energy Outlook 2013* (OECD, Paris, 2013)

International Energy Agency, *World Energy Outlook 2014* (OECD, Paris, 2014)

I. Kamien Morton, L. Nancy, N.L. Schwartz, Optimal maintenance and sale age for a machine subject to failure. Manag. Sci. **17**, 427–449 (1971)

N. Kleit Andrew, Did open access integrate natural gas markets? An arbitrage cost approach. J. Regul. Econ. **14**, 19–33 (1998)

R. La Porta, F. Lopez-de-Silanes, A. Shleifer, R. Vishny, Legal determinants of external finance. J. Financ. **52**(3), 1131–1150 (1997)
M.A. Levi, The geopolitics of natural gas. Natural gas in the United States. WP of James A. Baker III Institute for Public Policy, Rice University, 23 October, 2013
D. Maxwell, Z. Zhu, Natural gas prices, LNG transport costs, and the dynamics of LNG imports. Energy Econ. **33**, 217–226 (2011)
H.V. Rogers, J. Stern, Challenges to JCC pricing in Asian LNG markets, OIES Paper: NG 81, 2014
US Energy Information Administration, The global liquefied natural gas market: status and outlook, Report #: DOE/EIA-0637, Release Date: December 2003
C. von Hirschhausen, A. Neumann, Long-term contracts and asset specificity revisited: an empirical analysis of producer–importer relations in the natural gas industry. Rev. Ind. Organ. **32**, 131–143 (2008)
J.G. Weber, A decade of natural gas development: the makings of a resource curse? Resour. Energy Econ. **37**, 168–183 (2014)
Y. Yegorov, J. Dehnavi, Future of shale gas in China and its influence on the global markets for natural gas, Available at SSRN 2524357, 2014

The Deterministic Optimal Liquidation Problem

Pavol Brunovský, Margaréta Halická, and Mario Mitas

Dedicated to Vladimir Veliov on the occasion of his 65th birthday.

Abstract A one-dimensional free terminal time optimal control problem stemming from mathematical finance is studied. To find the optimal solution and prove its optimality the standard maximum principle procedure including Arrow's sufficiency theorem is combined with specific properties of the problem. Certain unexpected features of the solution are pointed out and discussed.

1 The Problem

In the papers Brunovský et al. (2013, in press) and Černý (1999), the following stochastic optimal control problem is studied:

For given $S(0) = S_0 > 0,\ Z(0) = Z_0 > 0$ maximize

$$J(v) = E\left(\int_0^{T(Z=0)} e^{-\rho t} v(t)(S(t) - \eta v(t))dt\right), \tag{1}$$

subject to

$$dS(t) = \lambda S(t)dt + \sigma S(t)dB(t), \tag{2}$$

$$dZ(t) = (rZ(t) - v(t))\,dt, \tag{3}$$

$$v(t) \geq 0 \tag{4}$$

P. Brunovský (✉) · M. Halická · M. Mitas
Department of Applied Mathematics and Statistics, Comenius University in Bratislava, Bratislava, Slovakia
e-mail: brunovsky@fmph.uniba.sk; halicka@fmph.uniba.sk; mitas.mario@gmail.com

G. Feichtinger et al. (eds.), *Control Systems and Mathematical Methods in Economics*, Lecture Notes in Economics and Mathematical Systems 687,
https://doi.org/10.1007/978-3-319-75169-6_13

with $B(t)$ the standard Brownian motion and $T(Z = 0)$ being the first arrival time of $Z(t)$ at 0. The problem models optimal liquidation of a given asset by its dominant owner. The variable Z stands for the amount of the asset the owner is in possession, S for its reservation price, v for the rate of selling, the objective function J representing the current value of the revenue the owner receives under the liquidation policy $v(t)$. The value of the asset appreciates with rate r, its price develops randomly with drift λ and future revenue is discounted by rate $\rho > 0$. The dominance of the owner is reflected by the term $-\eta v(t)$, $\eta > 0$, by which the actual price is diminished proportionally to the amount the owner sells. It represents temporary drop of the asset price due to increased supply.

In Černý (1999), the HJB (or, dynamic programming) equation of the problem is developed and discussed. In Brunovský et al. (2013) its solvability as well as properties of the solution are studied. It is shown that, in order for the problem to be solvable one needs that

$$\rho > r + \lambda. \tag{5}$$

This assumption means that the discount is strong enough to prevent the owner to hold the asset infinitely long. In Brunovský et al. (in press) the problem is viewed in a broader perspective and a survey of the rich literature going back to Almgren and Chriss (2000) is presented. Optimality of the solution of the HJB equation is proved and a numerical scheme is developed for the computation of the solution of the equation which is strongly degenerate.

The papers leave open the question about the nature of the endogenously determined terminal time which may in principle be finite as well as infinite. Of course, this is a question of random nature.

In this note we address the deterministic version of the problem, i.e. the problem with $\sigma = 0$. As in the general stochastic case, in a straightforward way one can check that the change of variables

$$x = \eta \frac{Z}{S}, \tag{6}$$

$$u = \eta \frac{v}{S} \tag{7}$$

reduces this problem to the following one-dimensional one:

Given $x_0 > 0$,

$$\text{maximize } \frac{S_0^2}{\eta} \int_0^{T(x=0)} e^{(2\lambda-\rho)t} u(t)(1 - u(t))dt \tag{8}$$

subject to

$$\dot{x}(t) = (r - \lambda)x(t) - u(t), \tag{9}$$

$$x(0) = x_0, \tag{10}$$

$$u(t) \geq 0, \tag{11}$$

with T the first t at which $x(t)$ vanishes if finite, $T = \infty$ otherwise. By denoting $a = \rho - 2\lambda$, $b = \lambda - r$ the number of parameters can be reduced. The constant factor S_0^2/η dropped, the problem (8)–(11) then reads as follows:

Given $x_0 > 0$,

$$\text{maximize } J(u, T) = \int_0^T e^{-at} u(t)(1 - u(t))dt \tag{12}$$

subject to (11), (10),

$$\dot{x}(t) = -bx(t) - u(t), \tag{13}$$

$$x(T) = 0 \text{ if } T < \infty, \tag{14}$$

$$x(t) > 0 \text{ for } 0 < t < T. \tag{15}$$

As pointed out by the anonymous referee, the substitution $u = -I$ turns (12)–(15) into a problem resembling the capital accumulation one (cf., e.g., Feichtinger et al. 2006).

The condition (5) now reads

$$a + b > 0, \tag{16}$$

which, from now on, will be our standing hypothesis.[1] A control/response/time triple (u, x, T) and its components will be called admissible if $T > 0$, $u(t)$ satisfies (11) and its response $x(t)$ satisfies (10) (13), (14) and (15).

Thus we have reduced the problem to a standard one with fixed terminal state and free horizon. This problem will be discussed in the sequel.

We now show that, along any admissible control the objective function is bounded. The bound means that, because of discount and the adverse effect of the sale on the price of the asset, the revenue cannot exceed the one from immediate sale for a price not affected by the latter. The bound will be used substantially in the proof of the main result.

[1] Because there is no need of the assumption $\rho > 0$ in the deterministic problem, we ignore it in the sequel.

Proposition 1 *Let (u, x, T) be an admissible triple. Then for every $t_0 \in [0, T]$ one has*

$$\int_{t_0}^{T} e^{-at} u(t)(1 - u(t))dt \le e^{-at_0} x(t_0).$$

Proof One has

$$\int_{t_0}^{T} e^{-at} u(t)(1 - u(t))dt \le \int_{t_0}^{T} e^{-at} u(t)dt$$
$$= \int_{t_0}^{T} e^{-at}(-\dot{x}(t) - bx(t))dt = -\int_{t_0}^{T} e^{-(a+b)t} \frac{d}{dt}(e^{bt} x(t))dt$$
$$= e^{-at_0} x(t_0) - e^{-aT} x(T) - (a + b) \int_{t_0}^{T} e^{-at} x(t)dt \le e^{-at_0} x(t_0).$$

□

2 The Result

The maximum of the integrand of the objective function is achieved at $u = \frac{1}{2}$. Yet, it would be a myopic policy to sell the asset at the corresponding rate, because it does not take into account future revenues. The theorem below shows that for certain values of the parameters, the optimal selling rate is smaller than the myopically optimal value while for others the latter is optimal. Also, the theorem implicitly determines the optimal liquidation time, control and the maximal value of the objective function.

Theorem 1 *Let $a \ne 0$, $b \ne 0$ and $a + 2b \ne 0$.*

(i) If either $b > 0$, or both $b < 0$ and $x_0 < -1/2b$, then the optimal liquidation time $\hat{T}$ is finite. It is the (unique) positive root of the implicit equation

$$x_0 = \frac{1}{2b}(e^{bT} - 1) - \frac{e^{-(a+b)T}}{2(a + 2b)}(e^{(a+2b)T} - 1). \tag{17}$$

The unique optimal control $\hat{u}(t)$ is given by

$$\hat{u}(t) = \frac{1}{2}(1 - e^{(a+b)(t-\hat{T})}) \tag{18}$$

and

$$\hat{J} = J(\hat{u}, \hat{T}) = \frac{1}{4a}(1 - e^{-a\hat{T}}) - \frac{e^{-(a+b)\hat{T}}}{4(a + 2b)}(e^{(a+2b)\hat{T}} - 1) \tag{19}$$

is the maximal value of J.

(ii) If both $b < 0$ and $x_0 \geq -1/2b$, then $\hat{T} = \infty$, $\hat{u}(t) \equiv \frac{1}{2}$ and $\hat{J} = \frac{1}{4a}$. The optimal trajectory $\hat{x}(t)$ satisfies

$$\hat{x}(t) \begin{cases} \equiv x_0, & \text{if } x_0 = -\frac{1}{2b} \\ \nearrow \infty, & \text{if } x_0 > -\frac{1}{2b}. \end{cases} \tag{20}$$

Note that the formulas (17), (19) are in accord with (A.24), (A.25) of Černý (1999). In the excluded cases $a = 0, b = 0, a + 2b = 0$ (no two of which can take place simultaneously because of (16)) the formulas (17), (19) remain valid if $(e^{AT} - 1)/A$ for $A = 0$ is understood as $\lim_{A \to 0}(e^{AT} - 1)/A = T$.

Proof Case (i)

The optimal triple $(\hat{u}(t), \hat{x}(t), \hat{T})$ has to satisfy the Pontryagin maximum principle for a problem with fixed terminal state and free time. In the current-value formulation (Sethi and Thompson 2006, Section 3.3), it reads:

There is a constant $\hat{\psi}^0 \geq 0$ and a solution $\hat{\psi}(t)$ of the adjoint equation

$$\dot{\psi} = (a + b)\psi \tag{21}$$

such that $(\hat{\psi}^0, \hat{\psi}(t)) \neq 0$,

$$H(\hat{x}(t), \hat{\psi}^0, \hat{\psi}(t), \hat{u}(t)) = \max_{u \geq 0} H(\hat{x}(t), \hat{\psi}^0, \hat{\psi}(t), u) \tag{22}$$

for all $0 \leq t \leq \hat{T}$ with

$$H(x, \psi^0, \psi, u) = \psi^0 u(1 - u) + \psi(-bx - u).$$

The terminal time being free, from Seierstad and Sydsaeter (1993, Theorem 11 of Chapter 2) we additionally obtain

$$H(\hat{x}(\hat{T}), \hat{\psi}^0, \hat{\psi}(\hat{T}), \hat{u}(\hat{T})) = H(0, \hat{\psi}^0, \hat{\psi}(\hat{T}), \hat{u}(\hat{T})) = 0. \tag{23}$$

We show $\hat{\psi}^0 \neq 0$. Suppose the contrary. Then, from (22) it follows that, for all t, $\hat{u}(t) = 0$, hence $\hat{J} = 0$ which is surely not optimal.

Because of homogeneity of H we can set $\hat{\psi}^0 = 1$. With this choice (22) implies

$$\hat{u}(t) = \frac{1}{2}[1 - \hat{\psi}(t)]^+, \tag{24}$$

where $A^+ = \max\{A, 0\}$. Because $\hat{x}(\hat{T}) = 0$, the condition (23) implies

$$\begin{aligned} 0 &= H(0, 1, \hat{\psi}(\hat{T}), \hat{u}(\hat{T})) = \hat{u}(\hat{T})(1 - \hat{u}(\hat{T}) - \hat{\psi}(\hat{T})) \\ &= \frac{1}{2}[1 - \hat{\psi}(\hat{T})]^+ \left(1 - \frac{1}{2}[1 - \hat{\psi}(\hat{T})]^+ - \hat{\psi}(\hat{T})\right) \\ &= \frac{1}{2}[1 - \hat{\psi}(\hat{T})]^+ \frac{1}{2}(1 - \hat{\psi}(\hat{T})), \end{aligned}$$

which holds if and only if

$$\hat{\psi}(\hat{T}) \geq 1. \tag{25}$$

We complete the proof in two steps. In Step 1 we show that for each x_0 satisfying the assumptions of Case (i) there exists a unique T and a unique solution $(x(t), \psi(t), u(t))$, $t \in [0, T]$ of (13), (21), (24), (10), such that (14), (15), (25) and that this solution satisfies

$$\psi(T) = 1. \tag{26}$$

Then, in the Step 2, we prove that the (u, x) component of this solution and the corresponding T is the optimal triple.

Step 1. First we prove that there is no solution of (13), (21), (24), (10), (14), (15), (25) not satisfying (26).

Indeed, suppose $\psi(T) > 1$ and let $t^* = \inf\{t \in [0, T] : \psi(s) > 1 \text{ for } t < s \leq T\}$. One has $t^* < T$, $x(t^*) > 0$ as well as $u(t) = 0$ for $t^* \leq t \leq T$. This means that $x(t)$ satisfies $\dot{x}(t) = -bx(t)$ for $t^* \leq t \leq T$, hence $x(T) = e^{-b(T-t^*)}x(t^*) > 0$ which contradicts (14) and proves the claim.

Suppose now that $((x(t), \psi(t), u(t))$, $t \in [0, T]$ satisfies (13), (21), (24), (14), (15) as well as (26) for some $T > 0$ to be considered as a parameter. Then, $\psi(t) = e^{(a+b)(t-T)} \leq 1$ for all t so

$$u(t) = \frac{1}{2}(1 - \psi(t)) = \frac{1}{2}(1 - e^{(a+b)(t-T)}). \tag{27}$$

We now show that the parameter T can be chosen in such a way that $u(t)$, $t \in [0, T]$ is an admissible control. Substituting for $u(t)$ into (13) and integrating the resulting equation from T to t taking into account (14), we obtain

$$x(t) = \frac{1}{2b}(e^{b(T-t)} - 1) + \frac{1}{2(a+2b)}(e^{b(T-t)} - e^{-(a+b)(T-t)}). \tag{28}$$

Denoting $\tau = T - t$ the "time to liquidation", $\xi(\tau) = x(T-\tau)$ we obtain

$$\xi(\tau) = \frac{1}{2b}(e^{b\tau} - 1) + \frac{1}{2(a+2b)}(e^{b\tau} - e^{-(a+b)\tau}). \tag{29}$$

The equality (10) holds if and only if

$$\xi(T) = x_0 \tag{30}$$

It can be readily checked that $\xi(0) = 0$ while, due to (16)

$$\lim_{\tau\to\infty} \xi(\tau) = \begin{cases} \infty \text{ if } b > 0 \\ -1/2b \text{ if } b < 0 \end{cases} \tag{31}$$

Therefore (30), and, consequently, (10), has a solution for each x_0 satisfying the hypotheses of Case(i)

To prove that this solution is unique observe that $\xi(\tau)$ satisfies

$$\begin{aligned} \dot{\xi} &= b\xi + \frac{1}{2}(1 - e^{(a+b)\tau}) \\ \xi(0) &= 0. \end{aligned}$$

Hence, $\xi(\tau)$ is strictly increasing for $\tau > 0$ and reaches each its value at a single τ.

Step 2. For given x_0, we label the unique solution of (13), (21), (24), (10), by hats and prove that $(\hat{u}, \hat{x}, \hat{T})$ is the optimal triple. To this end we employ the techniques of the Arrow sufficiency theorem (Sethi and Thompson 2006, Theorem 2.1) and Proposition 1. We can do so because the integrand of the objective function is concave in u and the right hand side of (12) is linear in both x and u.

We should prove that for any (u, x, T) triple, the following inequality holds

$$J(\hat{u}, \hat{T}) - J(u, T) \geq 0. \tag{32}$$

Since the admissible triple $(\hat{u}, \hat{x}, \hat{T})$ satisfies the conditions of the Pontryagin maximum principle with $\psi^0 = 1$, and $\hat{\psi}$, one has at any $t \in [0, \min\{T, \hat{T}\}]$

$$\begin{aligned} & e^{-at}[\hat{u}(1-\hat{u}) - u(1-u)] \\ &= e^{-at}\left[H(\hat{x}, 1, \hat{\psi}, \hat{u}) - H(\hat{x}, 1, \hat{\psi}, u) + \hat{\psi} b(x - \hat{x}) + \hat{\psi}(\dot{x} - \dot{\hat{x}})\right] \\ &\geq e^{-at}\hat{\psi} b(x - \hat{x}) + e^{-at}\hat{\psi}(\dot{x} - \dot{\hat{x}}) = \frac{d}{dt}\left(e^{-at}\hat{\psi}(x - \hat{x})\right), \end{aligned} \tag{33}$$

where we used the maximum condition (22), the adjoint equation (21) and

$$e^{-at}\hat{\psi}(t)b = \frac{d}{dt}(e^{-at}\hat{\psi}(t)).$$

To prove (32) for $T > \hat{T}$ we use (33) and Proposition 1 to obtain

$$J(\hat{u}, \hat{T}) - J(u, T) = \int_0^{\hat{T}} e^{-at}[\hat{u}(t)(1 - \hat{u}(t)]dt - \int_0^{T} e^{-at}[u(t)(1 - u(t))]dt$$

$$\geq \int_0^{\hat{T}} \frac{d}{dt}\left(e^{-at}\hat{\psi}(x - \hat{x})\right)dt - \int_{\hat{T}}^{T} e^{-at}u(t)(1 - u(t))]dt$$

$$\geq \hat{\psi}(\hat{T})(x(\hat{T}) - \hat{x}(\hat{T}) - (\hat{\psi}(0)(x(0) - \hat{x}(0) - e^{-a\hat{T}}\hat{\psi}(\hat{T})(x(\hat{T}) = 0.$$

If $T \leq \hat{T}$, we extend $u(t)$ to the interval $[0, \hat{T}]$ by $u(t) = 0$ for $t \in [T, \hat{T}]$, then

$$J(u, \hat{T}) = J(u, T). \tag{34}$$

Using (34) and (33) we obtain $J(\hat{u}, \hat{T}) - J(u, T) \geq 0$, which completes the proof of Case (i).

Case (ii)

Because of the standing hypothesis, $a > 0$. The integrand $e^{-at}u(t)(1 - u(t))$ reaches its maximum $\frac{1}{4}e^{-at}$ for $u(t) = \frac{1}{2}$. The function $x(t, 0)$ is nondecreasing in t for $x_0 \geq -1/2b$, hence it never vanishes. That is, the maximal revenue is achieved if, in (12), $T = \infty$ and $u(t) \equiv \frac{1}{2}$ yielding

$$J(u) = \frac{1}{4}\int_0^{\infty} e^{-at}dt = \frac{1}{4a} \tag{35}$$

which completes the proof of (ii). □

Transcribed into the original variables Theorem 1 reads as follows:
Let $\rho \neq 2\lambda$, $\lambda \neq r$, $\rho \neq 2r$. Then,

(i) If either $\lambda > r$, or both $\lambda < r$ and $Z_0/S_0 < \frac{1}{2\eta(r-\lambda)}$, then $\hat{T}$ is finite. It is the only positive root of the implicit equation

$$Z_0/S_0 = \frac{1}{2\eta(\lambda - r)}(e^{(\lambda-r)\hat{T}} - 1) + \frac{e^{(\lambda+r-\rho)\hat{T}}}{2\eta(\rho - 2r)}(e^{(\rho-2r)\hat{T}} - 1). \tag{36}$$

The optimal control $\hat{v}(t)$ is given by

$$\hat{v}(t) = \frac{S(t)}{2\eta}(1 - e^{(\rho-\lambda-r)(t-\hat{T})}) \tag{37}$$

and

$$\hat{J} = \frac{S_0^2}{\eta}\left[\frac{1}{4(\rho - 2\lambda)}(1 - e^{(2\lambda-\rho)\hat{T}}) - \frac{e^{(\lambda+r-\rho)\hat{T}}}{4(\rho - 2r)}(e^{(\rho-2r)\hat{T}} - 1)\right]. \tag{38}$$

(ii) If both $r > \lambda$ and $Z_0/S_0 \geq \frac{1}{2\eta(r-\lambda)}$, then $\hat{T} = \infty$, $\hat{J} = \frac{S_0^2}{4\eta(\rho-2\lambda)}$ and $\hat{v}(t) \equiv \frac{S(t)}{2\eta}$,

$$\frac{Z(t)}{S(t)} \begin{cases} \equiv \frac{Z_0}{S_0} \text{ if } Z_0 = \frac{1}{2\eta(r-\lambda)} S_0 \\ \nearrow \infty \text{ if } Z_0 > \frac{1}{2\eta(r-\lambda)} S_0. \end{cases} \tag{39}$$

3 Remarks

There are several unexpected features of the solution of the problem which we feel worth to be pointed out. To this end, let us call extremal a solution of the system of Eqs. (13), (21), (11), (10), (27) and extended extremal a solution of (13), (21), (10), (27) with (11) dropped. Observe that from (21) and (27) we obtain the differential equation

$$\dot{u} = -\frac{1}{2}(a+b)(1-2u) \tag{40}$$

for the control. Therefore, extended extremals can be alternatively represented by solutions of the system (13), (10), (40) in the (x, u)-space. Also, (27) implies that u decreases from $\frac{1}{2}$ to 0 for ψ increasing from 0 to 1. The trajectories of the extremals and the corresponding solutions of (13), (10), (40) are depicted in Figs. 1 and 3, those of extended extremals on Figs. 2 and 4, respectively. Parts of trajectories of the extended extremals which do not belong to extremals are represented by dashed lines.

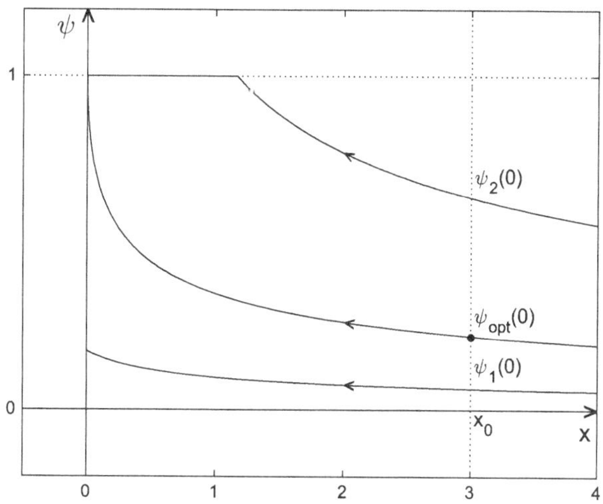

Fig. 1 Extremals in (x, ψ)-space for $a = -0.5$, $b = 1$

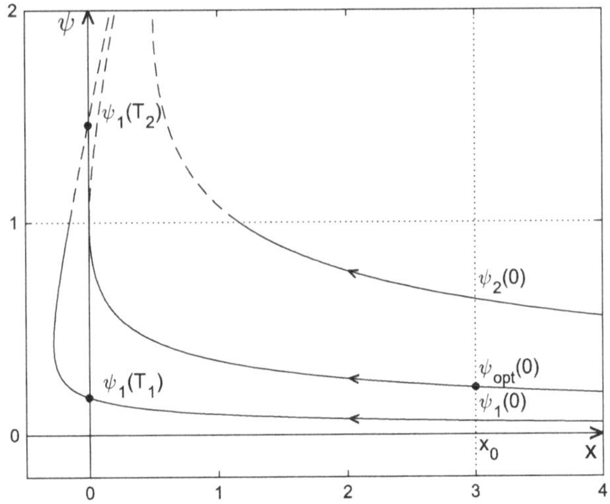

Fig. 2 Extended extremals in (x, ψ)-space for $a = -0.5,\ b = 1$. $\psi_1(0) < \psi_{opt}(0)$: the extended extremal meets $x = 0$ in two points $T_1 < \hat{T} < T_2$; $\psi_2(0) > \psi_{opt}(0)$: $x(t)$ never vanishes

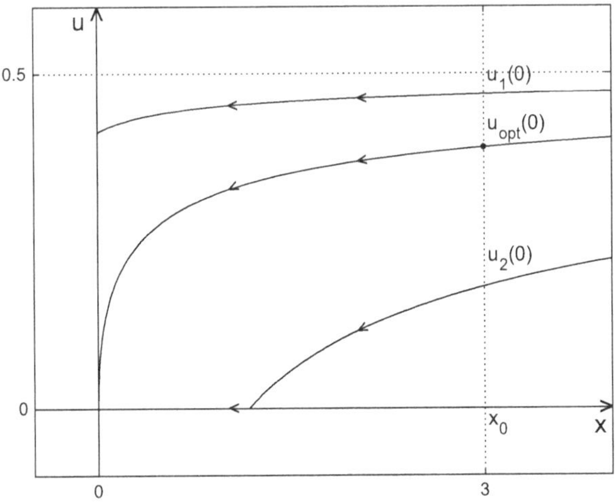

Fig. 3 Extremals in (x, u)-space for $a = -0.5,\ b = 1$

- As illustrated by Fig. 3, $\hat{u}(t) < \frac{1}{2}$ for all t. That is, in Case (i) it is always optimal to sell less than the myopically optimal amount $u = \frac{1}{2}$. Yet, somewhat counter-intuitively, at the beginning, when the inventory is large and the liquidation time is far, optimal control is closer to the myopic one than at the end, when the inventory is almost exhausted and the time to liquidation is small.
- Unlike extremals (Figs. 1 and 3), extended extremals (Figs. 2 and 4) admit purchase (short sale in finance) of the asset. As seen from Fig. 2 (and can easily be verified by elementary algebra), the value $\psi_{opt}(0)$ is the only value of $\psi(0)$ whose extended extremal satisfies (14) in a single T. For $\psi_2(0) > \psi_{opt}(0)$ (i.e. $u(0) < u_{opt}(0)$), (14) has no solution, while for $0 < \psi_1(0) < \psi_{opt}(0)$ (i.e. $1/2 > u(0) > u_{opt}(0)$), it has two roots $T_1 < T_2$. So, if we parametrize extended extremals by the terminal time T, then $T = \hat{T}$ maximizes $\psi(0)$ and

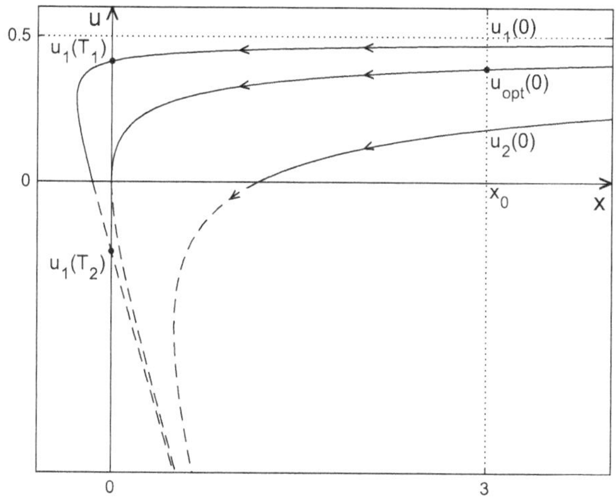

Fig. 4 Extended extremals in (x, u)-space for $a = -0.5,\ b = 1$. $u_1(0) > u_{opt}(0)$: the extended extremal meets $x = 0$ in two points $T_1 < \hat{T} < T_2$; $u_2(0) < u_{opt}(0)$: $x(t)$ never vanishes

minimizes $u(0)$. It can be shown by simulations that, for the extended extremal, the value of the objective function J with $T = T_2$ exceeds that with $T = T_1$. That is, condition (11) is substantial. This is not overly surprising: because of the standing assumption ((5) resp. (16)) the seller may increase his revenue by purchasing the asset earlier and selling it later to end up with zero inventory. The above conditions on $u(0)$ can of course be expressed in the original variables using (7). For example, condition $1/2 > u_1(0) > u_{opt}(0)$ is equivalent to the condition

$$\frac{1}{2\eta} S_0 > u_1(0) \frac{1}{\eta} S_0 = v(0) > u_{opt}(0) \frac{1}{\eta} S_0 = v_{opt}(0)$$

- It is tempting to try to use the maximum principle to reduce the search for optimal control to extremals represented by $u(0)$. This reduction leads to a constrained two-dimensional Lagrange problem with objective function J and constraint $x(T) = 0$. However, the Lagrange formalism fails because for $u(0) = u_{opt} = 1/2(1 - e^{-(a+b)T})$ both $\frac{\partial x(T)}{\partial x(0)}$ and $\frac{\partial x(T)}{\partial T}$ vanish. Interestingly, for the extended extremals, J has a point of inflexion at $\hat{T}$.
- Step 1 could alternatively be proved by examining the phase portrait of the linear system for extremals (13), (21), (27) or (13), (40) . However, depending on the sign of $a, b, a + 2b$, three types of phase portraits can occur, although the picture of trajectories of Figs. 1 and 2 looks qualitatively the same.

Acknowledgements The authors are thankful to the anonymous referee for enlightening comments. Research of the second author has been supported by VEGA Grants 1/0780/15 and 01/0062/18.

References

R. Almgren, N. Chriss, Optimal execution of portfolio transactions. J. Risk **3**, 5–39 (2000)

P. Brunovský, A. Černý, M. Winkler, A singular differential equation stemming from am optimal control problem in financial economics. Appl. Math. Optim. **68**, 255–274 (2013)

P. Brunovský, A. Černý, J. Komadel, Optimal trade execution under endogenou pressure to liquidate: theory and numerical solutions. Eur. J. Oper. Res. **264**, 1159–1171 (2018)

A. Černý, Current crises: introduction of spot speculators. Int. J. Fin. Econ. **4**, 75–90 (1999)

G. Feichtinger, R.F. Hartl, P.M. Kort, V.M. Veliov, Anticiaptiaon effects of technological progress on capital accumulation: a vintage capital approach. J. Econ. Theory **126**, 143–164 (2006)

A. Seierstad, K. Sydsaeter, *Optimal Control Theory with Economic Applications*, 2nd edn. (Elsevier, North-Holland, 1993)

P.S. Sethi, G.T. Thompson, *Optimal Control Theory*, 2nd edn. (Springer, Berlin, 2006)

Dynamic Models of the Firm with Green Energy and Goodwill

Herbert Dawid, Richard F. Hartl, and Peter M. Kort

Abstract This paper considers the effect of investment in solar panels on optimal dynamic firm behavior. To do so, an optimal control model is analyzed that has as state variables goodwill and green capital stock. Following current practice in companies like Tesla and Google, we take into account that the use of green energy has positive goodwill effects. As a solution, we find an optimal trajectory that overshoots before reaching a stable steady state.

1 Introduction

Vladimir Veliov is the author of several contributions in the field of dynamic models that take the environment into consideration; see e.g. Feichtinger et al. (2005), Georgiev et al. (2005), Moser et al. (2014). For this reason, the present contribution is centered around a dynamic model of the firm, in which the firm has the opportunity to produce goods using green energy, e.g. by using solar panels. This is a topical subject; see e.g. the recent article in The Economist (2017). Our model is inspired by Amigues et al. (2015), who present a model in which an energy provider appears that provides two different ways to produce energy, namely in the old fashioned way by burning oil and in the "green" way by investing in solar panels.

H. Dawid (✉)
Department of Business Administration and Economics and Center for Mathematical Economics, Bielefeld University, Bielefeld, Germany
e-mail: hdawid@wiwi.uni-bielefeld.de

R. F. Hartl
Department of Business Administration, Production and Operations Management, University of Vienna, Vienna, Austria

P. M. Kort
Department of Econometrics and Operations Research & Center, Tilburg University, Tilburg, The Netherlands

Department of Economics, University of Antwerp, Antwerp, Belgium

G. Feichtinger et al. (eds.), *Control Systems and Mathematical Methods in Economics*, Lecture Notes in Economics and Mathematical Systems 687,
https://doi.org/10.1007/978-3-319-75169-6_14

In particular, we design and analyze a dynamic model of the firm where the firm uses energy as an input to produce goods. In practice, some firms are already using solar energy and use this fact in public relations, e.g. Tesla or Google. For this reason the special feature of our model is that the use of green energy attributes to the development of the firm's goodwill.

We analyze two different models. In the first one, the firm is only able to use green energy. For this, a green capital stock needs to be built up. As a solution, we find a unique saddle point steady state to which the firm monotonically converges. In the second model we analyze a firm that can choose between two different sources of energy: traditional energy obtained from the energy market (modeled as a control) and green energy as in the first model. For this model, we also find a unique saddle point steady state, but the difference with the previous model is that convergence takes place in a non-monotonic way. Hence, our main conclusion is that the use of two different inputs, namely traditional energy and green energy results in an optimal trajectory that converges to a steady state, but that overshoots.

A number of dynamic models capturing the transition from traditional to renewable resources for energy production have been studied in the literature. Wirl (1991, 2008) studies the capacity buildup of renewable energy producers entering a market (initially) dominated by incumbent producers with market power, which rely on production with exhaustible resources. Tsur and Zemel (2011) are closer to our setup by studying the transition from fossil fuels to solar energy within a firm. They characterize scenarios under which solar energy is adopted and finally dominates the industry. Contrary to Tsur and Zemel (2011) the contribution by Amigues et al. (2015) takes into account that fossil resources are exhaustible and shows that under certain conditions the optimal transition is characterized by different phases, starting with exclusive use of the non-renewable resource, exhibiting then parallel use of both resources a final phase in which only renewable resources are used. The dynamics in all these contributions are driven by costs considerations of the planing firms, but do not consider any reputational effects. The main innovative contribution of our paper in this respect is that we analyze the dynamic implications of the interplay of cost and goodwill considerations for the transition from traditional to green energy.

The content of the paper is as follows: Sect. 2 describes the model. Section 3 analyzes the model where only green capital stock is used in the production process, whereas Sect. 4 looks at the extension where the production process has two possible inputs, green capital and traditional energy. Section 5 concludes and gives directions for future research.

2 The Model

The firm's production process has energy, E, as input, which results in an output $q(E)$. Energy is available from two different sources. There is the traditional way of producing energy by using e.g. fossil fuels. Usage of traditional energy is a control denoted by X. On the other hand, the firm can invest in a green capital stock, K, (e.g. solar panels) to produce energy. The total energy used is therefore

$$E = K + X.$$

The evolution of the green capital stock over time follows the traditional capital accumulation equation

$$\dot{K} = I - \delta K, \qquad K(0) = K_0$$

in which I is investment in the green capital stock, whereas δ is the depreciation rate.

Investing in the green capital stock has a positive side effect. Making such investment known to the public helps increasing the firm's goodwill, G. This can be done by advertising, a. The larger the green capital stock, the more goodwill increases by a given advertising expenditure. In this way, the goodwill dynamics becomes

$$\dot{G} = f(K)a - \delta_G G, \qquad G(0) = G_0$$

where the effectiveness of advertising, $f(K)$, is an increasing function of the green capital stock, and δ_G is the depreciation rate of goodwill.

The firm's output can be sold on the market. The output price increases in goodwill and decreases in quantity. Assuming a linear demand function we express the output price, p, by

$$p = \max[g(G) - \alpha q(E), 0]$$

where α is a positive constant, and $g(.)$ is an increasing function of goodwill G. In the following analytical treatment we will always assume that the price is strictly positive. Advertising, investment in green energy, and traditional energy use are costly. Advertising cost are denoted by $C_a(a)$, green energy investment costs are given by $C_s(I)$, and the unit cost of traditional energy is p_X.

Assuming that the firm maximizes the discounted cash flow stream, we arrive at the following dynamic model of the firm:

$$\max_{I,a,X} \int_0^\infty e^{-rt}\left[(g(G) - \alpha q(E))\, q(E) - C_a(a) - C_s(I) - p_X X\right] dt \quad (1)$$

$$\dot{G} = f(K)a - \delta_G G, \qquad G(0) = G_0 \quad (2)$$

$$\dot{K} = I - \delta K, \qquad K(0) = K_0 \quad (3)$$

$$E = K + X. \quad (4)$$

We rule out negative advertising and negative usage of traditional energy, i.e. $a \geq 0$ and $X \geq 0$ has to hold. With respect to investment in green capital we allow for (costly) disinvestment, however the non-negativity constraint for the stock applies, i.e. $K \geq 0$. Before we analyze this general model, we first look at a simplified variant, where the traditional energy use is abolished, i.e., $X = 0$.

3 Analysis of the Model with Only Clean Input

With $X = 0$ the dynamic model of the firm straightforwardly becomes

$$\max_{I,a} \int_0^\infty e^{-rt} \left[(g(G) - \alpha q(K))\, q(K) - C_a(a) - C_s(I) \right] dt$$

$$\dot{G} = f(K)\,a - \delta_G G, \qquad G(0) = G_0$$
$$\dot{K} = I - \delta K, \qquad K(0) = K_0, \; K \geq 0.$$

Applying the maximum principle, Grass et al. (2008), Feichtinger and Hartl (1986) we start out with setting up the Hamiltonian

$$H = (g(G) - \alpha q(K))\, q(K) - C_a(a) - C_s(I) + \lambda (f(K)\,a - \delta_G G) + \mu (I - \delta K) + \nu K.$$

This results in the following necessary optimality conditions:

$$H_I = 0 = -C_s'(I) + \mu$$
$$H_a = 0 = -C_a'(a) + \lambda f(K)$$
$$\dot{\lambda} = (r + \delta_G)\,\lambda - g'(G)\, q(K)$$
$$\dot{\mu} = (r + \delta)\,\mu - q'(K)\,(g(G) - 2\alpha q(K)) - \lambda a f'(K) - \nu.$$

In order to obtain some analytical results, we proceed the analysis with the following specifications:

$$C_a(a) = \frac{\beta}{2} a^2 \tag{5}$$

$$C_s(I) = \frac{\gamma}{2} I^2 \tag{6}$$

$$f(K) = K \tag{7}$$

$$q(K) = \eta K \tag{8}$$

$$g(G) = \theta G^\rho, \qquad \text{with } \theta > 0,\, 0 < \rho < 1. \tag{9}$$

The necessary optimality conditions now become

$$\begin{aligned}
H_I &= 0 = -\gamma I + \mu \\
H_a &= 0 = -\beta a + \lambda K \\
\dot{\lambda} &= (r + \delta_G)\,\lambda - \rho\theta G^{\rho-1}\eta K \\
\dot{\mu} &= (r + \delta)\,\mu - \eta\left(\theta G^{\rho} - 2\alpha\eta K\right) - \lambda a - \nu
\end{aligned}$$

with $\nu \geq 0$ and $\nu K = 0$. The optimal investment rate and advertising rate are given by

$$\begin{aligned}
I &= \frac{\mu}{\gamma} \\
a &= \frac{\lambda K}{\beta}.
\end{aligned}$$

This results in the canonical system

$$\dot{\lambda} = (r + \delta_G)\,\lambda - \rho\theta\eta G^{\rho-1} K \tag{10}$$

$$\dot{\mu} = (r + \delta)\,\mu - \eta\left(\theta G^{\rho} - 2\alpha\eta K\right) - \lambda^2 \frac{K}{\beta} - \nu \tag{11}$$

$$\dot{G} = \frac{\lambda K^2}{\beta} - \delta_G G, \qquad G(0) = G_0 \tag{12}$$

$$\dot{K} = \frac{\mu}{\gamma} - \delta K, \qquad K(0) = K_0. \tag{13}$$

Based on this canonical system we now prove the following important result:

Proposition 1 *For $\rho \neq 0.5$, the canonical system* (10)–(13) *admits a unique steady state.*

Proof Assuming that $K > 0$ we set $\nu = 0$. Hence, from (10), (12), and (13) we obtain that a possible steady state must satisfy

$$G = \frac{\lambda}{\beta\delta_G}\left(\frac{\mu}{\gamma\delta}\right)^2 \tag{14}$$

$$K = \frac{\mu}{\gamma\delta} \tag{15}$$

$$\lambda = \frac{\rho\theta\eta}{\gamma\delta\,(r + \delta_G)}\left(\frac{\lambda}{\beta\delta_G}\left(\frac{\mu}{\gamma\delta}\right)^2\right)^{\rho-1}\mu. \tag{16}$$

From (16) we obtain the following explicit expression for λ :

$$\lambda = \left(\frac{\rho\theta\eta\left(\gamma\delta\right)^{1-2\rho}\left(\beta\delta_G\right)^{1-\rho}}{\left(r+\delta_G\right)} \right)^{\frac{1}{2-\rho}} \mu^{\frac{2\rho-1}{2-\rho}}. \tag{17}$$

After substitution of (14), (15), and (17) into (11), and setting $\dot{\mu} = 0$, some tedious calculations lead to an equation for the steady state value of μ

$$\mu = \rho^{\frac{\rho}{2-4\rho}} \gamma\delta \left(\frac{\eta\theta}{r+\delta_G} \right)^{\frac{1}{1-2\rho}} \left(\beta\delta_G\right)^{\frac{-\rho}{2-4\rho}} \left(\frac{r+\delta_G+\rho\delta_G}{\left(r+\delta\right)\gamma\delta+2\alpha\eta^2} \right)^{\frac{2-\rho}{2-4\rho}}. \tag{18}$$

We can conclude that for $\rho \neq 0.5$, the steady state exists and is uniquely determined. Finally, it is easy to check that in the steady state indeed $K > 0$ holds and also that $a \geq 0$ is satisfied. ■

To undertake the stability analysis, use the characterization of stability properties of steady states of two-dimensional control problems provided in Dockner (1985) and Feichtinger et al. (1994). This characterization requires the calculation of the determinant of the Jacobian of the dynamical system and Dockner's K. This expression, which we denote by κ, is defined as

$$\kappa = \det \begin{pmatrix} \frac{\partial \dot{G}}{\partial G} & \frac{\partial \dot{G}}{\partial \lambda} \\ \frac{\partial \dot{\lambda}}{\partial G} & \frac{\partial \dot{\lambda}}{\partial \lambda} \end{pmatrix} + \det \begin{pmatrix} \frac{\partial \dot{K}}{\partial K} & \frac{\partial \dot{K}}{\partial \mu} \\ \frac{\partial \dot{\mu}}{\partial K} & \frac{\partial \dot{\mu}}{\partial \mu} \end{pmatrix} + 2\det \begin{pmatrix} \frac{\partial \dot{G}}{\partial K} & \frac{\partial \dot{G}}{\partial \mu} \\ \frac{\partial \dot{\lambda}}{\partial K} & \frac{\partial \dot{\lambda}}{\partial \mu} \end{pmatrix}. \tag{19}$$

The determinant of the Jacobian of the canonical system (10)–(13) is given by

$$\begin{aligned} \det J &= \det \begin{pmatrix} -\delta_G & \frac{2\lambda K}{\beta} & \frac{K^2}{\beta} & 0 \\ 0 & -\delta & 0 & \frac{1}{\gamma} \\ \left(1-\rho\right)\rho\theta\eta G^{\rho-2}K & -\rho\theta\eta G^{\rho-1} & r+\delta_G & 0 \\ -\eta\rho\theta G^{\rho-1} & 2\alpha\eta^2-\frac{\lambda^2}{\beta} & -2\lambda\frac{K}{\beta} & r+\delta \end{pmatrix} \\ &= \delta_G\left(\beta\left(\gamma\delta\left(r+\delta\right)+2\alpha\eta^2\right)-\lambda^2\right)\frac{r+\delta_G}{\beta\gamma} - 2\theta\eta\rho\frac{r+2\delta_G}{\beta\gamma}G^{\rho-1}\lambda K \\ &\quad -\frac{\theta^2\eta^2\rho^2}{\beta\gamma}G^{2\rho-2}K^2 + \theta\eta\rho\left(1-\rho\right)\frac{3\lambda^2+\beta\left(\gamma\delta\left(r+\delta\right)+2\alpha\eta^2\right)}{\beta^2\gamma}G^{\rho-2}K^3. \end{aligned}$$

Now we insert expressions for K and G in order to express everything in terms of λ and μ which yields

$$\begin{aligned}\det J &= \delta_G\beta\left(\gamma\delta\left(r+\delta\right)+2\alpha\eta^2\right)\frac{r+\delta_G}{\beta\gamma}\\ &-\delta_G\frac{r+\delta_G}{\beta\gamma}\lambda^2\\ &+\frac{\theta\eta\rho}{\beta^2\gamma}\left(\frac{\mu}{\gamma\delta}\right)^{2\rho-1}\left(\frac{1}{\beta\delta_G}\right)^{\rho-2}\left(3\left(1-\rho\right)-2\frac{r+2\delta_G}{\delta_G}\right)\lambda^\rho\\ &-\frac{\theta^2\eta^2\rho^2}{\beta\gamma}\left(\frac{\mu}{\gamma\delta}\right)^{4\rho-2}\left(\frac{1}{\beta\delta_G}\right)^{2\rho-2}\lambda^{2\rho-2}\\ &+\theta\eta\rho\left(1-\rho\right)\frac{\beta\left(\gamma\delta\left(r+\delta\right)+2\alpha\eta^2\right)}{\beta^2\gamma}\left(\frac{\mu}{\gamma\delta}\right)^{2\rho-1}\left(\frac{1}{\beta\delta_G}\right)^{\rho-2}\lambda^{\rho-2}.\end{aligned}$$

We plug in λ from (17) and, after some algebra, we obtain

$$\begin{aligned}\det J &= \frac{\gamma\delta\left(r+\delta\right)+2\alpha\eta^2}{\gamma}\delta_G\left(r+\delta_G\right)\left(2-\rho\right)\\ &-\delta_G\frac{1}{\beta\gamma}\left(\frac{1}{r+\delta_G}\right)^{\frac{\rho}{2-\rho}}\left(\rho\theta\eta\right)^{\frac{2}{2-\rho}}\left(\gamma\delta\right)^{\frac{2-4\rho}{2-\rho}}\left(\beta\delta_G\right)^{\frac{2-2\rho}{2-\rho}}\mu^{\frac{4\rho-2}{2-\rho}}\\ &+\frac{\left(\gamma\delta\right)^{\frac{2-4\rho}{2-\rho}}\left(\beta\delta_G\right)^{\frac{4-3\rho}{2-\rho}}}{\beta^2\gamma}\frac{\left(3\delta_G\left(1-\rho\right)-2\left(r+2\delta_G\right)\right)}{\delta_G\left(r+\delta_G\right)^{\frac{\rho}{2-\rho}}}\left(\rho\theta\eta\right)^{\frac{2}{2-\rho}}\mu^{\frac{4\rho-2}{2-\rho}}\\ &-\frac{1}{\beta\gamma}\left(\rho\theta\eta\right)^{\frac{2}{2-\rho}}\left(\gamma\delta\right)^{\frac{2-4\rho}{2-\rho}}\left(\beta\delta_G\right)^{\frac{2-2\rho}{2-\rho}}\left(\frac{1}{r+\delta_G}\right)^{\frac{2\rho-2}{2-\rho}}\mu^{\frac{4\rho-2}{2-\rho}}.\end{aligned}\tag{20}$$

Next we compute κ (see (19)) as

$$\begin{aligned}\kappa &= \det\begin{pmatrix}-\delta_G & \frac{K^2}{\beta}\\ \left(1-\rho\right)\rho\theta\eta G^{\rho-2}K & r+\delta_G\end{pmatrix}+\det\begin{pmatrix}-\delta & \frac{1}{\gamma}\\ 2\alpha\eta^2-\frac{\lambda^2}{\beta} & r+\delta\end{pmatrix}\\ &+2\det\begin{pmatrix}\frac{2\lambda K}{\beta} & 0\\ -\rho\theta\eta G^{\rho-1} & 0\end{pmatrix}.\end{aligned}$$

Now we can insert the steady state expressions for G, K, and λ to get

$$\begin{aligned}\kappa &= -\delta\left(r+\delta\right)-\frac{2\alpha\eta^2}{\gamma}-\left(2-\rho\right)\left(r+\delta_G\right)\delta_G\\ &+\frac{1}{\beta\gamma}\left(\frac{\rho\theta\eta\left(\gamma\delta\right)^{1-2\rho}\left(\beta\delta_G\right)^{1-\rho}}{\left(r+\delta_G\right)}\right)^{\frac{2}{2-\rho}}\mu^{\frac{4\rho-2}{2-\rho}}.\end{aligned}\tag{21}$$

In general, it is impossible to conclude whether $\det J$ and κ are positive or negative. For this reason, we proceed with a numerical analysis. For our benchmark case we set the parameter values as follows:

$$\beta = 10;\ \gamma = 5;\ \eta = 1;\ \theta = 0.6;\ \rho = 0.25;$$
$$\alpha = 0.15;\ \delta_G = 0.2;\ \delta = 0.2;\ r = 0.04. \tag{22}$$

In this scenario, we get

$$\begin{aligned} \mu &= 1.252 \\ \det J &= 0.005184 \\ \kappa &= -0.1734 \\ \det J - (\kappa/2)^2 &= -0.00233 \end{aligned}$$

From Table 1 (respectively Fig. 1) in Feichtinger et al. (1994), we conclude that the optimal trajectories monotonically converge to the steady state. We have carried out numerical calculations of the three key indicators $\det J$, κ and $\det J - (\kappa/2)^2$ for a wide range of parameter variations satisfying the relevant constraints and we found for all parameter settings the same signs as in our benchmark scenario, implying, at least locally, monotone convergence to the steady state.

In order to obtain deeper insights concerning the global dynamics under optimal investment, and also the dependence of the optimal investment from the states, we complement these findings about the uniqueness of the steady state and its stability properties with a numerical analysis. In particular, we determine the firm's optimal investment functions on the (relevant) state space and the induced dynamics starting from an initial condition in which both the stocks of goodwill and of green capital are zero. To obtain the optimal investment functions we rely on a collocation method, in which an approximation based on Chebychev polynomials of the value function is determined such that the Hamilton-Jacobi-Bellman equation associated with the control problem is solved on a suitably determined grid in the state space. The optimal investment functions are then determined from the first order conditions using this approximate value function for the problem. A more detailed description of the method can be found for example in Dawid et al. (2015).

Figure 1 shows that the firm's value is increasing in goodwill and the green capital stock. With respect to goodwill this is quite obvious, since this stock has a positive impact on the price. Concerning the green capital stock, it should be mentioned that for sufficiently small values of goodwill and sufficiently large values of K the induced output is above the monopoly output for the current market size, which means that current market profit of the firm could actually be increased by reducing its capital stock.[1] However, as can be seen in Fig. 1, even for these parts of the state space the value function is increasing in K, which is due to the fact, that,

[1] We assume that the capital stock is always fully used for production.

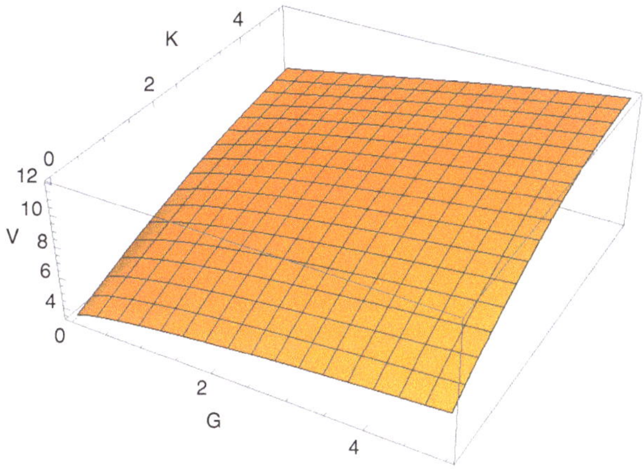

Fig. 1 Value of the firm, V, as a function of goodwill, G, and green capital stock, K

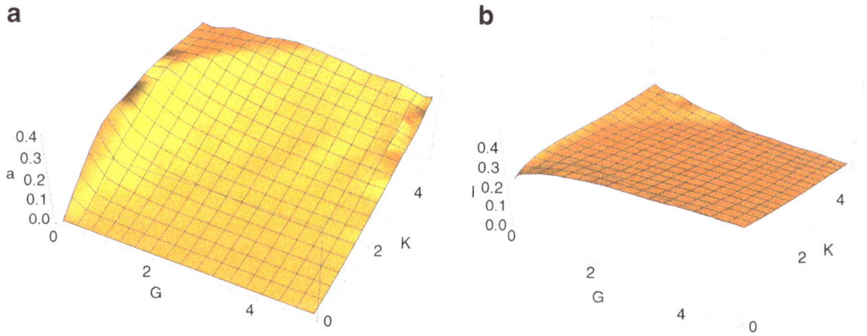

Fig. 2 Optimal (**a**) advertising and (**b**) and green investment as a function of goodwill, G, and green capital stock, K

first, the firm anticipates a future increase in the stock of goodwill, which will make a higher capital stock more profitable, and, second, a high stock of green capital always increases the effectiveness of the firm's advertising activities.

Figure 2a depicts how advertising depends on goodwill and the green capital stock. In most situations, advertising increases with green capital and decreases with goodwill. The former result can be explained by noting that advertising is more efficient when the green capital stock is large. The effect of goodwill on the output price is increasing but concave. This explains why advertising decreases with goodwill. The flat region of the investment function for combinations of small goodwill and a large stock of green capital can be explained by the observation that for these combinations, the non-negativity constraint of the price is strictly binding. Hence, the price of the product is zero and the instantaneous marginal effect of an increase in the stock of goodwill on the price is zero as well, regardless of the precise level of G and K. Therefore, the impact of a change in the value of these state variables on investment is small in this region.

Figure 2b illustrates how green investment depends on goodwill and the green capital stock. First of all, we have the standard capital accumulation result that investment decreases with the stock, which is called the flexible accelerator mechanism (Lucas 1967). In our framework this effect is driven by the decrease in the product price which is induced by a larger capital stock. It should be noted that there is also a second opposite effect in our model. Investment in K also increases the (future) effectiveness of advertising, thereby increasing future prices. The associated investment incentive is increasing in K for two reasons. First, a larger K induces a higher level of advertising and, second, it depends positively on the firm's output, which is determined by K. However, as Fig. 2b illustrates, in the model with $X = 0$ this effect is always dominated by the standard flexible accelerator mechanism and we have a negative dependence of green investment on K. The positive dependence of green investment on goodwill is due to the fact that the price of the product is positively affected by the goodwill stock, which increases incentives to expand output. Similarly to the observations for optimal advertising, we see that these effects are more or less eliminated in the part of the state space in which the current price is zero.

Figure 3 shows the state trajectory for zero initial stocks induced by these optimal investment functions. The figure shows a monotonic convergence to the steady state, which confirms our local stability analysis. From Fig. 4a we obtain that on this trajectory goodwill reaches its steady state after approximately twenty time units, whereas Fig. 4b shows faster convergence for the green capital stock.

Figure 5a depicts the advertising as a function of time. We see that some overshooting takes place here. This is because on the growth part of the trajectory, i.e. when $t < 10$, advertising gets more and more efficient as time passes because K goes up there. On the other hand, as shown in Fig. 2a, advertising decreases as the stock of goodwill goes up, because of the concave dependency of the market's reservation price from goodwill. This explains the decreasing convergence to the

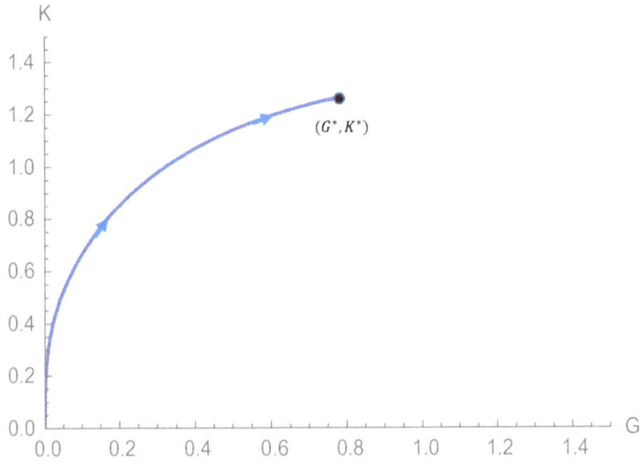

Fig. 3 The projection of the optimal trajectory, which starts at the origin, into the state space

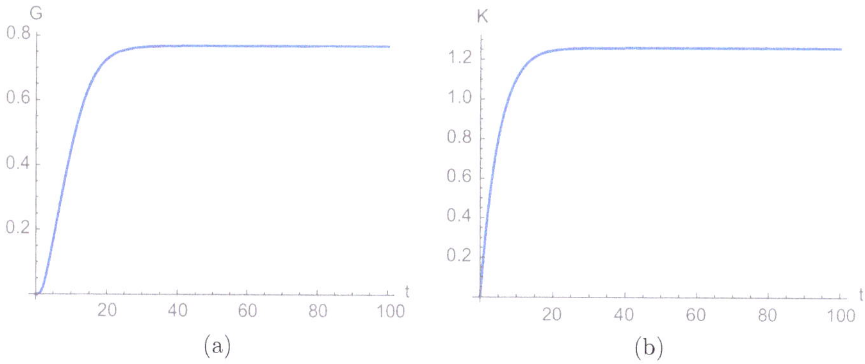

Fig. 4 The time path of (**a**) goodwill, G, and of (**b**) green capital stock K

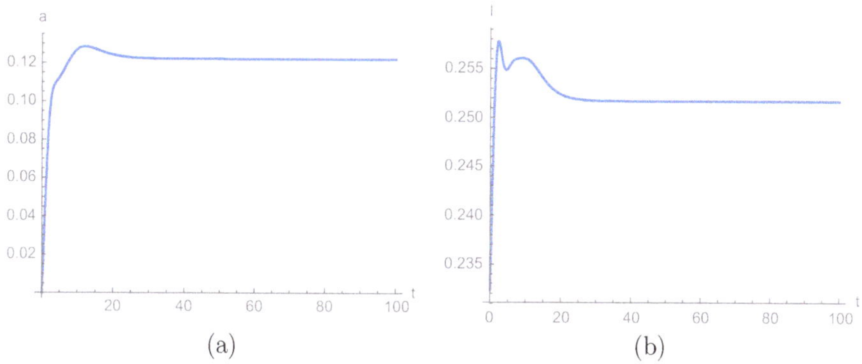

Fig. 5 The time path of (**a**) advertising, a, and of (**b**) investment in green capital I

steady state value for $t > 10$. Figure 5b shows investment in green capital as a function of time. Initially, investment incentives are large and mainly driven by the firm's goal to make its advertising activities more effective in increasing the goodwill stock. Once the goodwill stock has reached a certain positive level this effect becomes less important and the dominant mechanism is the flexible accelerator rule that implies that investment is decreasing in the corresponding capital stock. In Fig. 6 we see that the firm makes losses in the beginning. This is because the firm starts out with zero goodwill so that the initial output price is also zero. Hence the firm needs to invest in goodwill and green capital before it can produce in a profitable way. Instantaneous profits go up until the steady state is almost reached around time 20.

Overall, we conclude that, in the absence of an option for the firm to produce using traditional energy sources in addition to the green capital, some non-monotonicities of the investment paths might occur, but the dynamics of the two stocks seems to be globally characterized by monotone convergence to the unique steady state.

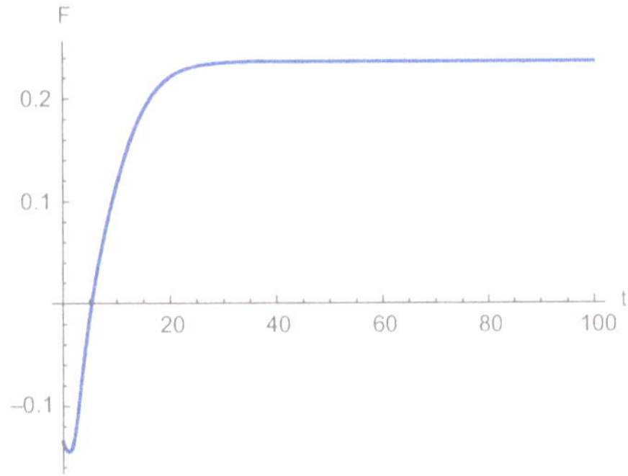

Fig. 6 The time path of the instantaneous profit function

4 Analysis of the Complete Model with Clean and Conventional Input

We now turn to the analysis of the complete model (1) to (4) where the firm has the additional choice of using a conventional input, X, on top of the green capital stock, K. To derive the optimality conditions we first define the Hamiltonian

$$\begin{aligned} H = &\left(g\left(G\right) - \alpha q\left(K + X\right)\right) q\left(K + X\right) - C_a\left(a\right) - C_s\left(I\right) - p_X X \\ &+ \lambda\left(f\left(K\right) a - \delta_G G\right) + \mu\left(I - \delta K\right) + \nu K. \end{aligned}$$

The necessary optimality conditions are

$$\begin{aligned} H_I &= 0 = -C_s'(I) + \mu \\ H_a &= 0 = -C_a'(a) + \lambda f\left(K\right) \\ H_X &= 0 = q'\left(K + X\right)\left(g\left(G\right) - 2\alpha q\left(K + X\right)\right) - p_X \\ \dot{\lambda} &= \left(r + \delta_G\right)\lambda - g'\left(G\right) q\left(K + X\right) \\ \dot{\mu} &= \left(r + \delta\right)\mu - q'\left(K + X\right)\left(g\left(G\right) - 2\alpha q\left(K + X\right)\right) - \lambda a f'\left(K\right) - \nu. \end{aligned}$$

With the special functions (5) to (9), we get

$$\begin{aligned} H_I &= 0 = -\gamma I + \mu \\ H_a &= 0 = -\beta a + \lambda K \\ H_X &= 0 = \eta\left(\theta G^{\rho} - 2\alpha\eta\left(K + X\right)\right) - p_X \\ \dot{\lambda} &= \left(r + \delta_G\right)\lambda - \rho\theta G^{\rho-1}\eta\left(K + X\right) \\ \dot{\mu} &= \left(r + \delta\right)\mu - \eta\left(\theta G^{\rho} - 2\alpha\eta\left(K + X\right)\right) - \lambda a - \nu \end{aligned}$$

Now all controls can be determined

$$I = \frac{\mu}{\gamma}$$
$$a = \frac{\lambda K}{\beta}$$
$$X = \max\left[\frac{\eta\theta G^{\rho} - p_X}{2\alpha\eta^2} - K, 0\right] \tag{23}$$

Since the calculations of the steady state and the corresponding stability analysis are even more involved than for the model in the previous section, we refrain from reporting them in detail here. The steady state values of the G and K are again given by (14) and (15). Using this, and numerically solving the system of equations $\dot{\lambda} = 0$ and $\dot{\mu} = 0$ allows to determine the steady states and the stability indicators also for this extended model. Using the same parameter values as in (22) and in addition

$$p_X = 0.175. \tag{24}$$

we obtain again a unique steady state in the relevant state space with the following stability indicators

$$\begin{aligned}\mu &= 1.2325\\ \lambda &= 0.99\\ \det J &= 0.00334\\ \kappa &= -0.09492\\ \det J - (\kappa/2)^2 &= 0.00109\end{aligned}$$

Using again Table 1 (respectively Fig. 1) in Feichtinger et al. (1994), this implies that we now have a (locally) stable steady state with complex eigenvalues, such that transient oscillations in the state dynamics occur before convergence to the steady state (we are in area b of Figure 1 in Feichtinger et al. (1994)).

Relying again on the collocation method to determine the value and optimal investment functions for this parameter setting, we observe in Fig. 7, that, as in the previous model, the value of the firm increases in both, goodwill, G, and green capital, K. Again, like before, as shown in Fig. 8a, in most situations, advertising, a, increases with green capital, K, and decreases with goodwill, G.

Figure 8b again illustrates how green investment depends on goodwill and the green capital stock. If we compare with Fig. 2b, we conclude that adding the input X has a considerable effect on the qualitative investment behavior. For relatively small values of the green capital stock investment incentives are now increasing in K. Intuitively, this can be explained by taking into account that the firm's choice of traditional production X depends negatively on K, such that, differently from the model considered in the previous section, the firm's output does not increase with K, as long as X is positive (see (23)). Hence, in this part of the state space the

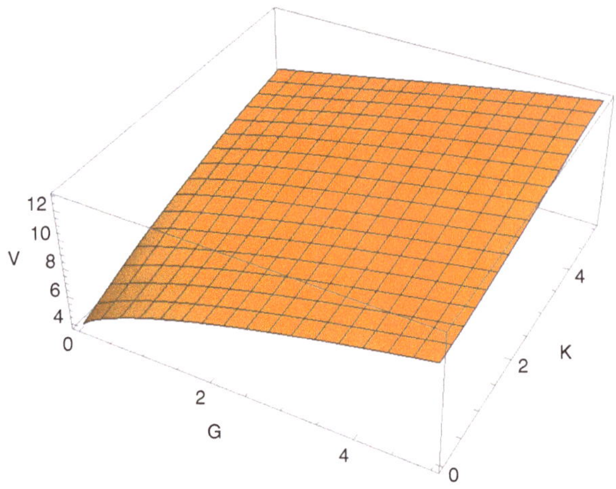

Fig. 7 Value of the firm, V, as a function of goodwill, G, and green capital stock, K

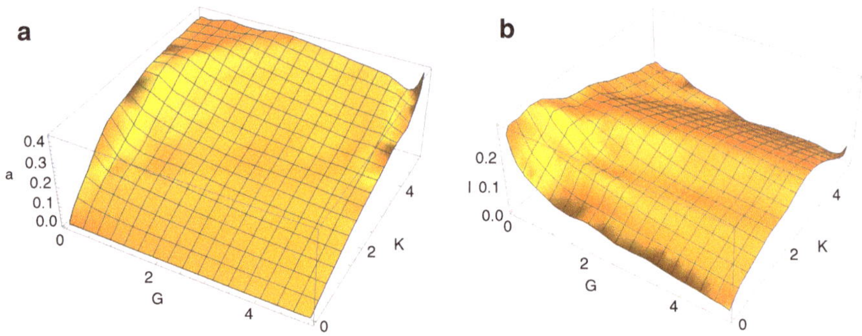

Fig. 8 Optimal (**a**) advertising and (**b**) and green investment as a function of goodwill, G, and green capital stock, K

standard price effect of an increase of K, which was the dominating force for the negative slope with respect to K in Fig. 2b disappears. The only remaining effect is the one that advertising increases with K and therefore the incentive to make advertising more effective also increases with K. If the green capital stock is so large that the firm does not use any traditional production, then essentially we have the same situation as in the model without traditional production and green investment depends negatively on K.

Unlike Fig. 3, in which we had monotonic convergence to the steady state, now for this model in Fig. 9 we see that the trajectory converges to the steady state in an non-monotonic way, which confirms the insights from our local stability analysis above.

The intuition can be best understood considering Figs. 10, 11, and 12. Initially (for $G(0) = 0$) market size is zero and the firm invests in K only to foster the

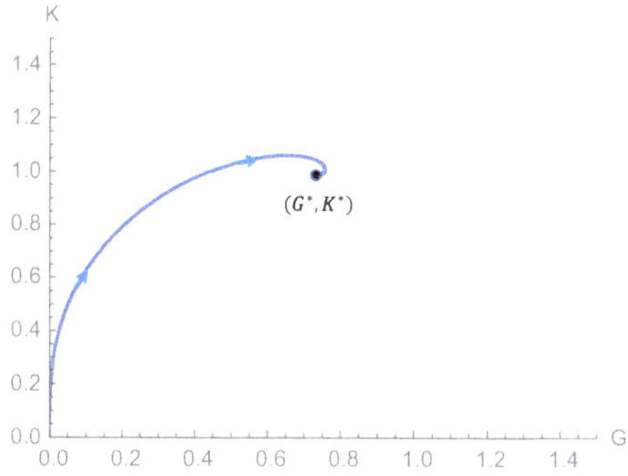

Fig. 9 The projection of the optimal trajectory, which starts at the origin, into the state space

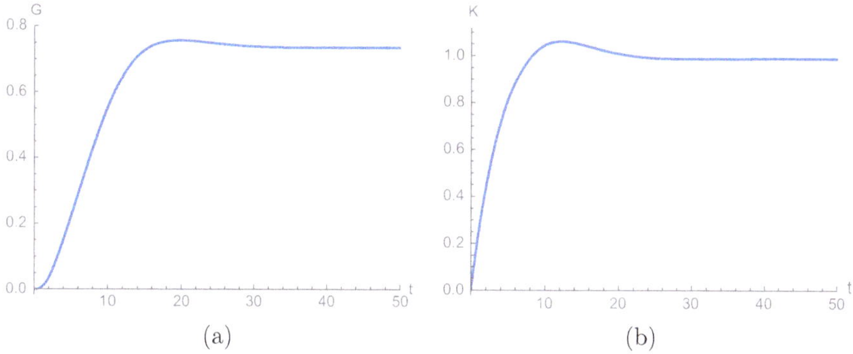

Fig. 10 The time path of (**a**) goodwill, G, and of (**b**) green capital stock K

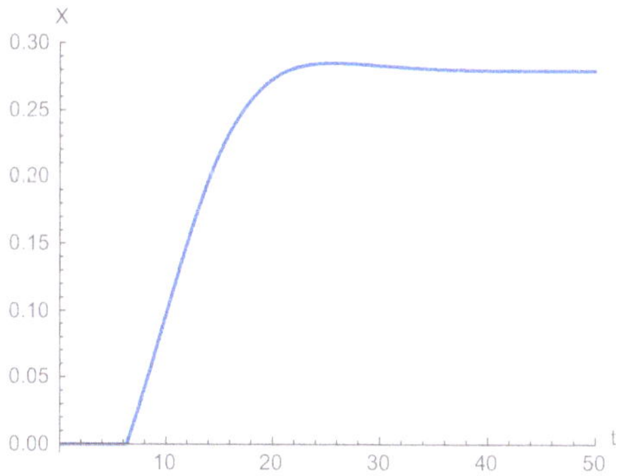

Fig. 11 The time path of usage of traditional input X

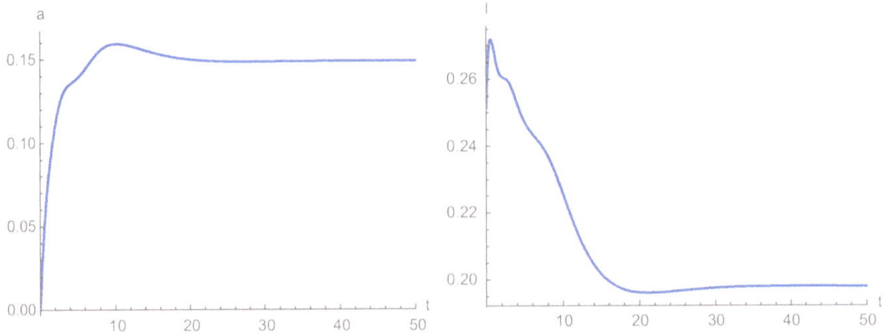

(a) (b)

Fig. 12 The time path of (**a**) advertising, a, and of (**b**) investment in green capital I

building up of goodwill (Fig. 12b). Once a positive stock of K has been generated, the firm increases investment in G (advertising a) (see Fig. 12a). The induced increase in the goodwill stock (Fig. 10a) then leads to decreasing investment in K over time (Fig. 12b). Initially, the K-stock is still sufficiently small to keep rising in spite of decreasing investment, but at approximately $t = 8$ the goodwill stock is sufficiently large such that the (inverse) demand for the product is sufficiently strong for the firm to start using traditional production as well as green production (see Fig. 11). The increase in production triggered by the increasing goodwill is now covered by an increase in traditional production rather than green production and traditional production also partly substitutes green production because the incentives to have green capital for fostering goodwill decreases. Hence, contrary to Fig. 4b, the green capital stock now eventually decreases and this has a negative effect on the effectiveness of advertising and induces the slight decrease in the stock of goodwill before the steady state is reached (Fig. 10a).

In Fig. 13a we see that the initial output price is also zero. This is because the firm starts out with zero goodwill. The firm invests in goodwill and green capital so that the price increases over time. Figure 13b is a direct consequence of panel (a). Because the output price is low in the beginning, the instantaneous profits is negative during an initial time period.

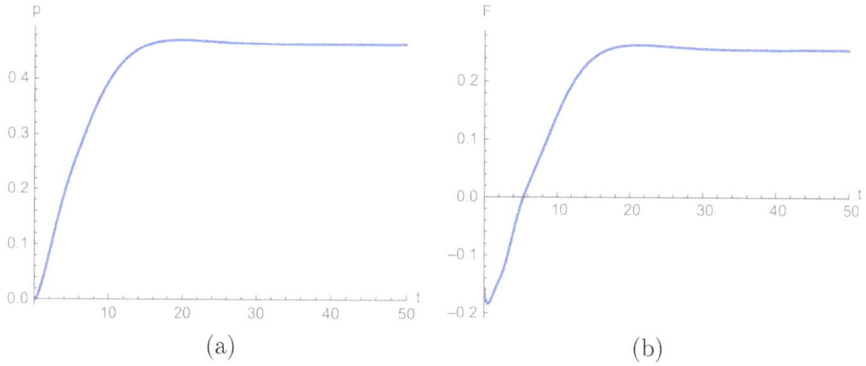

Fig. 13 The time path of (**a**) the output price, p, and of (**b**) the instantaneous profit

5 Conclusions

This contribution is a first analysis of the effect of investments in solar panels on dynamic firm behavior. The main result is that simultaneous use of traditional and green energy in the firm's production process leads to a trajectory that overshoots before converging towards a unique saddle point steady state. We plan to do future work in this area where, among others, we want to investigate the effect of government intervention in the form of subsidies and investment grants.

References

J.-P. Amigues, A.A. Le Kama, M. Moreaux, Equilibrium transitions from non-renewable energy to renewable energy under capacity constraints. J. Econ. Dyn. Control. **55**, 89–112 (2015)

H. Dawid, M. Keoula, M. Kopel, P.M. Kort, Product innovation incentives by an incumbent firm: a dynamic analysis. J. Econ. Behav. Organ. **117**, 411–438 (2015)

E.J. Dockner, Local stability analysis in optimal control problems with two state variables, in *Optimal Control Theory and Economic Analysis*, ed. by G. Feichtinger, vol. 2 (North-Holland, Amsterdam, 1985), pp. 89–103

G. Feichtinger, R.F. Hartl, Optimale Kontrolle ökonomischer Prozesse: Anwendungen des Maximumprinzips in den Wirtschaftswisssenschaften (de Gruyter, Berlin, 1986)

G. Feichtinger, A. Novak, F. Wirl, Limit cycles in intertemporal adjustment models: theory and applications. J. Econ. Dyn. Control. **18**, 353–380 (1994)

G. Feichtinger, R.F. Hartl, P.M. Kort, V.M. Veliov, Environmental policy, the porter hypothesis and the composition of capital: effects of learning and technological progress. J. Environ. Econ. Manag. **50**, 434–446 (2005)

K. Georgiev, S. Margenov, V.M. Veliov, Emission control in single species air pollution problems, in *Advances in Air Pollution Modeling for Environmental Security*, ed. by I. Fargo, K. Georgiev, A. Havasi. NATO Science Series, IV Earth and Environmental Sciences, vol. 54 (Springer, Berlin, 2005), pp. 219–228

D. Grass, J.P. Caulkins, G. Feichtinger, G. Tragler, D.A. Behrens, *Optimal Control of Nonlinear Processes* (Springer, Berlin, 2008)

R.E. Lucas, Jr., Optimal investment policy and the flexible accelerator. Int. Econ. Rev. **8**, 78–85 (1967)
E. Moser, W. Semmler, G. Tragler, V.M. Veliov, eds., *Dynamic Optimization in Environmental Economics*. Springer series Dynamic Modeling and Econometrics in Economics and Finance, vol. 15 (Springer, Berlin, 2014)
The Economist, Renewable energy: a world turned upside down. Feb 25th 2017
Y. Tsur, A. Zemel, On the dynamics of competing energy sources. Automatica **47**(1), 1357–1365 (2011)
F. Wirl, (Monopolistic) resource extraction and limit pricing: the market penetration of competitively produced synfuels. Environ. Resour. Econ. **1**, 157–178 (1991)
F. Wirl, Intertemporal inverstments into synfuels. Nat. Resour. Model. **21**(3), 466–488 (2008)

An Extended Integrated Assessment Model for Mitigation and Adaptation Policies on Climate Change

Willi Semmler, Helmut Maurer, and Anthony Bonen

Abstract We present an extended integrated assessment model (IAM) that explicitly solves for optimal climate financing policies. As with other IAMs, our approach ties economic activity with their externalities and feedback effects. We extend standard IAM methodologies to find the optimal allocation of infrastructure expenditure to carbon-neutral physical capital, climate change adaptation, and emissions mitigation. Optimal control solutions are obtained by discretizing the control problem and applying nonlinear programming methods. We demonstrate that the endogenously selected infrastructure shares out-perform fixed allocations by increasing consumption, private capital and tax revenue, while reducing public debt and CO_2 emissions. We find 92–95% of spending should be allocated to physical infrastructure with the remainder going to mitigation and adaptation, for which the major part is used for adaptation. Further, homotopic analysis is conducted on unobservable parameters. We show that adaptation expenditure increases with the

This paper was completed while Willi Semmler was a visiting scholar at the IIASA in July and August 2016. He would like to thank the IIASA for its hospitality. A related version of this paper was worked out while Willi Semmler was a visiting scholar at the IMF Research Department in April 2016. The authors would like to thank Prakash Loungani and his colleagues for many helpful insights.

W. Semmler
New School for Social Research, New York, NY, USA

International Institute for Applied Systems Analysis, Laxenburg, Austria

University of Bielefeld, Bielefeld, Germany

H. Maurer
Institute for Analysis and Numerics, University of Münster, Münster, Germany

A. Bonen (✉)
Labour Market Information Council, Ottawa, ON, Canada
e-mail: tony.bonen@lmic-cimt.ca

G. Feichtinger et al. (eds.), *Control Systems and Mathematical Methods in Economics*, Lecture Notes in Economics and Mathematical Systems 687,
https://doi.org/10.1007/978-3-319-75169-6_15

productive efficiency of non-renewables and emissions mitigation rises as its effect becomes nonlinear. The homotopic results demonstrate that our main findings are stable.

1 Introduction

Balancing the competing yet often complementary needs of climate change mitigation, adaptation and development is a complex policy goal (Bernard and Semmler 2015; IMF 2014, 2016). This paper presents a modelling framework for prioritizing funding to these three policy areas. Building on Bonen et al. (2016), we develop an extended integrated assessment model (IAM) that explicitly solves for the public funding allocation problem for climate change policy in the decision framework of a developing economy. Climate change policy is operationalized as the share of government expenditures made in support of carbon-neutral productivity-enhancing infrastructure, infrastructure that helps people adapt to the negative effects of a changing climate, and infrastructure used to mitigate carbon emissions. Depending on the parameterization, we find that between 92 and 95% of infrastructure expenditure should be allocated to productivity-enhancing infrastructure, 5–8% should be spent on adaptation, and the remainder on emissions mitigation. Productivity-enhancing infrastructure is prioritized as it increases the overall wealth in the country, thereby increasing the total capital available for the climate change adaptation and mitigation while increasing consumption and reducing government indebtedness.

Leading IAMs typically assume the economy's carbon intensity falls over time because of an exogenous 'back stop' of green technology. Our approach endogenizes carbon intensity by linking emissions to the extraction of a non-renewable resource (e.g., fossils fuels), and shows how renewable energy can be phased in through public-sector investment. This allows us to combine contemporary 'social cost of carbon' IAM approaches with the resource extraction models due to Hotelling (1931) and Pindyck (1978) as extended by Maurer and Semmler (2011). Thus, the IAM presented here extends the recent modelling advances that allow agents to respond to climate change by combining mitigation and adaptation actions (e.g., Ingham et al. 2005; Tol 2007; Lecoq and Zmarak 2007; Bosello 2008; de Bruin et al. 2009; Bréchet et al. 2013; Zemel 2015).

Computationally, IAMs represent complex dynamic systems that do not lend themselves to standard, closed-form solutions. Early iterations developed work-arounds such as forecasting economic growth trajectories in isolation and then using those output scenarios to generate emissions and/or temperature responses (Bonen et al. 2014). We avoid such simplifications by determining optimal control solutions for the full IAM system—a facet we believe to be crucial in accurately modelling economic-environmental interrelations. To this end, the optimal control problem is discretized on a fine grid which leads to a large-scale nonlinear programming problem (NLP) that can be conveniently formulated via the Mathematical Programming

Language (AMPL), *cf.* Fourer et al. (1993). AMPL can be linked to several efficient optimization solvers. In our computations, we use interior point optimization solver IPOPT (Wächter and Biegler 2006) that furnishes the control and state variables as well as the adjoint (co-state) variables. In this way, we are able to check whether we have found an *extremal solution* satisfying the necessary optimality conditions.

Employing AMPL enables us to advance the complexity—and thus realism—of the policymaker's action set. Under the initial parameterization, which is designed to match the stylized facts of a typical developing country, we find that 95% of funding should go toward productivity-enhancing investments, 5% to adaptation infrastructure, and none to emissions mitigation.[1] As expected, we show that allowing the optimizing policymaker to control the infrastructure expenditure allocations significantly improves social welfare relative to the case of fixed spending shares (a limitation other solution techniques would have to accept). Furthermore, we show that each constitutive element of social welfare is improved by the advancement: per capita consumption and private capital increase while public debt and CO_2 emissions fall relative to the fixed allocation scenario.

After demonstrating the superiority of expanding the policymaker's action set, we conduct a series of homotopic analyses to test both the model's stability and sensitivity of the main allocation results. First, the efficiency (viz. inverse of marginal cost) of fossil fuel energy is explored. We find that as fossil fuels become more efficient (cheaper for producers), the relative funding of productivity-enhancing infrastructure falls to 92% and the allocation to adaptation-focused infrastructure increases. Optimal mitigation, for the developing country, remains nil. Secondly, the concavity of the emissions effect of mitigation efforts is allowed to vary. Here we find that as mitigation's concavity increases, the impetus to reduce CO_2 emissions rises as the marginal return (at low mitigation levels) has a greater-than-proportional effect. Although allocations to emissions mitigation do not surpass 1.2%, social welfare monotonically increases with increased mitigation efforts. We also test welfare's sensitivity to intertemporal discounting. Our results here demonstrate the model conforms to the important theoretical insight that outcomes improve when policymakers reduce their discounting of the future. Crucially, improvements in terminal welfare are shown to flow from increased expenditure of emissions mitigating infrastructure.

The remainder of the paper is organized as follows. Section 2 presents the model and optimal control solution technique. Results are reported and discussed in Sect. 3. Section 4 concludes.

[1]We have also tested a specification in which these allocations are continuously updated in each time period, instead of being selected based on the initial expected social utility. There is little improvement in moving to this approach. In addition to reducing computational costs, the slight reduction in utility from optimally selecting a single set of allocations suggests that any loss of flexibility in guaranteeing long-term mitigation and adaptation funding is likely outweighed by the benefits of policy stability. Due to space constraints we do not present these results here.

2 Integrated Assessment Model as Optimal Control Problem

The integrated assessment model (IAM) has 5 state variables

$$X = (K, R, M, b, g) \in \mathbf{R}^5, \tag{1}$$

where K is private capital, R is the stock of the non-renewable resource, M is the atmospheric concentration of CO_2, b is the government's debt, and g is public capital. The dynamic system of the IAM is defined according to

$$\dot{K} = Y \cdot (\nu_1 g)^{\beta} - C - e_P - (\delta_K + n)K - u\,\psi R^{-\tau}, \tag{2}$$

$$\dot{R} = -u, \tag{3}$$

$$\dot{M} = \gamma\, u - \mu(M - \kappa\widetilde{M}) - \theta(\nu_3 \cdot g)^{\phi}, \tag{4}$$

$$\dot{b} = (\bar{r} - n)b - (1 - \alpha_1 - \alpha_2 - \alpha_3) \cdot e_P. \tag{5}$$

$$\dot{g} = \alpha_1 e_P + i_F - (\delta_g + n)g, \tag{6}$$

The control vector is given by

$$U = (C, e_P, u) \in \mathbf{R}^3, \tag{7}$$

where C denotes consumption, e_P is tax revenue, and u is the quantity of the resource R extracted each period.

The first dynamic $\dot{K}$ is the accumulation rate of private capital K that produces renewable energy and which drives output by the CES production function,[2]

$$Y(K, u) := A(A_K K + A_u u)^{\alpha} \tag{8}$$

where A is multifactor productivity, A_K and A_u are efficiency indices of private capital inputs K and (non-renewable) fossil fuel energy u, respectively. In (2), private-sector output Y is modified by the infrastructure share allocated to productivity enhancement $\nu_1 g$, for $\nu_1 \in [0, 1]$. This public-private interaction generates total output as $Y(\nu_1 g)^{\beta}$ from which the economy consumes C, pays taxes e_P, and is subject to physical δ_K and demographic n depreciation. The exponent β is the output elasticity of public infrastructure, $\nu_1 g$. The last term in (2) is the opportunity cost of extracting the non-renewable resource u, where ψ and τ are the scale and shape parameters that tie the marginal cost of u to the remaining stock of the resource à la Hotelling.

[2]For such a simplification of a production function see Acemoglu et al. (2012) and Greiner et al. (2014).

Equation (3) indicates the stock of the non-renewable resource R depletes by u units in each period.

The non-renewable resource emits carbon dioxide and thus increases the atmospheric concentration of CO_2 at the rate γ in Eq. (4). The stable level of CO_2 emissions is $\kappa > 1$ of the pre-industrial level $\widetilde{M}$, which is naturally re-absorbed into the ecosystem (e.g., oceanic reservoirs) at the rate μ. The last term in (4) is the reduction of per-period emissions $\dot{M}$ due to the allocation of $0 \le \nu_3 \le 1$ of infrastructure g to mitigation projects.

The last two dynamics are the accumulation of debt b and public capital g. In (5) public debt grows at the fixed interest rate $\bar{r}$, and is serviced with the share of tax revenue e_P not allocated respectively to capital accumulation α_1, social transfers α_2 or administrative overhead $\alpha_3 > 0$. Thus, $\alpha_4 \equiv 1 - \alpha_1 - \alpha_2 - \alpha_3$. Equation (6) says the stock of public capital, or total infrastructure, evolves according to the allocated tax revenue stream $\alpha_1 e_P$ and funds paid in from abroad, i_F. For developed countries $i_F = 0$, but may be positive for many developing countries. As with private capital, g depreciates by δ_g, and is adjusted for population growth n.

We assume throughout that the infrastructural allocations satisfy

$$\nu_k \ge 0 \quad (k = 1, 2, 3), \qquad \nu_1 + \nu_2 + \nu_3 = 1. \tag{9}$$

In later analyses, we either choose fixed values of ν_1, ν_2, ν_3 or we consider the allocations as additional optimization variables. All parameters in the dynamics (2)–(6) may be found in Table 1.

Using the state variable $X \in \mathbf{R}^5$ and control variable $U \in \mathbf{R}^3$, we write the dynamics (2)–(6) in compact form as

$$\dot{X}(t) = f(X(t), U(t)), \quad X(0) = X_0. \tag{10}$$

The initial state vector X_0 will be specified later. To this system we add the terminal constraint

$$K(T) = K_T \ge 0, \tag{11}$$

the control constraint

$$0 \le u(t) \le u_{max}, \tag{12}$$

and the pure state constraint

$$M(t) \le M_{max} \quad \forall\, t \in [0, T]. \tag{13}$$

The terminal constraint restricts the final level of the capital stock to a predetermined non-negative value, the control constraint prescribes an upper bound for the extraction rate, and finally the state constraint places a cap on the total level of CO_2 in the atmosphere in each period.

Table 1 Parameter values

Variable	Value	Definition
ρ	0.03	Pure discount rate
n	0.015	Population Growth Rate
η	0.1	Elasticity of transfers and public spending in utility
ϵ	1.1	Elasticity of CO_2-eq concentration in (dis)utility
ω	0.05	Elasticity of public capital used for adaptation in utility
σ	1.1	Intertemporal elasticity of instantaneous utility
A	$\in [1, 10]$	Total factor productivity
A_K	1	Efficiency index of private capital
A_u	$\in [50, 500]$	Efficiency index of the non-renewable resource
α	0.5	Output elasticity of privately-owned inputs, $A_k k + A_u u$
β	0.5	Output elasticity of public infrastructure, $\nu_1 g$
ψ	1	Scaling factor in marginal cost of resource extraction
τ	2	Exponential factor in marginal cost of resource extraction
δ_K	0.075	Depreciation rate of private capital
δ_g	0.05	Depreciation rate of public capital
i_F	0.05	Official development assistance earmarked for public infrastructure
α_1	0.1	Proportion of tax revenue allocated to new public capital
α_2	0.7	Proportion of tax revenue allocated to transfers and public consumption
α_3	0.1	Proportion of tax revenue allocated to administrative costs
$\bar{r}$	0.07	World interest rate (paid on public debt)
$\widetilde{M}$	1	Pre-industrial atmospheric concentration of greenhouse gases
γ	0.9	Fraction of greenhouse gas emissions not absorbed by the ocean
μ	0.01	Decay rate of greenhouse gases in atmosphere
κ	2	Atmospheric concentration stabilization ratio (relative to $\widetilde{M}$)
θ	0.01	Effectiveness of mitigation measures
ϕ	$\in [0.2, 1]$	Exponent in mitigation term $(\nu_3 g)^\phi$

Let us now define the objective functional, the social welfare functional. We maximize (viz. minimize the negative) the following functional over a given planning horizon $[0, T]$, where $T > 0$ denotes the terminal time:

$$W(T, X, U) = \int_0^T e^{-(\rho-n)t} \frac{\left(C\,(\alpha_2 e_P)^\eta \left(M - \widetilde{M}\right)^{-\epsilon} (\nu_2 g)^\omega\right)^{1-\sigma} - 1}{1-\sigma}\, dt\,. \tag{14}$$

The felicity (utility) function in (14) is isoelastic with four input components all in per capita terms: (1) consumption C; (2) the share $0 \leq \alpha_2 \leq 1$ of tax revenue e_P used for direct welfare enhancement (e.g., healthcare); (3) atmospheric concentration of CO_2 M above the pre-industrial level $\widetilde{M}$; and (4) the share

$0 \leq \nu_2 \leq 1$ of infrastructure g allocated to climate change adaptation. Restricting the exponents $\eta, \epsilon, \omega > 0$ ensures social expenditures and adaptation are utility enhancing, and that carbon emissions directly reduce utility. This approach differs from other models that map emissions to temperature changes and then to reduced productivity-*cum*-output. We believe the direct disutility approach better captures the wide ranging impacts of climate change that may include health impacts, ecological loss and heightened uncertainty, in addition to reduced productivity. Finally, note that the discount factor adjusts for the population growth rate n from the pure discount rate ρ as all values are normalized by the population.

To summarize, the IAM gives rise to an optimal control problem $OC(p)$, where the social welfare (14) is maximized subject to the dynamic constraints (10) and the terminal, control and state constraints (11)–(13). In this problem $OC(p)$, the notation p denotes a suitable parameter in Table 1 for which we shall conduct a sensitivity analysis in the next section.

A detailed discussion of the necessary optimality conditions of the Maximum Principle for optimal control problems with state constraints (*cf.* Hartl et al. 1995) is beyond the scope of this paper and will be given elsewhere.

3 Results

3.1 Discretization and Nonlinear Programming Methods

We choose the numerical approach "First Discretize then Optimize" to solve the optimal control problem $OC(p)$ defined in (10)–(14). The discretization of the control problem on a fine grid leads to a large-scale nonlinear programming problem (NLP) that can be conveniently formulated with the help of the Mathematical Programming Language AMPL (Fourer et al. 1993). AMPL can be linked to several powerful optimization solvers. We use the Interior-Point optimization solver IPOPT developed by Wächter and Biegler (2006). Details of discretization methods may be found in Betts (2010), Büskens and Maurer (2000), and Göllman and Maurer (2014). The subsequent computations for the terminal time $T = 25$ are performed with $N = 1000$ to $N = 5000$ grid points using the trapezoidal rule as integration method. Choosing the error tolerance $tol = 10^{-8}$ in IPOPT, we can expect that the state variables are correct up to 6 or 7 decimal digits. The Lagrange multipliers and adjoint variables are computed *a posteriori* by IPOPT thus enabling us to verify the necessary optimality conditions.

3.2 Parameter Values and Initial Conditions

The parameter values in the dynamics (2)–(5) are reported in Table 1. We set the initial conditions to

$$K(0) = 1.5,\ g(0) = 0.5,\ b(0) = 0.8,\ R(0) = 1.5,\ M(0) = 1.5,$$

and choose the terminal time terminal constraint as

$$T = 25, \quad K(T) = K_T = 3.$$

Furthermore, we restrict the extraction rate to

$$0 \leq u(t) \leq 0.1,\ \forall\, t \in [0, T].$$

We have considered the following two strategies for the allocations:

Strategy 1: Choose fixed values ν_1, ν_2, ν_3 satisfying (9).
Strategy 2: Consider ν_1, ν_2, ν_3 as optimization variables satisfying (9).

It would be also possible to treat $\nu_k = \nu_k(t), k = 1, 2, 3,$ as time-varying control variables. However, our computations show that this strategy improves only slightly on Strategy 2 and is computationally much more expensive. For that reason, we do not report those results here.

Strategy 1 selects the fixed values for the allocation of infrastructural investments, such that the majority of infrastructure enhances productivity and the remainder is evenly split between mitigation and adaptation. Specifically, we consider $\nu_1 = 0.6, \nu_2 = 0.2, \nu_3 = 0.2$. In the second and third strategies we endogenize these allocative shares as choice variables maximizing (14).

3.3 Fixed Versus Optimal Values of ν_1, ν_2, ν_3

Comparing state variable trajectories under Strategies 1 and 2 demonstrates the latter considerably improves on the former. In the first comparison we assume the economic efficiency of the non-renewable resource is low ($A_u = 50$)[3] and that CO_2 mitigation efforts exhibit constant marginal returns, $\phi = 1$. The trajectories for the three control variables (C, e_P, u) and five state variables (K, R, M, g, b) are plotted in Fig. 1. Under this parameterization, Strategy 2's optimal allocation is $\nu_1 = 0.95, \nu_2 = 0.05, \nu_3 = 0$. That is, no infrastructure expenditures are put

[3]By construction the efficiency index A_u should be larger than A_K as the former calibrates a flow input and the former a stock value.

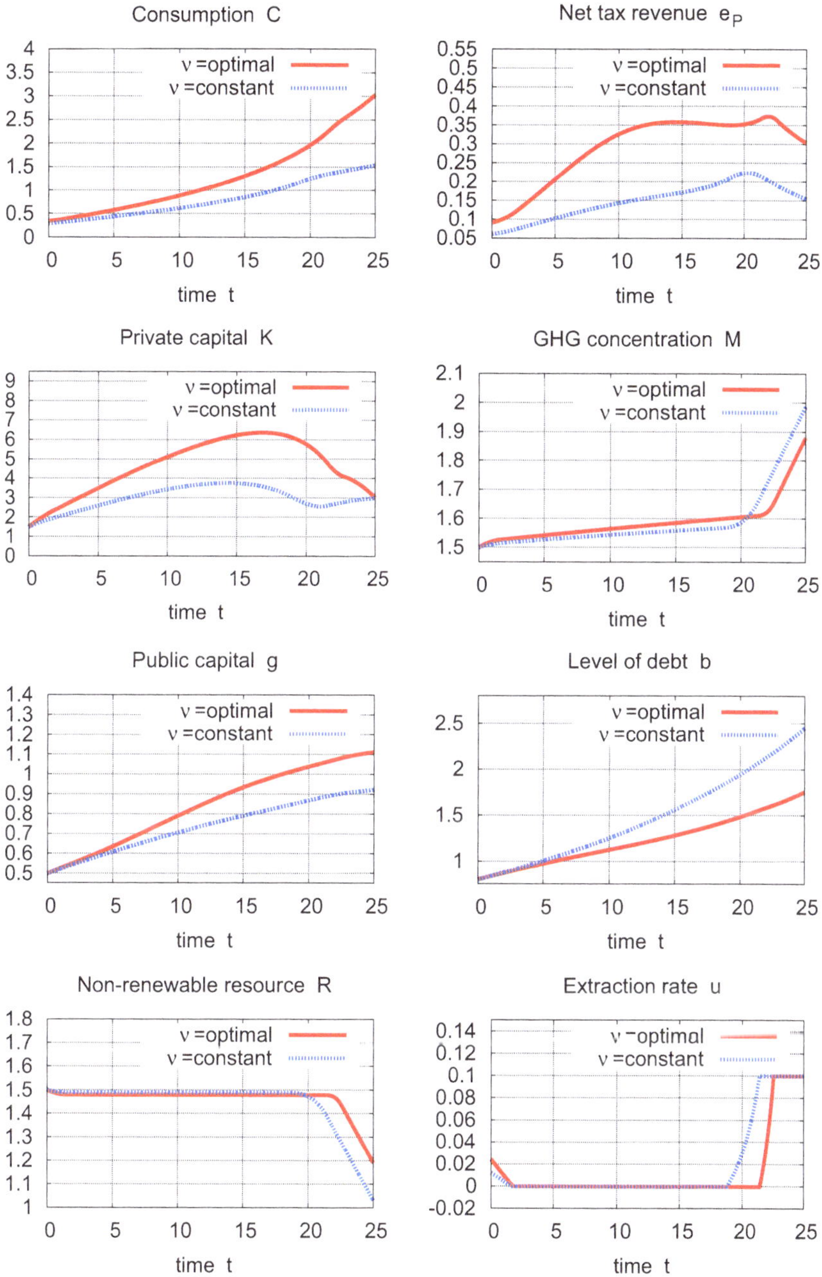

Fig. 1 Strategy 1 vs. 2, state and control variable trajectories. Strategy 1 (dashed blue) sets $\nu_1 = 0.6$, $\nu_2 = \nu_3 = 0.2$ and generates a final welfare value of $W(T) = -2.1006$. Strategy 2 (solid red) optimally selects $\nu_1 = 0.9534$, $\nu_2 = 0.04662$, $\nu_3 = 0$ and results in $W(T) = 5.1086$

toward mitigation and a mere 5% is allocated to adaptation.[4] The top four panels of Fig. 1 show this endogenous allocation, as compared to Strategy 1, results in higher per capita consumption, private capital accumulation and tax revenue in all periods, yet the final atmospheric CO_2 concentration is also lower. Although M is slightly lower under Strategy 1 through the first twenty periods, this abruptly reverses in the final periods when M grows exponentially. This seemingly odd result is explained by the trajectories in bottom four panels.

Under both strategies the per-period amount of non-renewable (and, here, inefficient) resource extracted is quickly pushed to zero so as to minimize the negative utility impact of CO_2 emissions. However, Strategy 1 over-allocates public infrastructure to mitigation efforts which generates suboptimal (climate-neutral) private capital accumulation. The low level of K in turn leads to less output and reduced tax revenue. Moreover, as the debt burden grows it begins to further dampen investment in K, which peaks in the fifteenth period. The falling per capita capital stock exhibits little impact until the terminal condition $K(t) = K_T$ begins to bite. From the twenty-first period onwards, preceding capital investment shortfalls are made up by shifting production to the inefficient non-renewable resource. The extracted amount u begins to ramp up from zero, reducing the stock R and generating CO_2 emissions.

Under Strategy 2 the peak in private capital comes at a delay and the terminal condition is not problematic since $K(t) > K_T$ for $3 < t < T$. Under this optimal allocation approach, overinvestment in mitigation infrastructure is avoided and the savings are put toward productivity enhancements. This generates a larger capital stock "buffer" allowing the economy to hold off the extraction of R. As in Strategy 1, maximum K is reached as the debt burden approaches 1.5, and tax revenue is redirected toward debt servicing. However, greater productivity and the lower stock of debt forestall this effect in Strategy 2. When extraction does begin in the twenty-second period, it merely reduces the *rate* at which K, the capital used for the production of green energy, falls toward K_T, rather than makes up for the previous investment shortfalls seen in Strategy 1. Again, the higher stock of private (green) capital has diminished the economy's reliance on the carbon-emitting non-renewable resource.

3.4 Homotopic Analysis of A_u

Many of the model parameters remain uncertain and/or unobservable. This limitation, common to all models, is particularly acute for IAMs due to the multifaceted feedback effects between economic decision-making and climatological impacts. To address the issue we apply homotopic parameter variation to $\mathbf{OCP}(p)$ for

[4]It is important to note that funding for renewable energy production is already captured through the variable K.

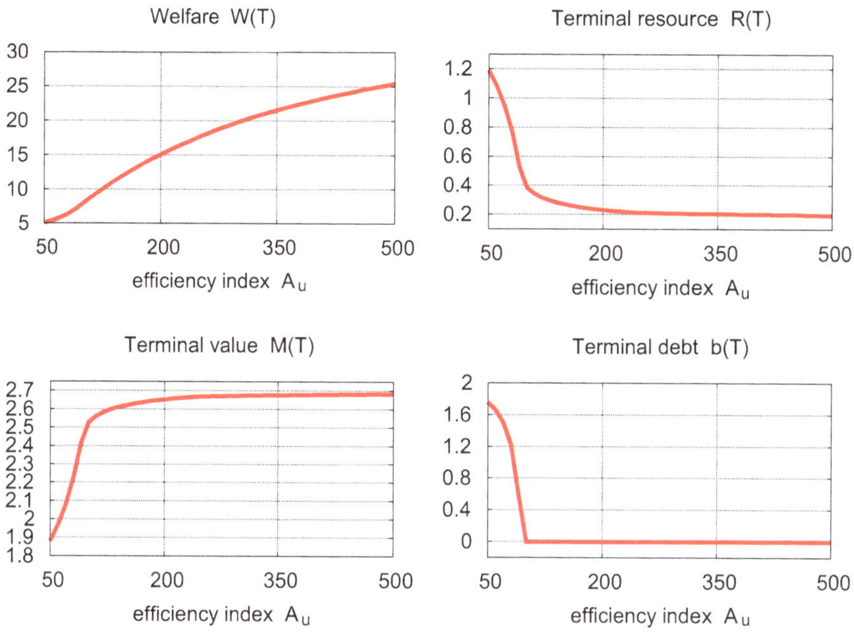

Fig. 2 Terminal states for homotopy $50 \le A_u \le 500$

several key parameters. In each case we use the optimal selection of infrastructure allocations ν_1, ν_2, ν_3 as they continue to outperform arbitrarily fixed values.

First, we consider scenarios in which the non-renewable resource—fossil fuel energy—generates output more efficiently than the generation of renewable energy by allowing A_u to range from a high of 500 down to 50 (as used in Sect. 3.3). Figure 2 plots the terminal values of welfare $W(T)$, CO_2 concentration $M(T)$, unextracted nonrenewable resource $R(T)$, and terminal debt $b(T)$. Unsurprisingly, welfare rises monotonically as the efficiency of this input is increased. Looked at the other way, welfare falls when fossil fuel energy becomes more costly to find and extract. The higher cost (viz. lower productive efficiency) of u decreases incentives to extract it, meaning the remaining stock of non-renewable resource rises from 0.2 for $A_u = 500$ to 1.2 at $A_u = 50$. At very low costs, the extraction rate is very inelastic, as shown by the slow increase in $R(T)$ between $A_u = 500$ and $A_u = 100$. After this point, the shift away from extraction rises rapidly as A_u halves from 100 to 50. This pattern of extraction maps inversely to CO_2 concentrations, which fall slowly as $A_u \to 100^+$, only to fall rapidly when extraction becomes sufficiently costly (which is calibrated here at $A_u = 100$).

The lower-right panel in Fig. 2 suggests why $R(T)$ rises in such a distinctly nonlinear fashion as A_u falls. At a low efficiency (high cost) of u, greater investment into K is supported through borrowed funds. For larger A_u, dependence on private capital K and productivity-enhancing infrastructure ν_1 is lower because the cheaper

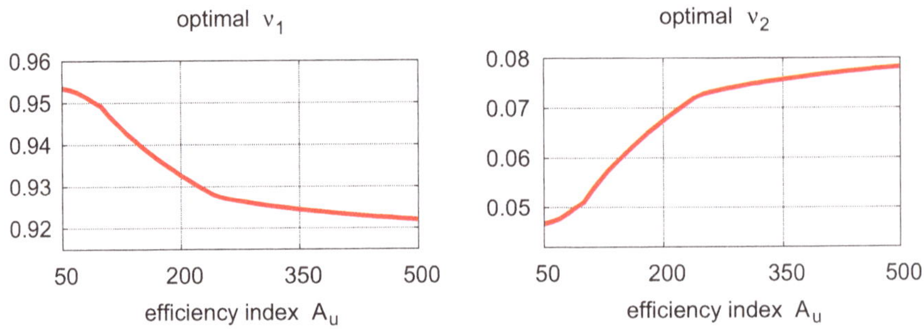

Fig. 3 Infrastructure allocations for homotopy $50 \leq A_u \leq 500$

non-renewable energy substitutes for carbon-neutral K. Figure 3 confirms this interpretation: the optimal allocation proportion ν_1 is 92% at $A_u = 500$ versus 95% for $A_u = 50$. In the former case, when extraction of the non-renewable resource is expensive, less infrastructure needs to be allocated toward adaptive projects: ν_2 falls from 8% to less than 5%. That said, the overall welfare outcome, is greater when A_u is large, in spite of the rise in M. Also implied by Fig. 3, $\nu_3 = 0$ for all values of A_u. Overall, the above case of $\nu_3 = 0$ is not likely to give realistic solutions since ν_3 enters the control problem linearly, which gives rise to the so-called 'bang-bang' problem.

3.5 Homotopic Analysis of ϕ

Since the result of no infrastructural investments put toward mitigation efforts is due to the linear relationship assumed by setting $\phi = 1$. Recall,

$$\dot{M} = \gamma u - \mu(M - \kappa \widetilde{M}) - \theta(\nu_3 \cdot g)^{\phi} \tag{4}$$

We now loosen this assumption of linearity to consider the mitigation exponent over the range $0.2 \leq \phi \leq 1$, which should be interpreted as the rate of diminishing returns to climate change mitigation efforts. Whereas $\nu_3 = 0$ for $\phi = 1$ (which is likely to be cause by the aforementioned 'bang-bang' problem), we obtain $\nu_3 > 0$ for $\phi \leq \phi_0 \approx 0.88$.

Figure 4 compares the optimal allocation of infrastructure expenditures toward productivity-enhancement ν_1, adaptation ν_2, and mitigation ν_3, as well comparing the final social welfare $W(T)$ at each value of ϕ. The results show that, as the rate of return to mitigation efforts diminishes, the impetus to reduce CO_2 emissions rises with ν_3 reaching 1.2% for $\phi = 0.2$. The rising mitigation share comes primarily at the (small) expense of traditional infrastructure, the allocation of g to which falls from 94% to just above 92.8%. The remaining difference ($\approx$ 0.1%) comes from reduced adaptation efforts. Note that as mitigation efforts are increased above nil,

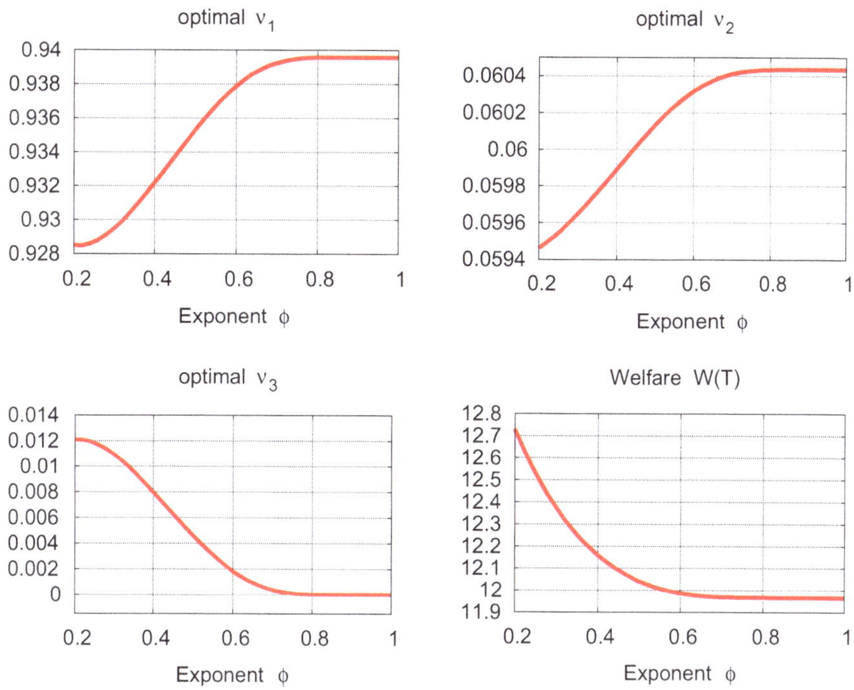

Fig. 4 Allocations and terminal welfare for homotopy $\phi \in [0.2, 1]$. The non-renewable resource's efficiency index is set at $A_u = 150$

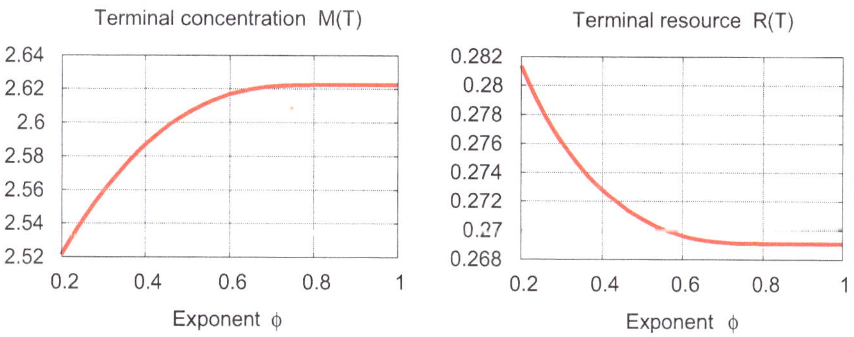

Fig. 5 Terminal resources and CO_2 for homotopy $\phi \in [0.2, 1]$. The non-renewable resource's efficiency index is set at $A_u = 150$

total social welfare increases by approximately 6%. Figure 5 confirms that as ϕ falls, the heightened mitigation effort helps reduce the final concentration of CO_2 in the atmosphere. Moreover, and corresponding to the latter result, the total amount of non-renewable resources extracted is lower ($R(T)$ higher) as ϕ falls.

3.6 Homotopic Analysis of A_u for $\phi = 0.2$

The unambiguous improvement to welfare and CO_2 concentration reduction for $\phi = 0.2$ found above assumed $A_u = 150$. To test whether the results from Sect. 3.5 were contingent on that efficiency index, we again perform a homotopy on A_u this time specifying a concave mitigation term in (4) at $\phi = 0.2$. As before we find that terminal welfare $W(T)$ increases when the efficiency of u falls (viz. the cost of extraction rises), infrastructural allocations to productivity ν_1 rise as adaptive efforts ν_2 fall (see Fig. 6). However, with $\phi = 0.2$ mitigation efforts ν_3 are no longer nil, although they remain between 1.0% and 1.7% of g. Interestingly, allocations mitigation are not monotonic over A_u. Over the 'high cost' range found in Sect. 3.4, $A_u \in [50, 100]$, ν_3 in Fig. 6 becomes increasingly desirable as extraction costs rise (A_u falls). For lower costs, $A_u > 100$, ν_3 falls as extraction costs increase (A_u falls) implying mitigation efforts must ramped up when fossil fuel energy is cheap in order to counter the increase in CO_2 emissions.

This interpretation of ν_3 is supported by the terminal states plotted in Fig. 7. The terminal atmospheric carbon concentrations rise rapidly over A_u (i.e., as extraction costs fall) and then stabilize above $A_u = 100$—aided in part by the increase in ν_3. Again, as the productive efficiency of u is increased, the extraction rate rises ($R(T)$ falls) nonlinearly and public debt becomes less relied upon as production shifts away from private capital toward non-renewable resources. Total infrastructure g

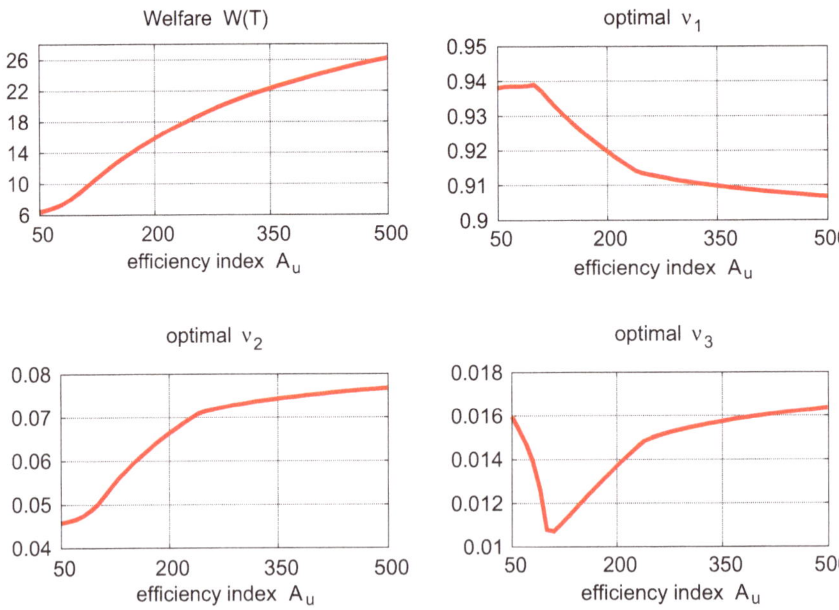

Fig. 6 Allocations and Welfare for homotopy $A_u \in [50, 500]$, $\phi = 0.2$

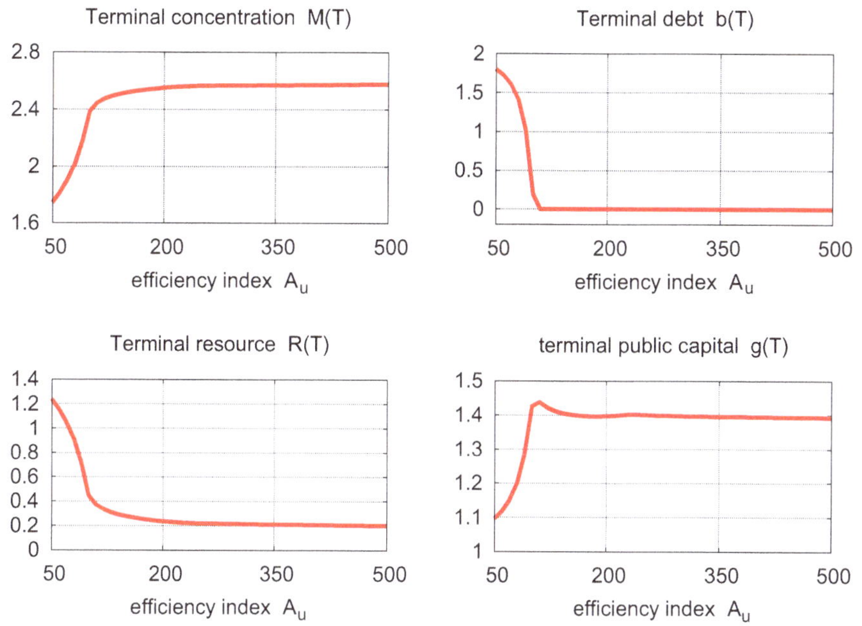

Fig. 7 Terminal states for homotopy $50 \leq A_u \leq 500$ for $\phi = 0.2$

also rises rapidly over the initial low range of A_u and then stabilizes for at values above 100.

Figure 8 shows the full trajectories of private capital K, consumption C, carbon concentrations M, the extraction rate u for three representative values of $A_u = 100, 200, 500$. In the extreme case of $A_u = 500$ private capital is driven to zero for the majority of periods between the initial and terminal points of K_0 and K_T, meaning production is driven entirely by the non-renewable resource. This result does not seem economically reasonable. The motivation to discard this parameterization is even stronger since the trajectories of M and u for $A_u = 500$ and $A_u - 200$ are nearly indistinguishable.

For an efficiency index of 150, K falls slightly from its initial value and fluctuates slightly before converging to K_T. Conversely, for $A_u = 100$, capital stock rises rapidly, peaks and then falls unevenly to K_T as was the case in Sect. 3.3 for $A_u = 50$, $\phi = 1$. As in Sect. 3.4, the extraction rate for $A_u = 100, 200$ reaches the maximal level near the end of the projection, with the less efficient scenario reaching the peak earlier. However, with $\phi = 0.2$ the lower efficiency index scenario now leads to a lower total and terminal CO_2 level as mitigation efforts are no longer held at zero.

Further trajectories for $\phi = 0.2$ are presented in Fig. 9. The total stock of infrastructure g is little changed under three A_u scenarios. As suggested by the trajectory of u in Fig. 8, the remaining stock of the non-renewable resource R is greatest for $A_u = 100$, but only by a small margin over the $A_u = 200$ scenario.

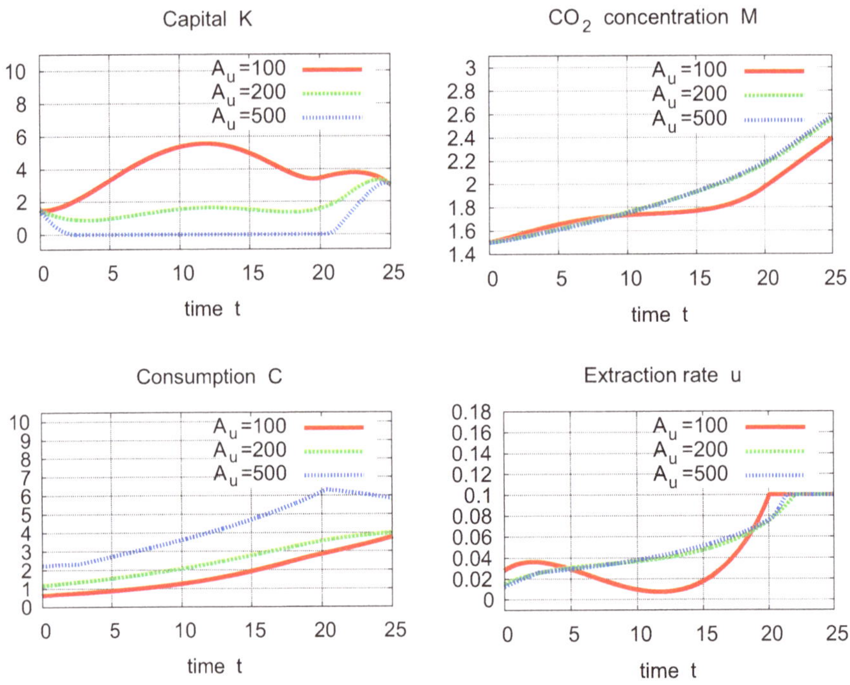

Fig. 8 Selected trajectories for $\phi = 0.2$ with $A_u = 100$, 200 and 500

Conversely, the tax revenue trajectory e_P fluctuates far more under $A_u = 100$ than the other scenarios. In the former case, e_P leads the fluctuations in u, falling before u rises and vice versa. This tendency supports the argument made above that greater reliance on the non-renewable resource reduces the need for fiscal deficits.

3.7 *Homotopic Analysis of ρ for $\phi = 0.2$*

Finally, we consider the homotopy of ρ, the pure discount rate. There has been much debate over the correct intertemporal discount rate that should be used in climate change economics (e.g., Stern 2007). While we do not weigh in on that debate here, it is informative to investigate the IAM results under various discount rate assumptions. Figure 10 shows that terminal welfare $W(T)$ falls smoothly as the discount on future outcomes rises. Although the falling allocation of infrastructure to mitigation ν_3 as ρ rises is expected, it is interesting to note that the shares of ν_1 and ν_2 move in opposite directions. In other words, the savings from ν_3 are not shared between productive infrastructure and adaptation. Instead, for higher discount rates, mitigation efforts are increased while ν_1 falls by a greater amount that ν_3.

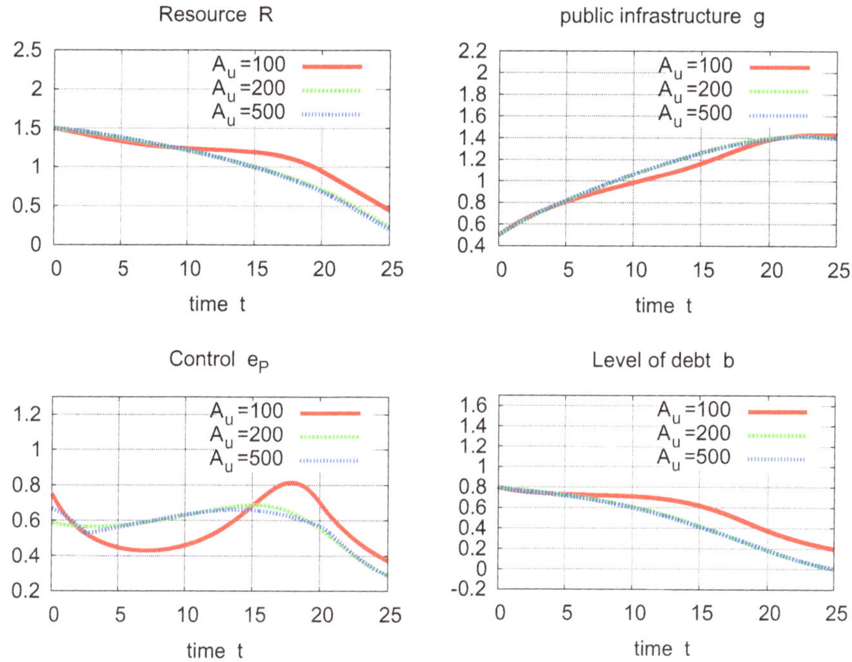

Fig. 9 Further trajectories for $\phi = 0.2$ with $A_u = 100, 200$ and 500

The reason for this behaviour is in Fig. 11. As the economy discounts future outcomes at a higher rate, the present cost of non-renewable resource extraction falls and thus the rate of extraction rises. The bottom panel in Fig. 11 indicates that indeed the remaining stock of non-renewable resource is driven down as ρ is increased. And, as in all other cases, when u rises the final stock of CO_2 concentration $M(T)$ rises. It is also notable that a higher discount rate is associated with a lower level of public infrastructure available to be used for any purpose. These results indicate that, indeed, the discount rate we choose to inform climate change policy can have a great effect on the trajectory ultimately followed.

4 Conclusion

Following a review of recent policy developments and modelling approaches to climate change economics, the paper developed an extended integrated assessment model explicitly accounting for the extraction of non-renewable resources and the phasing in of renewable energy. Another extension of the IAM framework is to include public sector policies concerning optimal decisions of both revenue and tax expenditures. Although the focus was on climate policy financing through

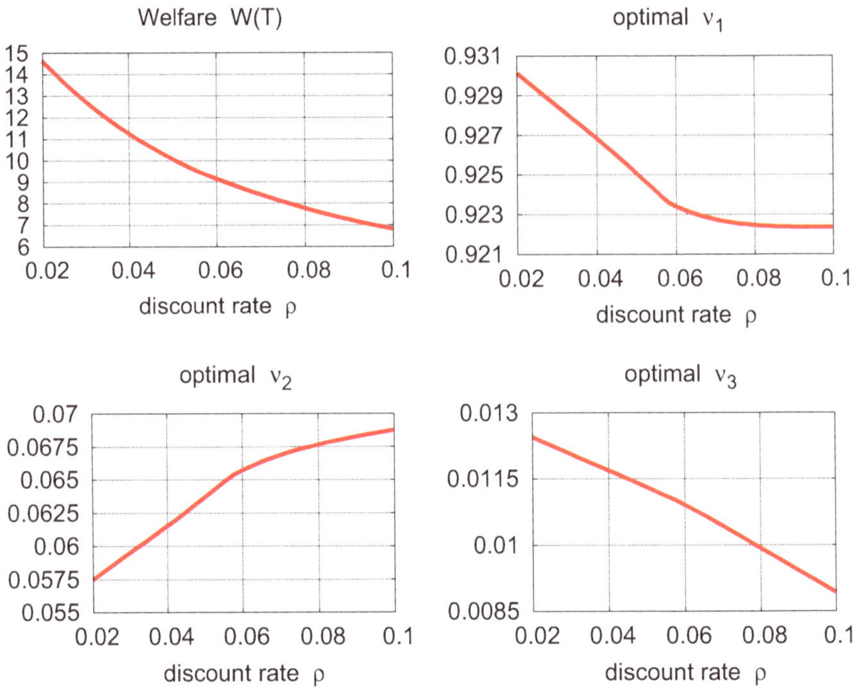

Fig. 10 Allocations and Welfare for homotopy $\rho \in [0.02, 0.1]$, $\phi = 0.2$

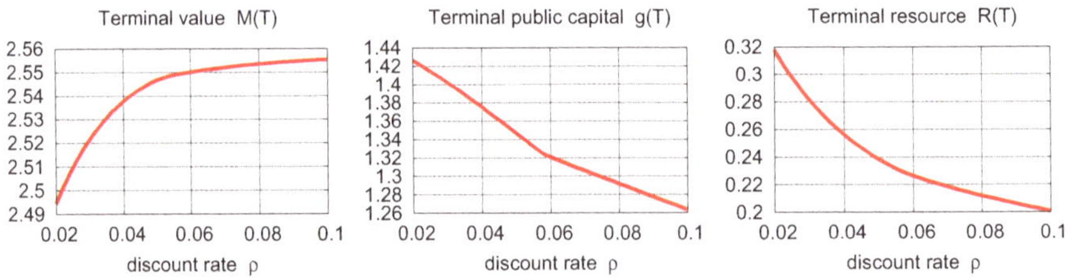

Fig. 11 Terminal states for homotopy $\rho \in [0.02, 0.1]$ for $\phi = 0.2$

taxation, future research could elaborate on the financing mechanisms through climate bonds.[5]

The IAM was solved using the AMPL algorithm which enabled us to maintain all of the system's nonlinearities and dynamic interactions. A particularly useful feature of this methodology is the ability to optimally determine the allocative variables ν_1, ν_2, ν_3 in order to indicate the best policy mix for addressing the challenges of climate change. In Sect. 3.3 we showed endogenously selected allo-

[5]In this context, a recent discussion of proposals for central banks to accept climate bonds as collateralizable securities is available in Flaherty et al. (2016).

cations consistently outperformed *ex ante* parameterizations. We then considered parameter homotopies under a strategy of optimally selecting the allocation shares to traditional, adaptive and climate change mitigating expenditures.

Given that green energy is already phased in through the accumulation of private capital, the model consistently found that over 90% of infrastructural investment should be geared toward productivity-enhancing investments. The phasing in of green energy is also supported by the fact that private capital enhancements $\nu_1 g$ are, by design, enhancements for carbon-neutral production. In other words, the model assumes that no public funds are used to directly support the extraction of CO_2-emitting resources.

Sections 3.4–3.6 consider the homotopy of A_u and ϕ, respectively the production efficiency index for the non-renewable resource and the exponent on mitigation efforts. The results demonstrated that greater efficiency of CO_2-generating resources incentivizes their use, thereby increasing carbon emissions. Increasing the input level of u also led to a reduced reliance on debt to finance ν_1. This result accords with the stylized fact that resource-dependent economies typically have large fiscal surpluses when primary products are in high demand. On the other hand as the efficiency of CO_2 generating energy declines, the results are reversed: more of this resource is left in the ground and cumulative CO_2 emissions are lower. The exponent ϕ proved to be crucial. As the concavity of mitigation efforts rose (lower ϕ), the level of mitigation efforts increased monotonically. One interpretation of this finding is that if mitigation is seen to be relatively inexpensive (i.e., fixed linear impacts on $\dot{M}$), then agents may continuously hold off on investing in mitigation.[6] We also considered the homotopy of ρ, the pure discount rate. As expected total social welfare was lower and CO_2 concentrations higher when, *ceteris paribus*, the discounting of future outcomes rose.

Overall, the IAM developed here is an advancement both in terms of the solution algorithm employed and in its use of novel dynamics. As mentioned, the modelling of non-renewable resource extraction and detailed public sector policies on climate change are new features in the IAM literature. In addition we have treated the damage function of climate change as impacting social welfare directly, as opposed to indirectly through reductions in the rate at which output is produced. While neither approach is perfect, we have employed the direct-utility impact version because we believe it is better able to capture the many physical, ecological and societal losses that may be induced by unabated climate change.

Finally, a necessary extension of the climate change policies studied here is consideration of the optimal financing sources, including policies for burden sharing. For example, standard IAMs place the cost and implementation burden of financing climate policies on the current generation. Indeed, the extended IAM developed here posits public sector financing of climate action through current tax revenues and expenditures. As an additional extension to the framework, we can

[6]Another issue is that when the control enters linearly, then the corresponding control variable (in this case mitigation effort) is driven to zero. This could be the result of a 'bang-bang' solution.

consider the extent to which climate policies can be funded by both a carbon tax and the issuing of climate bonds—the latter being repaid by future generations. For more specific work on this type of burden sharing between current and future generations, see Sachs (2014), Flaherty et al. (2016) and Gevorkyan et al. (2016).

References

L. Bernard, W. Semmler, *The Oxford Handbook of the Macroeconomics of Global Warming* (Oxford University Press, Oxford, 2015)

J.T. Betts, *Practical Methods for Optimal Control and Estimation Using Nonlinear Programming*, 2nd edn. (SIAM Publications, Philadelphia, 2010)

A. Bonen, P. Loungani, W. Semmler, S. Koch, Investing to mitigate and adapt to climate change: a framework model. IMF Working Paper, WP/16/164 (2016)

A. Bonen, W. Semmler, S. Klasen, Economic damages from climate change: a review of modeling approaches. SCEPA Working Paper 2014-03, Schwartz Center for Economic Policy Analysis (2014)

F. Bosello, Adaptation, mitigation and "green" R&D to combat climate change: insights from an empirical integrated assessment exercise. Working paper, Centro Euro-Mediterraneo Per I Cambiamenti Climatici (2008)

T. Bréchet, N. Hritoneko, Y. Yatsenko, Adaptation and mitigation in long-term climate policy. Environ. Resour. Econ. **55**(2), 217–243 (2013)

C. Büskens, H. Maurer, SQP methods for solving optimal control problems with control and state constraints: adjoint variables, sensitivity analysis and real-time controls. J. Comput. Appl. Math. **120**, 85–108 (2000)

K. de Bruin, R. Dellink, R. Tol, AD-DICE: an implementation of adaptation in the DICE model. Clim. Change **95**, 63–81 (2009)

M. Flaherty, A. Gevorkyan, S. Radpour, W. Semmler, Financing climate policies through climate bonds: a three stage model and empirics. Res. Int. Bus. Financ. **42**, 468–479 (2017)

R. Fourer, D.M. Gay, B.W. Kernighan, *AMPL: A Modeling Language for Mathematical Programming* (Duxbury Press, Brooks-Cole Publishing Company, North Scituate, 1993)

A. Gevorkyan, M. Flaherty, D. Heine, M. Mazzucato, S. Radpour, W. Semmler, Financing climate policies through carbon tax and climate bonds: a model and empirics. Manuscript, New School for Social Research (2016)

L. Göllman, H. Maurer, Theory and applications of optimal control problems with multiple time-delays. J. Ind. Manag. Optim. **10**, 413–441 (2014)

R.F. Hartl, S.P. Sethi, R.G. Vickson, A survey of the maximum principles for optimal control problems with state constraints. SIAM Rev. **37**, 181–218 (1995)

H. Hotelling, The economics of exhaustible resources. J. Polit. Econ. **39**(2), 137–175 (1931)

IMF, How much carbon pricing is in countries' own interest? The critical role of co-benefits. IMF Working Paper (2014)

IMF, After Paris: fiscal, macroeconomic, and financial implicatoins of climate change. IMF Staff Discussion Note (2016)

A. Ingham, J. Ma, A. Ulph, Can adaptation and mitigation be complements? Working Paper 79, Tyndall Centre for Climate Change Research (2005)

F. Lecoq, S. Zmarak, Balancing expenditures on mitigation and adaptation to climate change: an exploration of issues relevant for developing countries. Policy Research Working Paper 4299, World Bank (2007)

H. Maurer, W. Semmler, A model of oil discovery and extraction. Appl. Math Comput. **217**(13), 1163–1169 (2011)

R. Pindyck, Gains to producers from the cartelization of exhaustible reources. Rev. Econ. Stat. **60**(2), 238–251 (1978)

J. Sachs, Climate change and intergenerational well-being, in *The Oxford Handbook of the Macroeconomics of Global Warming*, ed. by L. Bernard, W. Semmler (Oxford University Press, Oxford, 2014), pp. 248–259

N. Stern, *The Economics of Climate Change: The Stern Review* (Cambridge University Press, Cambridge, 2007)

R. Tol, The double trade off between adaptation and mitigation for sea level rise: an application of FUND. Mitig. Adapt. Strateg. Glob. Chang. **12**(5), 741–753 (2007)

A. Wächter, L.T. Biegler, On the implementation of an interior-point filter line-search alogrithm for large-scale nonlinear programming. Math. Program. **106**, 25–57 (2006)

A. Zemel, Adaptation, mitigation and risk: an analytical approach. J. Econ. Dyn. Control. **51**, 133–147 (2015)

Part III
Population Dynamics and Spatial Models

Optimal Population Growth as an Endogenous Discounting Problem: The Ramsey Case

Raouf Boucekkine, Blanca Martínez, and J. Ramon Ruiz-Tamarit

Abstract This paper revisits the optimal population size problem in a continuous time Ramsey setting with costly child rearing and both intergenerational and intertemporal altruism. The social welfare functions considered range from the Millian to the Benthamite. When population growth is endogenized, the associated optimal control problem involves an endogenous *effective* discount rate depending on past and current population growth rates, which makes preferences intertemporally dependent. We tackle this problem by using an appropriate maximum principle. Then we study the stationary solutions (balanced growth paths) and show the existence of two admissible solutions except in the Millian case. We prove that only one is optimal. Comparative statics and transitional dynamics are numerically derived in the general case.

R. Boucekkine (✉)
Aix-Marseille University, CNRS, EHESS, Centrale Marseille, AMSE and IMRA, Marseille, France

Senior member, IUF, Marseille, France
e-mail: raouf.boucekkine@univ-amu.fr

B. Martínez
Department of Economic Analysis, Universidad Complutense de Madrid, Madrid, Spain

Instituto Complutense de Análisis Económico (ICAE), Madrid, Spain
e-mail: blmartin@ccee.ucm.es

J. R. Ruiz-Tamarit
Department of Economic Analysis, Universitat de València, València, Spain

IRES Department of Economics, Université Catholique de Louvain, Louvain-la-Neuve, Belgium
e-mail: ramon.ruiz@uv.es

G. Feichtinger et al. (eds.), *Control Systems and Mathematical Methods in Economics*, Lecture Notes in Economics and Mathematical Systems 687,
https://doi.org/10.1007/978-3-319-75169-6_16

1 Introduction

There is a growing body of the economic and population ethics literatures concerned with the demographic dimension of the sustainable growth debate (see for example, Arrow et al. 2004). Clearly, if sustainability implies coping with the needs of current generations without compromising those of the far future generations (i.e. the so-called Brundtland criterion), the question of sustainable demographic paths, namely those which can achieve the latter intergenerational fairness goal, comes easily into the story. Of course this question can be easily connected to current hot environment-oriented issues: *ceteris paribus*, larger populations are likely to pollute more and to deplete more quickly natural resources, which is a further strong threat on sustainable development (see Boucekkine et al. 2014b). Here, we strictly stick to the intergenerational fairness problem outlined by Arrow et al. (2004), and abstract from the environmental ingredients.

The key question is indeed a basic and recurrent one in population ethics: what is the optimal population size? By which social welfare ordering can we argue that the current European or world population is suboptimal or not? Clearly this question can be asked with or without the global warming threat. As recently outlined by Dasgupta (2005), the question of optimal population size traces back to antiquity. For example, Plato concluded that the number of citizens in the ideal city-state is 5040, arguing that it is divisible by every number up to ten and have as many as 59 divisors, which would allow for the population to "...suffice for purposes of war and every peacetime activity, all contracts for dealings, and for taxes and grants" (cited in Dasgupta 2005).

The early related economic literature is due to Edgeworth (1925): he claimed that the use of total utilitarianism (that's the Benthamite social welfare function) is highly problematic as it leads to choose a bigger population size (compared for example to the Millian social welfare function or average utilitarianism) with quite lower standard of living. Some recent inspections into this issue have reached the same conclusion. In particular, Nerlove et al. (1985), who examined the robustness of Edgeworth's claim to parental altruism within a simple static model, found that the claim still holds when the utility function of adults is increasing in the number of children and/or the utility of children. Interestingly enough, the latter economic literature was contemporaneous (and probably sympathetic) to a masterpiece of the population ethics literature, the Parfit's (1984) *Reasons and Persons* book. In particular, Parfit explicitly attributed to total utilitarianism the same unpleasant implication as Edgeworth 60 years ago, and he called it a *repugnant conclusion*.

The use of dynamic frameworks to assess the *repugnant conclusion* and its correlates traces back essentially to the 90s. Intriguingly, the settings considered were rooted in the endogenous growth literature, a strongly rising stream at that time. Palivos and Yip (1993) and Razin and Yuen (1995) are two excellent representatives of this literature. In particular, Palivos and Yip showed that Edgeworth's claim cannot hold in the framework of endogenous growth driven by an AK production function. The determination of the optimal population growth rate relies on the following trade-off: on one hand, the utility function depends explicitly on the demographic growth rate; on the other, the latter induces the standard linear dilution effect on capital accumulation, and therefore on economic growth. Palivos and Yip proved that in such a case the Benthamite criterion leads to a smaller population size and a higher growth rate of the economy provided the intertemporal elasticity of substitution is lower than one. More recently, Boucekkine and Fabbri (2013) have examined a more general AK framework with endogenous demographic growth allowing for any type of correlation between demographic and economic growth at equilibrium. This is made possible by considering a large class of dilution functions: in particular, following Blanchet (1988) who proved that such functions are nonlinear when accounting for the age structure of capital, Boucekkine and Fabbri found that the *repugnant conclusion* would more easily arise under non-monotonic dilution schemes.

In Boucekkine and Fabbri (2013) and Palivos and Yip (1993), human populations do not play any role in the production side of the economy since the assumed production function is AK. Clearly, if humans do not produce, a strong pro-natalist ingredient is lost. What if the size of population matters in the production function as in neoclassical growth? Things should be much more involved. In the extreme case where the production function is AN (that's only human capital matters), Boucekkine et al. (2014a) show that the results depend strongly on the humans' life span. If the life span is large enough, Parfit's *repugnant conclusion* for total utilitarianism does not hold: even more, all individuals of all generations will receive the same consumption, and therefore will enjoy the same welfare. If life spans are small enough, even the Benthamite social welfare function would legitimate finite time extinction.

This paper is concerned with the much more essential Ramsey version of the problem, that's the production function is neoclassical and both capital and labor (or human capital) are production factors. We shall not incorporate the finite life span assumption into the story as it has been already deeply explored in Boucekkine, Fabbri and Gozzi. Individuals have infinite lives and there are decreasing returns with respect to labor in the production sector. Time is continuous. To our knowledge, very few papers have tackled the optimal population problem within this frame. Perhaps the most popular contribution along this line is Barro and Sala-i-Martin (2004), section 9.2.2. Essentially, the very vast majority of papers dealing with endogenous fertility use overlapping generations and discrete time. The seminal contribution to this line of research is Barro and Becker (1989) and their fertility choice model within a Ramsey structure. Typically, agents live two periods (childhood and adulthood), the utility of children enters linearly the utility of parents but the degree of altruism is a decreasing function of the number of children. Barro and Sala-i-Martin's 2004 model (BSM hereafter) can be seen, roughly speaking, as a continuous time formulation of the latter seminal model. Accordingly, the counterpart of the number of children variable is the continuous growth rate of population, just like in the models surveyed above on the optimal population problem.

There are two main drawbacks in the BSM. First of all, BSM do not consider the case where the induced social welfare function is either Benthamite or Millian since their model is initially derived from Barro and Becker (1989) where the degree of altruism is a decreasing function of the number of children. Bringing the latter assumption to the continuous time model does not allow to tackle an important element of the optimal population debate. This simply reflects the fact that Barro and his coauthors have a clearly distinct focus. The second drawback is technical: as any Ramsey model, BSM's is not tractable. Moreover, endogenizing demographic growth brings more nonlinearity into the problem, favoring multiplicity of stationary equilibria and the like (a fact acknowledged, although not rigorously studied, in Barro and Becker 1989, pp. 489–490). These issues are left in the dark in BSM.

In this paper, we take seriously the specific implications of endogenous population growth. When we endogenize population growth decisions connecting the fertility choice with economic variables, the Ramsey growth model experiences indeed a drastic change in its structure. The standard discounted optimal control problem assumes that the instantaneous utility function depends on contemporary variables alone, and that the intertemporal welfare function discounts utility stream at a fixed exponential rate. However, the simple modification to an hyperbolic discount function causes systematic changes in decisions which are responsible of a time inconsistency in intertemporal choices. The same problem of time inconsistency appears when the intertemporal non-separability of preferences comes from an endogenous and variable discount function. In our model, the endogeneity of exponential population growth at a variable rate transforms the standard optimal control problem with a constant rate of time preference, into a new and nontrivial dynamic optimization problem, and one has to be cautious in the application of the Pontryagin's maximum principle. This is because the induced non-constant *effective* discount rate becomes endogenous, which makes preferences intertemporally dependent. From the literature on endogenous discounting (see for example, Obstfeld (1990), or more recently, Marin-Solano and Navas (2009)), we can implement a mathematical solution by introducing a new state variable representing the accumulated stock of impatience. Then, we can solve the transformed problem within the standard optimal control approach. The transformed is however far nontrivial as it involves a problem in higher dimension with a pure state constraint. We handle it using the appropriate approach described for example in Sethi and Thompson (2000). Furthermore, we study two sets of questions. One is related to the existence and uniqueness of stationary solutions (balanced growth paths): we show that two admissible steady states exist provided the social welfare function is not Millian; however, only one is proved to be optimal. Last but not least, we give some insight into the short term dynamics of the model. In particular, we numerically study the *optimal demographic transitions* in line with the typical imbalance effect analysis as designed in Boucekkine et al. (2008).

The article is organized as follows. Section 2 describes the model economy. Section 3 describes the optimal growth problem and applies an appropriate maximum principle to derive the set of necessary conditions. Section 4 analyses the long-run dynamics and characterize the balanced growth paths as described above, including comparative statics. Section 5 is numerical, it derives in particular some useful transitional dynamics. Section 6 concludes.

2 The Model

The model economy is a one sector closed economy. Output is obtained according to a neoclassical production function depending on the technological level, the physical capital stock and labor input. The latter, under the assumption of a fixed relation between labor supply and population, will be identified with the population stock. The aggregate production function, which comes from the direct summation of the individual production functions for many identical firms, is

$$Y(t) = A(t)^{1-\beta} K(t)^{\beta} N(t)^{1-\beta} . \tag{1}$$

In this function technical progress is assumed Harrod-neutral. Technological level, denoted by A, is exogenous and evolves according to the differential equation

$$\dot{A}(t) = xA(t) . \tag{2}$$

This is the standard law of motion for technology in Neoclassical growth theory, where it is assumed that technical progress arrives at a constant growth rate $x \geq 0$. The solution to the above equation implies that A increases monotonically according to the exponential form

$$A(t) = A_0 \exp(x(t - t_0)) , \tag{3}$$

where $A_0 = A(t_0) > 0$ is the initial technological level.

The economy is populated by many identical and infinitely lived agents. In this context, there is no point for differentiating between parents and children. Households face an infinite planning horizon, representing an immortal extended family where each member can be seen as a dynasty. Consequently, given that we focus on the link between demography and economic growth, trying to endogenize the demographic growth in a continuous time Ramsey model, we shall adapt the fertility conceptual schema from demographic theories to the requirements of our own model. From an aggregate point of view, we only have to deal with two demographic variables: population level and its variation. Population stock, denoted by N, is endogenously determined and it evolves according to the linear differential equation

$$\dot{N}(t) = n(t) N(t) , \tag{4}$$

where the rate of population growth $n(t)$ is a control variable.

The initial population stock is $N(t_0) = N_0 > 0$, and we assume that $n(t) \geq 0\, \forall t$. With respect to the individual preferences we assume that they are represented by a twice continuously differentiable instantaneous utility function, which depends positively on the current per capita consumption, and positively on the rate of population growth. The structure of our model allows for the existence of a long-run balanced growth path, defined as an allocation in which consumption per capita grows at a positive constant rate and the population growth rate is constant. We assume that the particular instantaneous utility function is of the form

$$U(c(t), n(t)) = \frac{c(t)^{1-\Phi}\, n(t)^{\epsilon(1-\Phi)}}{1-\Phi}. \tag{5}$$

In this function, the parameter $\Phi \left(\equiv \frac{c \cdot U_{cc}}{-U_c}\right)$ represents the inverse of the *conventional* intertemporal elasticity of substitution coefficient, which is constant and it is allowed to take values above or below unity, $0 < \Phi \lesseqgtr 1$; the parameter $\epsilon \left(\equiv \frac{U_n}{U_c} \frac{n}{c}\right)$ represents the weight of population changes in utility relative to the weight of consumption, and it is assumed positive but lower than one, $0 < \epsilon < 1$. According to the above parameter configuration we get $U_c > 0$, $U_n > 0$ and $U_{cc} < 0$, while we need $\Phi > \frac{\epsilon-1}{\epsilon}$ for $U_{nn} < 0$. However, the latter parameter constraint always holds for $\Phi > 0$ and $\epsilon < 1$. To ensure the strict concavity of the instantaneous utility function we assume that $\Phi > \frac{\epsilon}{1+\epsilon}$.[1]

In the present framework where population is endogenous because the stock N depends on the population growth rate n, which is currently decided by economic agents, we omit any population stock effect in the representation of individuals' preferences. We consider that people do not care about the population size N but only about the per capita number of offspring n. That is, the stock effect is not modeled entering the instantaneous utility function as a direct argument, but affecting other variables and functions in the model.

Finally, we introduce the aggregate resources constraint according to which output may be devoted to consumption, to capital accumulation or to rear population changes. Strictly speaking, here there are no parents rearing children but people looking after people. For the sake of simplicity we do not consider capital depreciation. Hence, net investment equals gross investment and the capital stock is governed by the differential equation

$$C(t) + \dot{K}(t) + bn(t)K(t) = Y(t). \tag{6}$$

The initial capital stock is $K(t_0) = K_0 > 0$. Adapting from Barro and Sala-i-Martin (2004), we assume that the per capita rearing cost is $b_0 + b\frac{K(t)}{N(t)}$, where

[1]This parameter constraint is a sufficient condition for the determinant of the Hessian matrix to be positive, but also implies that $\Phi > \frac{\epsilon-1}{\epsilon}$. Consequently, the Hessian matrix is negative definite, which corresponds to the standard sufficient condition for the utility function to be strictly concave.

$b_0 \geqslant 0$ and $b \geqslant 0$. This cost includes either purchases of market goods and services or the opportunity cost of time devoted to population rearing. Then, if we consider the real number representing the change in population size, total resources allocated to them are $\left(b_0 + b\frac{K(t)}{N(t)}\right)\overset{\bullet}{N}(t) = b_0 n(t) N(t) + bn(t) K(t)$. To obtain (6) we have simplified by setting $b_0 = 0$.

3 The Optimal Growth Problem

In the optimal growth problem, the benevolent planner has to consider the effect of population size on social welfare. In this setting, given that we are not particularly interested in the case $\Phi \rightarrow 1$ but rather in the most empirically relevant case in which $\Phi > 1$, we define the social welfare (which is the planner's objective function) as

$$W = \int_{t_0}^{+\infty} \frac{c^{1-\Phi} n^{\epsilon(1-\Phi)}}{1-\Phi} N^{\lambda} e^{-\rho(t-t_0)} dt \tag{7}$$

Parameter ρ is the positive social rate of discount or time preference. Parameter $\lambda \in [0, 1]$ contributes to specify social preferences, which are represented using a Millian, an intermediate, or a Benthamite intertemporal welfare function. In one extreme, when $\lambda = 0$ (average utilitarianism), the central planner maximizes per capita utility (average utility of consumption per capita). In the other, when $\lambda = 1$ (classical utilitarianism), the central planner maximizes total utility (the addition across total population of utilities of per capita consumption).[2]

The central planner's problem consists then in choosing the sequence $\{c(t), n(t), t \geq t_0\}$ that solves the optimization problem

$$\max_{\{K,N,c,n\}} \quad (7) \qquad s.t. \quad (1), (2), (4), \text{and } (6), \tag{8}$$

given $A(t_0) = A_0 > 0$, $K(t_0) = K_0 > 0$, and $N(t_0) = N_0 > 0$.

[2]The literature differentiates between two types of altruism depending on the two parameters ρ and λ. The first one is *intertemporal* altruism and depends on the discount rate applied to future population utility. The second one is *intergenerational* altruism and depends on the number of individuals which is taken into account each period. In particular, for representative and infinitely lived agent models, parameter λ controls for the degree of altruism towards total population including future generations. When agents are selfish the central planner maximizes W under $\lambda = 0$, and population size has no direct effect on the intertemporal utility. Instead, when agents are altruistic the central planner maximizes W under $\lambda = 1$, and the intertemporal utility function includes total population as a determinant.

Before solving the dynamic problem, we define the variables $\widetilde{c}(t) = \frac{c(t)}{A(t)}$ and $\widetilde{k}(t) = \frac{K(t)}{A(t)N(t)}$, which allow to write in per capita efficiency terms either the integrand and the dynamic resources constraint, $\dot{\widetilde{k}}(t) = \widetilde{k}(t)^{\beta} - \widetilde{c}(t) - (x + (1+b)\,n(t))\,\widetilde{k}(t)$.

In this context, solving Eq. (4) we get the following expression for the endogenous population size

$$N(t) = N_0 \exp\left(\int_{t_0}^{t} n(\tau)\,d\tau\right). \tag{9}$$

Hence, expressions (3) and (9) allow for the transformation

$$A(t)^{1-\Phi} N(t)^{\lambda} e^{-\rho(t-t_0)} = A_0^{1-\Phi} N_0^{\lambda} \exp\left(-\int_{t_0}^{t} (\rho - x(1-\Phi) - \lambda n(\tau))\,d\tau\right). \tag{10}$$

This term plays the role of a **variable discount factor** which also depends on past and current rates of population growth. So, adapting from Obstfeld (1990), Palivos et al. (1997), Ayong le Kama and Schubert (2007), and Schumacher (2011) who analyze optimal control problems extended to an **endogenous discounting** framework, we can define the accumulated stock of impatience as the non-negative

$$\Delta(t) = \int_{t_0}^{t} (\rho - x(1-\Phi) - \lambda n(\tau))\,d\tau = (\rho - x(1-\Phi))(t-t_0) - \lambda \int_{t_0}^{t} n(\tau)\,d\tau \geq 0, \tag{11}$$

where for obvious reasons $\Delta(t_0) = \Delta_0 = 0$.[3] This is a new state variable for which the motion equation reads

$$\dot{\Delta}(t) = \rho - x(1-\Phi) - \lambda n(t) \equiv \Theta(n(t)) \gtreqless 0. \tag{12}$$

That is, the **effective discount rate** is endogenous because of the endogeneity of population growth rates. Further, impatience is inversely proportional to the number of offspring, $\Theta'(\cdot) = -\lambda \leqslant 0$ and $\Theta''(\cdot) = 0$, except for the Millian case in which we recover the standard constant discount factor. This may happen because as population grows agents care more about the future, given that the increased population represents an investment having a positive impact on future welfare. The negative effect of population growth on the effective discount rate is greater

[3]The non-negativity of Δ might be replaced by a weaker constraint in line with Assumption 4 in Palivos et al. (1997) given the goal of a well-defined optimization problem.

as higher is the intergenerational altruism. Moreover, from (12) we get

$$\Theta(\cdot) \begin{cases} > 0 & \\ = 0 & \text{whenever} \\ < 0 & \end{cases} \begin{matrix} 0 < n(t) < \frac{\rho - x(1-\Phi)}{\lambda} \\ n(t) = \frac{\rho - x(1-\Phi)}{\lambda} \\ \frac{\rho - x(1-\Phi)}{\lambda} < n(t). \end{matrix} \tag{13}$$

A direct consequence of definition (11) is that the solution trajectory for population size may be rewritten as

$$N(t) = N_0 \exp\left(\frac{(\rho - x(1-\Phi))(t - t_0) - \Delta(t)}{\lambda}\right). \tag{14}$$

Overall, after introducing the new variable Δ, the intertemporal optimization problem becomes an **autonomous problem** without discounting and infinite planning horizon. According to Pittel (2002) and based on Marin-Solano and Navas (2009), due to the effective non-constant discount rate, the Pontryagin's maximum principle cannot be applied directly because intertemporally dependent preferences can create a time-consistency problem. We need this state variable to solve the problem within the standard optimal control approach, where it is no use distinguishing between present value and current value specifications. Here we follow Seierstad and Sydsaeter (1987), Chiang (1992), and Sethi and Thompson (2000).

Then, we can write the Hamiltonian function

$$\underset{\{\widetilde{c},n,q,\widetilde{k},\upsilon,\Delta\}}{H} = \frac{\widetilde{c}^{1-\Phi} n^{\epsilon(1-\Phi)} A_0^{1-\Phi} N_0^{\lambda} e^{-\Delta}}{1-\Phi} + q\left(\widetilde{k}^{\beta} - \widetilde{c} - (x + (1+b)n)\widetilde{k}\right)$$
$$+\upsilon(\rho - x(1-\Phi) - \lambda n). \tag{15}$$

Here $q \geqslant 0$ and $\upsilon \geqslant 0$ are the co-states for $\widetilde{k}$ and Δ respectively.[4] If we ignore the constraints involving only control variables, $\widetilde{c} \geqslant 0$ and $n \geqslant 0$, the first order necessary conditions arising from Pontryagin's Maximum principle are

$$q = \widetilde{c}^{-\Phi} n^{\epsilon(1-\Phi)} A_0^{1-\Phi} N_0^{\lambda} e^{-\Delta}, \tag{16}$$

$$q(1+b)\widetilde{k} = \epsilon \widetilde{c}^{1-\Phi} n^{\epsilon(1-\Phi)-1} A_0^{1-\Phi} N_0^{\lambda} e^{-\Delta} - \upsilon\lambda, \tag{17}$$

$$\dot{\widetilde{k}} = \widetilde{k}^{\beta} - \widetilde{c} - (x + (1+b)n)\widetilde{k}, \tag{18}$$

[4]The multiplier υ represents the marginal shadow value of relaxing the constraint (12). That is, the shadow price of (the accumulated stock of) impatience. As previously said, because of empirical reasons, we focus on the case in which $\Phi > 1$ and, consequently, $U(\widetilde{c}, n, \Delta) < 0$. Here we choose to write H in its canonical form with a positive sign preceding the (positive) multiplier υ.

$$\overset{\bullet}{q} = (x + (1+b)\,n)\,q - q\beta\widetilde{k}^{\beta-1}, \tag{19}$$

$$\overset{\bullet}{\Delta} = \rho - x\,(1-\Phi) - \lambda n, \tag{20}$$

$$\overset{\bullet}{\upsilon} = \frac{\widetilde{c}^{1-\Phi} n^{\epsilon(1-\Phi)}}{1-\Phi} A_0^{1-\Phi} N_0^{\lambda} e^{-\Delta}, \tag{21}$$

Finally, we also need the initial conditions A_0, N_0, K_0, $\widetilde{k}_0 = \frac{K_0}{A_0 N_0}$, and Δ_0, as well as the transversality conditions

$$\lim_{t\to+\infty} H\,(t) = 0, \tag{22}$$

$$\lim_{t\to+\infty} q\,(t) \geqslant 0 \text{ and } \lim_{t\to+\infty} q\,(t)\,\widetilde{k}\,(t) = 0, \tag{23}$$

$$\lim_{t\to+\infty} \upsilon\,(t) \geqslant 0 \text{ and } \lim_{t\to+\infty} \upsilon\,(t)\,\Delta\,(t) = 0. \tag{24}$$

The necessary conditions in the present dynamic optimization problem are also sufficient for a maximum because the Hamiltonian function satisfies the required concavity conditions [see Appendix 1]. Looking at (15) we can also check that H is autonomous. Consequently, along the optimal path H is constant and, given that our transversality condition (22) says that H eventually converges to zero, we conclude that

$$H = 0 \quad \forall t. \tag{25}$$

Next, given the solution to Eq. (20) as shown in (11), which assumes $\Delta\,(t) \geq 0$, as well as the finite values of the strictly concave function $\frac{\widetilde{c}^{1-\Phi} n^{\epsilon(1-\Phi)}}{1-\Phi} A_0^{1-\Phi} N_0^{\lambda}$, we can integrate (21) to obtain the expression[5]

$$\upsilon\,(t) = \upsilon\,(t_0) + \int_{t_0}^{t} \frac{\widetilde{c}^{1-\Phi} + n^{\epsilon(1-\Phi)}}{1-\Phi} A_0^{1-\Phi} N_0^{\lambda} e^{-\Delta} d\tau.$$

Then, substituting into the transversality condition (24) we get

$$\left(\upsilon\,(t_0) + \int_{t_0}^{+\infty} \frac{\widetilde{c}^{1-\Phi} n^{\epsilon(1-\Phi)} A_0^{1-\Phi} N_0^{\lambda} e^{-\Delta}}{1-\Phi} d\tau\right) \int_{t_0}^{+\infty} \Theta\,(n\,(\tau))\,d\tau = 0.$$

[5]The convergence of the objective integral (7) is shown in Nairay (1984), even for the case in which $\Phi < 1$, by proving that such a limit value exists and is finite.

This condition holds if and only if, for any $\lim_{t\to+\infty} \Delta(t)$ different from zero,

$$\upsilon(t_0) = \int_{t_0}^{+\infty} -\frac{\widetilde{c}^{1-\Phi} n^{\epsilon(1-\Phi)}}{1-\Phi} A_0^{1-\Phi} N_0^{\lambda} e^{-\Delta} d\tau.$$

Consequently, we conclude that the multiplier υ takes the value

$$\upsilon(t) = \int_{t}^{+\infty} -\frac{\widetilde{c}^{1-\Phi} n^{\epsilon(1-\Phi)}}{1-\Phi} A_0^{1-\Phi} N_0^{\lambda} e^{-\Delta} d\tau \quad \forall t \geq t_0. \tag{26}$$

Then, the first order conditions reduce to (16)–(19) together with (11), (26), the transversality conditions

$$0 = \frac{\widetilde{c}^{1-\Phi} n^{\epsilon(1-\Phi)}}{1-\Phi} A_0^{1-\Phi} N_0^{\lambda} e^{-\Delta} + q\dot{\widetilde{k}} + \upsilon(\rho - x(1-\Phi) - \lambda n), \tag{27}$$

$$\lim_{t\to+\infty} q(t) \geqslant 0 \text{ and } \lim_{t\to+\infty} q(t)\widetilde{k}(t) = 0, \tag{28}$$

$$0 < \left| \int_{t_0}^{+\infty} (\rho - x(1-\Phi) - \lambda n(t))\, dt \right|, \tag{29}$$

and the initial conditions A_0, N_0, and $\widetilde{k}_0 = \frac{K_0}{A_0 N_0}$.

Consider now Eqs. (16) and (17). As we have seen, gross product may be allocated to consumption, investment, or offspring. On the margin, goods must be equally valuable if they are consumed or accumulated as new physical capital. Namely, the marginal utility of consumption today must be equal to the current shadow price qe^{Δ} of physical capital (consumption tomorrow). Moreover, at equilibrium the marginal utility of population growth must be equal to the sum of the current implicit value of the *full* (rearing and dilution) marginal cost of increasing population $qe^{\Delta}(1+b)\widetilde{k}$, plus the current shadow value of the accumulated impatience scaled by the weight of the increased population in social welfare $\upsilon e^{\Delta}\lambda$. Taken together, these equations give the tangency condition

$$\frac{\epsilon\widetilde{c}}{n} = (1+b)\widetilde{k} + \lambda\frac{\upsilon}{q}, \tag{30}$$

which describes the optimal allocation between consumption goods and children. The marginal rate of substitution between n and c must be equal to the *full* marginal cost of increasing population plus the degree of intergenerational altruism times the relative (shadow) prices of impatience and physical capital.

Moreover, differentiating (16) with respect to time and substituting (19) and (20), we get the corresponding adapted version of the Ramsey rule,

$$\frac{\dot{\widetilde{c}}}{\widetilde{c}} = \frac{1}{\Phi}\left(\beta\widetilde{k}^{\beta-1} - (x+n+bn) - (\rho - x(1-\Phi) - \lambda n) + \epsilon(1-\Phi)\frac{\dot{n}}{n}\right), \tag{31}$$

where $\frac{1}{\Phi} = \frac{-U_c}{cU_{cc}}$, $\beta \widetilde{k}^{\beta-1} = f'\left(\widetilde{k}\right)$, $\rho - x(1-\Phi) - \lambda n = \Theta(n)$, and $\epsilon(1-\Phi) = \frac{nU_{cn}}{U_c}$. The growth rate of per capita consumption in efficiency units depends: i) positively on the net marginal productivity[6] of per capita capital in efficiency units; ii) negatively on the effective discount rate; as well as iii) on the rate of change of the population growth rate. For $\Phi > 1$ we get $U_{cn} < 0$, which implies that c and n are gross substitutes in utility. In this case their corresponding rates of growth are inversely related to each other. All the three above arguments are endogenous because of the endogeneity of the rate of population growth.

Finally, for the purpose of facilitating comparison with the exogenous discount rate model, the above Ramsey rule may be written as

$$\frac{\dot{\widetilde{c}}}{\widetilde{c}} = \frac{-U_c}{cU_{cc}} \left(f'\left(\widetilde{k}\right) - \rho - x\Phi - (1+b-\lambda)\, n + \frac{U_{cn}}{U_c} \dot{n} \right). \tag{32}$$

If the net return to capital exceeds the effective discount rate, agents would decide to invest now in physical capital leaving less resources to consumption and child rearing today. In the standard model with exogenous discount and constant rate of population growth, this would suffice to explain an increasing per capita consumption. However, in our model the expected increasing resources may allow for different combinations. Obviously, the additional future resources are available for the simultaneous growth of per capita consumption and population size, but this is not the only possibility given that the above expression still admits an increasing consumption with a decreasing population, or a decreasing consumption with an increasing population. In any case, all of them will produce more future welfare.

On the other hand, differentiating (17) with respect to time and substituting (18)–(21), we get the corresponding adapted version of the Meade rule (Dasgupta 1969; Constantinides 1988). This is a rule for the optimal population growth, which comes from the balance between the gains and losses due to the introduction of a new member into society.

$$\frac{\dot{n}}{n} = \frac{n}{\epsilon(1-\Phi)-1} \frac{(1+b)}{\epsilon \widetilde{c}} \left((1-\beta)\, \widetilde{k}^{\beta} - \widetilde{c} \right) + \frac{n}{\epsilon(1-\Phi)-1} \frac{\lambda}{\epsilon(1-\Phi)}$$
$$+ \frac{\rho - x(1-\Phi) - \lambda n}{\epsilon(1-\Phi)-1} - \frac{(1-\Phi)}{\epsilon(1-\Phi)-1} \frac{\dot{\widetilde{c}}}{\widetilde{c}}, \tag{33}$$

where $\frac{1}{\epsilon(1-\Phi)-1} = \frac{U_c}{nU_{cn}-U_c} < 0$, $(1-\beta)\, \widetilde{k}^{\beta} = f\left(\widetilde{k}\right) - \widetilde{k} f'\left(\widetilde{k}\right)$, $\rho - x(1-\Phi) - \lambda n = \Theta(n)$, and $1 - \Phi = \frac{U_c + cU_{cc}}{U_c} < 0$. The change in the population growth

[6] Even if we do not consider capital depreciation, the exogenous technical progress and the increase in the population size are the cause of a marginal dilution effect which adds to the corresponding marginal rearing cost to determine the net marginal productivity.

rate depends: (1) negatively on the difference between the marginal product of an additional person and his consumption measured in efficiency units; (2) positively on the degree of intergenerational altruism; (3) negatively on the effective discount rate; and (4) negatively on the growth rate of per capita consumption in efficiency units. It is worth noticing that again the changes in c and n are inversely related to each other.

Writing in terms of a more general specification, the Meade rule takes the following form

$$\dot{n} = \frac{n\left(U_c\right)^2}{nU_{cn}U_n - U_cU_n} A\left(1+b\right)\left(f\left(\widetilde{k}\right) - \widetilde{k}f'\left(\widetilde{k}\right) - \widetilde{c}\right) + \frac{n\left(U_c\right)^2}{n\left(U_{cn}\right)^2 - U_cU_{cn}}\lambda$$

$$+ \frac{nU_c}{nU_{cn} - U_c}\Theta\left(n\right) - \frac{n\left(U_c + cU_{cc}\right)}{nU_{cn} - U_c}\frac{\dot{\widetilde{c}}}{\widetilde{c}}. \tag{34}$$

According to our model, when the per capita consumption exceeds the marginal product of labor, there is an incentive for increasing the rate of population growth as well as the per capita consumption level. Moreover, for a given degree of altruism, if the effective discount rate is negative, the above incentive will be stronger.

4 The Balanced Growth Path and Comparative Statics

In the previous section we have solved the model for any exogenous and constant rate of technical progress. However, in the present section we assume, for the sake of simplicity, a constant technological level, that is x=0. Otherwise, we could not find most of the analytical long-run results associated with the balanced growth path. Hereafter we characterize the long term equilibria identifying the balanced growth path along which $\widetilde{c}$ and n are constant. In steady state $\dot{\widetilde{k}} = 0$, but given the transversality condition (27) we observe that $\dot{\Delta} = \rho - \lambda n^* \equiv \Theta\left(n^*\right) > 0$, which is compatible with the constraint (29). This implies, from (11), that

$$\Delta^* = \left(\rho - \lambda n^*\right)\left(t - t_0\right), \tag{35}$$

which makes q non-stationary according to (16). Consequently, we introduce a new variable $p = qe^{\Delta}$ and, hence, $\frac{\dot{p}}{p} = \frac{\dot{q}}{q} + \dot{\Delta}$. Now, in steady state $\dot{\widetilde{k}} = \dot{p} = 0$ and Eqs. (16)–(19) can be written as

$$p^* = \widetilde{c}^{*^{-\Phi}} n^{*^{\epsilon(1-\Phi)}} A_0^{1-\Phi} N_0^{\lambda}, \tag{36}$$

$$\left(1+b\right)\widetilde{k}^* = \frac{\epsilon\widetilde{c}^*}{n^*} - \left(\frac{\upsilon^*}{q^*}\right)\lambda, \tag{37}$$

$$\widetilde{k}^{*\beta} = \widetilde{c}^* + (1+b)\, n^* \widetilde{k}^*, \tag{38}$$

$$\beta \widetilde{k}^{*\beta-1} = \rho + (1+b-\lambda)\, n^*, \tag{39}$$

Moreover, from (26) or (27) we get

$$\upsilon^* = \frac{-\widetilde{c}^{*1-\Phi} n^{*\epsilon(1-\Phi)} A_0^{1-\Phi} N_0^{\lambda} e^{-\Delta^*}}{(1-\Phi)(\rho-\lambda n^*)} = \frac{-\widetilde{c}^* q^*}{(1-\Phi)(\rho-\lambda n^*)}. \tag{40}$$

These expressions allow us to directly obtain the stationary values $\widetilde{c}^*$, n^*, and $\widetilde{k}^*$ corresponding to the balanced growth path and, by substitution, all the remaining endogenous variables of the model. After some algebraic manipulations we get

$$\widetilde{k}^* = \left(\frac{\beta}{\rho + (1+b-\lambda)\, n^*}\right)^{\frac{1}{1-\beta}}, \tag{41}$$

$$\widetilde{c}^* = \frac{\beta^{\frac{\beta}{1-\beta}} \left(\rho + ((1-\beta)(1+b)-\lambda)\, n^*\right)}{\left(\rho + (1+b-\lambda)\, n^*\right)^{\frac{1}{1-\beta}}}, \tag{42}$$

$$\widetilde{y}^* = \left(\frac{\beta}{\rho + (1+b-\lambda)\, n^*}\right)^{\frac{\beta}{1-\beta}}, \tag{43}$$

where n^* corresponds to the roots of the second degree polynomial equation with real coefficients

$$\Psi_a n^{*2} + \Psi_b n^* + \Psi_c = 0, \tag{44}$$

These coefficients depend on the structural parameters of the model in the following way

$$\Psi_a(\lambda, \Phi, b, \beta, \epsilon) = \lambda\left(((1-\beta)(1+b)-\lambda)(\epsilon(1-\Phi)-1) - \beta(1-\Phi)(1+b)\right), \tag{45}$$

$$\Psi_b(\lambda, \Phi, b, \beta, \epsilon, \rho) = \rho(1-\Phi)(1+b)(\beta - \epsilon(1-\beta)) + \lambda(\epsilon(1-\Phi)2\rho - \rho), \tag{46}$$

$$\Psi_c(\Phi, \epsilon, \rho) = -\rho^2 \epsilon (1-\Phi). \tag{47}$$

The roots are:

$$n_1^* = \frac{-\Psi_b + \sqrt{D}}{2\psi_a}, \tag{48}$$

$$n_2^* = \frac{-\Psi_b - \sqrt{D}}{2\psi_a}, \tag{49}$$

where $D = \Psi_b^2 - 4\Psi_a\Psi_c$ is the discriminant. In case $D \geq 0$ the roots $n_1^\star$ and $n_2^\star$ are both real. Moreover, from (40), (26), and (13) we get $\Theta(n^*) > 0$, that is

$$0 < n_i^* < \frac{\rho}{\lambda} \qquad \forall i = \{1, 2\}. \tag{50}$$

4.1 *The Case of the Millian Welfare Function: λ = 0*

We first analyze the Millian case. When the central planner maximizes per capita utility (average utilitarianism), $\lambda = 0$ and population size has no direct effect on the intertemporal utility. It is easily checked that since $\Psi_a = 0$ for $\lambda = 0$, Eq. (44) has a unique solution given by,

$$n^* = \frac{\epsilon\rho}{(1+b)(\beta - \epsilon(1-\beta))} \tag{51}$$

which is positive for $\beta > \epsilon(1-\beta)$. Notice that under the Millian case Eq. (50) is always checked.

Substituting (51) in (41)–(43) we obtain the stationary values for $\widetilde{k}^*$, $\widetilde{c}^*$ and $\widetilde{y}^*$,

$$\widetilde{k}^* = \left(\frac{\beta - \epsilon(1-\beta)}{\rho(1+\epsilon)}\right)^{\frac{1}{1-\beta}}, \tag{52}$$

$$\widetilde{c}^* = \frac{-\rho(\epsilon - \beta(1+\epsilon))^{\frac{\beta}{1-\beta}}}{(-\rho(1+\epsilon))^{\frac{1}{1-\beta}}}, \tag{53}$$

$$\widetilde{y}^* = \left(\frac{\beta - \epsilon(1-\beta)}{\rho(1+\epsilon)}\right)^{\frac{\beta}{1-\beta}}, \tag{54}$$

Proposition 1 summarizes the associated comparative statics findings:

Proposition 1 *When $\beta > \epsilon(1-\beta)$:*
$\frac{\partial n^*}{\partial b} < 0,\ \frac{\partial n^*}{\partial \epsilon} > 0,\ \frac{\partial n^*}{\partial \rho} > 0,\ \frac{\partial n^*}{\partial \beta} < 0,\ \frac{\partial n^*}{\partial \Phi} = 0$
$\frac{\partial \widetilde{y}^*}{\partial b} = 0,\ \frac{\partial \widetilde{y}^*}{\partial \epsilon} < 0,\ \frac{\partial \widetilde{y}^*}{\partial \rho} < 0,\ \frac{\partial \widetilde{y}^*}{\partial \beta} > 0,\ \frac{\partial \widetilde{y}^*}{\partial \Phi} = 0$

Proposition 1 is trivially checked by taking the partial derivative with respect to the corresponding parameter in (51) and (54). In general, as we can see from the above two sets of partial derivative signs, the optimal long-run rate of population growth and the long-run per capita income level (measured in efficiency terms) are inversely correlated.

In particular, we can identify the following parameter-variable relationships. First, recall that b represents the opportunity cost of parental time devoted to child rearing. Then, an economy where parents experience a higher cost of offspring will optimally choose in the long-run a lower rate of population growth. Note that in Eq. (43) we can observe two different effects of a change in the per capita rearing costs on the per capita income level. First, an increase in b directly reduces the resources devoted to capital accumulation, which implies a lower long-run level of $\widetilde{y}^*$. Moreover, a higher b reduces the optimal population growth rate, which has an indirect positive effect on $\widetilde{y}^*$. These two effects are of opposite sign and, only in the Millian case, exactly compensate each other. Consequently, the per capita income level is independent of b.

Second, recall that ϵ represents the weight of children in utility relative to consumption. Then, a society with higher preference for children will optimally choose in the long-run a higher population rate of growth, and will experience a lower per capita income level. Moreover, recall that an economy showing a low ρ represents a patient society. Then, we find that in the long-run an impatient society will optimally choose a higher population rate of growth, and will reach a lower level of per capita income.

On the other hand, an economy with higher β is an economy with a technology implying a higher elasticity of output with respect to capital (higher capital share). Then, in the long-run, this economy will optimally choose a lower population rate of growth, and will have the benefit of a higher per capita income level. Finally, we observe that in the Millian case the inverse of the intertemporal elasticity of substitution in consumption, Φ, has no effect on the long run population growth rate or the level of income per capita.

4.2 The Case $\lambda \neq 0$

When $\lambda \neq 0$, two distinct balanced growth paths emerge. We show that both are admissible in the sense that the two associated demographic growth rates are positive.

Proposition 2 *Under the parameter constraints* $0 < \lambda \leqslant 1$, $\Phi > 1$, *and* $\epsilon < \frac{\beta}{1-\beta}$, *which imply* $\Psi_c > 0$ *and* $\Psi_b < 0$, *if* $\Psi_a > 0$ *then we get two real positive values* n_1^* *and* n_2^*, *which are different as long as* $D > 0$.

Proof We consider the most empirically relevant case in which $\Phi > 1$ and assume $\epsilon < \frac{\beta}{1-\beta}$. Under these assumptions we get $\Psi_c > 0$ and $\Psi_b < 0$. Given $D \geqslant 0$, if $\Psi_a > 0$, we have $\sqrt{D} < |\Psi_b|$, and we can express $\sqrt{D}$ as $\sqrt{D} = |\Psi_b| - \theta$, with $\theta > 0$. Taking into account all the above, it is straightforward to check that $n_1^\star = \frac{2|\Psi_b|+\theta}{2\Psi_a} > 0$ and $n_2^\star = \frac{\theta}{2\Psi_a} > 0$. ■

Remark 1 From Proposition 2 and using the expressions in Eqs. (48) and (49) we get

$$n_1^* > n_2^* > 0, \tag{55}$$

$$2\Psi_a n_1^* + \Psi_b = \sqrt{D} > 0, \tag{56}$$

$$2\Psi_a n_2^* + \Psi_b = -\sqrt{D} < 0. \tag{57}$$

Multiplicity of balanced growth paths cannot be a surprise in a model with endogenous fertility. In the seminal Barro and Becker (1989) discrete time OLG model, multiplicity is possible in the case where the cost of rearing children is large enough. In our model, the existence of two distinct solutions is generated under much milder parametric assumptions not related to the cost of rearing children. This said, the optimality analysis of the steady state solutions in our model does allow to eliminate one of the two candidates, as demonstrated here below.

Proposition 3 *Given the parameter constraints and the results shown in Proposition 2 and Remark 1, the two real positive values n_1^* and n_2^* are separated from each other by the root's limiting upper-bound, implying that condition (50) does not hold for both. That is, we get*

$$0 < n_2^* < \frac{\rho}{\lambda} < n_1^*. \tag{58}$$

Proof We consider the different combinations ordering the upper-bound and the two roots, and we conclude that only one of such orderings is feasible because it is the only that requires a compatible relationship between structural parameters.

First, $\frac{\rho}{\lambda} > n_1^* > n_2^*$. That is $\frac{\rho}{\lambda} > \frac{-\Psi_b+\sqrt{D}}{2\Psi_a}$, or $\lambda\sqrt{D} < 2\rho\Psi_a + \lambda\Psi_b$. Then, if $2\rho\Psi_a + \lambda\Psi_b \leqslant 0$ the previous inequality is incompatible because $D > 0$. Alternatively, if $2\rho\Psi_a + \lambda\Psi_b > 0$ we can square the two sides of the inequality getting the new inequality $-\lambda^2\Psi_c < \rho^2\Psi_a + \rho\lambda\Psi_b$. Using Eqs. (45)–(47) to transform into a constraint between structural parameters alone we get $0 < \lambda(1-\beta)(1+b)$, which is incompatible.

Second, $n_1^* > n_2^* > \frac{\rho}{\lambda}$. That is $\frac{-\Psi_b-\sqrt{D}}{2\Psi_a} > \frac{\rho}{\lambda}$, or $-\lambda\sqrt{D} > 2\rho\Psi_a + \lambda\Psi_b$. Then, if $2\rho\Psi_a + \lambda\Psi_b \geqslant 0$ the previous inequality is incompatible because $D > 0$. Alternatively, if $2\rho\Psi_a + \lambda\Psi_b < 0$ we can square the two sides of the inequality getting the new inequality $-\lambda^2\Psi_c < \rho^2\Psi_a + \rho\lambda\Psi_b$. Using Eqs. (45)–(47) to transform into a constraint between structural parameters alone we get $0 < -\lambda(1-\beta)(1+b)$, which is incompatible.

Therefore, $n_2^* < \frac{\rho}{\lambda} < n_1^*$ is the only case which is compatible with the signs of the coefficients Ψ_a, Ψ_b, and Ψ_c. ■

Remark 2 Given that $\lim_{\lambda\to 0}\Psi_a = 0$, $\lim_{\lambda\to 0}\Psi_b = \rho(1-\Phi)(1+b)(\beta-\epsilon(1-\beta)) < 0$, $\lim_{\lambda\to 0}\Psi_c = -\rho^2\epsilon(1-\Phi) > 0$, and consequently $\lim_{\lambda\to 0}\sqrt{D} = \left|\lim_{\lambda\to 0}\Psi_b\right| = -\rho(1-\Phi)(1+b)(\beta-\epsilon(1-\beta)) > 0$, we get

$$\lim_{\lambda\to 0} n_2^* = \lim_{\lambda\to 0}\frac{-\Psi_b-\sqrt{D}}{2\Psi_a} = \frac{\epsilon\rho}{(1+b)(\beta-\epsilon(1-\beta))} > 0. \tag{59}$$

which corresponds to the selfish case (see Eq. (51)).

4.2.1 Comparative Statics Results

Unfortunately, when $\lambda \neq 0$, the model becomes much more complex analytically speaking. Recall that in this case the endogenous discounting is active and preferences become intertemporally related. The comparative statics of the optimal steady state population growth rate, n_2^* become intractable in general. To get an immediate idea about it, let us consider the vector of parameters $\Omega = (\lambda, \Phi, b, \beta, \epsilon, \rho)$. Then, by successive differentiation of Eq. (44) with respect to the components of such a vector we get the general formula:

$$\frac{\partial n_2^*}{\partial\Omega} = \frac{-\left(n_2^*\right)^2\frac{\partial\Psi_a}{\partial\Omega} - n_2^*\frac{\partial\Psi_b}{\partial\Omega} - \frac{\partial\Psi_c}{\partial\Omega}}{2\Psi_a n_2^* + \Psi_b}. \tag{60}$$

Therefore, even though one can sign the terms $\frac{\partial\Psi_i}{\partial\Omega}$ where $i \in \{a, b, c\}$, which is not always the case indeed, this might be unlikely to do the job. We can establish the following partial results

$$\frac{\partial\Psi_a}{\partial\lambda} = \Psi_a - (\epsilon(1-\Phi)-1)\lambda > 0,$$

$$\frac{\partial\Psi_a}{\partial\Phi} = \lambda(-\epsilon((1-\beta)(1+b)-\lambda)+\beta(1+b)) > 0,$$

$$\frac{\partial\Psi_a}{\partial b} = \lambda((1-\beta)(\epsilon(1-\Phi)-1)-\beta(1-\Phi)) \lesseqgtr 0 \text{ depending on whether } \beta\Phi \lesseqgtr 1,$$

$$\frac{\partial\Psi_a}{\partial\beta} = -\lambda((1-\beta)(\epsilon(1-\Phi)-1)+(1-\Phi)(1+b)) > 0,$$

$$\frac{\partial\Psi_a}{\partial\epsilon} = \lambda((1-\beta)(1+b)-\lambda)(1-\Phi) < 0, \text{ because we assume } (1-\beta)(1+b)-2\lambda > 0,$$

$$\frac{\partial\Psi_b}{\partial\lambda} = \epsilon(1-\Phi)2\rho-\rho < 0,$$

$$\frac{\partial \Psi_b}{\partial \Phi} = -\rho (1+b)(\beta - \epsilon (1-\beta)) - \lambda \epsilon 2 \rho < 0,$$

$$\frac{\partial \Psi_b}{\partial b} = \rho (1-\Phi)(\beta - \epsilon (1-\beta)) < 0,$$

$$\frac{\partial \Psi_b}{\partial \beta} = \rho (1-\Phi)(1+b)(1+\epsilon) < 0,$$

$\frac{\partial \Psi_b}{\partial \epsilon} = -\rho (1-\Phi)((1-\beta)(1+b) - 2\lambda) > 0$, because we assume $(1-\beta)(1+b) - 2\lambda > 0$,

$$\frac{\partial \Psi_b}{\partial \rho} = (1-\Phi)(1+b)(\beta - \epsilon (1-\beta)) + \lambda \epsilon (1-\Phi) 2 - \lambda < 0,$$

$$\frac{\partial \Psi_c}{\partial \Phi} = \rho^2 \epsilon > 0,$$

$$\frac{\partial \Psi_c}{\partial \epsilon} = -\rho^2 (1-\Phi) > 0,$$

$$\frac{\partial \Psi_c}{\partial \rho} = -2\rho\epsilon (1-\Phi) > 0.$$

Unfortunately, all these properties do not allow us to sign the derivative $\frac{\partial n_2^*}{\partial \lambda}$, which is one of the important tasks we have to accomplish as increasing λ allows to move from the Millian to the Benthamite social welfare function. However, on gets after the appropriate substitutions

$$\frac{\partial n_2^*}{\partial \lambda} = \frac{-(n_2^*)^2 \frac{\partial \Psi_a}{\partial \lambda} - n_2^* \frac{\partial \Psi_b}{\partial \lambda} - \frac{\partial \Psi_c}{\partial \lambda}}{2\Psi_a n_2^* + \Psi_b} = \frac{-n_2^*}{2\Psi_a n_2^* + \Psi_b} \left(n_2^* \frac{\partial \Psi_a}{\partial \lambda} + \frac{\partial \Psi_b}{\partial \lambda} \right)$$

$$= \frac{-n_2^*}{2\Psi_a n_2^* + \Psi_b} \left(n_2^* \Psi_a - (\epsilon (1-\Phi) - 1) \left(n_2^* \lambda - \rho \right) + \rho \epsilon (1-\Phi) \right). \qquad (61)$$

Given (55), (57), and (58), we get

$$\frac{\partial n_2^*}{\partial \lambda} < 0 (> 0)$$

depending on whether

$$n_2^* \Psi_a < (>) (\epsilon (1-\Phi) - 1) \left(n_2^* \lambda - \rho \right) - \rho \epsilon (1-\Phi).$$

One can proceed in the same way for all the other comparative statics and identify sufficient conditions for the intended properties to hold. Unfortunately, it is not possible to extract sharp necessary and sufficient conditions. As a last example, consider the other altruism parameter ϵ and try to sign the derivative $\frac{\partial n_2^*}{\partial \epsilon}$, **which is expected to be strictly positive**:

$$\begin{aligned}\frac{\partial n_2^*}{\partial \epsilon} &= \frac{-(n_2^*)^2\frac{\partial \Psi_a}{\partial \epsilon} - n_2^*\frac{\partial \Psi_b}{\partial \epsilon} - \frac{\partial \Psi_c}{\partial \epsilon}}{2\Psi_a n_2^* + \Psi_b} \\ &= \frac{-n_2^*}{2\Psi_a n_2^* + \Psi_b}\left(n_2^*\frac{\partial \Psi_a}{\partial \epsilon} + \frac{\partial \Psi_b}{\partial \epsilon} + \frac{1}{n_2^*}\frac{\partial \Psi_c}{\partial \epsilon}\right) \\ &= \frac{-n_2^*}{2\Psi_a n_2^* + \Psi_b}\Bigg(\left(n_2^*\lambda - \rho\right)(1-\Phi)\left((1-\beta)(1+b) - 2\lambda\right) \\ &\quad + (1-\Phi)\frac{\left(n_2^*\lambda + \rho\right)\left(n_2^*\lambda - \rho\right)}{n_2^*}\Bigg). \end{aligned} \tag{62}$$

Consequently, if $(1-\beta)(1+b) - 2\lambda > 0$ (sufficient), then $\frac{\partial n_2^*}{\partial \epsilon} > 0$. Moreover, it also implies $(1-\beta)(1+b) - \lambda > 0$.

It appears clearly that when $\lambda \neq 0$, numerical exploration of the comparative statics properties is unavoidable. We shall do that together with the investigation of short-run dynamics.

5 Numerical Experiments

In this section we complement the previous analytical results with some numerical exercises. Since the comparative statics are analytically ambiguous in the general case, we present some numerical results in Sect. 5.1. Moreover, Sect. 5.2 is devoted to give some insight into the short term dynamics of the model. We consider the following parameter values: $\rho = 0.05$, $\beta = 0.36$ and $\Phi = 2$, which roughly conform to the standard values used in the literature (Caballé and Santos 1993; Canton and Meijdam 1997). Per capita rearing cost and the propensity to have children are given by $b = 1$, $\epsilon = 0.3$, according to Barro and Sala-i-Martin (2004) and de la Croix and Doepke (2003). Finally we assume A_0=1, $x = 0$ and fix $\lambda = 0.5$ as the benchmark value for the altruism parameter.

5.1 Comparative Statics

We first analyze how the long run population growth rate and the per capita consumption change as the degree of intertemporal altruism increases (Fig. 1).

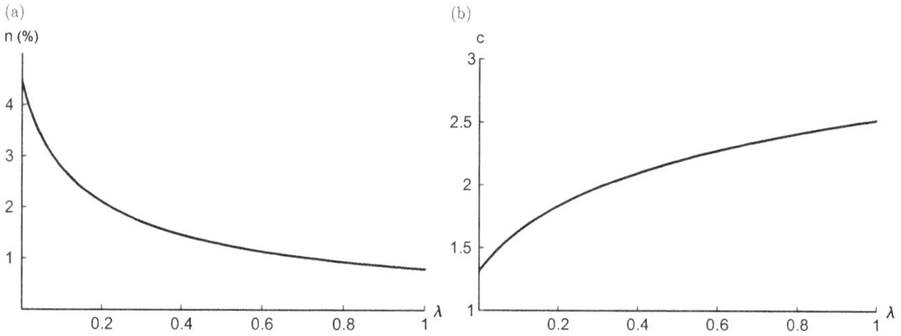

Fig. 1 Population growth rate and consumption per capita values as the altruism parameter λ changes. (**a**) Long run n as λ changes. (**b**) Long run c as λ changes

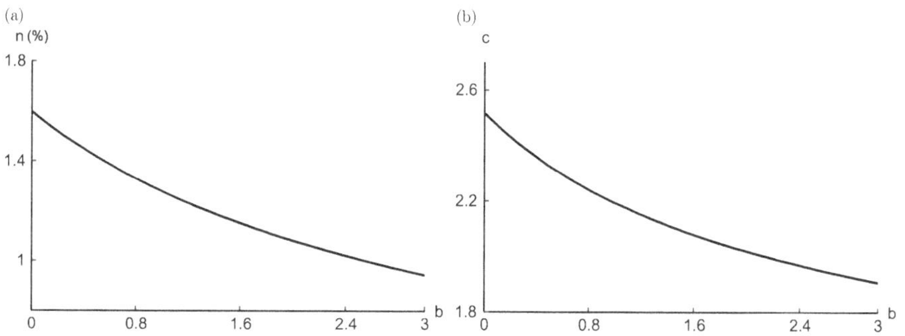

Fig. 2 Population growth rate and consumption per capita values as rearing cost b changes. (**a**) Long run n rate as b changes. (**b**) Long run c as b changes

One can see that consistently with Palivos and Yip, the population growth rate is decreasing with λ, that's the Benthamite social welfare function delivers the lowest demographic growth in the long-run. The fact that consumption per capita is at the same time increasing with λ allows also to conclude that either in the AK or in the Ramsey case (with decreasing returns to both human or physical capital) no *repugnant conclusion* arises under Benthamite preferences.[7] The Millian case shows in contrast the largest growth rate of population and the lowest per capita consumption. Nonetheless, in our calibrated model, the quantitative differences between the two extreme cases, though significant, can hardly lead us to conclude for any opposite *repugnant conclusion.*

Figure 2 shows a negative relationship between the child rearing cost and the stationary rate of population growth. The same result is obtained in the selfish case. However, the effect of higher child rearing costs on the long run income per capita depends on the degree of intertemporal altruism. We proved that in the Millian case

[7] We also obtain the same result for a wide range of reasonable parameter values.

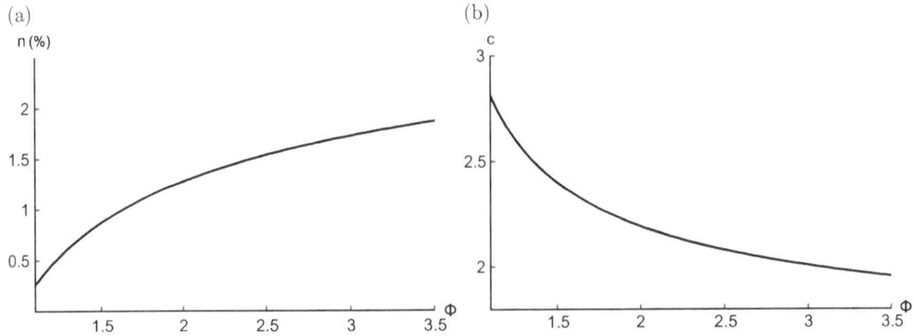

Fig. 3 Population rate and consumption per capita values as the inverse of the intertemporal elasticity of substitution Φ changes. (**a**) Long run n as Φ changes. (**b**) Long run c as Φ changes

$\lambda = 0$, the long run income per capita is independent of b. When $\lambda \neq 0$, numerical experiments show that the steady state income per capita decreases with the child rearing cost.

Finally we study the comparative statics with respect to the inverse of the intertemporal elasticity of substitution in consumption Φ. Under the Millian case (see Proposition 1), the long run values of the relevant variables of the model are independent of Φ. However, when $\lambda \neq 0$, an increase in the inverse of the intertemporal elasticity of substitution in consumption rises the population growth rate in the long run. As a consequence, income per capita decreases with Φ (Fig. 3).

5.2 *Short Run Dynamics*

We numerically study the optimal paths focusing on the typical imbalance effect analysis. We induce a transition process choosing $k_0 \neq \widetilde{k}^*$, and analyze two different situations, depending on the position of the economy, below or above the long run value of the capital stock per capita. In particular we set the initial condition $k_0 = 0.5\widetilde{k}^*$ and $k_0 = 1.5\widetilde{k}^*$ (Figs. 4 and 5).

The paths obtained can be roughly interpreted as optimal demographic transitions. When capital per capita is below the stationary value, capital is relatively scarce with respect to labor (or human capital). The economy starts investing massively in capital, and capital is gradually substituted for labor. As the process of substitution proceeds forward, the optimal population growth rate goes down leading to a kind of demographic transition (decreasing population growth rate, increasing consumption per capita) until convergence to the stationary equilibrium. It's worth pointing out that these dynamics can be interpreted as imbalance effect dynamics as depicted in Barro and Sala-i-Martin (2004), Chap. 5, or more recently in Boucekkine et al. (2008). Notice also the symmetry of optimal trajectories corresponding to initial relatively scarce capital and initial relatively abundant

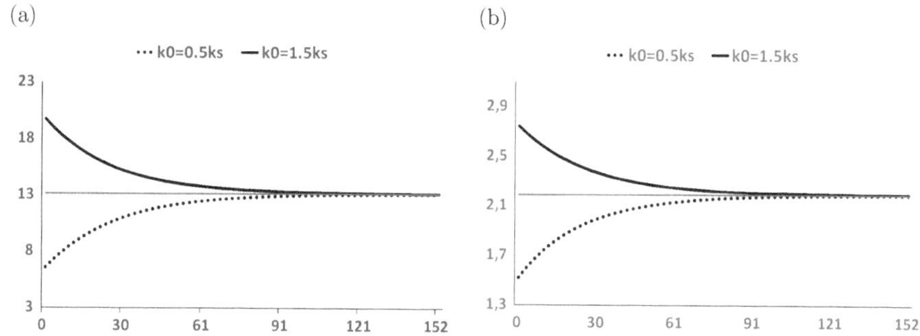

Fig. 4 Per capita physical capital stock and consumption. (**a**) Physical capital stock. (**b**) Per capita consumption

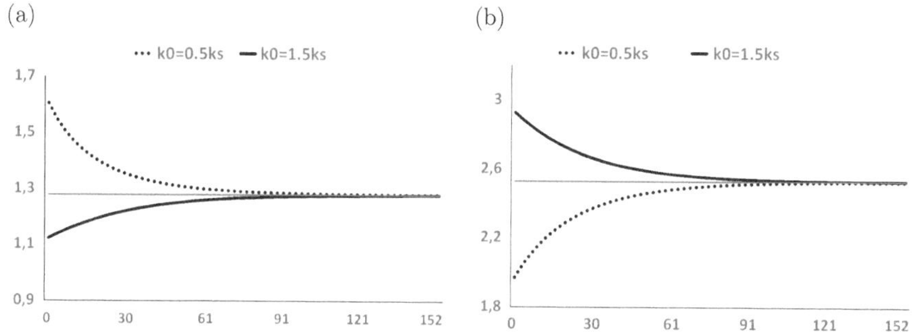

Fig. 5 Population growth rate and per capita income. (**a**) Population growth rate. (**b**) Per capita income

capital respectively. Symmetry here is granted because we have one production sector, the shape of imbalance effects are much less symmetric in two-sector models à la Lucas-Uzawa (see Boucekkine et al. 2008).

6 Conclusion

In this paper, we have studied an optimal population size problem of the Ramsey type in a continuous time framework. We have motivated this problem within the population ethics debate: the tricky question is not to derive optimal demographic paths but which social welfare functions are the most appropriate to cope with intergenerational fairness. We show that our framework does not go at odds with the early AK or AN literature dealing with these questions. In particular, the Benthamite criterion is shown to not deliver any *repugnant conclusion*, neither in the long run nor in the short run. Indeed, within the class of social welfare functions considered, the Benthamite criterion is the one which leads with the lowest stationary demographic growth and to the largest consumption per capita.

Our contribution is also methodological. We show that this type of problems with endogenous demography can be unambiguously connected to the class of optimal control problems with endogenous discounting, and should be therefore treated accordingly, that's with an appropriate version of the maximum principle. It goes without saying that this proviso applies *a fortiori* to any other extension of this model, for example if one has in mind to incorporate ecological concerns into the problem. Also notice, as already put forward by Ayong le Kama and Schubert (2007), that integrating those concerns would imply an additional channel of endogenous discounting: indeed, the planner could choose to also discount with respect to the ecological state, assigning the maximal weight to the situations of ecological emergency.

Acknowledgements Authors acknowledge the support of the Belgian research programmes ARC on Sustainability, the financial support from the Spanish Ministerio de Ciencia e Innovación, Project ECO2013-48884-C3-1-P and from the Spanish Ministerio de Economía, Industria y Competitividad, Project ECO2016-76818-C3-3-P.

Appendix 1

According to the Mangasarian's Sufficiency Theorem the necessary conditions of the Maximum Principle for an optimum are also sufficient if the Hamiltonian function $H\left(\widetilde{c}, n, q, \widetilde{k}, \upsilon, \Delta\right)$ given in (15) is concave in $\left(\widetilde{c}, n, \widetilde{k}, \Delta\right)$ jointly, under the proviso that the transversality conditions (23) and (24) hold. Here, the Hessian matrix associated to the Hamiltonian function may be written as follows

$$H_{essian} = (1-\Phi)\, H_{\Delta\Delta} \begin{pmatrix} \frac{-\Phi}{\widetilde{c}^2} & \frac{\epsilon(1-\Phi)}{\widetilde{c}n} & 0 & \frac{-1}{\widetilde{c}} \\ \frac{\epsilon(1-\Phi)}{\widetilde{c}n} & \frac{\epsilon(\epsilon(1-\Phi)-1)}{n^2} & \frac{-(1+b)}{\widetilde{c}} & \frac{-\epsilon}{n} \\ 0 & \frac{-(1+b)}{\widetilde{c}} & \frac{-(1-\beta)\beta}{\widetilde{c}\widetilde{k}^{2-\beta}} & 0 \\ \frac{-1}{\widetilde{c}} & \frac{-\epsilon}{n} & 0 & \frac{1}{1-\Phi} \end{pmatrix}, \tag{63}$$

where $(1-\Phi)\, H_{\Delta\Delta} = \widetilde{c}^{1-\Phi} n^{\epsilon(1-\Phi)} A_0^{1-\Phi} N_0^{\lambda} e^{-\Delta}$ is nonnegative.

A necessary and sufficient condition for $H\left(\widetilde{c}, n, q, \widetilde{k}, \upsilon, \Delta\right)$ to be concave in $\left(\widetilde{c}, n, \widetilde{k}, \Delta\right)$ is that the Hessian matrix is negative semidefinite. Moreover, a necessary and sufficient condition for a negative semidefinite H_{essian} is that the sign of the determinants known as principal minors accommodate to the following sequence: $\widetilde{D}_1 \leqslant 0$, $\widetilde{D}_2 \geqslant 0$, $\widetilde{D}_3 \leqslant 0$, and $\widetilde{D}_4 \equiv |H_{essian}| \geqslant 0$.

Given the parameter constraints assumed in this model, in particular $\Phi > 1$ and $1 > \epsilon(1-\Phi)$, it is easy to show that the required concavity conditions on the Hamiltonian function are satisfied if

$$\frac{\widetilde{c}}{\overline{k}} \frac{\widetilde{k}^{\beta}}{\widetilde{k}} \frac{1}{n^2} \geqslant \frac{(1+b)^2}{\epsilon(1-\beta)\beta}. \tag{64}$$

This condition imposes a stronger requirement on the degree of concavity of the production function. The above inequality, given that $f\left(\widetilde{k}\right) = \widetilde{k}^{\beta}$, may be rewritten as

$$f''\left(\widetilde{k}\right) \leqslant -\frac{(1+b)^2}{\epsilon}\frac{n^2}{\widetilde{c}} < 0. \tag{65}$$

In particular, given (41), (42) and (58), a sufficient condition for the required concavity condition (64) to be satisfied is that $1 > \beta\epsilon$, which always holds because of the assumed parameter configuration of the model.

Appendix 2: Volterra's Derivatives

In what follows we adapt to our model the analysis from Pittel (2002), appendix to Chap. 5. Recall that the particular instantaneous utility function is of the form

$$U\left(c\left(t\right), n\left(t\right)\right) = \frac{c\left(t\right)^{1-\Phi} n\left(t\right)^{\epsilon(1-\Phi)}}{1-\Phi} \tag{66}$$

whereas the welfare function takes the following intertemporal form

$$W\left(\widetilde{c}\left(t\right), n\left(t\right)\right) = \int_{t_0}^{+\infty} \frac{\widetilde{c}\left(t\right)^{1-\Phi} n\left(t\right)^{\epsilon(1-\Phi)} A_0^{1-\Phi} N_0^{\lambda} e^{-\Delta(t)}}{1-\Phi} dt \tag{67}$$

$$\Delta\left(t\right) = \int_{t_0}^{t} \Theta\left(n\left(s\right)\right) ds \tag{68}$$

$$\Theta\left(n\left(s\right)\right) = \rho - x\left(1-\Phi\right) - \lambda n\left(s\right) \tag{69}$$

Due to the structure of the exponential term, **intertemporal preferences are not time-additive**. Consequently, although the marginal utilities $U_c = c^{-\Phi} n^{\epsilon(1-\Phi)}$ and $U_n = \epsilon c^{1-\Phi} n^{\epsilon(1-\Phi)-1}$ in (66) represent the corresponding changes in utility at time t, with intertemporal preferences being **recursive** as in (67) we need the Volterra derivatives to determine the corresponding intertemporal marginal utilities. Changes in the determinants of W will have an impact on the current utility index but they can also affect the perception of future utility via the impact on the accumulated discount rates with which the future utility levels are discounted.

For the sake of simplicity we can write (67) in a more compact form

$$W = \int_{t_0}^{+\infty} F\left(\widetilde{c}\left(t\right), n\left(t\right)\right) e^{-\Delta(t)} dt = \int_{t_0}^{+\infty} F\left(\widetilde{c}\left(t\right), n\left(t\right)\right) \exp\left(-\int_{t_0}^{t} \Theta\left(n\left(s\right)\right) ds\right) dt \tag{70}$$

where

$$F\left(\widetilde{c}\left(t\right),n\left(t\right)\right)=\frac{\widetilde{c}\left(t\right)^{1-\Phi}n\left(t\right)^{\epsilon\left(1-\Phi\right)}A_{0}^{1-\Phi}N_{0}^{\lambda}}{1-\Phi} \tag{71}$$

On the other hand, from Eq. (26) we get

$$\begin{aligned}\upsilon\left(t\right)&=\int_{t}^{+\infty}-F\left(\widetilde{c}\left(\tau\right),n\left(\tau\right)\right)e^{-\Delta(\tau)}d\tau\\&=\int_{t}^{+\infty}-F\left(\widetilde{c}\left(\tau\right),n\left(\tau\right)\right)\exp\left(-\int_{t_0}^{\tau}\Theta\left(n\left(s\right)\right)ds\right)d\tau\end{aligned} \tag{72}$$

and

$$\begin{aligned}\upsilon\left(t\right)e^{\Delta(t)}&=\left(\int_{t}^{+\infty}-F\left(\widetilde{c}\left(\tau\right),n\left(\tau\right)\right)\exp\left(-\int_{t_0}^{\tau}\Theta\left(n\left(s\right)\right)ds\right)d\tau\right)\\&\quad\times\exp\left(\int_{t_0}^{t}\Theta\left(n\left(s\right)\right)ds\right)\end{aligned} \tag{73}$$

Then, we define the new variable

$$\omega\left(t\right)=-\upsilon\left(t\right)e^{\Delta(t)}=\int_{t}^{+\infty}F\left(\widetilde{c}\left(\tau\right),n\left(\tau\right)\right)\exp\left(-\int_{t}^{\tau}\Theta\left(n\left(s\right)\right)ds\right)d\tau \tag{74}$$

The Volterra derivative is used to determine the derivatives of the functional W near time t, which is supplied in (74).

Volterra derivative with respect to $\widetilde{c}$:

$$\frac{\partial F\left(\widetilde{c}\left(t\right),n\left(t\right)\right)}{\partial\widetilde{c}}\exp\left(-\int_{t_0}^{t}\Theta\left(n\left(s\right)\right)ds\right)=\widetilde{c}\left(t\right)^{-\Phi}n\left(t\right)^{\epsilon\left(1-\Phi\right)}A_{0}^{1-\Phi}N_{0}^{\lambda}e^{-\Delta(t)} \tag{75}$$

which corresponds to the right hand side of the first order condition (16).

Volterra derivative with respect to n:

$$\begin{aligned}&\left(\frac{\partial F\left(\widetilde{c}\left(t\right),n\left(t\right)\right)}{\partial n}-\frac{\partial\Theta\left(n\left(s\right)\right)}{\partial n}\omega\left(t\right)\right)\exp\left(-\int_{t_0}^{t}\Theta\left(n\left(s\right)\right)ds\right)=\\&\quad=\frac{\epsilon\widetilde{c}\left(t\right)^{1-\Phi}n\left(t\right)^{\epsilon\left(1-\Phi\right)}A_{0}^{1-\Phi}N_{0}^{\lambda}e^{-\Delta(t)}}{n\left(t\right)}-\lambda\upsilon\left(t\right)\end{aligned} \tag{76}$$

which corresponds to the right hand side of the first order condition (17).

References

K. Arrow, P. Dasgupta, L. Goulder, G. Daly et al., Are we consuming too much? J. Econ. Perspect. **18**, 147–172 (2004)

A. Ayong le Kama, K. Schubert, A note on the consequences of an endogenous discounting depending on the environmental quality. Macroecon. Dyn. **11**(2), 272–289 (2007)

R. Barro, G. Becker, Fertility choice in a model of economic growth. Econometrica **57**, 481–501 (1989)

R. Barro, X. Sala-i-Martin, *Economic Growth* (MIT Press, Boston, 2004)

D. Blanchet, Age structure and capital dilution effects in neoclassical growth models. J. Popul. Econ. **1**(3), 183–194 (1988)

R. Boucekkine, G. Fabbri, Assessing Parfit's repugnant conclusion within a canonical endogenous growth set-up. J. Popul. Econ. **26**, 751–767 (2013)

R. Boucekkine, B. Martínez, J.R. Ruiz-Tamarit, Note on global dynamics and imbalance effects in the Lucas-Uzawa model. Int. J. Econ. Theory **4**, 503–518 (2008)

R. Boucekkine, G. Fabbri, G. Gozzi, Egalitarianism under population change: age structure does matter. J. Math. Econ. **55**, 86–100 (2014a)

R. Boucekkine, B. Martínez, J.R. Ruiz-Tamarit, Optimal sustainable policies under pollution ceiling: the demographic side. Math. Model. Nat. Phenom. **9**(4), 38–64 (2014b)

J. Caballé, M.S. Santos, On endogenous growth with physical and human capital. J. Polit. Econ. **101**, 1042–1067 (1993)

E. Canton, L. Meijdam, Altruism and the macroeconomic effects of demographic changes. J. Popul. Econ. **10**(3), 317–334 (1997)

A.C. Chiang, *Elements of Dynamic Optimization* (McGraw-Hill, Singapore, 1992)

M.A. Constantinides, Optimal population growth and the social welfare function. East. Econ. J. **14**(2), 229–238 (1988)

P. Dasgupta, On the concept of optimum population. Rev. Econ. Stud. **36**, 295–318 (1969)

P. Dasgupta, Regarding optimum population. J. Polit. Philos. **13**, 414–442 (2005)

D. de la Croix, M. Doepke, Inequality and growth: why differential fertility matters. Am. Econ. Rev. **93**(4), 1091–1113 (2003)

F. Edgeworth, *Papers Relating to Political Economy*, vol. III (Macmillan, London, 1925)

J. Marin-Solano, J. Navas, Non-constant discounting in finite horizon: the free terminal time case. J. Econ. Dyn. Control **33**, 666–675 (2009)

A. Nairay, Asymptotic behavior and optimal properties of a consumption-investment model with variable time preference. J. Econ. Dyn. Control **7**(3), 283–313 (1984)

M. Nerlove, A. Razin, E. Sadka, Population size: individual choice and social optima. Q. J.Econ. **100**(2), 321–334 (1985)

M. Obstfeld, Intertemporal dependence, impatience, and dynamics. J. Monet. Econ. **26**, 45–75 (1990)

T. Palivos, C.K. Yip, Optimal population size and endogenous growth. Econ. Lett. **41**(1), 107–110 (1993)

Th. Palivos, P. Wang, J. Zhang, On the existence of balanced growth equilibrium. Int. Econ. Rev. **38**(1), 205–224 (1997)

D. Parfit, *Reasons and Persons* (Oxford University Press, Oxford, 1984)

K. Pittel, *Sustainability and Endogenous Growth* (Edward Elgar, Northampton, 2002)

A. Razin, C. Yuen, Utilitarian trade-off between population growth and income growth. J. Popul. Econ. **8**, 81–87 (1995)

I. Schumacher, Endogenous discounting and the domain of the felicity function. Econ. Model. **28**, 574–581 (2011)

A. Seierstad, K. Sydsaeter, *Optimal Control Theory with Economic Applications* (Elsevier Science, Amsterdam, 1987)

S.P. Sethi, G.L. Thompson *Optimal Control Theory. Applications to Management Science and Economics* (Springer, New York, 2000)

Does Demography Change Wealth Inequality?

Miguel Sánchez-Romero, Stefan Wrzaczek, Alexia Prskawetz, and Gustav Feichtinger

Abstract In this article, we investigate the effect of demography on wealth inequality. We propose an economic growth model with overlapping generations in which individuals are altruistic towards their children and differ with respect to the age of their parent. We denote the age gap between the parent and their child as generational gap. The introduction of the generational gap allows us to analyze wealth inequality not only across cohorts but also within cohorts. Our model predicts that a decline in fertility raises wealth inequality within cohorts and, simultaneously, it reduces inequality at the population level (across cohorts). In contrast, increases in life expectancy result in a non-monotonic effect on wealth inequality by age and across cohorts.

1 Introduction

The share of total wealth owned by the top 10% of the population has increased during the last decades in Britain, France, Sweden, and US (Piketty 2014), among many other countries. Inequality, in general, and wealth inequality, in particular, have become of main concern among policymakers and researchers, given that it can create political instability and prevent long-run economic growth (Alesina and Perotti 1996).

According to Piketty demography is one of the most important factors explaining wealth inequality. Specifically, demography can influence wealth inequality

M. Sánchez-Romero
Wittgenstein Centre for Demography and Global Human Capital (IIASA, VID/ÖAW, WU), Vienna Institute of Demography/Austrian Academy of Sciences, Vienna, Austria

S. Wrzaczek (✉) · A. Prskawetz · G. Feichtinger
Wittgenstein Centre for Demography and Global Human Capital (IIASA, VID/ÖAW, WU), Vienna Institute of Demography/Austrian Academy of Sciences, Vienna, Austria

Vienna University of Technology (TU Wien), Vienna, Austria
e-mail: stefan.wrzaczek@oeaw.ac.at

G. Feichtinger et al. (eds.), *Control Systems and Mathematical Methods in Economics*, Lecture Notes in Economics and Mathematical Systems 687,
https://doi.org/10.1007/978-3-319-75169-6_17

through: (1) changes in fertility, (2) changes in longevity, and (3) shifts in the age structure of the population. The first factors are related to the evolution of savings over the life cycle, the capital intensity of the economy (Bloom and Williamson 1998), and inheritances. Inherited wealth is frequently associated to the privileged and the lack of meritocracy, which prevents innovation (Piketty 2014). However, the effect of demography on inheritances (or bequeathed wealth) is not straightforward. On the one hand, if people live longer, wealth will be based largely on human capital (rather than on inheritances) and hence population aging will reduce wealth inequality. On the other hand, as rising longevity increases life cycle savings, wealth will be more unequally distributed across age groups and hence population aging will increase wealth inequality (Goldstein and Lee 2014). Also, the impact of fertility on wealth inequality is a priori ambiguous. If people have less children, the share of young individuals with low savings will fall and thus wealth inequality will be reduced (Vandenbroucke 2016). However, if people have less children, freeing up household's resources, savings will increase and, consequently, wealth inequality will rise.

Given the myriad of offsetting effects, in this article, we are interested in disentangling the effect that demography has on wealth inequality through bequests. In particular, should we expect greater or lower wealth inequality in the face of population aging? What is the effect of demography on intergenerational wealth transfers and how will these transfers change wealth inequality? Answering these two questions requires a well-founded economic and demographic model. However, many models give more importance to the economic assumptions and oversimplify the demographic assumptions. Indeed, in many economic models the population is divided into a finite number of age groups, typically only two, and childhood is either neglected or assumed to be optimally chosen by children according to their own budget constraint. As a consequence, much of the age variation that occurs in real life is neglected and, therefore, many of the results, which often depend strongly on such assumptions, can be flawed.

To improve our understanding of the effect of demography on wealth inequality, we introduce more realistic demographic features in a standard continuous-time overlapping generations (OLG) models. Specifically, we characterize individuals not only by their age and time (cohorts) dimensions, but also through the age gap between parents and children (from now on generational gap). The introduction of the generational gap, as a third dimension, helps to better analyze wealth inequality, since we consider explicitly the existence of different types of individuals that differ according to the time and quantity of inheritance received. As recently acknowledged by Alvaredo et al. (2017), this is an important feature for understanding the aggregate wealth accumulation process, given that it avoids the unrealistic results of the representative agent approach (Kotlikoff and Summers 1981; Kotlikoff 1988; Modigliani 1988).[1] As far as we know, this is the first time that heterogeneity by

[1] Alvaredo et al. (2017) cope with the representative agent approach by assuming a dual population model with 'savers' and 'rentiers'.

generational gap is introduced into a theoretical OLG model. Further, following Tobin (1967) and Lee (1980), we consider that individuals face a probabilistic lifespan and two distinguishable periods over their life cycle. A first period of dependency, or childhood, in which individuals rely on the consumption decisions made by parents and a second period, or adulthood, in which individuals build up their own household, have children, determine the consumption of the household, and save. To compare our results to the recent paper by Onder and Pestieau (2016), we also assume individuals save for retirement motives and for the joy of giving bequest. In line with empirical findings, we consider parents are altruistic and maximize the bequest per capita left to each child (see section 2.2.3 in Arrondel and Masson 2006). To keep the model computationally tractable, while maximizing the available demographic information, we assume each household head represents the average individual within a cohort with a given generational gap.[2] Thus, under a perfect annuity market and the joy of bequeathing, the fraction of wealth annuitized by each individual will be associated to the probability of having children.[3] As the macroeconomic framework we assume a small-open economy in which wage rates and interest rates are determined in international markets.

We should stress that in this paper we abstract from other important channels (besides fertility, mortality, and the age structure) through which demography may affect inequality. In particular, wealth inequality can also be driven by differences in the population composition; e.g., gender. For instance, Greenwood et al. (2014) show for the US that the increasing positive assortative mating together with the increasing labor force participation of women explains the increase in household income inequality from 1960 to 2005. Another important source of increasing inequality is the rising proportion of single parenthood, whose children generally face later in life a disadvantaged position (Chetty et al. 2016). However, moving from a one-sex model to a two-sex model requires many additional assumptions and complex interactions that would prevent us to obtain unique results. For this reason we opted for excluding population compositional effects. We restrict our analysis on wealth inequality and its relation to demographic factors. We therefore exclude the effect of labor income inequality and the effect of pension systems on our decision variables, which trigger inequality when there are compositional changes in the population (Greenwood et al. 2014; Chetty et al. 2016).

Given the non linear properties of our optimality conditions, we cannot analytically solve the model and we need to rely on numerical simulations. Our results are based on the following principle: individuals whose parents are older receive their bequest earlier and capitalize it over a longer period of time. Thus, demography causes wealth inequality (within cohorts) because individuals with a greater generational gap, or age gap between parents and children, have a higher

[2]A more realistic model, but also more complex and computationally burdensome, would be to assume a stochastic setting in which each individual also faces a probabilistic family size.

[3]At the individual level, this result is equivalent to say that only childless individuals annuitize their wealth, which is consistent with Yaari (1965) framework.

financial wealth over the life cycle. Moreover, wealth inequality diminishes with age given that wealthier individuals also have higher consumption.

The results are divided into the effect that demography has on wealth inequality at the individual level (within cohorts) and at the population level. At the individual level, our simulations suggest that a decline in fertility is associated to an increase in wealth inequality within each age group, while an increase in life expectancy has a non monotonic effect on wealth inequality at each age. At the population level, however, our simulations suggest that a decline in fertility reduces wealth inequality, while an increase in life expectancy has a small non monotonic effect on wealth inequality.

The paper is organized as follows: Sect. 2 introduces the demographic characteristics of each household, the different sources of wealth accumulation, and the preferences of the household head. In Sect. 3, we analytically solve the model and provide the intuition for wealth inequality. Section 4 introduces the population model and the aggregate physical capital wealth of the economy. In Sect. 5, we solve the model numerically and show how population aging reduces wealth inequality. We conclude and give suggestions for further research in Sect. 6.

2 Model

2.1 Households

The approach used here to represent the optimal household head's decisions over the life cycle borrows from the model proposed by Yaari (1965). We extend this model in two dimensions. First, by taking into consideration the average number of children that a cohort bears over the life cycle and, second, by introducing an additional time dimension in the analysis: the age difference between the parent and the child or generational gap. We denote with the letter l the generational gap.

Individuals face death and a changing family size along their life span based on age-specific mortality and fertility rates, which vary across birth cohorts. Let us denote age by x and the year of birth by τ. We represent the individual lifetime uncertainty by the survival function

$$S(x,\tau) = \exp\left\{-\int_0^x \mu(a,\tau)da\right\}, \tag{1}$$

where $S(x,\tau)$ is the probability that an individual born in year τ survives to age x. The probability of surviving satisfies that $S(0,\tau) = 1$ and $S(\omega,\tau) = 0$, where ω is the maximum age, and $\mu(a,\tau) \geq 0$ is the mortality hazard rate at age a by the same individual. To account for the fact that households are comprised of a household head and dependent children, while household heads may also have non dependent children living in a different household, we build the following two demographic measures:

$$n(x,\tau) = \int_0^x S(x-l,\tau+l)m(l,\tau)dl, \qquad \text{(children/heirs)} \tag{2}$$

$$h(x,\tau) = 1 + \int_{x-A}^x \delta(x-l)\frac{S(l,\tau)}{S(x,\tau)}m(l,\tau)S(x-l,\tau+l)dl, \quad \text{(household size)} \tag{3}$$

where $n(x,\tau)$ represents the total number of heirs, or offspring, of an individual born in year τ at age x. $h(x,\tau)$ is the size of the household measured in terms of equivalent adult consumers. The first term after the integral sign in Eq. (3), $\delta(x-l)$, reflects the equivalent scale of a child at age $x-l$ relative to the consumption of the household head. Notice that due to the existence of mortality risk households are formed not only by offspring but also by orphans. For simplicity, we assume orphans are raised by surviving household heads belonging to the same cohort, which is being taken into account through the fraction $\frac{S(l,\tau)}{S(x,\tau)}$.[4] The function $m(l,\tau)$ is the fertility rate of an individual born in year τ at age l and $S(x-l,\tau+l)$ is the survival probability of a child born in year $\tau+l$ to age $x-l$. Letter A represents the age at which individuals become independent and build up their own household.

Suppose there are three sources of wealth accumulation. Let us denote by $k(x,\tau,l)$ the wealth held by an individual born in year τ at age x whose parent is l years older. The first is the revaluation of the existing wealth via the sum of interests received on existing wealth, $rk(x,\tau,l)$, plus the mortality risk premiums received from annuitizing a fraction $\theta(x,\tau)$ of the total wealth, $\theta(x,\tau)\mu(x,\tau)k(x,\tau,l)$. Given a *laissez-faire* economy, we assume for consistency that $\theta(x,\tau)$ reflects the fraction of individuals at age x, born in year τ, who do not have children. This is equivalent to say that the fraction of individuals who have children, $1-\theta(x,\tau)$, transfer all their wealth to their heirs, whereas the fraction of individuals who do not have children, $\theta(x,\tau)$, fully annuitize their wealth. From Eq. (2) the probability of an individual born in year τ of not having children at age x is[5]

$$\theta(x,\tau) = \exp\{-n(x,\tau)\}. \tag{4}$$

[4] $\frac{S(x,\tau)}{S(l,\tau)}$ denotes the probability of individuals to survive up to age x given that they have been alive at l (at the age of childbearing). The reciprocal value thus divides the orphans to the surviving individuals of the same age-group. Note that $\frac{S(l,\tau)}{S(x,\tau)} = e^{\int_l^x \mu(a,\tau)\,da} \geq 1$ for $l \in [x-A,x]$.

[5] For the derivation of (4) note that the fraction of x-year old individuals without children is reduced at age t ($t \leq x$) by the fertility rate times the corresponding probability that the children survive up to age $x-t$ (corresponds to age x of the individual). These dynamics can be formalized as $\dot{\theta}(t,\tau) = -\theta(t,\tau)m(t,\tau)S(x-t,\tau)$ with $t \in [0,x]$ and $\theta(0,\tau) = 1$. The solution yields (4).

The second source of wealth accumulation is the bequest received from the death of the parent, which we denote by $B(x, \tau, l)$. We assume parents split their wealth among their surviving offspring equally.[6] The amount of wealth received as a bequest by an individual born in year τ at age x from a parent who is l years older is

$$B(x, \tau, l) = \mu(x + l, \tau - l)\frac{S(x + l, \tau - l)}{S(l, \tau - l)}k(x + l, \tau - l)\eta(x + l, \tau - l) \text{ for } x \leq \omega - l. \quad (5)$$

The first two terms on the right-hand side of Eq. (5) denote the probability that the parent dies at age $x + l$. The third term is the amount of wealth left by the parent at death, $k(x + l, \tau - l)$, and the last term is the fraction of wealth that corresponds to each offspring, which is given by

$$\eta(x + l, \tau - l) = \frac{1 - \theta(x + l, \tau - l)}{n(x + l, \tau - l)}. \quad (6)$$

The numerator of (6) is the probability that an individual born in year $\tau - l$ at age $x + l$ has at least one child (i.e. $1 - \theta$) and the denominator reflects the total number of living offspring (or siblings). For illustration, Fig. 1 shows the fraction of wealth annuitized as a function of the number of children within the cohort (θ) and the share of wealth that an individual receives as function of the number of offspring from the parent (η). From Fig. 1 we can see that an individual belonging to a cohort with an average of two offspring per adult ($n = 2$) annuitizes around 13.5% of her wealth and leaves around 43% of her wealth to each offspring. If an individual belongs to

Fig. 1 Fraction of annuitized wealth (θ) and fraction of wealth received according to the number of children within the cohort (η)

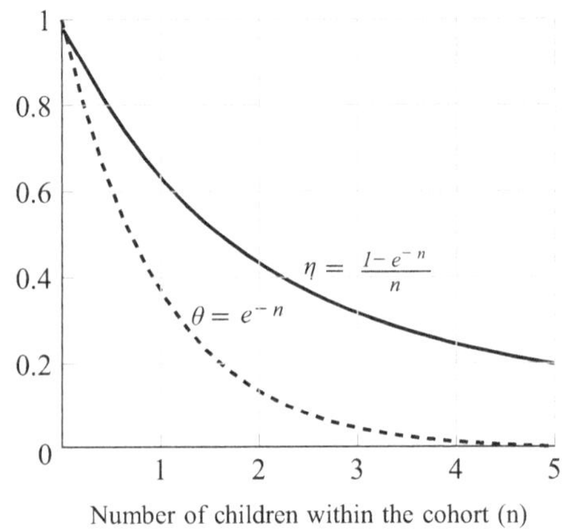

[6]If the bequest is received before reaching the age A, we assume the household head raising the orphan commits to invest the inheritance and to pass it on to the orphan once the child reaches the age A.

a cohort with an average of four offspring per adult ($n = 4$), she annuitizes around 2% of her wealth and leaves close to 25% of her wealth per child.[7]

Comparing (4) to (6), and as illustrated in Fig. 1, it is worth stressing that for any number of children $n > 0$ the fraction of wealth received (η) is always greater than the fraction of wealth annuitized (θ).

The third source of wealth accumulation is savings out of labor income. We denote by $y(x, \tau)$ the labor income generated by the household head born in year τ at age x. The total consumption of a household run by a household head born in year τ at age x, whose parent is l years older, is denoted by $c(x, \tau, l)$. Let the labor income of an individual born in year τ at age x be given by $\Gamma(\tau + x)y(x)$, where $\Gamma(\tau + x)$ is the labor-augmenting technological progress and $y(x)$ is the labor income profile which has an invariant shape over time with respect to age.

Adding all three sources of wealth accumulation we can represent the average change in wealth of a cohort born in year τ at age x, whose parent is l years older, as follows:

$$\frac{\partial k(x, \tau, l)}{\partial x} = \begin{cases} [r + \theta(x, \tau)\mu(x, \tau)]k(x, \tau, l) + B(x, \tau, l) & \text{for } x < A, \\ \\ [r + \theta(x, \tau)\mu(x, \tau)]k(x, \tau, l) + B(x, \tau, l) & \\ +y(x, \tau) - c(x, \tau, l) & \text{for } x \geq A. \end{cases} \tag{7}$$

The first period in Eq. (7) corresponds to the accumulation of the average wealth before age A, while the second period represents the accumulation of the average wealth during adulthood. As it is standard, we assume individuals are born with zero wealth, i.e. $k(0, \tau, l) = 0$.

2.2 Preferences

Suppose household heads have additively separable preferences that are represented by isoelastic functions U (that satisfy the Inada conditions: $U' > 0$, $U'' < 0$, with U being continuously differentiable, $U'(0) = \infty$, and $U'(\infty) = 0$) of their own consumption and the average bequest left at death to each child. Assuming no subjective discounting, the expected utility of a household head born in year τ, whose parent is l years older (*generational gap*), is

$$EU(c) = \int_A^{\omega} \frac{S(x, \tau)}{S(A, \tau)} \left\{ U\left(\frac{c(x, \tau, l)}{h(x, \tau)}\right) + \alpha\mu(x, \tau)U\left(\eta(x, \tau)k(x, \tau, l)\right) \right\} dx. \tag{8}$$

[7]One should notice that by assuming a one-sex model, we are forced to unrealistically double the proportion of wealth annuitized. However, this is not the case for the fraction of wealth received, since the higher number of siblings, as a result of using a two-sex model, is offset by the fact that individuals receive the bequest from two parents.

Note that Eq. (8) extends Yaari (1965) expected utility by introducing the household size and the number of offspring. Parameter $\alpha \geq 0$ controls for the degree of altruism towards the children; that is, the subjective weighting for giving bequest. The second term inside the brackets represents the fact that the household head not only takes into account the amount of wealth bequeathed to each offspring, $\eta(x,\tau)k(x,\tau,l)$, but also the time at which the bequest is given, $\frac{S(x,\tau)}{S(A,\tau)}\mu(x,\tau)$. This last demographic function is the probability of dying at age x, conditional on having survived to age A, for the cohort born in year τ.

It is worth noting that the instantaneous utility functions used for consumption and for bequest are the same in (8). This specification guarantees the existence of a stable consumption path once that we consider an exogenous productivity growth in a general equilibrium model.

3 Analytical Solution

The consumption path c that maximizes the expected utility (8) subject to the constraint (7) is the one that solves the Hamiltonian[8]

$$\mathcal{H}(k,c,\lambda,x) = \tilde{S}U(c/h) + \alpha\mu\tilde{S}U(\eta k) + \lambda([r+\theta\mu]k + B + y - c), \tag{9}$$

where λ is the adjoint variable related to k and $\tilde{S}$ denotes the probability of survival conditional on being alive at age A. We obtain the following first order condition (FOC)

$$\mathcal{H}_c = \tilde{S}[h]^{-1}U'(c/h) - \lambda \overset{!}{=} 0. \tag{10}$$

Equation (10) determines the optimal consumption path. The dynamics of the adjoint variable is given by

$$\frac{\partial\lambda}{\partial x} = -[r+\theta\mu]\lambda - \alpha\mu\tilde{S}\eta U'(\eta k). \tag{11}$$

Suppose from now on $U(c) = \log(c)$. Then, the optimal consumption path can be characterized by the system of equations:

$$\frac{\partial\lambda}{\partial x} = -[r+\theta\mu]\lambda - \alpha\mu\tilde{S}/k, \tag{12a}$$

$$\frac{\partial k}{\partial x} = [r+\theta\mu]k + B + y - \tilde{S}/\lambda, \tag{12b}$$

[8]Every pair $(c(\cdot),k(\cdot))$ that fulfills the necessary optimality conditions (10)–(11) are a unique optimal solution of the household problem (9), since the Mangasarian sufficiency conditions (see Theorem 3.29 in Grass et al. 2008) are fulfilled. Note that the discontinuity of the dynamics of k at A is no contradiction since the time horizon of the household is $[A,\omega]$. The behaviour during the period $[0,A)$is assumed to be determined by the parents.

and the boundary conditions

$$k(0, \tau, l) = 0 \text{ and } k(\omega, \tau, l) = 0. \tag{12c}$$

The above system of equations is clearly non-linear. Nevertheless, analyzing the dynamics of the average consumption per adult we can give some insight about the implications of introducing children into the model. Differentiating (10) with respect to x and using (11) we obtain the dynamics of consumption

$$\frac{1}{c}\frac{\partial c}{\partial x} = r + \mu \left\{ \alpha \frac{c}{k} - (1-\theta) \right\}. \tag{13}$$

The first term in the right-hand side of (13) is the standard Euler condition when individuals purchase actuarially fair annuities and there is no subjective discount factor. The second term reflects the degree of impatience that arises from having children, whose size depends on the degree of altruism towards the children. For a positive second term, household heads become more patient. This is because household heads reduce their consumption in the present and postpone it to later periods. However, if the second term is negative, then household heads become more impatient.

Except for the trivial case that $\alpha = 0$, a priori, nothing guarantees that the second term in the right-hand side of (13) will be either positive or negative. Indeed, given that the set of variables $\{c, k, n\}$ evolve over the life cycle, we can expect periods in which household heads are patient and periods in which household heads are impatient. To illustrate this point, we show in Fig. 2 the degree of impatience at a given age x as a function of the number of children and the value of $\alpha \frac{c}{k}$, which is a weighted measure of the consumption to capital ratio. The negative relationship between the ratio $\frac{c}{k}$ and the number of children, n, is embedded in the expected utility. Indeed, we have from (8) that individuals maximize the value function

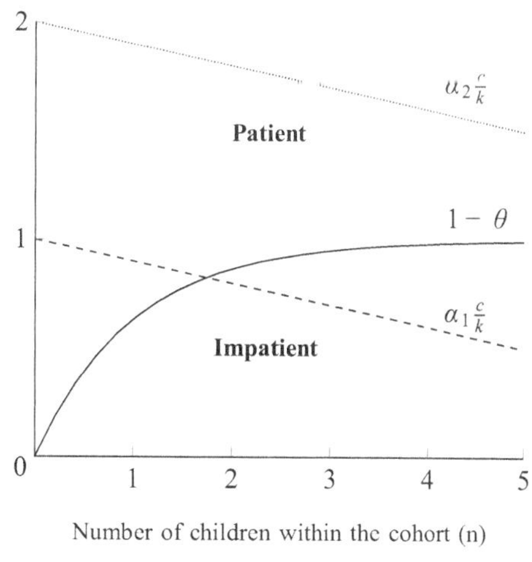

Fig. 2 Degree of impatience by number of children within the cohort

according to the bequest received per offspring. Thereby, an increase in the number of children, n, leads to a fall in consumption, and hence an increase in savings, in order to maintain the same level of bequest per offspring. For any value of $\alpha\frac{c}{k}$ above $1-\theta$ individuals become more patient (see the white area). In contrast, values of $\alpha\frac{c}{k}$ below $1-\theta$ imply that individuals become impatient (see the gray area).

Bequest also affects household head's behavior—i.e. consumption and saving—through a wealth effect. The wealth effect refers to changes in consumption due to changes in lifetime income. Assuming a fixed labor income profile for all individuals, then consumption and savings heterogeneity within a birth cohort can only be explained by the difference between the bequest given and the bequest received. This is clearly seen analyzing the lifetime budget constraint at birth. From (7) and using the boundary conditions (12c), the lifetime budget constraint of an individual born in year τ whose parent is l years older is[9]

$$\int_A^\omega e^{-rx} S(x,\tau)c(x,\tau,l)dx = \int_A^\omega e^{-rx} S(x,\tau)y(x,\tau)dx + T_B(0,\tau,l), \tag{14}$$

where $T_B(0,\tau,l)$ is the *bequest wealth* at birth; that is, the expected present value at birth of the difference between the bequest received and the bequest given

$$T_B(0,\tau,l) = \int_0^\omega e^{-rx} S(x,\tau)\left\{B(x,\tau,l) - \mu(x,\tau)[1-\theta(x,\tau)]k(x,\tau,l)\right\}dx. \tag{15}$$

A positive (resp. negative) $T_B(0,\tau,l)$ implies that individuals will not only consume more (resp. less) than they expect to earn over their lifetime from work, but they will also start at age A with a higher wealth. Multiplying (15) by $e^{rA}/S(A,\tau)$ gives

$$k(A,\tau,l) = \frac{e^{rA}}{S(A,\tau)} T_B(0,\tau,l) = T_B(A,\tau,l). \tag{16}$$

Therefore, wealth inequality at age A is explained by differences in bequest wealth. Moreover, under the assumption that all individuals born at time t face the same fertility and mortality rates, differences in bequest wealth at birth are explained by the generational gap. In order to clearly see whether a greater generational gap positively impacts on bequest wealth, we differentiate $B(x,l)$ with respect to l, see Eq. (5). Note that for expositional clarity we get rid of the time of birth. Thus, we have

$$\frac{\partial B(x,l)}{\partial l} = B(x,l)\left(\frac{\frac{\partial\mu(x+l)}{\partial l}}{\mu(x+l)} - [\mu(x+l)-\mu(l)] + \frac{\frac{\partial k(x+l)}{\partial l}}{k(x+l)} + \frac{\frac{\partial\eta(x+l)}{\partial l}}{\eta(x+l)}\right). \tag{17}$$

[9]In order to get (14) it is necessary to add and subtract $\mu(x,\tau)k(x,\tau,l)$ in (7).

According to (17) the impact on the bequest received of an increase in the generational gap depends on three components: (1) the evolution of the mortality rate of parents; (2) the rate of change in the financial wealth profile of parents; and (3) the rate of change in the share of wealth that each surviving heir receives. Note that the first component and the third component are frequently neglected in models that assume a constant mortality hazard rate and a fixed number of heirs. First, given that the mortality rate is an increasing function with respect to age when individuals are adults, the sum of the first three components inside the parenthesis is initially positive and it turns negative as the individual ages.[10] As a consequence, ceteris paribus (2) and (3), bequest received initially rises with an increase in the generational gap, since the senescence rate is greater than the probability of dying of the parent, and then it falls due to the increasing mortality rate of the parent. With respect to the second term in (17) that reflects the rate of change in the financial wealth profile, we know from the life cycle theory of saving that the financial wealth will be hump shaped with age. Therefore, an increase in the generational gap will reduce the effect of the rate of change in the financial wealth on bequest. And third, we have from (4) and (6) that the sign of the rate of change in the share of wealth that each surviving heir receives is equal to the sign of the inverse of the rate of change in the total number of heirs

$$\operatorname{sign}\left[\frac{\frac{\partial \eta(x+l)}{\partial l}}{\eta(x+l)}\right] = -\operatorname{sign}\left[\frac{\frac{\partial n(x+l)}{\partial l}}{n(x+l)}\right]$$

$$= \operatorname{sign}\left[\int_0^{x+l} \mu(x+l-a)\frac{S(x+l-a)m(a)}{\int_0^{x+l} S(x+l-z)m(z)dz}da - \frac{m(x+l)}{n(x+l)}\right]. \tag{18}$$

Equation (18) implies that when the number of heirs starts declining (resp. increasing) a greater generational gap l will have a positive (resp. negative) impact on the bequest received. Hence, when n increases (18) will decrease, whereas when later in life n does not change, the mortality effect (as represented in the first term of Eq. (18)) dominates and thereby the rate of change in the share of wealth that each surviving heir receives increases.

To illustrate the combined effect of the three components in (17), or the effect of increasing the generational gap on bequest, Fig. 3 shows the average bequest received over the life cycle of two representative individuals, that belong to the same birth cohort but differ with respect to the age of the parent, who are 22 and 45 years

[10] Assuming a Gompertz-Makeham law of mortality, with $\mu(x) = ae^{bx} + c$ where $a, b, c > 0$, the sum of the first three components inside the parenthesis of (17), denoted by $f(x)$, is approximately equal to $b - ae^{bl}(e^{bx} - 1)$. Thus, we have that $f(x) > 0$ for $x < x_0$ and $f(x) \leq 0$ for $x \geq x_0$, with $x_0 > 0$.

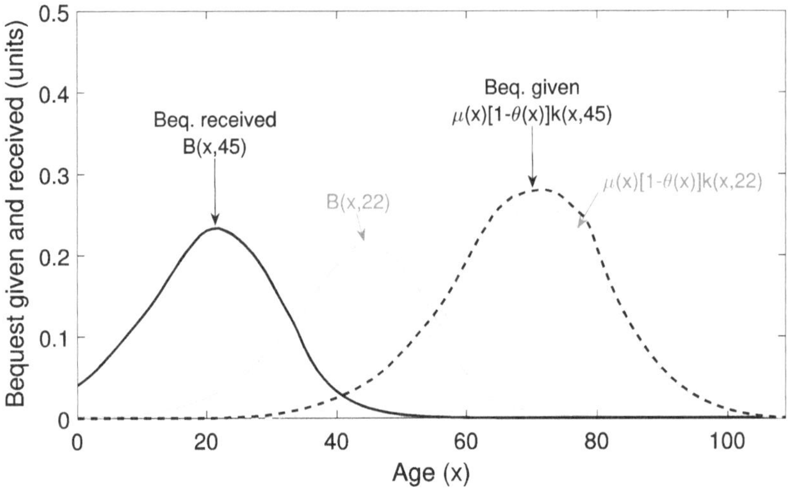

Fig. 3 Per capita bequest given (dashed) and received (solid) by generational gap. *Notes*: Units relative to the average labor income between ages 30 and 49. The first term in $B(x, l)$ and $k(x, l)$ denotes the age of the individual, x, while the second term denotes the generational gap, l. Both bequest profiles are derived using an annual interest rate of 3%, and fertility and mortality rates with an average TFR of 2.5 and a life expectancy of 65 years

older, respectively. Comparing both bequest profiles, we observe that individuals with older parents receive their bequest earlier. This demographic characteristic is key for explaining wealth inequality since those individuals who receive their bequest at young ages can capitalize it over a longer period of time. Moreover, the area below each bequest profile is fairly similar. Thus, wealth differences between individuals belonging to the same birth cohort are driven by the age at which the bequest is received, and not by the amount of bequest left.

To see the impact on financial wealth of receiving the bequest early in life, Fig. 4 shows the bequest wealth and the financial wealth by generational gap. In Fig. 4a it can be seen how bequest wealth at age A rises as the generational gap increases. Given that the area below the bequest profiles in Fig. 3 are roughly equal, the positive relationship between the bequest wealth, see (15), and the generational gap is explained by two discounting factors: the survival probability $S(x)$ and the interest rate r. The higher the value of r and the lower the survival probability $S(x)$, the more the bequest is discounted, and hence the lower the bequest wealth. Figure 4b shows how the financial wealth rises with increases in the generational gap between ages 25 and 65. The positive relationship between financial wealth and the generational gap persists over the life cycle, although the difference diminishes as individuals age.

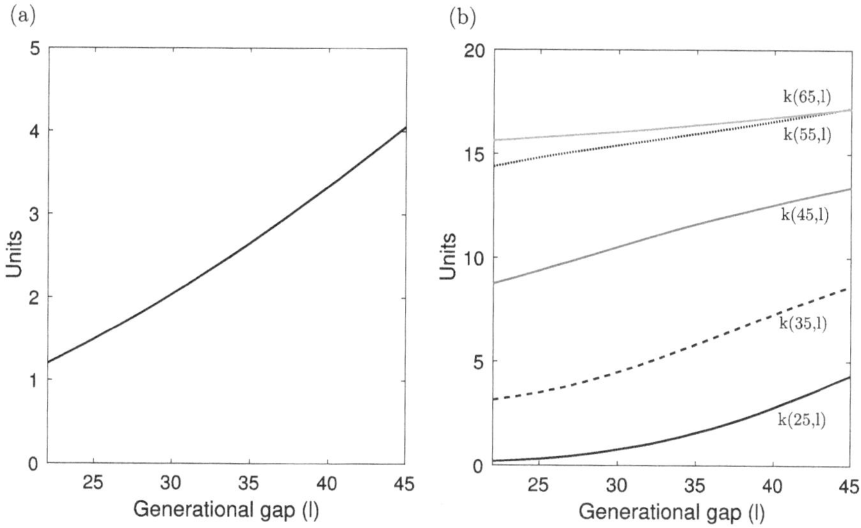

Fig. 4 Impact of the generational gap on bequest wealth and financial wealth. (**a**) Bequest wealth, $T_B(A,l)$. (**b**) Financial wealth, $k(x,l)$. *Notes*: Units relative to the average labor income between ages 30 and 49. First age of making decisions, A, is set at 22. Like Fig. 3 all profiles are derived using an annual interest rate of 3%, age-specific fertility rates with an average TFR of 2.5, and mortality rates with a life expectancy of 65 years

4 Aggregation

Before aggregating the optimal life cycle decisions of our individuals and obtaining the aggregate measures, it is necessary to know the age distribution of the population. Under the assumption of a closed population, we will consider the same demographic information as explained at the household level (i.e. age-specific fertility and mortality rates). We will also consider that all individuals born at time t face the same fertility and mortality rates.

4.1 Demography

Let us denote by $N(0,\tau,l)$ the total number of births at time τ born from a parent of age l. Hence, the number of people born at time τ, from a parent l years older, that survive to age $x = t - \tau$, with $t \geq \tau$, is

$$N(t-\tau,\tau,l) = S(t-\tau,\tau)N(0,\tau,l). \tag{19}$$

From (19) we can calculate the size of the population of age $x = t - \tau$ at time t as

$$N(t-\tau,\tau) = \int_0^{\omega} N(t-\tau,\tau,l)dl. \tag{20}$$

Using the age-specific fertility rate, we can then calculate the birth sequence from a parent of age $x = t - \tau$ at time t as

$$N(0, t, t - \tau) = m(t - \tau, \tau)N(t - \tau, \tau). \tag{21}$$

Integrating over all ages of parents, and using (20) and (21), the renewal equation becomes

$$\int_{t-\omega}^{t} N(0, t, t - \tau)d\tau = \int_{t-\omega}^{t} m(t - \tau, \tau)S(t - \tau, \tau)\left\{\int_{0}^{\omega} N(0, \tau, l)dl\right\}d\tau. \tag{22}$$

Note that if we substitute $\int_{t-\omega}^{t} N(0, t, t - \tau)d\tau$ for $N(0, t)$ and $\int_{0}^{\omega} N(0, \tau, l)dl$ for $N(0, \tau)$, Eq. (22) is the standard renewal equation.

Integrating over all birth cohorts and generational gaps, the size of the population at time t is

$$N(t) = \int_{t-\omega}^{t}\int_{0}^{\omega} N(t - \tau, \tau, l)dld\tau \tag{23}$$

and the dynamics of our population in year t is

$$\frac{dN(t)}{dt} = \int_{t-\omega}^{t}\left\{\left[m(t - \tau, \tau) - \mu(t - \tau, \tau)\right]\int_{0}^{\omega} N(t - \tau, \tau, l)dl\right\}d\tau, \tag{24}$$

which is also the standard balancing equation of a closed population.

4.2 Aggregate Physical Capital Wealth

To obtain the law of motion for aggregate physical capital, we define aggregate stock of physical capital at time t, $K(t)$, aggregate consumption at time t, $C(t)$, and aggregate labor income at time t, $Y_l(t)$, by integrating the financial wealth, consumption, and labor income over all ages (i.e. $x = t - \tau$) and generational gaps

$$K(t) = \int_{t-\omega}^{t}\int_{0}^{\omega} k(t - \tau, \tau, l)N(t - \tau, \tau, l)dld\tau, \tag{25}$$

$$C(t) = \int_{t-\omega}^{t-A}\int_{0}^{\omega} c(t - \tau, \tau, l)N(t - \tau, \tau, l)dld\tau, \tag{26}$$

$$Y_l(t) = \int_{t-\omega}^{t-A}\int_{0}^{\omega} y(t - \tau, \tau)N(t - \tau, \tau, l)dld\tau. \tag{27}$$

Differentiating (25) with respect to time yields

$$\frac{dK(t)}{dt} = \int_{t-\omega}^{t}\int_{0}^{\omega} \frac{\partial k(t-\tau,\tau,l)}{\partial t} N(t-\tau,\tau,l)dld\tau$$
$$+ \int_{t-\omega}^{t}\int_{0}^{\omega} k(t-\tau,\tau,l)\frac{\partial N(t-\tau,\tau,l)}{\partial t} dld\tau$$
$$+ \int_{0}^{\omega} k(0,\tau,l)N(0,\tau,l)dl - \int_{0}^{\omega} k(\omega,\tau,l)N(\omega,\tau,l)dl. \qquad (28)$$

From (7) and given the boundary conditions $k(0,\tau,l) = k(\omega,\tau,l) = 0$, it follows

$$\frac{dK(t)}{dt} = \int_{t-\omega}^{t}\int_{0}^{\omega} \big([r + \theta(t-\tau,\tau)\mu(t-\tau,\tau)]k(t-\tau,\tau,l)$$
$$+ B(t-\tau,\tau,l)\big)N(t-\tau,\tau,l)dld\tau$$
$$+ \int_{t-\omega}^{t-A}\int_{0}^{\omega} \big(y(t-\tau,\tau) - c(t-\tau,\tau,l)\big)N(t-\tau,\tau,l)dld\tau$$
$$- \int_{t-\omega}^{t} \mu(t-\tau,\tau)\int_{0}^{\omega} k(t-\tau,\tau,l)N(t-\tau,\tau,l)dld\tau. \qquad (29)$$

From (25)–(27) and after rearranging terms, we have

$$\frac{dK(t)}{dt} = rK(t) + Y_l(t) - C(t)$$
$$+ \int_{t-\omega}^{t}\int_{0}^{\omega} B(t-\tau,\tau,l)N(t-\tau,\tau,l)dld\tau$$
$$- \int_{t-\omega}^{t} \mu(t-\tau,\tau)[1-\theta(t-\tau,\tau)]\int_{0}^{\omega} k(t-\tau,\tau,l)N(t-\tau,\tau,l)dld\tau, \qquad (30)$$

where the sum of the last two components of (30), which represents respectively the total bequest received and the total bequest given, is equal to zero. See a detailed proof in appendix section "Total Bequest Given Equals Total Bequest Received". Note that (30) is the standard law of motion of the aggregate stock of capital in a closed economy. Hence, as it should be expected, the aggregate stock of capital increases over time when the total income generated in the economy—asset income plus labor income—exceeds total consumption.

5 Impact of Demography on Wealth Inequality

5.1 Data

Demographic Data In order to understand the impact of demography on wealth inequality, we will set up scenarios with different combinations of life expectancies (LEs) and total fertility rates (TFR). To have a clear result—without mixing changes in the age-shape with changes in levels—, we calculate the life expectancy and the total fertility rates using the same underlying vital rates as follows[11]

$$\mathrm{LE} = \int_0^{\omega} e^{-\int_0^x \beta_E \tilde{\mu}(a)da} dx, \quad (31)$$

$$\mathrm{TFR} = \int_0^{\omega} \beta_F \tilde{m}(x) dx, \quad (32)$$

where $\tilde{\mu}(x)$ and $\tilde{m}(x)$ are, respectively, the average mortality hazard rate and the average fertility rate at age x observed in the selected years 1900, 1925, 1950, 1975, and 2000 in the US; and where β_E and β_F are, respectively, scaling factors that adjust the mortality and fertility profiles so as to match the life expectancy to ages ranging from 50 to 90 years old, and the total fertility rate from 1.5 to 5 births per woman. The population size of each age group will be derived by applying (31) and (32) to the population model presented in Sect. 4.1.

Economic Data To complement our model with economic information, we assume individuals become independent and set their own household at the age of 22 (A) and live for a maximum of 110 years (ω). In order to generate in our model realistic age-specific saving and thus wealth profiles, we take the labor income per capita in the US in year 2003 from the National Transfer Accounts project (Lee and Mason 2011). Figure 5 shows the labor income profile normalized to the average labor income between ages 30–49. To have clear results about the effect of demography on the accumulation of capital, we assume the same labor income profile in all our simulations, i.e. $y(x, \tau) = \Gamma(\tau + x)y(x)$ with $\Gamma(\tau + x) = 1$ for all τ, x.

The interest rate, r, is set at 3%, which roughly corresponds in a Cobb-Douglas production function to a capital-output ratio of 4 with capital share of 0.33 and a depreciation rate of 5% (Piketty and Zucman 2014). Finally, we set the degree of altruism towards the children, α, at 1.5 in order to get an average capital-to-output

[11] Along the demographic transition age-specific fertility and mortality rates have not only decreased, but also changed in shape. Ceteris paribus, changes in the age shape of both demographic rates have an important impact on the generational gap (mean-age of childbearing) and the mortality variance. Hence, these two demographic processes also influence the bequest received and hence wealth inequality. However, given the limited space, we opt for investigating in this article only changes in level.

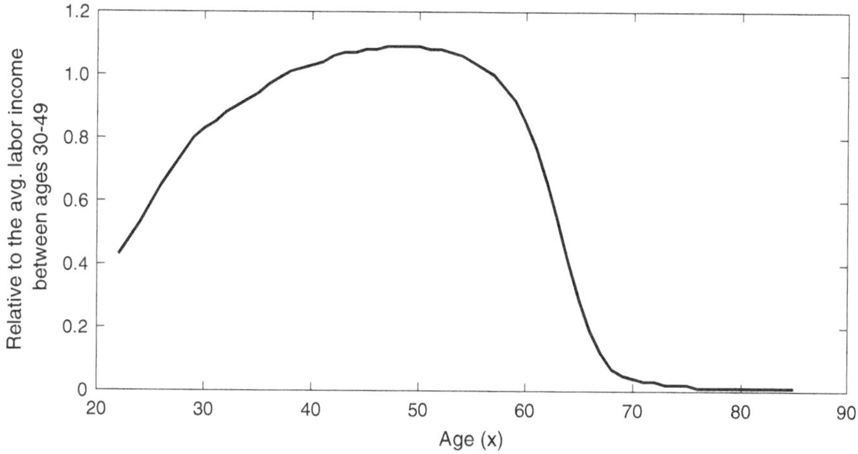

Fig. 5 Labor income per capita in USA, 2003. *Source*: www.ntaccounts.org

ratio of 4 in the median case scenario with a LE of 65 years and a TFR of 2.5 children per women.

5.2 *Wealth Inequality*

Our model works for any population structure and dynamics, however the results in this section are derived for a stable population. The model accounts for two sources of wealth inequality. At the cohort level, the inequality caused by the age at which individuals receive bequest and, at the aggregate level, the inequality caused by the age distribution of the population. First, we will focus on within cohort inequality and we will analyze the same effect across cohorts.

To analyze wealth inequality we will use the coefficient of variation

$$c[\boldsymbol{k}] = \frac{\sqrt{\mathrm{V}[\boldsymbol{k}]}}{\mathrm{E}[\boldsymbol{k}]}, \tag{33}$$

where $\mathrm{V}[\boldsymbol{k}]$ is the variance of financial wealth and $\mathrm{E}[\boldsymbol{k}]$ is the average financial wealth.

5.2.1 Within Cohorts

A novel characteristic of our model is the possibility of analyzing within cohort inequality driven by different fertility and mortality patterns. As we explained in Sect. 3, wealth inequality within the same birth cohort is caused by the fact that individuals receive their bequest at different ages (see Fig. 3). Children with older parents receive, on average, their bequest sooner than children with younger parents. As a consequence, the former group of children invest their inheritance over a longer period of time than the latter group, which has the potential to create large differences in wealth between individuals belonging to the same birth cohort.

To calculate wealth inequality within birth cohorts, we will use the coefficient of variation, $c_C[\boldsymbol{k}(x)] = \sqrt{\mathrm{V}_C[\boldsymbol{k}(x)]}/\,\mathrm{E}_C[\boldsymbol{k}(x)]$, using the following definitions:

$$\mathrm{V}_C[\boldsymbol{k}(x)] = \int_0^{\omega} \left(k(x,l) - \mathrm{E}_C[\boldsymbol{k}(x)]\right)^2 \frac{N(x,l)}{N(x)} dl, \tag{34}$$

$$\mathrm{E}_C[\boldsymbol{k}(x)] = \int_0^{\omega} k(x,l) \frac{N(x,l)}{N(x)} dl, \tag{35}$$

where $\mathrm{V}_C[\boldsymbol{k}(x)]$ is the variance of the financial wealth at age x and $\mathrm{E}_C[\boldsymbol{k}(x)]$ is the average financial wealth at age x.

From Sects. 2 and 3, we can identify four direct effects of an **increase in fertility** on the financial wealth profile and bequest. First, an increase in fertility lowers the bequest received given that the fraction of wealth that corresponds to each offspring falls (see Eqs. (5)–(6)). Second, since the proportion of childless individuals, that are assumed to fully annuitize their wealth, declines with the number of offspring, an increase in fertility is also accompanied with a reduction in the average interest on wealth due to the lower proportion of wealth annuitized (see Eq. (7)). Third, the increase in the number of children leads individuals to become more impatience (see Eq. (13) and Fig. 2). And, fourth, since the bequest given increases with the number of offspring, an increase in fertility leads to a decline in bequest wealth, i.e. T_B (see Eq. (15)). Notice that the sum of the third and fourth effect reduces consumption early in life and increase it at old age. Thus, an increase in fertility implies a rise in savings early in life, which yields a higher average financial wealth, i.e. $\mathrm{E}_C[\boldsymbol{k}(x)]$. In contrast, an increase in fertility reduces the variance of the financial wealth because of the fall in the bequest received. Therefore, an increase (resp. a fall) in fertility unambiguously leads to a reduction (resp. an increase) in wealth inequality within cohorts, given that the average financial wealth increases (resp. falls) and the variance decreases (resp. increases).

An **increase in life expectancy** has two major effects. First, a rise in life expectancy shifts the average age at death of any individual born in year τ towards older ages—i.e., $\mu(x,\tau)S(x,\tau)$ peaks at older ages—. Second, according to the life cycle model, an increase in life expectancy also increases savings for retirement motive in order to finance the additional years lived during retirement. As a consequence, an increase in life expectancy yields an increase in the average financial wealth. Combining both the demographic effect and the economic effect, we have that an increase in life expectancy not only raises the expected bequest, due to the increase in savings for retirement motive, but it also postpones the age at which the bequest is received (see Eq. (5)), which will imply an increase and a shift in the variance of financial wealth towards older ages.

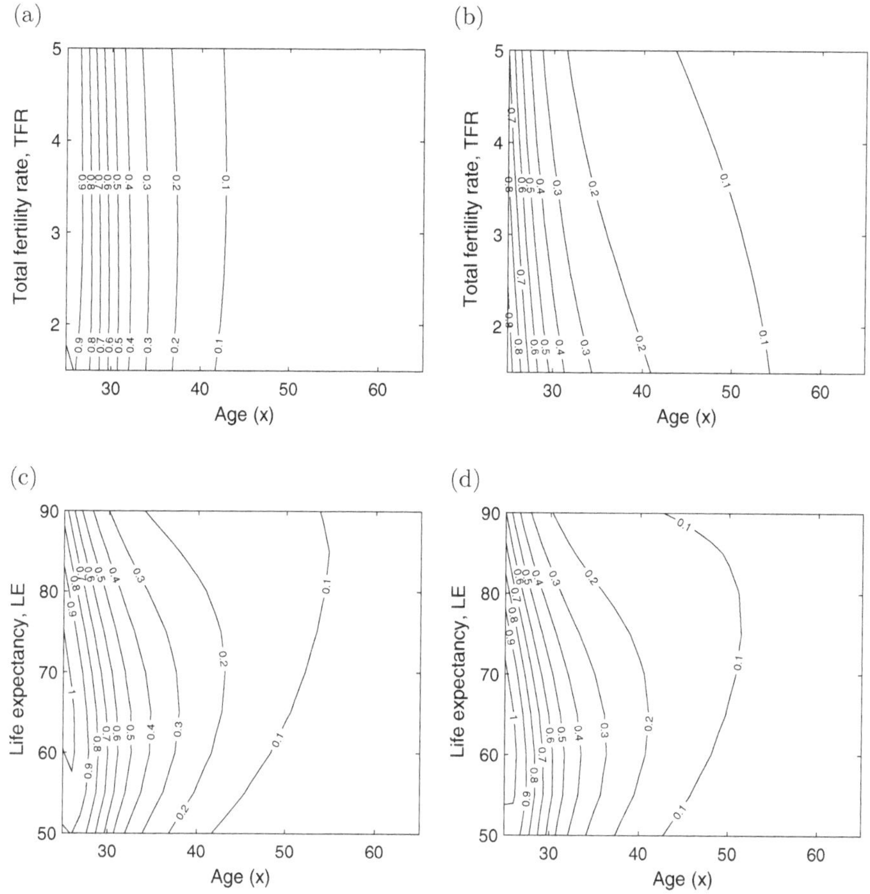

Fig. 6 Impact of changes in life expectancy (LE) and fertility (TFR) on financial wealth inequality at selected ages. (**a**) Fixed LE = 50. (**b**) Fixed LE = 80. (**c**) Fixed TFR = 1.5. (**d**) Fixed TFR = 3.0

Figure 6 shows the impact of changes in life expectancy and fertility on wealth inequality over a cohort's life cycle. The contour lines denote the combination of age and either life expectancy or fertility values that provide the same level of financial wealth inequality. Higher contour values indicate a higher wealth inequality.

Our simulations show three important results. First, in Fig. 6a,b it can be seen, for a given age x, that a decline (resp. increase) in fertility has no effect on wealth inequality when the LE is low. In contrast, when the LE is 80 years, a decline in fertility raises long run wealth inequality, while an increase in fertility reduces long run wealth inequality. The latter result is similar to that recently obtained by Onder and Pestieau (2016). As already explained, the greater wealth inequality associated to a fall in fertility is caused by a simultaneous decline in the average financial wealth and an increase in the variance. Second, in Fig. 6c,d it can be seen, for a given age x, that the effect of an increase in life expectancy on wealth inequality is non monotonic. In particular, when the initial life expectancy is low an increase in LE raises wealth inequality, while an increase in life expectancy reduces wealth inequality when the initial LE is high. The non monotonic behavior arises because of the postponement towards older ages of the bequest received. In particular, the highest variance is reached at the age equal to the difference between the life expectancy and the average generational gap. Thus, for instance, for a life expectancy equal to 65 and a generational gap of 25, we can see in Fig. 6d that the highest wealth inequality is reached around age 40. And third, it is worth noting that in all panels in Fig. 6 the wealth inequality decreases with age when the only source of inequality is driven by the generational gap. This is because the average financial wealth increases over the life cycle. This result implies, when all individuals face the same fertility and mortality, that the persistence in wealth inequality over the life cycle is not due to the generational gap.

5.2.2 Population

Our model also allows to analyze the population wealth inequality taking into consideration a realistic demographic structure. To calculate the wealth inequality of the whole population (N) through the coefficient of variation, $c_N[\boldsymbol{k}] = \sqrt{\mathrm{V}_N[\boldsymbol{k}]}/\mathrm{E}_N[\boldsymbol{k}]$, we use the following definitions:

$$\mathrm{V}_N[\boldsymbol{k}] = \int_0^{\omega}\int_0^{\omega}(k(x,l) - \mathrm{E}_N[\boldsymbol{k}])^2 \frac{N(x,l)}{N} dl dx, \tag{36}$$

$$\mathrm{E}_N[\boldsymbol{k}] = K/N, \tag{37}$$

where $\mathrm{V}_N[\boldsymbol{k}]$ is the variance of financial wealth of the population, $\mathrm{E}_N[\boldsymbol{k}]$ is the average financial wealth of the population, K is the total assets held by the population, and N is the total population size.

The impact of demography on population wealth inequality is explained by the inequality both within and across cohorts. Recall that in Sect. 5.2.1 we have dealt with wealth inequality within cohorts. The impact of changes in fertility and mortality on the financial wealth across age groups can be well accounted through the change in the mean-age of the population. Figure 7 shows the mean-age of the population that results from combining the alternative mortality and fertility rates using the population model introduced in Sect. 4.1. On the one side, we can see in Fig. 7 that for a given LE the mean-age of the population increases with lower fertility levels. For instance, with a LE of 60 years, a decline in TFR from 4 to 1.5 raises the mean-age of the population from 25 to 40 years. On the other side, for a given TFR, the mean-age of the population increases with higher life expectancy levels. However, it should be noted that the mean-age of the population is affected to a lower extent by a change in LE than by a change in TFR. For example, with

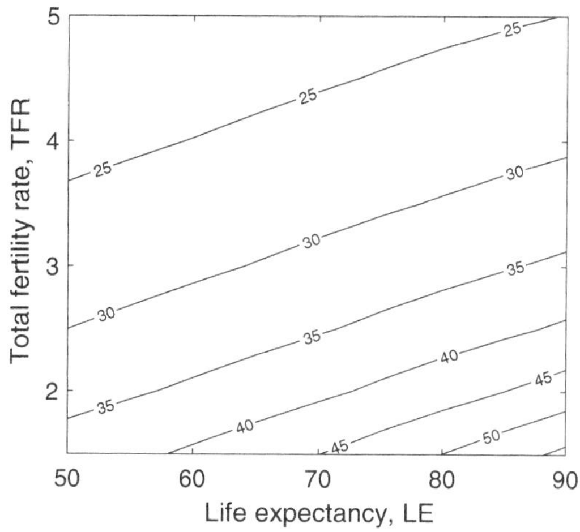

Fig. 7 Mean-age of the population by total fertility rate and life expectancy

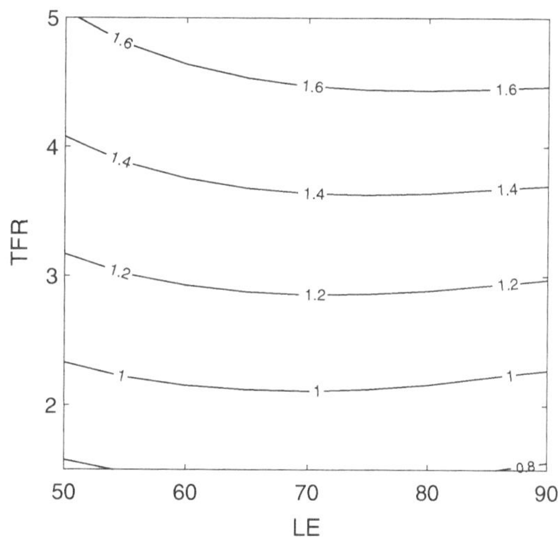

Fig. 8 Impact of changes in life expectancy (LE) and fertility (TFR) on financial wealth inequality

a TFR close to population replacement, an increase in LE from 50 to 90 raises the mean-age of the population from 35 to 45 years.

Combining the results in Sect. 5.2.1 with Fig. 7, we can explain the impact of changes in fertility and mortality on the total wealth inequality. Our simulations indicate that a decline in fertility decreases the financial wealth inequality, regardless the life expectancy value (see Fig. 8). This result is explained by the fact that, although a decline in fertility yields an increase in the financial inequality within cohorts, the mean-age of the population increases significantly, as Fig. 7 shows, which is associated to lower financial wealth inequality (see Fig. 6). Figure 8 also indicates that the effect of a change in life expectancy on financial wealth inequality depends on the mortality level on the one side, and its impact on the financial wealth inequality which is rather small on the other. In particular, for a TFR level of 3, an increase in life expectancy, when LE is below 70 years, yields an increase in financial wealth inequality. In contrast, an increase in life expectancy above age 70 reduces the financial wealth inequality.[12] Applying the same strategy as with fertility, looking at Fig. 6c it can be seen that an increase in LE has a non monotonic effect on wealth inequality within cohorts, which is reinforced with the increase in the mean-age of the population.

6 Conclusions

In this paper we have estimated the impact that changes in fertility and life expectancy have on wealth inequality. We have implemented a small-open economy model populated by overlapping generations (OLG) in which individuals are characterized by three time dimensions: age, cohort (time), and generational gap (the age gap between the parent and the child). The last time dimension allows us to demographically generate wealth inequality within cohorts, since individuals receive their bequest at different ages, and to consider explicitly the fact that in reality people in the same cohort have different wealth trajectories.

Our results suggest that a decline in fertility increases wealth inequality, at the individual level. This is explained by the fact that the decline in fertility increases the proportion of bequest received, which raises the difference in financial wealth between those individuals who have already received their bequest and those individuals whose parents are still alive. At the population level, however, a fall in fertility reduces wealth inequality. This result is explained by two factors. First, wealth inequality diminishes with age given that individuals become wealthier with age. Second, the fall in fertility increases the share of older individuals. The combined effect of these two factors at the population level dominates the higher inequality at the individual level. Moreover, our simulations show that changes in life expectancy have a small non monotonic effect on wealth inequality.

The results should be interpreted with caution, since our model is based on key, but necessary, assumptions in order to focus on the effect of demography on wealth inequality. In particular, in this paper we have abstracted from changes in the population composition, such as gender, and from non demographic sources of inequality, such as income inequality, which might be positively correlated with wealth inequality at the individual level. However, even in this particular case, Goldstein and Lee (2014) have shown for the US, see Table 2, that a decline in fertility leads to a reduction in wealth inequality.

To conclude the paper, it is worth mentioning that the current version of the model is ready for analyzing the impact of demography over the demographic

[12]The age threshold from which the effect of mortality on financial wealth inequality changes shouldn't be considered as a fixed age. Indeed, using different age-specific mortality rates would result in a shift in the age threshold.

transition, and not just under different steady-state scenarios. Nonetheless, this model should be considered as a first step towards a full demographic-economic model that will allow us to better understand the consequences of population aging on inequality. In this direction, we plan to extend the model in order to analyze the effect that alternative transfer systems, such as the pension system, might have on our results. Moreover, we believe it is necessary to consider the impact of demographic heterogeneity by socio-economic status, such as education, on inequality, and to consider the effect of compositional changes in the population.

Acknowledgment We would like to acknowledge the comments and suggestions given by two anonymous referees. This project has received funding from the European Union's Seventh Framework Program for research, technological development and demonstration under grant agreement no. 613247: "Ageing Europe: An application of National Transfer Accounts (NTA) for explaining and projecting trends in public finances".

Appendix 1: Number of Offspring and the Fraction of Wealth Received

Number of Offspring Similar to the foundation of the stable population theory by Lotka (1939), we will consider the total number of heirs of an individual at age x is a convolution of the fertility function with the survival probability. To simplify the notation and without loss of generality, in this proof, we get rid of the time at birth.

Let $\vartheta(l)$ be the number of children born from an individual of age l. Let us assume that $\vartheta(l)$ is an independently distributed random variable defined on the non-negative integers. Further let $z_i(a)$, with $i = 0, \ldots, \vartheta(l)$, denote the ith child of age a. Let $z_i(a)$ be a Bernoulli distributed independent random variable with probability $S(a)$ (i.e. $S(a)$ denotes the probability that the child survives up to age a). Then $\mathcal{Z}(a, l)$ defining the total number of children of exact age a born from an individual of age l can be defined as

$$\mathcal{Z}(a, l) = \sum\nolimits_{i=0}^{\vartheta(l)} z_i(a). \tag{38}$$

Assuming $\vartheta(l)$ is distributed at the population level according to a Poisson with parameter $m(l)$, where $m(l)$ is the age-specific fertility rate at the exact age l. Then, the distribution of $\mathcal{Z}(a, l)$ is a convolution of the fertility function with the survival probably

$$\begin{aligned} P\{\mathcal{Z}(a, l) = \sigma\} &= P\left\{\sum\nolimits_{i=1}^{\vartheta(l)} z_i(a) = \sigma\right\} \\ &= \sum\nolimits_{n=1}^{\infty} P\left\{\sum\nolimits_{i=1}^{n} z_i(a) = \sigma\right\} P\{\vartheta(l) = n\}. \end{aligned} \tag{39}$$

To find the distribution that results from (39) we use the characteristic function of $\mathcal{Z}(a,l)$ as follows

$$\begin{aligned}
\varphi_{\mathcal{Z}(a,l)}(t) &= \mathrm{E}\left[e^{it\mathcal{Z}(a,l)}\right] = \mathrm{E}\left[e^{it\sum_{j=0}^{\vartheta(l)} z_j(a)}\right] \\
&= \sum_{n=0}^{\infty} \mathrm{E}\left[e^{it\sum_{j=0}^{n} z_j(a)}\right] \mathrm{P}\{\vartheta(l)=n\} \\
&= \sum_{n=0}^{\infty} \left(\prod_{j=0}^{n} \mathrm{E}\left[e^{itz_j(a)}\right]\right) \mathrm{P}\{\vartheta(l)=n\} \\
&= \sum_{n=0}^{\infty} \left(\varphi_{z_j(a)}(t)\right)^n \mathrm{P}\{\vartheta(l)=n\} = \sum_{n=0}^{\infty} \frac{\left(\varphi_{z_j(a)}(t)m(l)\right)^n e^{-m(l)}}{n!} \\
&= e^{-m(l)} \sum_{n=0}^{\infty} \frac{\left(\varphi_{z_j(a)}(t)m(l)\right)^n}{n!} = e^{-m(l)} e^{\varphi_{z_j(a)}(t)m(l)} = e^{m(l)\left(\varphi_{z_j(a)}(t)-1\right)}.
\end{aligned} \tag{40}$$

Given that the characteristic function of the Bernoulli distribution is

$$\varphi_{z_j(a)}(t) = \mathrm{E}\left[e^{itz_j(a)}\right] = e^{it} S(a) + 1 - S(a). \tag{41}$$

Substituting (41) in (40) gives

$$\varphi_{\mathcal{Z}(a,l)}(t) = e^{m(l)S(a)(e^{it}-1)}, \tag{42}$$

which is the characteristic function of a Poisson process. Thus, we have

$$\mathcal{Z}(a,l) \overset{d}{\sim} \mathrm{Po}(m(l)S(a)). \tag{43}$$

If we define the random variable total number of children of an individual with exact age x as

$$\mathcal{N}(x) = \int_0^x \mathcal{Z}(x-l,l)dl. \tag{44}$$

Then, from probability theory we have that the sum of Poisson processes is also a Poisson process. Hence, from (43) and (44) it follows

$$\mathcal{N}(x) \overset{d}{\sim} \mathrm{Po}\left(\int_0^x m(l)S(x-l)dl\right), \tag{45}$$

where the mean of $\mathcal{N}(x)$ is equal to the total number of heirs, see Eq. (2), given that $n(x) = \int_0^x m(l)S(x-l)dl$.

Fraction of Wealth Received In order to derive (6) we must answer the question: what will be the expected fraction of wealth that corresponds to an individual if the parent dies at the exact age x? If the parent leaves one monetary unit as bequest, the wealth will be equally split between our individual and all the remaining siblings. From (45) we know that the number of offspring from a parent of x is a random number distributed according to a Poisson process. Therefore, the expected number of offspring from a parent of exact age x is, in this case, given by

$$\begin{aligned} E\left[\mathcal{N}(x)|\mathcal{N}(x) \geq 1\right] &= \frac{1}{1 - P\left\{\mathcal{N}(x) = 0\right\}} \sum_{\sigma=1}^{\infty} \sigma P\left\{\mathcal{N}(x) = \sigma\right\} \\ &= \frac{1}{1 - e^{-n(x)}} \sum_{\sigma=1}^{\infty} \sigma \frac{[n(x)]^{\sigma} e^{-n(x)}}{\sigma!} \\ &= \frac{n(x)}{1 - e^{-n(x)}}. \end{aligned} \tag{46}$$

Given that each unit of wealth will be split equally among the expected number of offspring, the fraction of wealth received by any offspring from a parent of age x becomes the inverse of (46),

$$\eta(x) = \frac{1 - e^{-n(x)}}{n(x)}, \tag{47}$$

which coincides with (6).

Appendix 2: Total Bequest Given Equals Total Bequest Received

For convenience we integrate with respect to age and generational gaps. Let us define all wealth transfers given at time t as

$$\int_0^{\omega} \mu(x, t-x) \int_0^{\omega} N(x, t-x, l) k(x, t-x, l) \left[1 - \theta(x, t-x)\right] dl dx. \tag{48}$$

Equation (48) is the integral over all ages and generational gaps of the capital left as bequest at each age x in year t by individuals whose parents were l years older at the time of birth, i.e. $k(x, t-x, l)\left[1 - \theta(x, t-x)\right]$, times the total number of people dying with those demographic characteristics, $\mu(x, t-x)N(x, t-x, l)$. Also, let us define all wealth transfers received as

$$\int_0^{\omega}\int_0^{\omega-x} B(x, t-x, l)N(x, t-x, l)dldx. \tag{49}$$

Equation (49) is the integral over all ages and generational gaps of the average bequest received at age x in year t from the death of their parent at age $x+l$. From (5) we have that the bequest received only takes non zero values for $l \in [0, w-x)$.

In order to prove that all wealth transfers given equal all wealth transfers received, we show that by substituting terms in (49) we get (48). The inverse is completely analogous and we leave it to the reader.

Proof First, by substituting (19) and (21) in (49), we get

$$\int_0^{\omega}\int_0^{\omega-x} B(x, t-x, l)S(x, t-x)m(l, t-x-l)N(l, t-x-l)dldx. \tag{50}$$

Using (5), defining $a = x + l$, and rearranging terms in (50), we have

$$\int_0^{\omega}\int_x^{\omega} \mu(a, t-a)N(a, t-a)k(a, t-a)\eta(a, t-a)S(x, t-x)m(a-x, t-a)dadx. \tag{51}$$

Changing the order of integration in (51) and leaving outside of the inner integral those variables that do not depend on x gives

$$\int_0^{\omega} \mu(a, t-a)N(a, t-a)k(a, t-a)\eta(a, t-a)\int_0^{a} S(x, t-x)m(a-x, t-a)dxda. \tag{52}$$

Expressing the inner integral of (52) in terms of the generational gap $l = a - x$ gives

$$\int_0^{\omega} \mu(a, t-a)N(a, t-a)k(a, t-a)\eta(a, t-a)\int_0^{a} S(a-l, t-a+l)m(l, t-a)dlda. \tag{53}$$

According to (2), the inner integral of (53) equals the average number of births of an individual born in year $t-a$ at age a. Then, using (6) in (53), we have

$$\int_0^{\omega} \mu(a, t-a)N(a, t-a)k(a, t-a)[1-\theta(a, t-a)]da. \tag{54}$$

Next, using the fact that $N(a, t-a)k(a, t-a) = \int_0^{\omega} N(a, t-a, l)k(a, t-a, l)dl$ in (54), and assuming $a = x$, we have

$$\int_0^{\omega} \mu(x, t-x)\int_0^{\omega} N(x, t-x, l)k(x, t-x, l)[1-\theta(x, t-x)]dldx, \tag{55}$$

which is equivalent to (48). □

References

A. Alesina, R. Perotti, Income distribution, political instability, and investment. Eur. Econ. Rev. **40**(6), 1203–1228 (1996)

F. Alvaredo, B. Garbinti, T. Piketty, On the share of inheritance in aggregate wealth: Europe and the USA, 1900–2010. Economica **84**, 239–260 (2017)

L. Arrondel, A. Masson, Altruism, exchange or indirect reciprocity: what do the data on family transfers show? Handb. Econ. Giving, Altruism and Reciprocity **2**, 971–1053 (2006)

D.E. Bloom, J.G. Williamson, Demographic transitions and economic miracles in emerging Asia. World Bank Econ. Rev. **12**(3), 419–455 (1998)

R. Chetty, N. Hendren, L.F. Katz, The effects of exposure to better neighborhoods on children: new evidence from the moving to opportunity experiment. Am. Econ. Rev. **106**(4): 855–902 (2016)

J.R. Goldstein, R.D. Lee, How large are the effects of population aging on economic inequality? Vienna Yearb. Popul. Res. **12**, 193–209 (2014)

D. Grass, J.P. Caulkins, G. Feichtinger, G. Tragler, D.A. Behrens, *Optimal Control of Nonlinear Processes: With Applications in Drugs, Corruption and Terror* (Springer, Heidelberg, 2008)

D. Greenwood, N. Guner, G. Kocharkov, C. Santos, Marry your like: assortative mating and income inequality. Am. Econ. Rev. Pap. Proc. **104**(5), 348–353 (2014)

L.J. Kotlikoff, Intergenerational transfers and savings. J. Econ. Perspect. **2**(2), 41–58 (1988)

L.J. Kotlikoff, L.H. Summers, The role of intergenerational transfers in aggregate capital accumulation. J. Polit. Econ. **89**(4), 706–732 (1981)

R.D. Lee, Age structure intergenerational transfers and economic growth: an overview. Rev. Écon. **31**(6), 1129–1156 (1980)

R.D. Lee, A. Mason (eds.), *Population Aging and the Generational Economy: A Global Perspective* (Edward Elgar, UK, 2011)

A.J. Lotka, *Théorie analytique des associations biologiques: Analyse démographique avec application particulière à l'espèce humaine*. Actualités Scientifiques et Industrielles, No. 780. Paris, Hermann et Cie (1939)

F. Modigliani, The role of intergenerational transfers and life cycle saving in the accumulation of wealth. J. Econ. Perspect. **2**(2), 15–40 (1988)

H. Onder, P. Pestieau, *Inherited wealth and demographic aging*. World Bank Group, Policy Research Working Paper 7739 (2016)

T. Piketty, *Capital in the Twenty-First Century* (Belknap Press of Harvard University Press, London, 2014)

T. Piketty, G. Zucman, Capital is back: wealth-income ratios in rich countries, 1700–2010. Q. J. Econ. **129**(3), 1155–1210 (2014)

J. Tobin, Life cycle saving and balanced economic growth. In: *Ten Economic Studies in the Tradition of Irving Fisher* ed. by W. Fellner, et al. (Wiley, New York, 1967), pp. 231–256

G. Vandenbroucke, Aging and wealth inequality in a neoclassical growth model. Review **98**(1), 61–80 (2016)

M.E. Yaari, Uncertain lifetime, life insurance, and the theory of the consumer. Rev. Econ. Stud. **32**, 137–150 (1965)

Stability Analysis of a New E-rumor Model

Séverine Bernard, Ténissia Cesar, and Alain Pietrus

Abstract The emergence of new communication tools leads us to have public discussions on social networks. These public spaces of exchange are firmly established in our societies but are strong sensors of both human behaviors and collective feelings. The propagation of rumors, namely e-rumor, is very fast and is one of the most dangerous for a society because it can destabilize financial, political and economical markets. Many mathematical models, based on the epidemic ones, have been constructed in order to understand this multi-dimensional diffusion process driven mainly by socio-psychological elements. In this paper, we propose a new model of propagation of rumor taking into account the different possible changes of classes of the individuals of a social network. With this new model, we point out admissible equilibrium states and the conditions of their stability, as well as the criteria of persistence of this model.

1 Introduction

For a long time, rumors were transmitted by word of mouth from one to another, then by newspapers, radio, television,... Nowadays, social networks are the main media responsible for spreading rumors. These public spaces of exchange are firmly established in our societies but are strong sensors of both human behaviors and collective feelings. Since it can rapidly jeopardise the public opinion and the economic and financial markets, it is important to understand the transmission rules of a rumor in order to stop or control it, which has yet been done by many mathematicians, computers scientists, economists, sociologists,...

S. Bernard · T. Cesar · A. Pietrus (✉)
Université des Antilles, LAboratoire de Mathématiques Informatique et Applications EA4540, Pointe-à-Pitre cedex, FWI, Guadeloupe
e-mail: Severine.Bernard@univ-antilles.fr; Tenissia.Cesar@univ-antilles.fr; Alain.Pietrus@univ-antilles.fr

G. Feichtinger et al. (eds.), *Control Systems and Mathematical Methods in Economics*, Lecture Notes in Economics and Mathematical Systems 687,
https://doi.org/10.1007/978-3-319-75169-6_18

Many mathematical models, based on the epidemic ones, have been constructed in order to understand this multi-dimensional diffusion process driven mainly by socio-psychological elements. In the first mathematical approaches, the three individual classes of susceptible, infected and recovered, traditionally considered in the contagion process, have been directly transposed in the rumor context as ignorant, spreader and stifler, that is an individual who knows the rumor but does not spread it for the time being, as for example in Daley and Kendall (1964), Daley and Kendall (1965), Dietz (1967), Maki (1973), Rapoport (1953), Rapoport et al. (1953), and Rapoport and Rebhun (1952).

But rumor propagation has a very high specificity compared to epidemics and it is difficult to model the problem with simple transmission rules that are not able to reflect the complexity of individuals behaviors. For example, in Stattner et al. (2015), the authors focussed on diffusion phenomena that occur on Twitter and showed a strong heterogeneity in the individuals behaviors. Other works, as Collard et al. (2015), try to model the psychological mechanism involved in the decision for a person to become or not a spreader.

Our first approaches was based on the model of Huang and Jin (2011) in which the authors divided the group of stiflers into two subcategories, those who accept and those who are under the rumor, and tested a random and then a targeted immunization strategy. In Bernard et al. (2015) and Bernard et al. (2016), the model is exactly the same as the one of Huang and Jin (2011) in which Bernard et al. put all the parameters depending on the time, which seems more realist to model the phenomenon since the transmission rules can change from one moment to the next.

In their first approach Bernard et al. (2015), the authors choose as control the rate for which an ignorant becomes a spreader after having met the last one in order to minimize the densities of spreaders and ignorants also, since an ignorant node is susceptible to become a spreader one at any time. In this case, the authors showed existence and gave the characterization of the optimal parameter, with the help of optimal control theory and some numerical simulations have been done to strengthen their theoretical study, showing that, in the optimal case, most of the individuals are stiflers.

In the second work (Bernard et al. 2016), the chosen control is the rate for which a spreader becomes a stifler who accepts the rumor after having met a spreader or a stifler and the minimized objective function leads to reduce as much as possible the number of spreaders and keep the largest possible number of individuals as ignorants. An optimal control approach leads to characterize the optimal control. The results of Bernard et al. (2016) seem to be better than the ones of Bernard et al. (2015) since a stifler can change position along the time and can become a spreader without having met a spreader, whereas an ignorant can become a spreader only after having met a spreader.

In the following, as it has been done for an epidemic model in Hansen and Day (2011), some realistic external actions has been added in the last rumor model and controlled separately in Bernard et al. (2017) and simultaneously in Bernard et al. (2018) with the same objective of spread reduction. These actions may be considered as the probability of persons who learn a disclaimer of a given rumor and as the isolation of the more active spreaders in order to avoid their contact with other nodes.

But in these works, the models do not take into account the different possible changes of classes of the individuals of a social network. In this state of minds, we propose in Sect. 2 a new model of propagation of e-rumor in which the opposing stiflers are those who know the rumor and will never spread it and the accepting stiflers are those who know the rumor and will stay an accepting stifler or become a spreader or an opposing stifler at any time. This model has been inspired by the construction of a mathematical model for the spread of two political parties in Misra (2012). The difficulty is that we consider a population divided into four subcategories for the rumor, contrary to Misra (2012) in which there are three classes, the voters and the members of the two political parties. Consequently, the computations are a little bit more complicated and the use of Maple is sometimes necessary.

There are important contributions about conditions for the optimality of a limit cycle in dynamic models done in Dockner and Feichtinger (1991), Feichtinger et al. (1994), Grass et al. (2008), and Wirl (2007) with applications in particular in social sciences. The model that we propose can be reduced in a system of three equations, and for this reason, we can not use the works of Feichtinger et al. (1994) and Grass et al. (2008) in which necessary conditions of an optimal scalar control are stated for a nonlinear problem of two state equations and the classification of equilibria is established by using Hopf bifurcation theorem adapted to optimal control problem. But in a parallel work, we construct another model of propagation of rumor consisting in dividing the population of a social network only into three categories, the ignorants, the spreaders and the stiflers, opposing and accepting, and taking into account the different possible changes of classes. Consequently, we obtain a system of two state equations for which we associate a nonlinear control problem with a scalar control. In this context, we can use the works of Feichtinger et al. (1994) and Grass et al. (2008) in order to classify the equilibria.

The stability analysis begin with the research of the admissible equilibrium states in Sect. 3, which consists in solving a system of three equations and the conditions for which these last ones are non negative since we work with densities of population. To continue this analysis, we point out the conditions of their stability, the fact that there is no limit cycle between the spreaders and the stiflers and the coexistence criteria of these different classes of individuals in Sect. 4. Most of the reasonings done in this work use classical results on dynamical systems and we refer the reader to Robinson (2001) for an overview and Cherkas and Grin (2010), Freedman and Moson (1990), and Freedman and Waltman (1984) for more details.

2 The New e-rumor Mathematical Model

Let N be the total number of persons and μ the rate at which an individual enters or leaves the social network. This population is divided into four groups: the one of ignorants, the one of spreaders, the one of accepting stiflers and the one of opposing stiflers. In the following, we denote by I the number of ignorants, S the number of spreaders, RA the number of accepting stiflers and RU the number of opposing stiflers. We assume that an ignorant can become a spreader, an accepting stifler or an opposing stifler with rate β_1, β_2 or β_3 respectively. A spreader can become an accepting stifler or an opposing stifler with rate θ_1 or α respectively. Moreover, an accepting stifler can become a spreader or an opposing stifler with rate θ_2 or λ respectively. Note that an opposing stifler can not change class but can just leave the social network, as all the individuals of the three other groups, with rate μ. All these transmission rules are synthetized in the following diagram

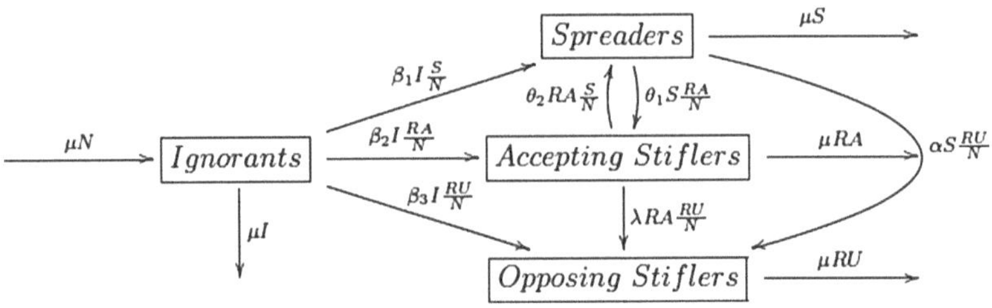

and can be written by the mathematical system of ordinary differential equations

$$
\begin{cases}
\dfrac{dI}{dt}(t) = \mu N - \beta_1 \dfrac{I(t)S(t)}{N} - \beta_2 \dfrac{I(t)RA(t)}{N} - \beta_3 \dfrac{I(t)RU(t)}{N} - \mu I(t), \\[2ex]
\dfrac{dS}{dt}(t) = \beta_1 \dfrac{I(t)S(t)}{N} - \theta_1 \dfrac{S(t)RA(t)}{N} - \alpha \dfrac{S(t)RU(t)}{N} + \theta_2 \dfrac{RA(t)S(t)}{N} - \mu S(t), \\[2ex]
\dfrac{dRA}{dt}(t) = \beta_2 \dfrac{I(t)RA(t)}{N} + \theta_1 \dfrac{S(t)RA(t)}{N} - \theta_2 \dfrac{RA(t)S(t)}{N} - \lambda \dfrac{RA(t)RU(t)}{N} - \mu RA(t), \\[2ex]
\dfrac{dRU}{dt}(t) = \beta_3 \dfrac{I(t)RU(t)}{N} + \lambda \dfrac{RA(t)RU(t)}{N} + \alpha \dfrac{S(t)RU(t)}{N} - \mu RU(t),
\end{cases}
\tag{1}
$$

with μ, β_1, β_2, β_3, θ_1, θ_2, α and λ strictly non negative real numbers. By setting $\theta = \theta_1 - \theta_2$ and $i = \frac{I}{N}$, $s = \frac{S}{N}$, $ra = \frac{RA}{N}$ and $ru = \frac{RU}{N}$, one obtains

$$\begin{cases} \dfrac{di}{dt}(t) = \mu - \beta_1 i(t)s(t) - \beta_2 i(t)ra(t) - \beta_3 i(t)ru(t) - \mu i(t), \\ \\ \dfrac{ds}{dt}(t) = \beta_1 i(t)s(t) - \theta s(t)ra(t) - \alpha s(t)ru(t) - \mu s(t), \\ \\ \dfrac{dra}{dt}(t) = \beta_2 i(t)ra(t) + \theta s(t)ra(t) - \lambda ra(t)ru(t) - \mu ra(t), \\ \\ \dfrac{dru}{dt}(t) = \beta_3 i(t)ru(t) + \lambda ra(t)ru(t) + \alpha s(t)ru(t) - \mu ru(t). \end{cases} \tag{2}$$

Moreover, $I + S + RA + RU = N$ so $i + s + ra + ru = 1$ and we can only study the following system of three ordinary differential equations

$$\begin{cases} \dfrac{ds}{dt}(t) = \beta_1(1 - s(t) - ra(t) - ru(t))s(t) - \theta s(t)ra(t) - \alpha s(t)ru(t) - \mu s(t), \\ \\ \dfrac{dra}{dt}(t) = \beta_2(1 - s(t) - ra(t) - ru(t))ra(t) + \theta s(t)ra(t) - \lambda ra(t)ru(t) - \mu ra(t), \\ \\ \dfrac{dru}{dt}(t) = \beta_3(1 - s(t) - ra(t) - ru(t))ru(t) + \lambda ra(t)ru(t) + \alpha s(t)ru(t) - \mu ru(t), \end{cases} \tag{3}$$

and obtain i with $i(t) = 1 - s(t) - ra(t) - ru(t)$. In the following, we assume that $\theta > 0$ but the study will be the same in the contrary case.

3 Equilibrium States

The study of equilibrium states leads us to solve the following system of three equations

$$\begin{cases} \beta_1(1 - s(t) - ra(t) - ru(t))s(t) - \theta s(t)ra(t) - \alpha s(t)ru(t) - \mu s(t) = 0, \\ \beta_2(1 - s(t) - ra(t) - ru(t))ra(t) + \theta s(t)ra(t) - \lambda ra(t)ru(t) - \mu ra(t) = 0, \\ \beta_3(1 - s(t) - ra(t) - ru(t))ru(t) + \lambda ra(t)ru(t) + \alpha s(t)ru(t) - \mu ru(t) = 0, \end{cases} \tag{4}$$

keeping in mind that an admissible equilibrium state is a triplet (s, ra, ru) of non negative solutions of these three equations. Some solutions are easy to obtain but not all so we use Maple to solve the previous system and to obtain eight solutions. We write the solution as a triplet (s, ra, ru). The first obvious one is the state

$E_1 = (0, 0, 0)$. The three following solutions are

$$E_2 = \left(\frac{\beta_1 - \mu}{\beta_1}, 0, 0\right), E_3 = \left(0, \frac{\beta_2 - \mu}{\beta_2}, 0\right), E_4 = \left(0, 0, \frac{\beta_3 - \mu}{\beta_3}\right),$$

which are admissible equilibrium states if and only if

$$\beta_1 > \mu, \beta_2 > \mu \text{ and } \beta_3 > \mu, \tag{5}$$

respectively. Another solution is

$$E_5 = \left(\frac{\mu(\beta_1 - \beta_2) - \theta(\beta_2 - \mu)}{\theta(\theta + \beta_1 - \beta_2)}, \frac{\theta(\beta_1 - \mu) - \mu(\beta_1 - \beta_2)}{\theta(\theta + \beta_1 - \beta_2)}, 0\right),$$

which is admissible if

$$\begin{cases} \theta + \beta_1 - \beta_2 > 0, \\ \mu(\beta_1 - \beta_2) - \theta(\beta_2 - \mu) > 0, \\ \theta(\beta_1 - \mu) - \mu(\beta_1 - \beta_2) > 0, \end{cases} \text{ or } \begin{cases} \theta + \beta_1 - \beta_2 < 0, \\ \mu(\beta_1 - \beta_2) - \theta(\beta_2 - \mu) < 0, \\ \theta(\beta_1 - \mu) - \mu(\beta_1 - \beta_2) < 0. \end{cases} \tag{6}$$

Let us remark that, if

$$\begin{cases} \theta + \beta_1 - \beta_2 < 0, \\ \mu(\beta_1 - \beta_2) - \theta(\beta_2 - \mu) < 0, \\ \theta(\beta_1 - \mu) - \mu(\beta_1 - \beta_2) < 0, \end{cases}$$

then $\beta_1 < \mu$ and this is not compatible with the condition of admissibility of E_2. For this reason, we say that E_5 is admissible if and only if

$$\begin{cases} \theta + \beta_1 - \beta_2 > 0, \\ \mu(\beta_1 - \beta_2) - \theta(\beta_2 - \mu) > 0, \\ \theta(\beta_1 - \mu) - \mu(\beta_1 - \beta_2) > 0. \end{cases} \tag{7}$$

In a same way, the solutions

$$E_6 = \left(\frac{\mu(\beta_1 - \beta_3) - \alpha(\beta_3 - \mu)}{\alpha(\alpha + \beta_1 - \beta_3)}, 0, \frac{\alpha(\beta_1 - \mu) - \mu(\beta_1 - \beta_3)}{\alpha(\alpha + \beta_1 - \beta_3)}\right)$$

and

$$E_7 = \left(0, \frac{\mu(\beta_2 - \beta_3) - \lambda(\beta_3 - \mu)}{\lambda(\lambda + \beta_2 - \beta_3)}, \frac{\lambda(\beta_2 - \mu) - \mu(\beta_2 - \beta_3)}{\lambda(\lambda + \beta_2 - \beta_3)}\right)$$

are admissible if and only if

$$\begin{cases} \alpha+\beta_1-\beta_3>0, \\ \mu(\beta_1-\beta_3)-\alpha(\beta_3-\mu)>0, \\ \alpha(\beta_1-\mu)-\mu(\beta_1-\beta_3)>0, \end{cases} \text{ or } \begin{cases} \alpha+\beta_1-\beta_3<0, \\ \mu(\beta_1-\beta_3)-\alpha(\beta_3-\mu)<0, \\ \alpha(\beta_1-\mu)-\mu(\beta_1-\beta_3)<0, \end{cases} \tag{8}$$

and

$$\begin{cases} \lambda+\beta_2-\beta_3>0, \\ \mu(\beta_2-\beta_3)-\lambda(\beta_3-\mu)>0, \\ \lambda(\beta_2-\mu)-\mu(\beta_2-\beta_3)>0, \end{cases} \text{ or } \begin{cases} \lambda+\beta_2-\beta_3<0, \\ \mu(\beta_2-\beta_3)-\lambda(\beta_3-\mu)<0, \\ \lambda(\beta_2-\mu)-\mu(\beta_2-\beta_3)<0, \end{cases} \tag{9}$$

respectively. By compatibility with the conditions (5), taken for the admissibility of E_2 and E_3, we can say that E_6 and E_7 are admissible if and only if

$$\begin{cases} \alpha+\beta_1-\beta_3>0, \\ \mu(\beta_1-\beta_3)-\alpha(\beta_3-\mu)>0, \\ \alpha(\beta_1-\mu)-\mu(\beta_1-\beta_3)>0, \end{cases} \text{ and } \begin{cases} \lambda+\beta_2-\beta_3>0, \\ \mu(\beta_2-\beta_3)-\lambda(\beta_3-\mu)>0, \\ \lambda(\beta_2-\mu)-\mu(\beta_2-\beta_3)>0, \end{cases} \tag{10}$$

respectively. For the last solution E_8, one has

$$\begin{cases} s=\dfrac{\mu(\alpha-\lambda-\theta)(\beta_2-\beta_3+\lambda)-\lambda(\alpha\beta_2-\lambda\beta_1-\theta\beta_3)}{(\alpha-\lambda-\theta)(\alpha\beta_2-\lambda\beta_1-\theta\beta_3)}, \\ ra=\dfrac{\mu(\alpha-\lambda-\theta)(\beta_3-\beta_1-\alpha)+\alpha(\alpha\beta_2-\lambda\beta_1-\theta\beta_3)}{(\alpha-\lambda-\theta)(\alpha\beta_2-\lambda\beta_1-\theta\beta_3)}, \\ ru=\dfrac{\mu(\alpha-\lambda-\theta)(\beta_1-\beta_2+\theta)-\theta(\alpha\beta_2-\lambda\beta_1-\theta\beta_3)}{(\alpha-\lambda-\theta)(\alpha\beta_2-\lambda\beta_1-\theta\beta_3)}. \end{cases} \tag{11}$$

If $\alpha\beta_2-\lambda\beta_1-\theta\beta_3$ and $\alpha-\lambda-\theta$ have the same sign, the solution E_8 is admissible if and only if

$$\begin{cases} \mu(\alpha-\lambda-\theta)(\beta_2-\beta_3+\lambda)>\lambda(\alpha\beta_2-\lambda\beta_1-\theta\beta_3), \\ \alpha(\alpha\beta_2-\lambda\beta_1-\theta\beta_3)>\mu(\alpha-\lambda-\theta)(\beta_1-\beta_3+\alpha), \\ \mu(\alpha-\lambda-\theta)(\beta_1-\beta_2+\theta)>\theta(\alpha\beta_2-\lambda\beta_1-\theta\beta_3), \end{cases} \tag{12}$$

and if $\alpha\beta_2-\lambda\beta_1-\theta\beta_3$ and $\alpha-\lambda-\theta$ have different signs, the solution E_8 is admissible if and only if

$$\begin{cases} \mu(\alpha-\lambda-\theta)(\beta_2-\beta_3+\lambda)<\lambda(\alpha\beta_2-\lambda\beta_1-\theta\beta_3), \\ \alpha(\alpha\beta_2-\lambda\beta_1-\theta\beta_3)<\mu(\alpha-\lambda-\theta)(\beta_1-\beta_3+\alpha), \\ \mu(\alpha-\lambda-\theta)(\beta_1-\beta_2+\theta)<\theta(\alpha\beta_2-\lambda\beta_1-\theta\beta_3). \end{cases} \tag{13}$$

Let us remark that, if $\alpha\beta_2 - \lambda\beta_1 - \theta\beta_3 > 0$, $\alpha - \lambda - \theta < 0$ and (10) holds, then

$$(\alpha - \lambda - \theta)(\alpha\beta_2 - \lambda\beta_1 - \theta\beta_3) < 0$$

and

$$\mu(\alpha - \lambda - \theta)(\beta_3 - \beta_1 - \alpha) + \alpha(\alpha\beta_2 - \lambda\beta_1 - \theta\beta_3) > 0.$$

In this case, ra is negative and this solution E_8 is not admissible.

In a same way, if $\alpha\beta_2 - \lambda\beta_1 - \theta\beta_3 < 0$, $\alpha - \lambda - \theta > 0$ and (10) holds, then

$$(\alpha - \lambda - \theta)(\alpha\beta_2 - \lambda\beta_1 - \theta\beta_3) < 0$$

and

$$\mu(\alpha - \lambda - \theta)(\beta_2 - \beta_3 + \lambda) - \lambda(\alpha\beta_2 - \lambda\beta_1 - \theta\beta_3) > 0.$$

In this case, s is negative and this solution E_8 is not admissible. For these reasons, we say that E_8 is admissible if and only if (12) holds.

4 Stability Analysis

Taking all the hypothesis (5), (7), (10) and (12), in order all the eight equilibrium states to be admissible, we are going now to determine the nature of all of them. For this, we need to know the sign of the eigenvalues of the Jacobian matrix associated to System (3) and calculated at each equilibrium state. Let us remark that it is possible not to assume all these conditions (5), (7), (10) and (12) but in this case, we will have less admissible equilibrium states and more or less stable nodes. So it is possible to consider different combinations of sign for (5), (7), (10) and (12) but the reasoning will not change really.

The Jacobian matrix associated to system (3) is

$$J = \begin{pmatrix} \beta_1(1 - 2s - ra - ru) - \theta ra - \alpha ru - \mu, -\beta_1 s - \theta s, -\beta_1 s - \alpha s \\ -\beta_2 ra + \theta ra, \beta_2(1 - s - 2ra - ru) + \theta s - \lambda ru - \mu, -\beta_2 ra - \lambda ra \\ -\beta_3 ru + \alpha ru, -\beta_3 ru + \lambda ru, \beta_3(1 - s - ra - 2ru) + \lambda ra + \alpha s - \mu \end{pmatrix}.$$

Let us set J_i this Jacobian matrix calculated at the equilibrium state E_i, for $i = 1, \ldots, 8$. One has

$$J_1 = \begin{pmatrix} \beta_1 - \mu & 0 & 0 \\ 0 & \beta_2 - \mu & 0 \\ 0 & 0 & \beta_3 - \mu \end{pmatrix}$$

and the hypothesis done to have the admissibility of E_2, E_3 and E_4 imply that the eigenvalues of J_1 are all strictly non negative, which leads us to say the equilibrium state E_1 is an unstable node.

Remark 1 If $\beta_1 < \mu$, $\beta_2 < \mu$ and $\beta_3 < \mu$ then E_2, E_3 and E_4 are not admissible equilibrium states but E_1 is locally asymptotically stable.

Moreover,

$$J_2 = \begin{pmatrix} \mu - \beta_1 & \frac{(\beta_1+\theta)(\mu-\beta_1)}{\beta_1} & \frac{(\beta_1+\alpha)(\mu-\beta_1)}{\beta_1} \\ 0 & \frac{\theta(\beta_1-\mu)-\mu(\beta_1-\beta_2)}{\beta_1} & 0 \\ 0 & 0 & \frac{\alpha(\beta_1-\mu)-\mu(\beta_1-\beta_3)}{\beta_1} \end{pmatrix}.$$

The eigenvalues of J_2 are $v_2^1 = \mu - \beta_1 < 0$, $v_2^2 = \frac{\theta(\beta_1-\mu)-\mu(\beta_1-\beta_2)}{\beta_1} > 0$ and $v_2^3 = \frac{\alpha(\beta_1-\mu)-\mu(\beta_1-\beta_3)}{\beta_1} > 0$ due to the conditions taken to have the admissibility of the equilibrium states E_2, E_5 and E_6 respectively. Consequently E_2 is unstable and it will be the same for the equilibrium states E_3 and E_4.

Remark 2 If $\theta(\beta_1 - \mu) - \mu(\beta_1 - \beta_2) < 0$ and $\alpha(\beta_1 - \mu) - \mu(\beta_1 - \beta_3) < 0$ with $\beta_1 - \beta_2 < 0$ and $\beta_1 - \beta_3 < 0$ then $\beta_1 - \mu < 0$ and E_2 is not an admissible equilibrium state. But if $\theta(\beta_1-\mu)-\mu(\beta_1-\beta_2) < 0$ and $\alpha(\beta_1-\mu)-\mu(\beta_1-\beta_3) < 0$ with $\beta_1 - \beta_2 > 0$ and $\beta_1 - \beta_3 > 0$ and if $\beta_1 - \mu > 0$ then E_5 and E_6 are not admissible equilibrium states but E_2 is a locally asymptotically stable one.

For the three Jacobian matrix J_5, J_6 and J_7, the reasoning will be the same for the three ones so we only give the details for J_6 for example. In fact J_6 is in the form

$$J_6 = \begin{pmatrix} \beta_1(1-2s-ru)-\alpha ru-\mu & -(\beta_1+\theta)s & -(\beta_1+\alpha)s \\ 0 & \beta_2(1-s-ru)+\theta s-\lambda ru-\mu & 0 \\ (\alpha-\beta_3)ru & (\lambda-\beta_3)ru & \beta_3(1-s-2ru)+\alpha s-\mu \end{pmatrix}.$$

It is clear that one of the eigenvalues of J_6 is

$$v_6^1 = \beta_2(1-s-ru) + \theta s - \lambda ru - \mu$$

calculated with $s = \frac{\mu(\beta_1-\beta_3)-\alpha(\beta_3-\mu)}{\alpha(\alpha+\beta_1-\beta_3)}$ and $ru = \frac{\alpha(\beta_1-\mu)-\mu(\beta_1-\beta_3)}{\alpha(\alpha+\beta_1-\beta_3)}$, which gives, after some computations,

$$v_6^1 = \frac{\mu(\alpha-\lambda-\theta)(\beta_3-\beta_1-\alpha)+\alpha(\alpha\beta_2-\lambda\beta_1-\theta\beta_3)}{\alpha(\alpha+\beta_1-\beta_3)},$$

and this eigenvalue is strictly non negative due to the conditions done for the admissibility of E_6 and E_8. This is sufficient to claim that E_6 is unstable, and it is the same for E_5 and E_7.

Remark 3 We can discuss also here about different possibilities of signs for the quantities $\mu(\alpha-\lambda-\theta)(\beta_3-\beta_1-\alpha)+\alpha(\alpha\beta_2-\lambda\beta_1-\theta\beta_3)$ and $\alpha(\alpha+\beta_1-\beta_3)$ and find in some cases the non admissibility of the eight equilibrium states and the locally asymptotically stability of E_6 but the reasoning is the same as previously.

It remains to determine the nature of the equilibrium point E_8. In fact,

$$J_8 = \begin{pmatrix} -\beta_1 s & -(\beta_1+\theta)s & -(\beta_1+\alpha)s \\ (\theta-\beta_2)ra & -\beta_2 ra & -(\beta_2+\lambda)ra \\ (\alpha-\beta_3)ru & (\lambda-\beta_3)ru & -\beta_3 ru \end{pmatrix},$$

with (s, ra, ru) defined in (11). We did not succeed to compute the eigenvalues of J_8, even with the help of Maple, because of the number of parameters and the fact that J_8 has no nil terms, contrary to the other matrix J_i, for $i = 1, \ldots, 7$. So we decided to use a code that generates randomly some values for the parameters $\beta_1, \beta_2, \beta_3, \alpha, \lambda, \theta$ and μ satisfying all the previous conditions giving the admissibility of all the equilibrium states E_i, for $i = 1, \ldots, 8$. With these different parameters, as for example

$$\begin{cases} \mu = 0.54264808887658644, \\ \alpha = 0.52664122046531703, \\ \lambda = 0.67619411213999614, \\ \theta = 0.64496066886789249, \\ \beta_1 = 0.83583066279073104, \\ \beta_2 = 0.65780618725161588, \\ \beta_3 = 0.58579971570993117, \end{cases}$$

we can easily compute the eigenvalues of this particular

$$J_8 = \begin{pmatrix} -0.15443654970694631 & 0 & 0 \\ 0 & -0.026890980865398251 & -0.026813306878478306 \\ 0 & -0.026813306878478306 & -0.026890980865398251 \end{pmatrix}$$

and see that they are all negative, which leads us to claim that the equilibrium state E_8 is locally asymptotically stable. Let us remark that with this choice of parameters, we find again the unstability of the seven first equilibrium points E_i, for $i = 1, \ldots, 7$.

Theorem 1 *There is no limit cycle between the density of spreaders and the two densities of stiflers.*

Proof Let B be the $\mathscr{C}^1$-function on $]0, +\infty[^3$ defined by

$$B(s, ra, ru) = \frac{1}{s.ra.ru}$$

and let us set

$$\begin{cases} f(s, ra, ru) = \beta_1(1 - s - ra - ru)s - \theta s.ra - \alpha s.ru - \mu s, \\ g(s, ra, ru) = \beta_2(1 - s - ra - ru)ra + \theta s.ra - \lambda ra.ru - \mu ra, \\ h(s, ra, ru) = \beta_3(1 - s - ra - ru)ru + \lambda ra.ru + \alpha s.ru - \mu ru, \end{cases}$$

the functions defined from the right hand sides of the equations of System (3). Then we have criterion of Bendixson-Dulac (see Cherkas and Grin (2010) for example).

$$\frac{\partial}{\partial s}(fB) + \frac{\partial}{\partial ra}(gB) + \frac{\partial}{\partial ru}(hB) = -\frac{\beta_1}{ra.ru} - \frac{\beta_2}{s.ru} - \frac{\beta_3}{s.ra}$$

□

There exists various forms of persistence and a among them (see Freedman and Moson (1990) for example). By noting that System (3) is a Kolmogorov type one and many results exist on the persistence of this type of system in particular for predator-prey ones, as in Freedman and Moson (1990) and Freedman and Waltman (1984) for example, we can establish the following result.

Theorem 2 *If the conditions (5), (7), (10) and (12), taken for the admissibility of all the equilibrium states, are satisfied then System (3) is uniformly persistent.*

Proof Let us set $\sigma(s, ra, ru) = s^p ra^q ru^v$, with $p, q, v > 0$. Then σ is a non negative function on $]0, +\infty[^3$. Moreover, the function defined by

$$\psi(s, ra, ru) = \frac{1}{\sigma(s, ra, ru)} \frac{d\sigma}{dt}$$

satisfies

$$\psi(s, ra, ru) = p\left[\beta_1(1 - s - ra - ru) - \theta ra - \alpha ru - \mu\right]$$
$$+ q\left[\beta_2(1 - s - ra - ru) + \theta s - \lambda ru - \mu\right] + v\left[\beta_3(1 - s - ra - ru) + \lambda ra + \alpha s - \mu\right].$$

Since there is no limit cycle, to show the uniform persistence, it is sufficient to show that $\psi(E_i) > 0$, for all $i = 1, \ldots, 7$. But

$$\psi(E_0) = p(\beta_1 - \mu) + q(\beta_2 - \mu) + v(\beta_3 - \mu),$$

is strictly non negative if the conditions of admissibility of E_2, E_3 and E_4 are satisfied. Moreover,

$$\psi(E_2) = \frac{q}{\beta_1}\left[\theta(\beta_1 - \mu) - \mu(\beta_1 - \beta_2)\right] + \frac{v}{\beta_1}\left[\alpha(\beta_1 - \mu) - \mu(\beta_1 - \beta_3)\right] > 0$$

from the conditions taken for the admissibility of E_5 and E_6. It will be the same for $\psi(E_3)$ and $\psi(E_4)$. And

$$\psi(E_5) = \frac{v}{\theta(\theta+\beta_1-\beta_2)}\left[\mu(\alpha-\lambda-\theta)(\beta_1-\beta_2+\theta)-\theta(\alpha\beta_2-\lambda\beta_1-\theta\beta_3)\right]$$

is strictly non negative when the conditions of admissibility of E_5 and E_8 are satisfied. It will be the same for E_6 and E_7, which completes the proof. □

Remark 4 The reasoning would be the same for different signs conditions for (5), (7), (10) and (12) and the proof would be faster since we would have less admissible equilibrium states and less unstable ones.

Theorem 3 *Each admissible and locally asymptotically stable equilibrium state is globally asymptotically stable.*

Proof Let $\Gamma(t) = (s(t), ra(t), ru(t))$ be any non trivial periodic orbit of System (3) with period $T > 0$. Let us show that

$$\int_0^T tr\,(J(s(t), ra(t), ru(t)))\,dt$$

is strictly negative. One has

$$\begin{aligned} tr\,(J(s(t), ra(t), ru(t))) = {} & \beta_1(1-2s-ra-ru)-\theta ra-\alpha ru-\mu+\beta_2(1-s-2ra-ru) \\ & +\theta s-\lambda ru-\mu+\beta_3(1-s-ra-2ru)+\lambda ra+\alpha s-\mu. \end{aligned}$$

But

$$\int_0^T \left(\beta_1(1-s-ra-ru)-\theta ra-\alpha ru-\mu\right)dt = \int_0^T \frac{\frac{ds}{dt}(t)}{s(t)}dt = 0,$$

due to the periodicity. In a same way,

$$\int_0^T \left(\beta_2(1-s-ra-ru)+\theta s-\lambda ru-\mu\right)dt = \int_0^T \frac{\frac{dra}{dt}(t)}{ra(t)}dt = 0$$

and

$$\int_0^T \left(\beta_3(1-s-ra-ru)+\lambda ra+\alpha s-\mu\right)dt = \int_0^T \frac{\frac{dru}{dt}(t)}{ru(t)}dt = 0.$$

Consequently,

$$\int_0^T tr\,(J(s(t), ra(t), ru(t)))\,dt = \int_0^T \left(-\beta_1 s(t)-\beta_2 ra(t)-\beta_3 ru(t)\right)dt$$

is strictly negative. In order to complete the proof, we just have to notice that this computation does not depend on the conditions of admissibility of the equilibrium states so is valid for all admissible and locally asymptotically stable equilibrium state. □

5 Concluding Remarks

In this work, we introduced a new model for the propagation of a rumor taking into account the different changes of classes of the individuals of a social network. We point out conditions of coexistence of each of these classes. Another approach would be to say for example in which conditions one or two of the three groups will die out, in particular the ones of the spreaders and the accepting stiflers.

It will be interesting in a following to study the sociological, economical and psychological elements involved in the decision for a person to become or not a spreader himself. A first approach could be a stochastical one because the rumor diffusion process is mainly driven by socio-psychological elements and it is in this direction that we have to look for.

References

S. Bernard, G. Bouza, A. Piétrus, An optimal control approach for e-rumor. Rev. Invest. Oper. **36**(2), 108–114 (2015)

S. Bernard, G. Bouza, A. Piétrus, An e-rumour model with control on the spreaders. C. R. Acad. Bulg. Sci. **69**(11), 1407–1414 (2016)

S. Bernard, T. César, A. Piétrus, Some actions to control e-rumor. e-J. Caribb. Acad. Sci. **9**(1), 1–8 (2017)

S. Bernard, T. César, A. Piétrus, Spreading rumors and external actions, in *Large-Scale Scientific Computing (LSSC 2017)*. Lecture Notes in Computer Science, vol. 10665 (Springer, 2018), pp. 193–200

L.A. Cherkas, A.A. Grin, Bendixson-Dulac criterion and reduction to global uniqueness in the problem of estimating the number of limit cycles. Differ. Equ. **46**(1), 61–69 (2010)

M. Collard, P. Collard, L. Brisson, E. Stattner, Rumor spreading modeling: profusion versus scarcity, in *ASONAM 2015: IEEE/ACM International Conference on Advances in Social Networks Analysis and Mining*, Paris, France (2015), pp. 1547–1554

D.J. Daley, D.G. Kendall, Epidemics and rumors. Nature **204**, 11–18 (1964)

D.J. Daley, D.G. Kendall, Stochastic rumours. IMA J. Appl. Math. **1**, 42–55 (1965)

K. Dietz, Epidemics and rumours: a survey. J. Royal Soc. A **130**(4), 505–528 (1967)

E.J. Dockner, G. Feichtinger, On the optimality of limit cycles in dynamic economic systems. J. Econ. **53**(1), 31–50 (1991)

G. Feichtinger, A. Novak, F. Wirl, Limit cycles in intertemporal adjustment models. J. Econ. Dyn. Control **18**, 353–380 (1994)

H.I. Freedman, P. Moson, Persistence definitions and their connections. Proc. Am. Math. Soc. **109**(4), 1025–1033 (1990)

H.I. Freedman, P. Waltman, Persistence in models of three species interacting predator-prey populations. Math. Biosci. **68**, 213–231 (1984)

D. Grass, J.P. Caulkins, G. Feichtinger, G. Tragler, D.A. Behrens, *Optimal Control of Nonlinear Processes, with Applications in Drugs, Corruption and Terror* (Springer, Berlin, 2008)

E. Hansen, T. Day, Optimal control of epidemics with limited resources. J. Math. Biol. **62**, 423–451 (2011)

J. Huang, X. Jin, Preventing rumor spreading on small-world networks. J. Syst. Sci. Complex. **24**, 449–456 (2011)

D. Maki, *Mathematical Models and Applications, with Emphasis on Social, Life, and Management Sciences* (Prentice Hall College Division, Englewood Cliffs, 1973)

A.K. Misra, A simple mathematical model for the spread of two political parties. Nonlinear Anal. Modell. Control **17**(3), 343–354 (2012)

A. Rapoport, Spread of information through a population with socio-structural bias. I: assumption of transitivity. Bull. Math. Biophys. **15**, 523–533 (1953)

A. Rapoport, Spread of information through a population with socio-structural bias. II: various models with partial transitivity. Bull. Math. Biophys. **15**, 535–546 (1953)

A. Rapoport, L.I. Rebhun, On the mathematical theory of rumor spread. Bull. Math. Biophys. **14**, 375–383 (1952)

J.C. Robinson, *Infinite-Dimensional Dynamical Systems: An Introduction to Dissipative Parabolic PDEs and the Theory of Global Attractors*. Cambridge Texts in Applied Mathematics (Cambridge University Press, Cambridge, 2001)

E. Stattner, R. Eugenie, M. Collard, How do we spread on Twitter, in *9th IEEE International Conference on Research Challenges in Information Science RCIS*, Athens, Greece (2015), pp. 334–341

F. Wirl, Social interactions within a dynamic competitive economy. J. Optim. Theory Appl. **133** 385–400 (2007)

Optimal Spatiotemporal Management of Resources and Economic Activities Under Pollution Externalities

Anastasios Xepapadeas and Athanasios N. Yannacopoulos

Abstract We propose a non-local spatial model for the study of the effects of pollution externalities on optimal resource management. The effects of the spatial variation and non-locality on the optimal state and the optimal policy are studied.

1 Introduction

Resource management—renewable or exhaustible resources—is usually analyzed in terms of dynamic models where the resource stock is a state variable that evolves in time, and harvesting or extraction per unit time is a control variable. The evolution of the state variables under the influence of resource growth functions and harvesting or extraction is modeled in general by dynamical systems consisting of nonlinear ordinary differential equations (ODEs). Pollution management problems are dynamic when pollution has stock and not flow characteristics (e.g. accumulation of phosphorus in a lake that may cause eutrophication, accumulation of airborne particles and pollutants from combustion creating "brown clouds"). The system's evolution is described in this case by dynamical systems of ODEs.

However, variables describing the state of an environmental system such as resources (renewable or exhaustible), pollutants, greenhouse gases (GHGs), heat, and precipitation have a profound spatial dimension in addition to their temporal dimension. This is because:

(i) Resources or pollutants are harvested, extracted, emitted, or abated in a specific location or locations.
(ii) The impacts of environmental variables, whether beneficial or detrimental, have a strong spatial dimension. Polar amplification suggests that the temperature increases faster in the Poles than the Equator because of the surface albedo

A. Xepapadeas · A. N. Yannacopoulos (✉)
Athens University of Business and Economics, Athens, Greece
e-mail: ayannaco@aueb.gr

G. Feichtinger et al. (eds.), *Control Systems and Mathematical Methods in Economics*, Lecture Notes in Economics and Mathematical Systems 687,
https://doi.org/10.1007/978-3-319-75169-6_19

feedback and heat transfer polarwards. Polar amplification is an established natural phenomenon since the recorded temperature anomaly is consistently higher at the Poles that the Equator and results in the loss of sea ice and land ice. Loss of sea ice could be beneficial because of the potential opening of new shipping lanes and the potential opening of access to previously inaccessible natural resources and fossil fuel reserves Loss of land ice may cause damages to lower latitudes because of sea level rise. Thus there is strong spatial dimension in climate change policies. The Atmospheric Brown Clouds (ABC) can be regarded as re‡ecting the spatial structure of air pollution. As stated in a recent UNEP study (Ramanathan et al. 2008), ABC consist of particles (or primary aerosols) and pollutant gases, such as nitrogen oxides (NOx), carbon monoxide (CO), sulphur dioxide (SO2), ammonia (NH3), and hundreds of organic gases and acids. ABC plumes which result from the combustion of biofuels from indoors; biomass burning outdoors and fossil fuels, are found in all densely inhabited regions and oceanic regions downwind of populated continents.[1] Fisheries crashes have a profound spatial dimension e.g. the Peruvian coastal anchovies fisheries crash on the 1970s, or the collapse of the Atlantic northwest cod fishery of the cost of Newfoundland in early 1990s.

(iii) There is transport of environmental state variables across geographical space due to natural processes.

- Energy balance climate models (EBCMs) explicitly account for the transport of heat across the globe from the Equator to the Poles (e.g. North et al. 1981)
- There is horizontal heat and moisture transport across the globe in more general EBCMs.
- Air-borne contaminants are transported in the atmosphere from the source of emissions due to turbulent eddy motion and wind.
- Renewable resources move in a given spatial domain.

When forward-looking optimizing economics agents that take decisions regarding resource management or emissions ignore transport effects, they essentially ignore the impact of their own actions on the utility or products of agents located at different sites. This is a spatial externality, which is not internalized. Therefore efficient policy seeking to maximize social welfare should involve mechanisms to internalize spatial spillovers, along with potential temporal spillovers.

It should be noted that although the spatial dimension is important in resource management not much research in the spatial aspect of environmental and resource economics has been undertaken, although there are notable exceptions in several cases such as: Spatially dependent taxes (e.g. Xabadia et al. 2004, Goetz and Zilberman 2007); Spatial resource models and spatial fishery models (e.g. Wilen 2007, Smith et al. 2009, Desmet and Rossi-Hansberg 2010, Brock et al. 2014b,a,

[1]Five regional ABC hotspots around the world have been identified: (i) East Asia, (ii) Indo-Gangetic Plain in South Asia, (iii) Southeast Asia, (iv) Southern Africa; and (v) the Amazon Basin.

Behringer and Upmann 2014, Camacho and Perez-Barahona 2015); Spatial models of climate and the economy (e.g. Brock et al. 2013, Brock et al. 2014c, Hassler and Krusell 2012, Desmet and Rossi-Hansberg 2015); Spatial growth models (e.g. Boucekinne et al. (2009), Boucekinne et al. (2013), Fabbri (2016), Camacho and Perez-Barahona (2016) and references therein).

The lack of substantial literature incorporating spatial issues in environmental and resource economics can be attributed to the technical difficulties involved when the mathematics of optimal control theory is extended to infinite dimensional state spaces that naturally emerge when optimization takes place in a spatiotemporal domains. Exceptions try to overcome the mathematical complication by imposing a certain structure to the problem that allows simplifications and sometimes closed form solutions. However, the importance of transport phenomena in environmental and resource economics, and the need to design regulation for internalizing spatial externalities emerging from these transport phenomena, make it necessary to extend dynamic optimization methods into spatial settings.

In this context we study dynamic optimization for the joint management of resources and pollution when pollution affects resource growth and when spatial transport phenomena both for the resources and the pollution are present. We present approaches that deal with dynamic optimization in infinite dimensional spaces which can be used as tools in environmental and resource economics, along with examples of their application. We also present methods which can be used to study the emergence of spatial patterns in dynamic optimizations models.

Our methods draw on the celebrated Turing difusion induced instability but the nature of the instability discussed here is fundamentally diferent from Turing's mechanism since it applies to forward-optimization models, and to a saddle point solution. We believe that this approach provides the tools to analyze a wide range of problems with explicit spatial structure which are very often encountered in environmental and resource economics.

2 The Model

Consider a two sector economy, consisting of an sector producing output which generates emissions (pollution) and a harvesting sector. The economy exists in a 1 finite spatial domain D and evolves in time. Thus the pair (x, t) denotes the spatial location $x \in D$ at time $t \in \mathbb{R}_+$. The harvesting sector specializes in a single species of a renewable resource and is localized on a subset $D_h \subset D$. The biomass of the resource is modelled by a density function $v : \mathbb{R}_+ \times D \to \mathbb{R}$, where $v(t, x)$ determines biomass at spatial location x and time t, and is such that $v(t, x) \geq 0$ if $x \in D_e$ while $v(t, x) = 0$ if $x \notin D_e$. The industrial sector is localized on a subset $D_e \subset D$ such that $D_e \cap D_h = \emptyset$, and without loss of generality we may assume that $D_e \cup D_h = D$. The industrial sector generates externalities in the form of emissions which are accumulated as pollution stock S. The pollution stock is generated at D_e but it effects through spatial transport also D_h i.e., the whole of the spatial domain. Without loss of generality and for the sake of visualisation one may consider $D = [0, L]$, $D_h = [0, L_0]$ and $D_e = [L_0, L]$, for $L_0 < L$.

The spatio-temporal evolution of the biomass is given by the partial differential equation (PDE):

$$\frac{\partial}{\partial t} v(t,x) = f(v(t,x), K(t)) + d_1 \Delta v(t,x) - H(t,x)\mathbf{1}_{D_h}(x), \quad (t,x) \in \mathbb{R}_+ \times D_h,$$
$$v(t,x) = 0 \quad (t,x) \in \mathbb{R}_+ \times D \setminus D_h, \tag{1}$$

where Δ is the Laplacian operator, which models spatial transport of the biomass (e.g., in Cartesian coordinates and one dimension $\Delta = \frac{\partial^2}{\partial x^2}$), f is a nonlinear function modelling the biological dynamics of the species, K is the carrying capacity of the environment and H is a harvesting function. Harvesting is assumed to take place on a open subset $D_h \subset D$, and the indicator function $\mathbf{1}_{D_h}$ ($\mathbf{1}_{D_h}(x) = 1$ if $x \in D_h$, $\mathbf{1}_{D_h}(x) = 0$ if $x \notin D_h$) is modelling the localized nature of harvesting. A good choice for the harvesting function H is to choose $H(t,x) = h(t,x)\, v(t,x)$ where $h(t,x) \in (0,1)$ is a function which gives the fraction of the biomass which is harvested at position x and time t. A typical example for the function f is the logistic function $f(v, K) = av\left(1 - \frac{v}{K}\right)$. The term $d_1 \Delta v$ models the spatial transport of the species in terms of a diffusion mechanism, whereas h is the harvesting term. The carrying capacity of the environment is not considered to be a constant but rather a varying quantity which depends on spatial location and time modelled by a function $K : D \times \mathbb{R}_+ \to \mathbb{R}_+$. Depending on the level of complexity of the model we may discard the spatial dependence on K and leave only the temporal dependence. Concerning the boundary conditions for the biomass, we may consider homogeneous boundary conditions for all boundaries, a choice which is consistent with our assumption that v vanishes on $D \setminus D_h$, i.e. that it also vanishes on the internal boundary between D_e and D_h. From the modelling point of view we may assume that the biomass is exterminated once it hits the boundary with the industrial region. Other boundary conditions such as eg. Neumann boundary conditions may be accomodated.

The output producing sector of the economy interacts with the harvesting sector via the carrying capacity K. Although in the majority of renewable resource management models, carrying capacity is regraded as fixed, in reality as noted by Arrow et al. (1995, p.520)

> Carrying capacities in nature are not fixed, static, or simple relations. They are contingent on technology, preferences, and the structure of production and consumption. They are also contingent on the everchanging state of interactions between the physical and biotic environment.

In terms of the present model the output producing sector generates pollution externalities through emissions which have a negative effect on the harvesting sector by decreasing the carrying capacity K thus reducing the biomass and having a negative overall effect on the harvesting. This can be for example the case of a lake or a closed sea (e.g. the Baltic) where agricultural pollution through phosporous deposits in the water affects the carying capacity of lakes' or the sea's fisheries.

Emissions are generated by economic agents (or producers) at different locations $x \in D_e$, where by $s(t,x)$ we denote the emissions at time t and location x. Emissions generate benefits locally through a benefit function B, in particular, $B(x, s(t,x))$ is the benefit generated from emissions at point $(t,x) \in \mathbb{R}_+ \times D_e$. The benefit function is assumed to be increasing and strictly concave.

The emissions aggregate and create locally (in D_e) an emissions stock $S(t,x)$, which may however be transported beyond D_e, on the whole of D and thus affect the biomass stock evolution at D_h, through the dependence of the carrying capacity on the spatial distribution of the emissions stock S. The emissions stock displays a spatio temporal dynamics of the form

$$\frac{\partial}{\partial t} S(t,x) = d_2 \Delta S(t,x) - \delta S(t,x) \mathbf{1}_{D_e}(x) + s(t,x) \mathbf{1}_{D_e}(x), \quad (t,x) \in \mathbb{R}_+ \times D, \tag{2}$$

where $d_2 \Delta S$ models the spatial transport of the emissions and the term $-\delta S \mathbf{1}_{D_e}$ models the absorption of emissions and the tendency to restore to its initial state. This term is assumed to act only over the domain D_e, where abatement procedures which operate in the background and are not modelled here may be undertaken. We assume that pollution in the water body depreciates very slowly, because, for example phosphorus loading are absorbed in sediments and remain there for a very their for a long time with very slow recycling to water (Grass et al. 2017). The emissions s, which contributes to the emissions aggregate S, is localized to an open subset $D_e \subset D$ of the spatial domain, and this localization effect is modelled in terms of the indicator function $\mathbf{1}_{D_e}$ of the subset D_e, ($\mathbf{1}_{D_e}(x) = 1$ if $x \in D_e$ and $\mathbf{1}_{D_e}(x) = 0$ if $x \notin D_e$). The spatial domain D_e models the positions in D where industrial or other activity generates pollution externalities, that are localized in certain parts of the globe. We assume that $D_h \cap D_e = \emptyset$ and $D_h \cup D_e = D$. Note that emissions stock S, on the contrary to s, is not localized but global, through the spatial transport of the effects of emissions, modelled here by the Laplacian term $d_2 \Delta S(t,x)$. The PDE (2) must be complemented with appropriate boundary conditions, which can either be assumed to be Dirichlet, Neumann or periodic boundary conditions.

The carrying capacity is negatively affected by the emissions stock S. This effect depends on some weighted spatial average of the total stock i.e. is assumed to depend on

$$(\mathcal{T}S)(t) = \int_D k(x') \mathbf{1}_{D_h}(x') S(t,x') dx',$$

where k is a kernel function providing the weights for the spatial average. It is very reasonable to assume that the kernel function k only has support on D_h, i.e. it is a function of the form $k \mathbf{1}_{D_h}$, as we expect only the pollution stock over the region occupied by the biomass to affect the carrying capacity. The effect of the total emissions stock on the carrying capacity is then modelled by

$$K(t) = K((\mathcal{T}S)(t)) = K(\int_D k(x')\mathbf{1}_{D_h}(x')S(t,x')dx'), \tag{3}$$

where $K : \mathbb{R}_+ \to \mathbb{R}_+$ is a decreasing function which is positive and bounded away from zero.

Then we get a coupled system of the form

$$\begin{aligned} \frac{\partial}{\partial t}v &= f(v, K(\mathcal{T}(S)) + d_1\Delta v - hv, \text{ in } D_h \\ v &= 0, \text{ in } D \setminus D_h \\ \frac{\partial}{\partial t}S &= d_2\Delta S - \delta S\mathbf{1}_{D_e} + s\mathbf{1}_{D_e}, \text{ in } D \end{aligned} \tag{4}$$

where e.g. $f(v, S) = \alpha v\left(1 - \frac{v}{K(\mathcal{T}(S))}\right)$, with $K(\mathcal{T}(S))$ as in (3). Our state equations is a coupled nonlocal reaction diffusion system, with homogeneous Dirichlet boundary conditions for v and either Dirichlet, Neumann or periodic boundary conditions for S. The variables (v, S) are the state variables and the variables (h, s) are the control variables.

The decision maker (social planner or regulator) seeks to maximize the total net benefits from both sectors, the harvesting sector and the output producing sector. The benefits from the harvesting sector depends on h and is a concave function of $U(hv)$ of total harvesting at each spatial point.[2] The net benefits from the output producing sector are represented a concave function of emissions s, $B(s)$. At the level of the economy we also assume that the social planner takes into account external damages to the economy by the pollution stock $S(x,t)$. This is modeled by a convex damage function $Z(S)$. These damages could include adverse effects on the population living on land, and uses of the amenities of the water body. Thus the decision maker chooses (h, s) so as to maximize

$$J(h,s) = \int_0^\infty \int_D e^{-rt}\Big(U(hv)\mathbf{1}_{D_h} + B(s)\mathbf{1}_{D_e} - Z(S)\Big)dxdt, \tag{5}$$

under the dynamic constraints (4).

The control variable h is constrained to take values in $[0, 1]$ and the control variable s in constrained to take values in $[0, \bar{s}]$ where $\bar{s}$ is the maximum possible pollution rate allowed by the production capacity of the economy. Let us define by C the set of admissible controls. It can be shown that the state Eq. (4) is well posed for any admissible choice of control procedure $(h, s) \in C$.

[2]To simplify the exposition we assume no harvesting costs.

3 A Necessary Condition in Terms of the Pontryagin Maximum Principle

In this section, assuming the existence of a solution to this optimal control system, we derive a necessary condition that allows for the identification of the optimal path and the optimal control procedure.

In order to express the maximum principle we need to define the adjoint variables (p, q), which are solutions of the backward system

$$\begin{aligned}
&\frac{\partial p}{\partial t} = -d_1 \Delta p - F_1 p + rp + U'(hv)h, \quad \text{in } \mathbb{R}_+ \times D_h, \\
&p = 0, \quad \text{in } \mathbb{R}_+ \times (D \setminus D_h), \\
&\frac{\partial q}{\partial t} = -d_2 \Delta q - \mathcal{T}^*(F_2 p) + (\delta \mathbf{1}_{D_e} + r)q - Z'(S), \quad \text{in } \mathbb{R}_+ \times D,
\end{aligned} \tag{6}$$

where $\mathcal{T}^*$ is the adjoint operator of $\mathcal{T}$, and the functions $F_0, F_1, F_2 : \mathbb{R}_+ \times D \to \mathbb{R}$ are defined by

$$\begin{aligned}
&F_0(t,x) := f_v(v(t,x), K(\mathcal{T}(S))(t)), \quad (t,x) \in \mathbb{R}_+ \times D_h, \\
&F_1(t,x) := F_0(t,x) - h(t,x), \quad (t,x) \in \mathbb{R}_+ \times D_h, \\
&F_2(t,x) := f_K(v(t,x), K(\mathcal{T}(S))(t))K'(\mathcal{T}(S)(t)), \quad (t,x) \in \mathbb{R}_+ \times D_h, \\
&F_0(t,x) = F_1(t,x) = F_2(t,x) = 0, \quad (t,x) \in \mathbb{R}_+ \times (D \setminus D_h),
\end{aligned} \tag{7}$$

for some choice of functions (h, s, v, S) satisfying the state Eq. (4), and the term $\mathcal{T}^*(F_2 p)$ is defined as

$$\left(\mathcal{T}^*(F_2 p)\right)(t,x) = \left(\int_{D_h} F_2(t,x')p(t,x')dx'\right) k(x)\mathbf{1}_{D_h}(x). \tag{8}$$

This system is solved for homogeneous Dirichlet boundary conditions for p, Dirichlet, Neumann or periodic boundary conditions for q and with a transversality condition of the form

$$\lim_{t\to\infty} e^{-rt} \int_D (v(t,x)p(t,x) + S(t,x)q(t,x))dx = 0, \quad a.e. \, t \in (0,\infty).$$

Without loss of generality let us assume that q satisfies periodic boundary conditions.

The following proposition provides the maximum principle.

Proposition 1 *Let* (v, S) *be and optimal path and* (h, s) *be the corresponding optimal control protocol. Then, there exists a pair of processes* (p, q) *which along with the quadruple* (v, S, h, s) *satisfy the set of forward-backward PDEs* (4)–(6)

along with the maximality condition

$$\int_D [(U'(hv) + p)v\eta \mathbf{1}_{D_h} + (B'(s) - q)\sigma \mathbf{1}_{D_e}]dx \leq 0, \quad \forall\ (\eta, \sigma) \in C_0, \tag{9}$$

where C_0 *is the relative interior of* C.

Proof See Appendix 1. ■

One can in principle reduce the co-state variables (p, q) from the above system as follows: Choose any pair of functions (u, S) and define the mapping $(u, S) \mapsto (p, q)$ where (p, q) is the solution of the adjoint system for the chosen pair. Denote $(p, q) = \mathcal{G}(u, S)$, where $\mathcal{G}$ is a functional mapping the (functional) parameters of the adjoint system to its solution. Then by substituting $(p, q) = \mathcal{G}(u, S)$ into the state system we obtain a system in terms only of the state equations, where the mapping $\mathcal{G}$ may be used to construct the relevant control procedure. Since the construction of the map $\mathcal{G}$ is not an easy task, we prefer to work with the full system provided by the maximum principle.

Remark 2 An equivalent way of expressing the maximum principle is by defining a Hamiltonian function, in terms of the adjoint variables $(\mathfrak{p}, \mathfrak{q}) = (-p, -q)$, and by defining the Hamiltonian function

$$\mathfrak{H}(h, s) = \int_D \Big\{(U(hv)\mathbf{1}_{D_h} + B(s)\mathbf{1}_{D_e} - D(S))$$

$$+\mathfrak{p}(d_1\Delta v + f(v, K(\mathcal{T}S)) - hv)\mathbf{1}_{D_h} + \mathfrak{q}(d_2\Delta S - \delta S\,\mathbf{1}_{D_e} + s\,\mathbf{1}_{D_e})\Big\}dx.$$

It can be seen that condition (9) is a maximization condition for this Hamiltonian function, thus allowing us to derive the forward-backward PDE system (4)–(6) in terms of the Pontryagin maximum principle. This allows us to understand $\mathfrak{p}$ as the shadow price for the resource biomass and $\mathfrak{q}$ as the shadow cost for pollution stock.

To simplify the exposition a little, assume that the functions U and B are such that the maximum is attained for $h > 0$ and $s > 0$ so that for any (η, σ) we may consider $\epsilon(\eta, \sigma)$ and $-\epsilon(\eta, \sigma)$ as a viable perturbation in (9), leading to the equivalent condition

$$U'(hv) + p = 0,$$

$$B'(s) - q = 0.$$

Under this assumption, and defining by I_1 the inverse function of U' and I_2 the inverse function of B', we can express the maximality condition as

$$hv = I_1(-p),$$

$$s = I_2(q).$$

This allows to express the necessary condition in terms of (v, S, p, q) only, and characterize the optimal path and the optimal control procedure in terms of the solution of the forward-backward PDE system

$$\begin{aligned}
&\frac{\partial v}{\partial t} = d_1 \Delta v + f(v, K(\mathcal{T}(S)) - I_1(-p), \quad \text{in } D_h \\
&\frac{\partial S}{\partial t} = d_2 \Delta S - \delta S \mathbf{1}_{D_e} + I_2(q) \mathbf{1}_{D_e}, \quad \text{in } D \\
&\frac{\partial p}{\partial t} = -d_1 \Delta p + (r - F_0) p, \quad \text{in } D_h \\
&\frac{\partial q}{\partial t} = -d_2 \Delta q - \mathcal{T}^*(F_2\, p) + (r + \delta \mathbf{1}_{D_e}) q - Z'(S), \quad \text{in } D \\
&v = p = 0, \quad \text{in } D \setminus D_h
\end{aligned} \tag{10}$$

with the functions F_0, F_2 defined as in (7) and the term $\mathcal{T}^*(F_2\, p)$ as in (8). This is a nonlinear PDE with nonlocal terms, which must be supplemented with appropriate boundary conditions for all the variables, initial conditions for the state variables (v, S) and the transversality condition for the adjoint variables (p, q).

It is highly unlikely that the system (10) can be solved analytically, in closed form and we need to resort to numerical techniques. Furthermore, the treatment of the problem of solvability such forward-backward integrodifferential PDE systems requires the development of abstract techniques, a task which is beyond the scope of the present work. Note that even in the absence of the non local terms systems of the type (10) display subtleties as to their global behaviour, and sensitivity with respect to the transversality condition (see for instance Boucekinne et al. (2009) where the possibility of ill posedness in the sense of Hadamard for a similar system without non local terms is reported, see also Ballestra (2016)). Such behaviour has motivated certain researchers (see e.g. Boucekinne et al. (2016) and references therein) to work with the dynamic programming formulation of the control problem (in the absence of non local effects), using the Hamilton-Jacobi equation for the value function. This approach bypasses the problem of ill posedness by endogenizing the choice of transversality condition for the adjoint variable, and requires the use of tools from infinite dimensional analysis (the HJ equation is a PDE on an infinite dimensional Hilbert space). While equivalent, the Pontryagin approach has a more dynamic flavour and is thus better suited for understanding the dynamical behaviour of the system when deviating from the optimal state (which is the main focus of this paper). On the other hand, the dynamic programming approach is ideal for obtaining feedback optimal laws, but unfortunately (as with the Pontryagin approach) examples of analytic solutions are very rare (see e.g. Boucekinne et al. (2016) for the AK system).

The above discussion incites the need for further study of system (10) in order to clarify whether such effects are present here. The answer to these questions would be important for the uniqueness of the maximizer and the stability of calculation of the optimal path.

However, even before clarifying these questions there is still a lot that can be said from the local analysis point of view, that will provide important information on the qualitative nature of the solutions. By local we mean that we focus our attention around a particular optimal state which a solution of system (10). If the system happens to be ill posed (i.e. if there are possibly more that one solutions of (10) we focus our attention to the particular optimal state that either the system has locked in by chance or the by manipulation of the boundary conditions the decision maker has forced it to lock in. Such issues will be the main concern of the remaining part of the paper.

4 The Steady State Solution and Its Properties

We are now interested in special solutions of the optimally controlled system, for which there is no time dependence, which will be here after called the steady state solution. These can be understood as fixed points of the forward-backward infinite dimensional dynamical system which is presented in (10), and is the infinite dimensional analogue of a saddle point. Such a saddle point however, in principle will have still a spatial structure, i.e., it will be a function of space. A steady state will be the solution of the system of non-local partial differential equations (integro-differential equations)

$$
\begin{aligned}
&0 = d_1 \Delta v + f(v, K(\mathcal{T}(S)) - I_1(-p), \quad \text{on } D_h\\
&0 = d_2 \Delta S - \delta S \mathbf{1}_{D_e} + I_2(q)\mathbf{1}_{D_e}, \quad \text{on } D\\
&0 = -d_1 \Delta p + (r - f_v(v, K(\mathcal{T}(S)))p, \quad \text{on } D_h\\
&0 = -d_2 \Delta q - \mathcal{T}^*(f_K(v, K(\mathcal{T}(S))K'(\mathcal{T}(S))p) + (r + \delta \mathbf{1}_{D_e})q - Z'(S), \quad \text{on } D,\\
&v = p = 0, \quad \text{on } D \setminus D_h
\end{aligned}
\tag{11}
$$

with homogeneous Neumann boundary conditions, where the actions of the operators $\mathcal{T}$ and $\mathcal{T}^*$ on any function u (see Appendix A) are defined as

$$
(\mathcal{T}u)(x) = \int_{D_h} k(x')u(x')dx',
$$

$$
(\mathcal{T}^*u)(x) = \left(\int_{D_h} u(x')dx' \right) k(x)\mathbf{1}_{D_h}(x).
$$

Note that in general, when $\mathcal{T}$ acts on any function it delivers a constant, whereas when $\mathcal{T}^*$ acts on any function it delivers a constant multiple of the spatially varying function k which is the averaging kerne, localized on D_hl. In the special case where $k = k_0$, a constant, both $\mathcal{T}$ and $\mathcal{T}^*$ deliver a constant when acting on any function. This special case, corresponds to the case where the effect of pollution on the carrying capacity of the species is obtained by an unweighted average of the pollution stock over the whole spatial domain D (i.e. all regions contribute equally on the effect of pollution on the carrying capacity).

Upon defining the constants

$$\begin{aligned} S_a &= \int_{D_h} k(x)S(x)dx, \\ K_a &= K(S_a), \quad K'_a = K'(S_a), \\ \Lambda_a &= \int_{D_h} f_K(v(x), K_a)K'_a p(x)dx, \end{aligned} \tag{12}$$

system (11) can be expressed as

$$\begin{aligned} &0 = d_1 \Delta v + f(v, K_a) - I_1(-p), \quad \text{on } D_h \\ &0 = d_2 \Delta S - \delta S \mathbf{1}_{D_e} + I_2(q)\mathbf{1}_{D_e}, \quad \text{on } D \\ &0 = -d_1 \Delta p + (r - f_v(v, K_a))p, \quad \text{on } D_h \\ &0 = -d_2 \Delta q - \Lambda_a k \mathbf{1}_{D_h} + (r + \delta \mathbf{1}_{D_e})q - Z'(S), \quad \text{on } D, \\ &v = p = 0, \quad \text{on } D \setminus D_h. \end{aligned} \tag{13}$$

This is a nonlocal system, which in principle can by solved as follows: Fix the constants S_a, Λ_a and for this choice solve the resulting nonlinear elliptic PDE (13) with the appropriate boundary conditions, providing a quadruple (v, S, p, q) depending on the choice of S_a, Λ_a. Using this solution return to (12) and evaluate the new values $\bar{S}_a$ and $\bar{\Lambda}_a$ of the constants as provided by the solution of (13). This defines the map $M : \mathbb{R}_+ \times \mathbb{R} \to \mathbb{R}_+ \times \mathbb{R}$, with action $(S_a, \Lambda_a) \mapsto (\bar{S}_a, \bar{\Lambda}_a)$. If (S_a^*, Λ_a^*) is a fixed point of this map, then the solution of the system (13) corresponding to the choice of the parameters (S_a^*, Λ_a^*) is the solution we seek for. Clearly, the procedure we have described above, can be turned to a proof of existence of solution to the nonlocal elliptic problem (11), and may also allow us to construct a numerical method for the solution of this problem.

A particularly interesting steady state is the so called flat steady state, which corresponds to a solution of (10) which is spatially independent. The means we look for a solution of (10) such that

$$\begin{aligned} u(x) = u_h, \quad p(x) &= p_h, \quad \text{on } D_h, \\ u(x) = p(x) &= 0, \quad \text{on } D \setminus D_h, \\ S(x) &= S, \quad \text{on } D, \\ q(x) &= q_e, \quad \text{on } D_e, \\ q(x) &= q_h, \quad \text{on } D_h. \end{aligned} \tag{14}$$

Note that we assume continuity of the pollution stock over the whole of the domain D. With minor adjustements we could assume different levels pollution stock over D_e and D_h, e.g. S_e and S_h respectively. Clearly, by a quick inspection of the system the solution of (10) such a solution is only feasible if $k(x) = k$ a constant, and we will make this assumption from now on.

We first address the question of when such a solution may exist. We claim that such a solution exists when the averaging kernel k is constant $k = k_0$ for all $x \in D$. If $k = k(x)$, i.e. if the operator $\mathcal{T}$ corresponds to a weighted averaging then we claim that a flat steady state of (10) cannot exist.

Proposition 3 *A spatially independent steady state optimal solution exists if and only if $k(x) = k$ is a constant. In this case, it is unique and is given in terms of the solution ν of the algebraic equation*

$$\nu = \frac{1}{r+\delta} Z'\left(\frac{1}{\delta} I_2(\nu)\right)$$

as

$$S = \frac{1}{\delta} I_2(\nu), \quad v_h = I_3(r; K_a),$$

$$-p_h = U'(f(v_h, K_a)), \quad q_e = \nu, \quad q_h = \frac{\Lambda_a k}{r} + \frac{1}{r} Z'(S),$$

where $I_3(y; K)$ is the inverse function of $f_v(v, K)$ with respect to the first argument (i.e. $f_v(I_3(y; K), K) = y$), whereas

$$K_a = K(\bar{k}S), \quad \bar{k} = k\,|D_h|,$$
$$\Lambda_a = f_K(v_h, K_a)K'((\bar{k}S)|D_h|.$$

and $|D_h|$ is the Lebesgue measure of D_h.[3]

The optimal state is maintained by the control strategy

$$h = \frac{f(v_h, K_a)}{v_h}, \quad s = \delta S.$$

Proof The proof is given in Appendix 2 ■

The following remark is useful: The optimal steady states v_h and S are determined by the golden rules

$$f_v(v_h, K_a) = r, \quad \delta S = s.$$

[3] $|D| = \mathcal{L}(D)$ is the length of D if $D \subset \mathbb{R}$, the area of D if $D \subset \mathbb{R}^2$ and the volume of D if $D \subset \mathbb{R}^3$.

The above rules are a sensible suggestion, since it requires that the pollution rate should equal the natural rate by which the pollution stock decays, where as the rate of regeneration of the natural resource f_v should equal the discount rate r. This naturally holds for specific values of v_h and K_a (hence S). Note also that the shadow price of pollution is different over D_h and over D_e.

Example 4 (The Logistic Growth Model) In order to provide a concrete example of this flat steady state solution, consider the case where the function f corresponds to logistic growth,

$$f(v, K) = \alpha v \left(1 - \frac{v}{K}\right),$$

and the carrying capacity K depends on the global pollution stock in terms of an exponential decay law as

$$K = K_m \exp\left(-\beta \mathcal{T}(S)\right), \quad \beta > 0.$$

In the above, K_m is the maximal carrying capacity, which decreases as the pollution stock increases ($\mathcal{T}$ is a positive operator). We furthermore assume that the utility, the benefit and the damage function are Cobb-Douglas and of the form are all logarithmic and of the form

$$U(y) = \frac{y^{1-\gamma_1}}{1-\gamma_1}, \quad B(y) = \frac{y^{1-\gamma_2}}{1-\gamma_2}, \quad Z(y) = \frac{y^{1-\gamma_3}}{1-\gamma_3}, \quad \gamma_i < 1.$$

In this case we see that $I_i(y) = y^{-\gamma_i}$, $i = 1, 2$, so that the flat steady state is parametrized in terms of the solution of the algebraic equation

$$\nu = \delta^{\gamma_3} \nu^{\gamma_3/\gamma_2}.$$

This admits the trivial solution $\nu = 0$ which is discarded[4] and the non trivial solution

$$\nu = \delta^{\frac{\gamma_2\gamma_3}{\gamma_2-\gamma_3}}.$$

Then,

$$S = \delta^{-\frac{\gamma_3}{\gamma_2-\gamma_3}}, \quad v_h = \frac{K_a}{2a}(a - r),$$

where

$$K_a = K_m \exp\left(- \beta\, k\, |D_h| \delta^{-\frac{\gamma_3}{\gamma_2-\gamma_3}} \right)$$

[4]It leads to solutions which are spurious and non sensical, as the lead to infinite stock of pollution S.

5 Local Behaviour Near a Steady State and Spatial Pattern Formation

We now consider a solution $\bar{z} := (\bar{v}, \bar{S}, \bar{p}, \bar{q})$ of the time independent problem (11). This a time independent optimal solution **not necessarily** spatially independent (an infinite dimensional analogue of a saddle point). The stable and the unstable manifolds of this saddle point can be obtained through linearization of the full time dependent problem (10).

Consider a solution of the full time dependent system (10) of the form $Z = \bar{z} + \epsilon z$ for some small ϵ, where $z = (v, S, p, q)$ denote the deviation from the time independent optimal solution $\bar{z} = (\bar{v}, \bar{S}, \bar{p}, \bar{q})$. The spatio-temporal evolution of $z = (v, S, p, q)$ is approximated for small enough ϵ by the linearized version of (10) which can be expressed in the compact form

$$z' = Az + Lz + Nz, \tag{15}$$

where $z = (v, S, p, q)^{tr}$ and

$$A = \begin{pmatrix} d_1 \mathbf{1}_{D_h} \Delta_h & 0 & 0 & 0 \\ 0 & d_2 \Delta & 0 & 0 \\ 0 & 0 & -d_1 \mathbf{1}_{D_h} \Delta_h & 0 \\ 0 & 0 & 0 & -d_2 \Delta \end{pmatrix}$$

$$L = \begin{pmatrix} L_1 \mathbf{1}_{D_h} & 0 & L_2 \mathbf{1}_{D_h} & 0 \\ 0 & L_3 \mathbf{1}_{D_e} & 0 & L_4 \mathbf{1}_{D_e} \\ L_5 \mathbf{1}_{D_h} & 0 & (L_6 + r) \mathbf{1}_{D_h} & 0 \\ 0 & L_7 & 0 & L_8 \mathbf{1}_{D_h} + r \end{pmatrix}$$

$$N = \begin{pmatrix} 0 & N_1 \mathbf{1}_{D_h} A_h(kI) & 0 & 0 \\ 0 & 0 & 0 & 0 \\ 0 & -N_2 \mathbf{1}_{D_h} A_h(kI) & 0 & 0 \\ -k\mathbf{1}_{D_h} A_h(N_2 I) & -A_h(N_3) k \mathbf{1}_{D_h} A_h(kI) & -k \mathbf{1}_{D_h} A_h(N_1 I) & 0 \end{pmatrix}$$

where Δ_h, Δ denote the Laplacian operator with Neumann boundary conditions on D_h, and D respectively, $A_h : L^2(D) \to \mathbb{R}$, is the averaging operator over D_h defined by its action on any function u as

$$A_h(u) = \int_{D_h} u(x) dx,$$

and

$$
\begin{aligned}
L_1(x) = f_v(\bar{z}),\ \ L_2(x) = -I_1'(-\bar{p}),\\
L_3(x) = -\delta,\ \ L_4(x) = I_2'(\bar{q}),\\
L_5(x) = -\bar{p} f_{vv}(\bar{z}),\ \ L_6(x) = -f_v(\bar{z}),\\
L_7(x) = -Z''(\bar{z}),\ \ L_8(x) = \delta,
\end{aligned}
$$

and

$$
\begin{aligned}
N_1(x) = f_K(\bar{z})K'(\bar{z}),\\
N_2(x) = \bar{p} f_{vK}(\bar{z})K'(\bar{z}),\\
N_3(x) = \bar{p} f_{KK}(\bar{z})(K'(\bar{z}))^2 + \bar{p} f_K(\bar{z})K''(\bar{z}).
\end{aligned}
$$

This is a linear system consisting of a local part $A + L$ and a nonlocal part N. Recall, that v is identically zero over D_e. If $\bar{z}$ is a **spatially dependent** solution of the time independent problem (10) then the operator $A + L$ is a local operator with spatially dependent coefficients, while N is a nonlocal (integral) operator again with spatially dependent coefficients.

An important question that arises is the stability of the steady state $\bar{z}$. This is an optimal procedure leading to the optimal state $(\bar{v}, \bar{S})$ which is supported by an optimal control policy $\bar{h} = \frac{1}{\bar{v}} I_1(-\bar{p})$ and $\bar{s} = I_2(\bar{q})$. Is that steady state stable, i.e., what happens if we deviate from $\bar{z} = (\bar{v}, \bar{S}, \bar{p}, \bar{q})$ (hence equivalently from $(\bar{v}, \bar{S}, h^{(0)}, s^{(0)})$) by a small deviation that will lead to new path $\bar{z} + \epsilon z = (\bar{v} + \epsilon v, \bar{S} + \epsilon S, \bar{p} + \epsilon p, \bar{q} + \epsilon q)$.

The stability of the steady state $\bar{z}$ is determined by the properties of the spectrum of the linear operator $S := A + L + N$. In particular the steady state $\bar{z}$ is stable if the real part of essential spectrum of S is negative and unstable otherwise. In particular we need to consider the eigenvalue problem for the operator S

$$\lambda z = (A + L + N)z. \tag{16}$$

The eigenmodes that correspond to eigenvalues with positive real part are unstable eigenmodes, whereas the eigemodes that correspond to eigenvalues with negative real part are stable eigenmodes. This provides a generalization of the familiar picture of a saddle point, usually encountered in finite dimensional problems, however, care must be taken with subtle technical issues arising from the infinite dimensional nature of the dynamics. The spectral expansion used is essentially a Galerkin approximation of the linearized system, using a basis for the expansion which diagonalizes the operator and thus reveals in the most transparent fashion the dynamics of the system. Such spectral analysis has been used also in spatial economic optimal control problems by Boucekinne et al. (2016) or Fabbri (2016) for the treatment of the HJ equation of the AK model, using as expansion basis the eigenfunction of the operator $A + L$ for L constant and A being either the Laplace or the Laplace-Beltrami operator. As the problem we consider is nonlinear, in the absence of non local terms, our approach can be considered as a collection of local (in phase space) AK models. However, the nonlocal terms N which are present in (16) are expected to introduce further complications and phenomena.

The treatment of the nonlocal eigenvalue problem is not a very easy task. We note that if $k = 0$ i.e. if the carrying capacity K is unaffected by the pollution stock that the nonlocal operator $N = 0$, thus reducing (16) to a local problem. In this limit the existence of stable and unstable eigenmodes can be obtained using perturbation theory for linear operators.

In the remaining of this section we sketch a numerical procedure that will allow us to approximate certain parts of the spectrum of the operator S and determine conditions for the onset of a pattern formation instability in the case where $\bar{z}$ is spatially independent. If $\bar{z}$ is **spatially independent**, i.e. corresponds to a flat steady state then all the operators, local and nonlocal are constant coefficient operators, and the analysis simplifies considerably. Note that this happens in the case where k is a constant, so that for any function u, we have that $A_h(ku) = kA_h(u)$, and since the functions N_i are also constant we have that $A_h(N_i u) = N_i A_h(u)$, $i = 1, 2, 3$. In this case, the nonlocal operator N simplifies to

$$N = \begin{pmatrix} 0 & N_1 k \mathbf{1}_{D_h} A_h & 0 & 0 \\ 0 & 0 & 0 & 0 \\ 0 & -N_2 k \mathbf{1}_{D_h} A_h & 0 & 0 \\ -kN_2 \mathbf{1}_{D_h} A_h & -N_3 |D_h| k^2 \mathbf{1}_{D_h} A_h & -kN_1 \mathbf{1}_{D_h} A_h & 0 \end{pmatrix}$$

In this case we may use a Galerkin approximation for the solution of the linearized problem, using a properly selected orthonormal basis.

Let $\{\phi_n, \lambda_n\}$, $n \in \mathbb{N}_0 := \{0, 1, 2, \cdots\}$, be the solutions to the eigenvalue problem

$$-\Delta\phi_n = \lambda_n \phi_n, \quad \text{in } D,$$
$$\nabla\phi_n \cdot \eta = 0, \quad \text{on } \partial D,$$

and $\{\psi_n, \mu_n\}$, $n \in \mathbb{N} := \{1, 2, \cdots\}$, be the solutions to the eigenvalue problem

$$-\Delta\psi_n = \mu_n \psi_n, \quad \text{in } D_h,$$
$$\nabla\psi_n \cdot \eta = 0, \quad \text{on } \partial D_h.$$

The $\{\phi_n\}$ are orthogonal in the sense that $\int_D \phi_n \phi_m dx = \delta_{nm}$. The same holds for $\{\phi_n\}$ are orthogonal in the sense that $\int_{D_h} \psi_n \psi_m dx = \delta_{nm}$. The eigenfunctions ϕ_1 and ψ_1 correspond to the constant functions $\phi_1(x) = \frac{1}{\sqrt{D}}$ and $\psi_1(x) = \frac{1}{\sqrt{D_h}}$ respectively with eigenvalues $\lambda_1 = \mu_1 = 0$. We also define

$$b_{nm} = \int_{D_e} \phi_n \phi_m dx, \quad c_{nm} = \int_{D_h} \psi_n \phi_m dx.$$

Note however that $b_{nm} = \int_{D_e} \phi_n \phi_m dx \neq \delta_{nm}$ and $c_{nm} = \int_{D_h} \psi_n \phi_m dx \neq \delta_{nm}$, so that the (infinite) matrices $B = (b_{nm})$ and $C = (c_{nm})$ are not diagonal, with B being symmetric and C non symmetric. The coefficients of the matrices B and C depend only on the geometry of the domains D_h and D_e, and in principle can be calculated via quadrature.

The sets $\{\phi_n\}$, and $\{\psi_n\}$ form complete orthonormal systems for $L^2(D)$ and $L^2(D_h)$ so that expansions for $z = (v, S, p, q)$ of the form

$$v = \sum_{n \in \mathbb{N}_0} v_n \psi_n, \quad S = \sum_{n \in \mathbb{N}} S_n \phi_n, \quad p = \sum_{n \in \mathbb{N}} p_n \psi_n, \quad q = \sum_{n \in \mathbb{N}} q_n \phi_n, \tag{17}$$

$$v_n = \int_{D_h} v \psi_n dx, \quad S_n = \int_D S \phi_n dx \quad p_n = \int_{D_h} p \psi_n dx, \quad q_n = \int_D q \phi_n dx,$$

are guaranteed. The action of the operator L is diagonalized in this basis, thus rendering it to an excellent choice for the treatment of the problem. The action of the operator A_h on v and p is very simple in terms of this expansion since,

$$A_h(v) = |D_h|^{1/2} v_1, \quad A_h(p) = |D_h|^{1/2} p_1,$$

whereas,

$$A_h(S) = |D_h|^{1/2} \sum_{n \in \mathbb{N}} c_{1n} S_n.$$

We now use the expansions (17) in (15), to perform a Galerkin approximation of this system of nonlocal PDEs, and transform it into a countable system of ODEs for the expansion coefficients (v_n, S_n, p_n, q_n), $n \in \mathbb{N}_0$. We substitute the expansions (17) in (15), multiply the first and third component with ψ_n and integrate over all D_h, multiply the second and fourth component with ϕ_n and integrated over all D, and using the orthogonality of ψ_n over D_h and ϕ_n over D, we obtain the system of ODEs

$$\begin{aligned}
\frac{d}{dt} v_n &= (-d_1 \mu_n + L_1) v_n + L_2 p_n + k N_1 \delta_{n,1} \sum_m c_{1m} S_m, \\
\frac{d}{dt} S_n &= -d_2 \lambda_n S_n + L_3 \sum_m b_{nm} S_m + L_4 \sum_m b_{nm} q_m, \\
\frac{d}{dt} p_n &= (d_1 \mu_n + L_6) p_n + L_5 v_n - \bar{k} N_2 \delta_{n,1} \sum_m c_{1m} S_m, \\
\frac{d}{dt} q_n &= (d_2 \lambda_n + L_8) q_n + L_7 S_n - \bar{k} N_2 c_{1n} v_1 \\
&\quad - \bar{k}^2 N_3 c_{1,n} \sum_m c_{1m} S_m - \bar{k} N_2 c_{1n} p_1
\end{aligned} \tag{18}$$

This is an infinite dimensional ODE system that must be expressed in more compact form. We will use the following notation: Define the vector $z = (z_n) \in \ell^2$ with

$$z_{4n-3} = z_{4(n-1)+1} = v_n, \quad z_{4n-2} = z_{4(n-1)+2} = S_n,$$
$$z_{4n-1} = z_{4(n-1)+3} = p_n, \quad z_{4n} = z_{4(n-1)+4} = q_n.$$

We also introduce the notation

$$\mathfrak{r}(\ell) = mod(\ell, 4),$$
$$\mathfrak{d}(\ell) = \begin{cases} \frac{1}{4}(\ell - \mathfrak{r}(\ell)) + 1 & \text{if } \mathfrak{r}(\ell) \neq 0, \\ \frac{1}{4}(\ell - \mathfrak{r}(\ell)) & \text{if } \mathfrak{r}(\ell) = 0 \end{cases}$$

This notation allows us to express

$$\sum_m c_{1m} S_m = \sum_m c_{1m} z_{4m-2} = \sum_\ell c_{1,\mathfrak{d}(\ell)} \delta_{\mathfrak{r}(\ell),2} z_\ell,$$
$$\sum_m b_{nm} S_m = \sum_m b_{nm} z_{4m-2} = \sum_\ell b_{n,\mathfrak{d}(\ell)} \delta_{\mathfrak{r}(\ell),2} z_\ell,$$
$$\sum_m b_{nm} q_m = \sum_m b_{nm} z_{4m} = \sum_\ell b_{n,\mathfrak{d}(\ell)} \delta_{\mathfrak{r}(\ell),0} z_\ell.$$

We may therefore express (18) as

$$\frac{d}{dt} z_{4m-3} = (-d_1 \mu_m + L_1) z_{4m-3} + L_2 z_{4m-1} + \bar{k} N_1 \delta_{m,1} \sum_\ell c_{1,\mathfrak{d}(\ell)} \delta_{\mathfrak{r}(\ell),2} z_\ell,$$
$$\frac{d}{dt} z_{4m-2} = -d_2 \lambda_m z_{4m-2} + L_3 \sum_\ell b_{m,\mathfrak{d}(\ell)} \delta_{\mathfrak{r}(\ell),2} z_\ell + L_4 \sum_\ell b_{m,\mathfrak{d}(\ell)} \delta_{\mathfrak{r}(\ell),0} z_\ell,$$
$$\frac{d}{dt} z_{4m-1} = (d_1 \mu_m + L_6) z_{4m-1} + L_5 z_{4m-3} - \bar{k} N_2 \delta_{m,1} \sum_\ell c_{1,\mathfrak{d}(\ell)} \delta_{\mathfrak{r}(\ell),2} z_\ell,$$
$$\begin{aligned} \frac{d}{dt} z_{4m} &= (d_2 \lambda_m + L_8) z_{4m} + L_7 z_{4m-2} - \bar{k} N_2 c_{1m} z_1 \\ &\quad - \bar{k}^2 N_3 c_{1,m} \sum_\ell c_{1,\mathfrak{d}(\ell)} \delta_{\mathfrak{r}(\ell),2} z_\ell - \bar{k} N_2 c_{1m} z_3, \end{aligned} \tag{19}$$

which is further expressed in the compact form

$$\frac{d}{dt} z_n = \sum_\ell K_{n\ell} z_\ell, \quad n \in \mathbb{N},$$

where $K = (K_{n\ell})$ is the infinite dimensional matrix with elements

$$
\begin{aligned}
K_{n\ell} &= (-d_1\mu_m + L_1)\delta_{\ell,n} + L_2\delta_{\ell,n+2} + \bar{k}N_1\delta_{m,1}c_{1,\mathfrak{d}(\ell)}\delta_{\mathfrak{r}(\ell),2}, \quad n = 4m-3,\\
K_{n,\ell} &= -d_2\lambda_m\delta_{\ell,n} + L_3 b_{m,\mathfrak{d}(\ell)}\delta_{\mathfrak{r}(\ell),2} + L_4 b_{m,\mathfrak{d}(\ell)}\delta_{\mathfrak{r}(\ell),0}, \quad n = 4m-2,\\
K_{n,\ell} &= (d_1\mu_m + L_6)\delta_{\ell,n} + L_5\delta_{\ell,n-2} - \bar{k}N_2\delta_{m,1}c_{1,\mathfrak{d}(\ell)}\delta_{\mathfrak{r}(\ell),2}, \quad n = 4m-1,\\
K_{n,\ell} &= (d_2\lambda_m + L_8)\delta_{\ell,n} + L_7\delta_{\ell,n-1} - \bar{k}N_2 c_{1m}\delta_{\ell,1} - \bar{k}N_2 c_{1,m}\delta_{\ell,3}\\
&\quad -\bar{k}^2 N_3 c_{1,m}c_{1,\mathfrak{d}(\ell)}\delta_{\mathfrak{r}(\ell),2}, \quad n = 4m
\end{aligned}
\tag{20}
$$

or equivalently,

$$
\begin{aligned}
K_{n\ell} = &\left\{(-d_1\mu_{\mathfrak{d}(n)} + L_1)\delta_{\ell,n} + L_2\delta_{\ell,n+2} + \bar{k}N_1\delta_{\mathfrak{d}(n),1}c_{1,\mathfrak{d}(\ell)}\delta_{\mathfrak{r}(\ell),2}\right\}\delta_{\mathfrak{r}(n),1)}\\
&+\left\{-d_2\lambda_{\mathfrak{d}(n)}\delta_{\ell,n} + L_3 b_{\mathfrak{d}(n),\mathfrak{d}(\ell)}\delta_{\mathfrak{r}(\ell),2} + L_4 b_{\mathfrak{d}(n),\mathfrak{d}(\ell)}\delta_{\mathfrak{r}(\ell),0}\right\}\delta_{\mathfrak{r}(n),2)}\\
&+\left\{(d_1\mu_{\mathfrak{d}(n)} + L_6)\delta_{\ell,n} + L_5\delta_{\ell,n-2} - \bar{k}N_2\delta_{\mathfrak{d}(n),1}c_{1,\mathfrak{d}(\ell)}\delta_{\mathfrak{r}(\ell),2}\right\}\delta_{\mathfrak{r}(n),3}\\
&+\left\{(d_2\lambda_{\mathfrak{d}(n)} + L_8)\delta_{\ell,n} + L_7\delta_{\ell,n-1} - \bar{k}N_2 c_{1\mathfrak{d}(n)}\delta_{\ell,1} - \bar{k}N_2 c_{1,\mathfrak{d}(n)}\delta_{\ell,3}\right.\\
&\left.\quad -\bar{k}^2 N_3 c_{1,\mathfrak{d}(n)}c_{1,\mathfrak{d}(\ell)}\delta_{\mathfrak{r}(\ell),2}\right\}\delta_{\mathfrak{r}(n),0}
\end{aligned}
\tag{21}
$$

By truncating the infinite matrix K into a finite dimensional one K_N we may approximate the infinite dimensional problem in terms of a Galerkin approximation. Furthermore, the eigenvalue problem can be approximated by the eigenvalue problem

$$(\lambda I - K_N)z = 0.$$

If this problem admits eigenvalues with positive real part then the flat optimal solution may develop a spatial pattern formation instability. If on the other hand the eigenvalues admit negative real part then the flat optimal solution will be stable, in the sense that any spatially dependent perturbation from the flat optimal steady state will be suppressed.

It should be noted that the spatial pattern formation instability we identify in this paper is different from the original diffusion induced Turing instability. The difference is that Turing instability emerges in a non-optimizing reaction diffusion system and destabilizes a stable flat steady state, while the optimal spatial instability emerges in an optimally controlled spatial system, with the instability occurring in state-costate space. Since in economic terms the states represents quantities and the costates represents the shadow prices of the quantities, the optimal instability occurs in the quantity-price space. Thus, the emergence of the patterns suggests that the optimal spatial distribution of prices and quantities will be spatially heterogeneous.

The second remark relates to the characteristics of the instability in terms of the steady state of the optimally controlled system. The standard optimal control problem with discounting, exhibits a steady state which is either a saddle point or completely unstable. When saddle-point stability occurs, for a range of initial values of the state variables which depend on the structure of the problem, the system can be controlled to the steady state along the stable manifold, by appropriate choice of initial values for the costate variables. When a spatial perturbation around the steady state destabilizes the stable manifold, this means that the steady state is unstable under spatial perturbations, so that the spatial pattern does not die out with passage of time, but grows. The system, therefore, will not return to the flat steady state, but it will generate spatial patterns. Provided that rate of growth of the spatial instability is less than the discount rate, the instability will not be explosive. As the system moves away from the steady state the linearization will not apply, nonlinear dynamics will take over and they are expected to confine the system to an optimal spatial pattern in the price-quantity space.

6 Conclusion

Transport phenomena, local or non-local, are very closely associated with environmental and resource systems. Spatial transport in renewable resources, air or water pollution, and heat transfer towards the Poles, are some of the issues that should be taken into account in environmental and resource management. Most of the times we tend to ignore these issues and design policies assuming that spatially homogeneity is a good approximation and spatial heterogeneities are not that important. However, this might not be the case.

In this paper we model spatial interactions in a coupled system of a renewable resource and industrial pollution and study the optimal control of this system, in the sense of attaining an optimal solution for a regulator Our results suggests that optimal policies may have a spatial structure and that the emergence of Turing type optimal instability indicates the potential existence of optimal agglomerations in the environmental systems. Optimal agglomerations imply that spatially homogenous policies which are derived when spatial interactions and transport phenomena are ignored may not be the optimal policies, but instead spatial structured policies are required. The study of spatial interactions in urban and regional economics, mainly through non-local interactions, has recently led to very interesting results on the structure of cities and agglomeration dynamics. In this paper we provide analytical tools to study the qualitatively similar issues of spatial patterns and agglomeration dynamics associated with environmental and resource systems by introducing spatial transport phenomena which are an empirically relevant. The type of research presented here could therefore provide new insights into the spatial dimension of environmental policy. Spatially differentiated instruments—

price or quantities—zoning systems, reserves and no-take areas could be cases where spatially heterogenous policies are appropriate.

Acknowledgements The authors wish to thank the anonymous referee for helpful and insightful comments.

Athanasios Yannacopoulos acknowledges partial support from an AUEB Innovative Research Grant funded by the AUEB Research Center.

Anastasios Xepapadeas acknowledges partial support from AUEB Innovative Research Grant funded by the AUEB Research Center.

Appendix 1: Proof of Proposition 1

Consider the problem of maximizing the functional J defined in (5) under the dynamic constraints defined by the nonlocal nonlinear PDEs (4). Let $\mathcal{U}$ be the set of admissible controls (h, s), and assume that $(h_*\mathbf{1}_{D_h}, s_*\mathbf{1}_{D_e}) \in \mathcal{U}$ is a maximizer of the functional $J : \mathcal{U} \to \mathbb{R}$ under the stated constraints. Consider any perturbation of the optimal protocol $(u_* + \epsilon u, s_* + \epsilon s)$ chosen so that $(u_* + \epsilon u, s_* + \epsilon s) \in \mathcal{U}$. The new control protocol is not optimal so it leads to a path for the state equation which is not optimal. Let us denote the optimal path by (v_*, S_*), which is nothing but the solution of the state Eq. (4) when $h = h_*$ and $s = s_*$. The adoption of the perturbed protocol $(u_* + \epsilon u, s_* + \epsilon s)$ will lead to a new path (non-optimal in general) which we will denote by $(v_* + \epsilon v, S_* + \epsilon S)$ and will be the solution of system (4) when we substitute the control procedure $(u_* + \epsilon u, s_* + \epsilon s)$. We are interested in small deviations, from the optimal path, so we will assume that $\epsilon \to 0$. Under technical regularity conditions for the state Eq. (4) (typically smoothness of the nonlinearities) for small ϵ, the evolution of the deviation from the optimal path is given by the solution of the linearized system

$$\begin{aligned}
\frac{\partial}{\partial t} v &= d_1 \Delta v + (f_v(v_*, K(\mathcal{T}(S_*))) - h_*\mathbf{1}_{D_h})v \\
&\quad + f_K(v_*, K(\mathcal{T}(S_*)))K'(\mathcal{T}(S_*))\mathcal{T}(S) - v_* h \mathbf{1}_{D_h}, \quad \text{in } \mathbb{R}_+ \times D_h, \\
v &= 0 \quad \text{in } \mathbb{R}_+ \times (D \setminus D_h), \\
\frac{\partial}{\partial t} S &= d_2 \Delta S - \delta\, S \mathbf{1}_{D_e} + s \mathbf{1}_{D_e}, \quad \text{in } \mathbb{R}_+ \times D
\end{aligned} \tag{22}$$

with Dirichlet boundary conditions for v and periodic boundary conditions for S, and initial conditions $v(0, x) = 0$ and $S(0, x) = 0$. This is an approximation of the deviation from the optimal path, which under smoothness assumptions can be shown to be a good approximation up to $O(\epsilon^2)$. To simplify the notation we will define the functions $F_1, F_2 : \mathbb{R}_+ \times D \to \mathbb{R}$ by

$$\begin{aligned} F_1(t,x) &:= f_v(v_*(t,x), K(\mathcal{T}(S_*))(t)) - h_*(t,x), \quad (t,x) \in \mathbb{R}_+ \times D_h, \\ F_2(t,x) &:= f_K(v_*(t,x), K(\mathcal{T}(S_*))(t))K'(\mathcal{T}(S_*)(t)), \quad (t,x) \in \mathbb{R}_+ \times D_h, \\ F_1(t,x) &= F_2(t,x) = 0, \quad (t,x) \in \mathbb{R}_+ \times (D \setminus D_h). \end{aligned} \tag{23}$$

and express (22) as

$$\begin{aligned} &\frac{\partial}{\partial t} v = d_1 \Delta v + F_1 v + F_2\, \mathcal{T}(S) - v_* h \mathbf{1}_{D_h}, \ \text{ in } \mathbb{R}_+ \times D_h, \\ &\quad v = 0 \ \text{ in } \mathbb{R}_+ \times (D \setminus D_h), \\ &\frac{\partial}{\partial t} S = d_2 \Delta S - \delta\, S \mathbf{1}_{D_e} + s \mathbf{1}_{D_e}, \ \text{ in } \mathbb{R}_+ \times D \end{aligned} \tag{24}$$

We furthermore, assume smoothness assumptions for U, B and Z and we calculate $J(h_* + \epsilon h, s_* + \epsilon s)$ which using the Taylor approximation around (h_*, s_*) we obtain that up to $O(\epsilon)$,

$$\begin{aligned} &\frac{1}{\epsilon}(J(h_* + \epsilon h, s_* + \epsilon s) - J(h_*, s_*)) \\ &\quad = \int_0^\infty e^{-rt} \int_D \Big\{ U'(h_* v_*) v_* h \mathbf{1}_{D_h} + B'(s_*) s \mathbf{1}_{D_e} \Big\} dx dt \\ &\quad + \int_0^\infty e^{-rt} \int_D \Big\{ U'(h_* v_*) h_* v \mathbf{1}_{D_h} - Z'(S_*) S \Big\} dx dt. \end{aligned} \tag{25}$$

The first integral in the above expression is in a desirable form since it contains only the optimal quadruple (h_*, s_*, v_*, S_*) and the arbitrary perturbation (h, s). The second integral however is not so nice, as it contains the optimal path (h_*, s_*, v_*, S_*) again but now the deviation (v, S) from the optimal path, rather than the deviation of the policy (h, s). Clearly, (v, S) is connected to (h, s) through the solution of the linearized system (22) and this is dependence is furnished in general via the action of the solution operator whose action is defined as $(h, s) \mapsto (v, S)$. Our aim is to make this connection explicit and this is accomplished by the construction of the adjoint system.

To motivate the construction of the adjoint system, we introduce two auxiliary variables (p, q) (one for each state variable) and introduce the auxiliary functional

$$\bar{I} = \int_D (v\, p + S\, q) dx = \int_D (v\, p\, \mathbf{1}_{D_h} + S\, q) dx.$$

Since this functional involves only integration over space, its value depends on time and taking the time derivative and substituting $(\frac{\partial}{\partial t} v, \frac{\partial}{\partial t} S)$, by their evolution given by the linearized system (22), and noting that v vanishes identically on $D \setminus D_h$ we obtain that

$$\frac{d\bar{I}}{dt} = \int_D \Big\{ \Big(d_1 \Delta v + F_1 v + F_2 \mathcal{T} S - v_* h \Big) \mathbf{1}_{D_h} p \tag{26}$$

$$+ \Big(d_2 \Delta S - \delta \, S \mathbf{1}_{D_e} + s \mathbf{1}_{D_e} \Big) q + v \mathbf{1}_{D_h} \frac{\partial p}{\partial t} + S \frac{\partial q}{\partial t} \Big\} dx$$

We use the Green's formulae

$$\int_{D_h} \Delta v \, p dx = \int_{D_h} \Delta p \, v dx,$$

$$\int_D \Delta S \, q dx = \int_D \Delta q \, S dx,$$

as well as the definition of the adjoint operator of $\mathcal{T}$,

$$\int_D F_2 p \mathcal{T}(S) dx = \int_D \mathcal{T}^*(F_2 p) S dx,$$

to bring the expression (26) into the equivalent form,

$$\frac{d\bar{I}}{dt} = \int_D \Big(d_1 \Delta p + F_1 p + \frac{\partial p}{\partial t} \Big) v \mathbf{1}_{D_h} dx + \int_D \Big(d_2 \Delta q - \delta \, q \mathbf{1}_{D_e} + \mathcal{T}^*(F_2 p) + \frac{\partial q}{\partial t} \Big) S dx$$

$$+ \int_D \Big(- v_* h \mathbf{1}_{D_h} p + s \mathbf{1}_{D_e} q \Big) dx.$$

Note that given any function $g : \mathbb{R}_+ \times D \to \mathbb{R}$ the operator $\mathcal{T}^*$ acting on g defines a new function $G : \mathbb{R}_+ \times D \to \mathbb{R}$, denoted by $G = \mathcal{T}^*(g)$ such that

$$G(t,x) = \Big(\mathcal{T}^*(g) \Big)(t,x) = \Big(\int_{D_h} g(t,x') dx' \Big) \; k(x) \mathbf{1}_{D_h}(x),$$

i.e. provides a copy of the function k localized in D_h, multiplied by a factor depending only on t (and the choice of the function g), equal to $C(t) := \int_{D_h} g(t,x') dx'$. The notation $\mathcal{T}^*(F_2 p)$ means that the operator $\mathcal{T}^*$ is acting on the function $g = F_2 p$ taken by the pointwise product of the functions F_2 and p, therefore,

$$\Big(\mathcal{T}^*(F_2 p) \Big)(t,x) = \Big(\int_{D_h} F_2(t,x') p(t,x') dx' \Big) \; k(x) \mathbf{1}_{D_h}(x). \tag{27}$$

We have so far left (p, q) completely unspecified. We now assume that (p, q) are solutions of the evolution equations

$$\begin{aligned} d_1 \Delta p + F_1 p + \frac{\partial p}{\partial t} &= G_1, \quad \text{in } \mathbb{R}_+ \times D_h, \\ p &= 0, \quad \text{in } \mathbb{R}_+ \times (D \setminus D_h), \\ d_2 \Delta q - \delta q \mathbf{1}_{D_e} + \mathcal{T}^*(F_2 p) + \frac{\partial q}{\partial t} &= G_2, \quad \text{in } \mathbb{R}_+ \times D, \end{aligned}$$

with Dirichlet boundary conditions for p, periodic for q and (G_1, G_2) to be specified shortly. This yields,

$$\frac{d\bar{I}}{dt} = \int_D (G_1 v \mathbf{1}_{D_h} + G_2 S) dx + \int_D (-v_* h \mathbf{1}_{D_h} p + s \mathbf{1}_{D_e} q) dx. \tag{28}$$

Since (G_1, G_2) are left unspecified, we will choose them in such a way that $\int_D (G_1 v \mathbf{1}_{D_h} + G_2 S) dx$ resembles or reproduces the second problematic integral in (25). To this end we choose

$$\begin{aligned} G_1 &= \Big(U'(h_* v_*) h_* + r\, p \Big) \mathbf{1}_{D_h}, \\ G_2 &= -Z'(S_*) + r\, q. \end{aligned}$$

For this choice, (28) becomes

$$\frac{d\bar{I}}{dt} = r\bar{I} + \int_D \Big(U'(h_* v_*) h_* \mathbf{1}_{D_h} v - Z'(S_*) S \Big) dx + \int_D (-v_* h \mathbf{1}_{D_h} p + s \mathbf{1}_{D_e} q) dx,$$

where we used the definition of $\bar{I}$, and integrating over time between $t = 0$ and $t = T$ (for T arbitrary) we get after taking the limit as $T \to \infty$ that,

$$\begin{aligned} \lim_{T \to \infty} e^{-rT} \bar{I}(T) - \bar{I}(0) = &\int_0^\infty e^{-rt} \int_D \Big(U'(h_* v_*) h_* \mathbf{1}_{D_h} v - Z'(S_*) S \Big) dx\, dt \\ &+ \int_0^\infty e^{-rt} \int_D (-v_* h \mathbf{1}_{D_h} p + s \mathbf{1}_{D_e} q) dx\, dt. \end{aligned} \tag{29}$$

By the definition of the functional $\bar{I}$ and the chosen initial conditions for (22), $I(0) = 0$. If we choose

$$\lim_{T \to \infty} e^{-rT} \bar{I}(T) := \lim_{T \to \infty} e^{-rT} \int_D (v(T, x) p(T, x) + S(T, x) q(T, x)) dx = 0, \tag{30}$$

then (29) yields,

$$\int_0^\infty e^{-rt}\int_D \Big(U'(h_* v_*)h_* \mathbf{1}_{D_h} v - Z'(S_*)S\Big)dx\,dt \tag{31}$$
$$= -\int_0^\infty e^{-rt}\int_D (-v_* h\mathbf{1}_{D_h} p + s\mathbf{1}_{D_e} q)dx\,dt,$$

so that we have managed to express the problematic second integral in (25) in terms of the variation of the optimal protocol (h, s) at the cost of introducing the adjoint variables (p, q).

Combining (25) with (31) we conclude that

$$\frac{1}{\epsilon}(J(h_* + \epsilon h, s_* + \epsilon s) - J(h_*, s_*))$$
$$= \int_0^\infty e^{-rt}\int_D \Big\{\Big(U'(h_* v_*)v_* - v_* p\Big)h\mathbf{1}_{D_h} + \Big(B'(s_*) - q\Big)s\mathbf{1}_{D_e}\Big\}dxdt,$$

and since $J(h_*, s_*)$ is assumed to be a maximum, for any admissible (h, v), and any $\epsilon > 0$ we have that $\frac{1}{\epsilon}(J(h_* + \epsilon h, s_* + \epsilon s) - J(h_*, s_*)) \leq 0$, hence we conclude that

$$\int_0^\infty e^{-rt}\int_D \Big\{\Big(U'(h_* v_*)v_* - v_* p\Big)h\mathbf{1}_{D_h} + \Big(B'(s_*) - q\Big)s\mathbf{1}_{D_e}\Big\}dxdt \leq 0, \;\; \forall\,(h, s) \tag{32}$$

This holding for any (h, s) can be seen to hold a.e. on D and this can be recognized as an optimality condition. To make this point more clear assume, for the time being, that any variation $(h, s) \in L^2(D) \times L^2(D)$ is possible (meaning that we may take as variations $(h, 0)$ and $(-h, 0)$ for any $h \in L^2(D)$, as well as $(0, s)$ and $(0, -s)$ for any $s \in L^2(D)$). This of course implies that $h_* > 0$ and $s_* > 0$. Then condition (32) yields,

$$v_*(U'(h_* v_*) + p) = 0, \;\text{ in } D_h, \tag{33}$$
$$B'(s_*) - q = 0, \;\text{ in } D_e,$$

with the above holding a.e., which can be recognized as the first order conditions for the maximization of

$$\bar{H}_1(h) := U(hv) + vhp,$$
$$\bar{H}_2(s) := B(s) - qs,$$

with respect to h and s in D_h and D_e respectively. These are considered are static optimization problems, since the optimality conditions hold for any (fixed) (t, x) a.e. on $[0, T] \times D$. In the general case where constraints are to be taken into account on the admissible set for the allowed (h, s), we recognize (32) as the condition for maximization of $\bar{H}_1$ and $\bar{H}_2$ with respect to h and s subject to the constraints that $(h_* + \epsilon h, s_* + \epsilon s) \in \mathcal{U}$. Note that the optimality condition (32) (or the simplest special case version (33) connects the adjoint variables (p, q) with the optimal policy (h_*, s_*).

We therefore conclude that if (v_*, S_*, h_*, s_*) is an optimal quadruple, then it is connected with the solution (p, q) for the adjoint system

$$\frac{\partial p}{\partial t} = -d_1 \Delta p - F_1 p + U'(h_* v_*) h_* + r\, p,$$
$$\frac{\partial q}{\partial t} = -d_2 \Delta q - \mathcal{T}^*(F_2 p) - Z'(S_*) + (r + \delta\, \mathbf{1}_{D_e})\, q,$$

with $\mathcal{T}^*(F_2 p)$ defined as in (27), subject to Dirichlet boundary conditions for p and periodic boundary conditions for q and the transversality condition (30), (v_*, S_*) is the solution of the state Eq. (4) for $h = h_*$, $s = s_*$, and that $(v_*, S_*, p, q, h_*, s_*)$ must be connected through the optimality condition (32).

Appendix 2: Proof of Proposition 3

If k is not a constant it is straightforward to see that a flat solution cannot exist, since the fourth Eq. (13) is inconsistent. We therefore consider the case where $k(x) = k$ a constant. In this case and seeking a solution of the form (14), the system of Eq. (13) simplifies to the following system of algebraic equations

$$\begin{aligned}
0 &= f(v_h, K_a) - I_1(-p_h), \\
0 &= -\delta S + I_2(q_e) \\
0 &= (r - f_v(v_h, K_a)) p_h, \\
0 &= (r + \delta) q_e - Z'(S), \\
0 &= -\Lambda_a k + r q_h - Z'(S),
\end{aligned} \tag{34}$$

that can be used for the determination of the unknown constants v_h, p_h, S, q_e, q_h. The second equation gives $S = \frac{1}{\delta} I_2(q_e)$, which upon substitution to the fourth leads to

$$q_e = \frac{1}{r + \delta} Z' \left(\frac{1}{\delta} I_2(q_e) \right),$$

which is an algebraic equation that can be solved to obtain the optimal level q_e^*, and from that obtain $S^* = \frac{1}{\delta} I_2(q_e^*)$. Having obtained S^* we now have the values of the constants K_a which is

$$K_a^* = K(\bar{k}S^*), \quad \bar{k} = k\,\mathcal{L}(D_h),$$

where $\mathcal{L}(D_h)$ is the Lebesgue measure of D_h.

We now consider the third equation that yields, $f_v(v, K_a^*) = r$ which can be solved to obtain $v_h^* = \Phi(S^*) := I_3(r, K_a^*)$ where I_3 is the inverse of $f_v(v; K_a)$ with respect to the argument v (which is of course dependent on K_a and S as a parameter). The first equation then allows us to calculate p_h as

$$-p_h^* = U'(f(v_h^*, K_a^*),$$

(recall that I_1 is the inverse of U'). We are now able to calculate Λ_a, which is

$$\Lambda_a^* = f_K(v_h^*, K_a^*)K'((\bar{k}S^*)\mathcal{L}(D_h),$$

and finally the fifth equation provides

$$q_h^* = \frac{\Lambda_a^* k}{r} + \frac{1}{r} Z'(S^*).$$

That implies, that the optimal state is given by the following control strategy

$$\begin{aligned} h^* &= \frac{f(v_h^*, K_a^*)}{v_h^*}, \\ s^* &= \delta S^*. \end{aligned} \tag{35}$$

References

K. Arrow, B. Bolin, R. Costanza, P. Dasgupta, C. Folke, C.S. Holling, B.-O. Jansson, S. Levin, K.-G. Maler, C. Perrings, D. Pimentel, Economic growth, carrying capacity, and the environment. Science **268**, 520–521 (1995)

L.V. Ballestra, The spatial AK model and the Pontryagin maximum principle. J. Math. Econ. **67**, 87–94 (2016)

S. Behringer, T. Upmann, Optimal harvesting of a spatial renewable resource. J. Econ. Dyn. Control **42**, 105–120 (2014)

R. Boucekinne, C. Camacho, B. Zhou, Bridging the gap between growth theory and the new economic geography: the spatial Ramsey model. Macroecon. Dyn. **13**, 20–45 (2009)

R. Boucekinne, C. Camacho, G. Fabbri, Spatial dynamics and convergence: the spatial AK model. J. Econ. Theory **148**, 2719–2736 (2013)

R. Boucekinne, G. Fabbri, S. Federico, F. Gozzi, Growth and agglomeration in the heterogeneous space: a generalized AK approach (2016, Preprint)

W.A. Brock, G. Engstrom, D. Grass, A. Xepapadeas, Energy balance climate models and general equilibrium optimal mitigation policies. J. Econ. Dyn. Control **37**(12), 2371–2396 (2013)

W.A. Brock, A. Xepapadeas, A. Yannacopoulos, Optimal control in space and time and the management of environmental resources. Ann. Rev. Resour. Econ. **6**(1), 33–68 (2014a)

W.A. Brock, A. Xepapadeas, A.N. Yannacopoulos, Robust control of a spatially distributed commercial fishery, in *Dynamic Optimization in Environmental Economics* (Springer, New York, 2014b), pp. 215–241

W.A. Brock, A. Xepapadeas, A.N. Yannacopoulos, Spatial externalities and agglomeration in a competitive industry. J. Econ. Dyn. Control **42**, 143–174 (2014c)

C. Camacho, A. Perez-Barahona, Land use dynamics and the environment. J. Econ. Dyn. Control **52**, 96–118 (2015)

C. Camacho, A. Perez-Barahona, The diffusion of economic activity across space: a new approach. J. Econ. Lit. **62**, R11 (2016)

K. Desmet, E. Rossi-Hansberg, On spatial dynamics. J. Reg. Sci. **50**(1), 43–63 (2010)

K. Desmet, E. Rossi-Hansberg, On the spatial economic impact of global warming. J. Urban Econ. **88**, 16–37 (2015)

G. Fabbri, Geographical structure and convergence: a note on geometry in spatial growth models. J. Econ. Theory **162**, 114–136 (2016)

R.U. Goetz, D. Zilberman, The economics of land-use regulation in the presence of an externality: a dynamic approach. Optim. Control Appl. Methods **28**(1), 21–43 (2007)

D. Grass, A. Xepapadeas, A. de Zeeuw, Optimal management of ecosystem services with pollution traps: the lake model revisited. J. Assoc. Environ. Resource Econ. **4**(4), 1121–1154 (2017)

J. Hassler, P. Krusell, Economics and climate change: integrated assessment in a multi-region world. J. Eur. Econ. Assoc. **10**(5), 974–1000 (2012)

G.R. North, R.F. Cahalan, J.A. Coakley, Energy balance climate models. Rev. Geophys. **19**(1), 91–121 (1981)

V. Ramanathan, H. Rodhe, M. Agrawal, H. Akimoto, M. Auffhammer, U.K. Chopra, et al., *Atmospheric Brown Clouds: Regional Assessment Report with Focus on Asia* (Scientific Publications, 2008)

M.D. Smith, J.N. Sanchirico, J.E. Wilen, The economics of spatial-dynamic processes: applications to renewable resources. J. Environ. Econ. Manag. **57**(1), 104–121 (2009)

J.E. Wilen, Economics of spatial-dynamic processes. Am. J. Agric. Econ. **89**(5), 1134–1144 (2007)

A. Xabadia, R. Goetz, D. Zilberman, Optimal dynamic pricing of water in the presence of waterlogging and spatial heterogeneity of land. Water Resour. Res. **40**(7), W07S02 (2004)

Some Regional Control Problems for Population Dynamics

Laura-Iulia Aniţa, Sebastian Aniţa, Vincenzo Capasso, and Ana-Maria Moşneagu

Abstract This paper deals with some control problems related to structured population dynamics with diffusion. Firstly, we investigate the regional control for an optimal harvesting problem (the control acts in a subregion ω of the whole domain Ω). Using the necessary optimality conditions, for a fixed ω, we get the structure of the harvesting effort which gives the maximum harvest; with this optimal effort we investigate the best choice of the subregion ω in order to maximize the harvest. We introduce an iterative numerical method to increase the total harvest at each iteration by changing the subregion where the effort acts. Numerical tests are used to illustrate the effectiveness of the theoretical results. We also consider the problem of eradication of an age-structured pest population dynamics with diffusion and logistic term, which is a zero-stabilization problem with constraints. We derive a necessary condition and a sufficient condition for zero-stabilizability. We formulate a related optimal control problem which takes into account the cost of intervention in the subregion ω.

L.-I. Aniţa
Faculty of Physics, "Alexandru Ioan Cuza" University of Iaşi, Iaşi, Romania
e-mail: lianita@uaic.ro

S. Aniţa (✉)
Faculty of Mathematics, "Alexandru Ioan Cuza" University of Iaşi, Iaşi, Romania

"Octav Mayer" Institute of Mathematics of the Romanian Academy, Iaşi, Romania
e-mail: sanita@uaic.ro

V. Capasso
ADAMSS (Centre for Advanced Applied Mathematical and Statistical Sciences), Universitá degli Studi di Milano, Milan, Italy
e-mail: vincenzo.capasso@unimi.it

A.-M. Moşneagu
Faculty of Mathematics, "Alexandru Ioan Cuza" University of Iaşi, Iaşi, Romania
e-mail: anamaria.mosneagu@uaic.ro

G. Feichtinger et al. (eds.), *Control Systems and Mathematical Methods in Economics*, Lecture Notes in Economics and Mathematical Systems 687,
https://doi.org/10.1007/978-3-319-75169-6_20

1 Introduction

An extensive literature was developed for the optimal harvesting problems of population dynamics (e.g. Aniţa et al. 2013, Aniţa 2000, Aniţa et al. 2011, Aniţa and Capasso 2009, Aniţa et al. 2016, Arnăutu and Moşneagu 2015, Barbu 1994, Belyakov and Veliov 2015, Fister and Lenhart 2006, Gurtin and Murphy 1981a,b, He 2006, Hritonenko and Yatsenko 2005, Luo 2007, Luo et al. 2004, Murphy and Smith 1990, Zhao et al. 2005, Zhao et al. 2006). In this paper we firstly remind an optimal harvesting problem for a spatially structured population with diffusion which has been introduced in Aniţa et al. (2016). For spatially structured harvesting problems it has usually been taken into consideration an effort that acts in the whole habitat Ω (see for example Aniţa 2000). Instead here we consider the case in which the effort is localized in a suitably chosen subregion ω of Ω. In addition to the problem of finding the magnitude of the control to act on a given subdomain ω, the most important task will be to identify an optimal subregion ω, where the control acts, in order to maximize the harvest. To this aim, at first we have derived necessary optimality conditions for the situation when the support of the control is fixed; as a fall out we have obtained information concerning the structure of the optimal control. Hence we have taken into account this structure to investigate the optimal subregion ω where the control is localized, by taking into account the cost paid for harvesting in ω. Here we have adapted some shape optimization methods, based on the level set method. These results have been previously presented in Aniţa et al. (2016). In this paper we consider also the problem of eradication of an age-structured pest population dynamics with diffusion and logistic term. We consider a related optimal control problem which can be again investigated by means of the level set method.

We consider the following population dynamics model with diffusion. A single population species is free to move in an isolated habitat $\Omega \subset \mathbb{R}^2$, with Ω a bounded domain with a sufficiently smooth boundary:

$$\begin{cases} \partial_t y(x,t) - d\Delta y(x,t) = a(x)y(x,t) - \chi_\omega(x)u(x,t)y(x,t), & (x,t) \in Q_T \\ \partial_\nu y(x,t) = 0, & (x,t) \in \Sigma_T \\ y(x,0) = y_0(x), & x \in \Omega, \end{cases} \tag{1}$$

where $Q_T = \Omega \times (0,T)$, $\Sigma_T = \partial\Omega \times (0,T)$, $T > 0$, $y = y(x,t)$ is the population density at position $x \in \overline{\Omega}$ and time $t \in [0,T]$, while $y_0(x)$ is the initial population density. Here $a(x)$ denotes the natural growth rate of the population, and $d \in (0,+\infty)$ is the diffusion coefficient. No-flux boundary conditions are considered.

In System (1), $u(x,t)$ represents the harvesting effort (control), bounded and localized in the subdomain $\omega \subset \Omega$ (χ_ω is the characteristic function of ω). The term $\chi_\omega(x)u(x,t)y(x,t)$ represents the rate of the harvested population at position

$x \in \Omega$ and time $t \in [0, T]$. Actually, for any $x \in \Omega \setminus \omega$ and time $t \in [0, T]$ we have that $\chi_\omega(x)u(x,t)y(x,t) = 0$.

The following hypotheses are considered:

(H1) $a \in L^\infty(\Omega)$;
(H2) $y_0 \in L^\infty(\Omega), \quad y_0(x) \geqslant 0 \quad \text{a.e. } x \in \Omega \text{ with } \|y_0\|_{L^\infty(\Omega)} > 0.$

We consider a related optimal harvesting problem

$$Maximize \int_0^T \int_\omega u(x,t)y^u(x,t)dx\,dt, \tag{2}$$

subject to $u \in K_\omega$, where $K_\omega = \{w \in L^\infty(\omega \times (0,T));\ 0 \leq w(x,t) \leq L \text{ a.e. in } \omega \times (0,T)\}$. Here $L > 0$ is a constant and y^u is the solution to (1) corresponding to a harvesting effort $u \in K_\omega$.

The existence result of an optimal control for Problem (2) follows Aniţa et al. (2011) or Arnăutu and Moşneagu (2015).

Theorem 1 *Problem (2) admits at least one optimal control.*

We denote by p the adjoint state, i.e. p satisfies

$$\begin{cases} \partial_t p(x,t) + d\Delta p(x,t) = -a(x)p(x,t) & \\ \qquad\qquad + \chi_\omega(x)u^*(x,t)(1+p(x,t)), & (x,t) \in Q_T \\ \partial_\nu p(x,t) = 0, & (x,t) \in \Sigma_T \\ p(x,T) = 0, & x \in \Omega, \end{cases} \tag{3}$$

where (u^*, y^{u^*}) is an optimal pair for (2). For the construction of the adjoint problems in optimal control theory we refer to Barbu (1994). Concerning the first order necessary optimality conditions it can be proved the following result (as in Aniţa et al. 2011 and Arnăutu and Moşneagu 2015):

Theorem 2 *If (u^*, y^{u^*}) is an optimal pair for Problem (2) and if p is the solution of Problem (3), then we have:*

$$u^*(x,t) = \begin{cases} 0, & 1 + p(x,t) < 0 \\ L, & 1 + p(x,t) > 0 \end{cases} \quad a.e.\ (x,t) \in \omega \times (0,T).$$

In Sect. 2 we will treat the regional harvesting problem as a shape optimization problem. We remind that the geometry of a set ω can be characterized in terms of its Minkowski functionals. There are three such functionals and these are proportional to the area, the perimeter and the Euler-Poincaré characteristic. In this paper we control the shape of ω as follows: by the length of the boundary of ω, and by the area of ω.

We shall use the implicit interface representation to control the shape of the 2D domain ω. Therefore, the boundary of a domain is defined as the isocontour of some function φ (see Delfour and Zolesio 2011 or Osher and Fedkiw 2003). By using the level set method, we introduce a level set function $\varphi : \overline{\Omega} \to \mathbb{R}$ such that $\omega = \{x \in \Omega;\ \varphi(x) > 0 \text{ a.e.}\}$ and $\partial\omega = \{x \in \Omega : \varphi(x) = 0 \text{ a.e.}\}$ (the boundary is defined as the zero level set of φ). We will then manipulate ω implicitly, through the function φ. This function φ is assumed to take positive values inside the region delimited by the curve $\partial\omega$ and negative values outside.

If φ is the implicit function of ω, in order to integrate over ω a function f defined over the whole Ω we may write $\int_\Omega f(x)H(\varphi(x))dx$, where we have used the Heaviside function $H : \mathbb{R} \to \{0, 1\}$, such that

$$H(z) = \begin{cases} 1, & \text{if} \quad z \geq 0 \\ 0, & \text{if} \quad z < 0. \end{cases}$$

If φ is sufficiently smooth, the directional derivative of the Heaviside function in the normal direction at a point $x \in \partial\omega$ is given by $H'(\varphi(x))|\nabla\varphi(x)|$, and by using the usual Dirac Delta δ on $\mathbb{R}$, we have $\delta(\varphi(x))|\nabla\varphi(x)|$. If we need to integrate over $\partial\omega$ a function f defined over the whole Ω we may write $\int_\Omega f(x)\delta(\varphi(x))|\nabla\varphi(x)|dx$.

We find the derivative of the optimal cost value with respect to the implicit function φ of the subregion ω. In order to improve the region where the control acts we derive a conceptual iterative algorithm based on these theoretical results. We also present the numerical implementation of this conceptual algorithm and some numerical tests. Basically, the theoretical results in Sect. 2 have been obtained in Aniţa et al. (2016). Here we give some additional details concerning the numerical scheme and its implementation. Further we present here some new numerical tests.

In Sect. 3 we treat the problem of eradication of an age-structured pest population with diffusion, which is a zero-stabilization problem with constraints. We derive a necessary condition and a sufficient condition of zero-stabilization. We consider a related optimal control problem which takes into account the cost paid by acting in the subregion ω. We formulate this optimal control problem by means of the level set method. The results in this section are new.

We may mention that one might rephrase the above control problem in terms of suitable measures, as done e.g. in Bressan et al. (2013); indeed they have treated the problem of existence for a generic optimal harvesting effort, with constraints. Our approach allows us to take advantage of the meaning and of the specific structure of our system.

For optimal control related to biological models see Lenhart and Workman (2007), and references therein (for an updated introduction and literature one may refer to Lenhart 2014). For basic results and methods in shape optimization we refer to Bucur and Buttazzo (2002), Delfour and Zolesio (2011), Henrot and Pierre (2005), Osher and Fedkiw (2003), Sethian (1999), Sokolowski and Zolesio (1992).

2 An Iterative Method to Localize an Optimal Subdomain ω Where the Control Acts

Here we intend to use the level set method in order to obtain the optimal subregion ω where the control is localized. Consider $\varphi : \overline{\Omega} \to \mathbb{R}$ the implicit function of ω, the subregion of Ω where the control acts.

We rewrite the optimal control problem (2) such that will include both the magnitude of the harvesting effort $u \in K_\omega$, and the choice of the subdomain ω with respect to its implicit function φ:

$$\underset{\varphi}{Maximize}\ \underset{u\in K_\omega}{Maximize}\left\{\int_0^T\int_\omega u(x,t)y^u(x,t)dx\,dt\right.$$
$$\left.-\alpha\ length(\partial\omega)-\beta\ area(\omega)\right\},$$

where y^u is the solution to (1) corresponding to a harvesting effort $u \in K_\omega$ and α, β are positive constants. α length$(\partial\omega) + \beta$ area(ω) represents the cost paid to harvest in the subregion ω. Here $\varphi : \overline{\Omega} \to \mathbb{R}$ is a smooth function and $\omega = \{x \in \Omega;\ \varphi(x) > 0\}$.

By using De Giorgi's formula for the length (perimeter) of a set and assuming that φ is smooth, the optimal control problem becomes

$$\underset{\varphi}{Maximize}\ \underset{u\in K_\omega}{Maximize}\left\{\int_0^T\int_\omega u(x,t)y^u(x,t)dx\,dt\right.$$
$$\left.-\alpha\int_\Omega \delta(\varphi(x))|\nabla\varphi(x)|dx-\beta\int_\Omega H(\varphi(x))dx\right\},$$

where $\varphi : \overline{\Omega} \to \mathbb{R}$ is a smooth function and $\omega = \{x \in \Omega;\ \varphi(x) > 0\}$.

We have now two maximization problems: firstly, for a fixed φ (and implicitly, ω) we have to find the structure of the harvesting effort which gives the maximum harvest, as a function of φ (or ω); secondly, using this structure of the optimal control we investigate the optimal choice of the subregion ω with respect to its implicit function φ in order to maximize the harvest. Since for a fixed φ we have that $\alpha \int_\Omega \delta(\varphi(x))|\nabla\varphi(x)|dx + \beta \int_\Omega H(\varphi(x))dx$ is a constant it means that firstly we have to investigate problem (2).

For any arbitrary but fixed φ, we denote by $(u^*_\varphi, y^*_\varphi)$ an optimal pair for the harvesting problem (2). Next we have to investigate the following optimal control problem:

$$\underset{\varphi}{Maximize}\left\{\int_0^T\int_\omega u^*_\varphi(x,t)y^*_\varphi(x,t)dx\,dt\right.$$
$$\left.-\alpha\int_\Omega \delta(\varphi(x))|\nabla\varphi(x)|dx-\beta\int_\Omega H(\varphi(x))dx\right\},$$

where y^*_φ is the solution to

$$\begin{cases} \partial_t y(x,t) - d\Delta y(x,t) = a(x)y(x,t) - H(\varphi(x))u^*_\varphi(x,t)y(x,t), & (x,t) \in Q_T \\ \partial_\nu y(x,t) = 0, & (x,t) \in \Sigma_T \\ y(x,0) = y_0(x), & x \in \Omega. \end{cases}$$

Note that $H(\varphi)$ represents the characteristic function of ω.

Assume that the hypotheses (H1, H2) are satisfied. We denote by p_φ the adjoint state. From Theorem 2, the optimal control is given by

$$u^*_\varphi(x,t) = \begin{cases} 0, & 1 + p_\varphi(x,t) < 0 \\ L, & 1 + p_\varphi(x,t) > 0 \end{cases} \tag{4}$$

a.e. $(x,t) \in \omega \times (0,T)$, where p_φ is the solution to (3).

By multiplying (1) by p_φ and (3) by y^*_φ, and integrating both of them on Q_T we obtain:

$$\int_0^T \int_\Omega [\partial_t y^*_\varphi p_\varphi + y^*_\varphi \partial_t p_\varphi] dx\, dt + \int_0^T \int_\Omega [-d\Delta y^*_\varphi p_\varphi + d\Delta p_\varphi y^*_\varphi] dx\, dt +$$

$$+ \int_0^T \int_\Omega [-a(x) y^*_\varphi p_\varphi + a(x) y^*_\varphi p_\varphi] dx\, dt =$$

$$= \int_0^T \int_\Omega [-H(\varphi(x))u^*_\varphi y^*_\varphi p_\varphi + H(\varphi(x))u^*_\varphi y^*_\varphi + H(\varphi(x))u^*_\varphi y^*_\varphi p_\varphi] dx\, dt.$$

This means that

$$-\int_\Omega y_0(x) p_\varphi(x,0) dx = \int_0^T \int_\Omega H(\varphi(x)) u^*_\varphi(x,t) y^*_\varphi(x,t) dx\, dt$$

and therefore

$$\int_0^T \int_\omega u^*_\varphi(x,t) y^*_\varphi(x,t) dx\, dt = -\int_\Omega y_0(x) p_\varphi(x,0) dx$$

Our problem of optimal harvesting becomes a problem of minimizing another functional with respect to the implicit function of ω. Therefore, we may rewrite the optimal problem as

$$\underset{\varphi}{Minimize} \left\{ \int_\Omega y_0(x) p_\varphi(x,0) dx + \alpha \int_\Omega \delta(\varphi(x)) |\nabla \varphi(x)| dx + \beta \int_\Omega H(\varphi(x)) dx \right\}.$$

By using (3) and (4) we get that p_φ is the solution to

$$\begin{cases} \partial_t p + d\Delta p = -a(x)p + LH(\varphi(x))(1+p)H(1+p), & (x,t) \in Q_T \\ \partial_\nu p(x,t) = 0, & (x,t) \in \Sigma_T \\ p(x,T) = 0, & x \in \Omega. \end{cases}$$

We shall adapt some shape optimization techniques to treat this last harvesting problem (see also Chan and Vese 2001, Getreuer et al. 2012). As usual, we will approximate this problem by the following one, where the Heaviside function H is substituted by its mollified version $H_\varepsilon(t) = \frac{1}{2}\left(1 + \frac{2}{\pi}\arctan\left(\frac{t}{\varepsilon}\right)\right)$, and its derivative by the mollified function $\delta_\varepsilon(t) = \frac{\varepsilon}{\pi(\varepsilon^2+t^2)}$.

Therefore, for a small but fixed $\varepsilon > 0$, the harvesting problem to be investigated is:

$$\underset{\varphi}{Minimize}\; J(\varphi),$$

where $\varphi : \overline{\Omega} \longrightarrow \mathbb{R}$ is a smooth function,

$$J(\varphi) = \int_\Omega y_0(x)p_\varphi(x,0)dx + \alpha \int_\Omega \delta_\varepsilon(\varphi(x))|\nabla\varphi(x)|dx + \beta \int_\Omega H_\varepsilon(\varphi(x))dx,$$

and $p_\varphi = p_\varphi(x,t)$ is the solution to

$$\begin{cases} \partial_t p + d\Delta p = -a(x)p + LH_\varepsilon(\varphi(x))(1+p)H_\varepsilon(1+p), & (x,t) \in Q_T \\ \partial_\nu p(x,t) = 0, & (x,t) \in \Sigma_T \\ p(x,T) = 0, & x \in \Omega. \end{cases} \tag{5}$$

In the following we derive the directional derivative of J (see Aniţa et al. 2016).

Theorem 3 *For any smooth functions $\varphi, \psi : \overline{\Omega} \longrightarrow \mathbb{R}$ we have that*

$$dJ(\varphi)(\psi) = \int_\Omega \delta_\varepsilon(\varphi(x))[-\alpha \ \mathrm{div}\left(\frac{\nabla\varphi(x)}{|\nabla\varphi(x)|}\right) + \beta$$

$$-L\int_0^T (1+p_\varphi(x,t))H_\varepsilon(1+p_\varphi(x,t))r_\varphi(x,t)dt]\psi(x)dx + \alpha\int_{\partial\Omega} \frac{\delta_\varepsilon(\varphi(x))}{|\nabla\varphi(x)|}\partial_\nu\varphi(x)\psi(x)d\sigma,$$

where r_φ is the solution to

$$\begin{cases} \partial_t r - d\Delta r = a(x)r - LH_\varepsilon(\varphi(x))H_\varepsilon(1+p_\varphi)r \\ \qquad -LH_\varepsilon(\varphi(x))(1+p_\varphi)\delta_\varepsilon(1+p_\varphi)r, & (x,t) \in Q_T \\ \partial_\nu r(x,t) = 0, & (x,t) \in \Sigma_T \\ r(x,0) = y_0(x), & x \in \Omega. \end{cases} \tag{6}$$

Proof For the sake of clarity we give here the proof (see Aniţa et al. 2016). It is possible to prove that

$$\frac{1}{\theta}[p_{\varphi+\theta\psi} - p_\varphi] \to q_\psi \quad \text{in } C([0,T]; L^\infty(\Omega)),$$

as $\theta \to 0$, where q_ψ is the solution to the problem

$$\begin{cases} \partial_t q + d\Delta q = -a(x)q + L\delta_\varepsilon(\varphi(x))(1+p_\varphi)H_\varepsilon(1+p_\varphi)\psi(x) & \\ \qquad +LH_\varepsilon(\varphi(x))H_\varepsilon(1+p_\varphi)q + LH_\varepsilon(\varphi(x))(1+p_\varphi)\delta_\varepsilon(1+p_\varphi)q, & (x,t) \in Q_T \\ \partial_\nu q(x,t) = 0, & (x,t) \in \Sigma_T \\ q(x,T) = 0, & x \in \Omega. \end{cases} \tag{7}$$

For any φ we have that

$$\begin{aligned} \lim_{\theta\to 0}\frac{1}{\theta}[J(\varphi+\theta\psi) - J(\varphi)] &= \int_\Omega y_0(x)q_\psi(x,0)dx + \alpha\int_\Omega \delta'_\varepsilon(\varphi(x))\psi(x)|\nabla\varphi(x)|dx \\ &+ \alpha\int_\Omega \delta_\varepsilon(\varphi(x))\frac{\nabla\varphi(x)\cdot\nabla\psi(x)}{|\nabla\varphi(x)|}dx + \beta\int_\Omega \delta_\varepsilon(\varphi(x))\psi(x)dx \\ &= \int_\Omega y_0(x)q_\psi(x,0)dx + \alpha\int_\Omega \operatorname{div}(\delta_\varepsilon(\varphi(x))\frac{\nabla\varphi(x)}{|\nabla\varphi(x)|}\psi(x))dx \\ &- \alpha\int_\Omega \delta_\varepsilon(\varphi(x))\psi(x)\operatorname{div}\left(\frac{\nabla\varphi(x)}{|\nabla\varphi(x)|}\right)dx + \beta\int_\Omega \delta_\varepsilon(\varphi(x))\psi(x)dx. \end{aligned}$$

Using Gauss-Ostrogradski's formula we get

$$\begin{aligned} dJ(\varphi)(\psi) &= \int_\Omega y_0(x)q_\psi(x,0)dx - \alpha\int_\Omega \delta_\varepsilon(\varphi(x))\operatorname{div}\left(\frac{\nabla\varphi(x)}{|\nabla\varphi(x)|}\right)\psi(x)dx \\ &+ \beta\int_\Omega \delta_\varepsilon(\varphi(x))\psi(x)dx + \alpha\int_{\partial\Omega}\frac{\delta_\varepsilon(\varphi(x))}{|\nabla\varphi(x)|}\partial_\nu\varphi(x)\psi(x)d\sigma. \end{aligned} \tag{8}$$

By multiplying the first equation in (7) by r_φ and integrating over Q_T, using (6) we get that

$$\int_\Omega y_0(x)q_\psi(x,0)dx = -\int_0^T\int_\Omega Lr_\varphi(x,t)\delta_\varepsilon(\varphi(x))(1+p_\varphi(x,t))H_\varepsilon(1+p_\varphi(x,t))\psi(x)dx\, dt. \tag{9}$$

Now, from (8) and (9), we get the conclusion of Theorem 3. □

Let us remark that the gradient descent with respect to φ is

$$\begin{cases} \partial_\theta \varphi(x,\theta) = \delta_\varepsilon(\varphi(x,\theta))[\alpha \operatorname{div}\left(\frac{\nabla\varphi(x,\theta)}{|\nabla\varphi(x,\theta)|}\right) - \beta \\ \qquad + L\int_0^T (1+p_\varphi(x,t))H_\varepsilon(1+p_\varphi(x,t))r_\varphi(x,t)dt], \ x \in \Omega,\ \theta > 0 \\ \frac{\delta_\varepsilon(\varphi(x,\theta))}{|\nabla\varphi(x,\theta)|}\partial_\nu \varphi(x,\theta) = 0, \qquad x \in \partial\Omega,\ \theta > 0. \end{cases} \tag{10}$$

(θ is an artificial time).

Numerical Implementation

From Theorem 3 we derive the following conceptual iterative algorithm, a semi-implicit gradient descent method, to improve at each step the region where the harvesting effort acts in order to obtain a smaller value for J.

STEP 0: set $n := 0$, $J^{(0)} := 10^6$ and $\theta_0 > 0$ a small constant
initialize $\varphi^{(0)} = \varphi^{(0)}(x,0)$

STEP 1: compute $p^{(n+1)}$ the solution of (5) corresponding to $\varphi^{(n)}(\cdot,0)$

compute $$J^{(n+1)} = \int_\Omega y_0(x)p^{(n+1)}(x,0)dx + \alpha\int_\Omega \delta_\varepsilon(\varphi^{(n)}(x,0))|\nabla\varphi^{(n)}(x,0)|dx + \beta\int_\Omega H_\varepsilon(\varphi^{(n)}(x,0))dx\,.$$

Step 2: if $\left|J^{(n+1)} - J^{(n)}\right| < \varepsilon_1$ or $J^{(n+1)} \geq J^{(n)}$ then **STOP**
else go to **Step 3**.

Step 3: compute $r^{(n+1)}$ the solution of problem (6) corresponding to
$\varphi^{(n)}(\cdot,0)$ and $p^{(n+1)}$.

Step 4: compute $\varphi^{(n+1)}$ using (10) and the initial condition
$\varphi^{(n+1)}(x,0) = \varphi^{(n)}(x,\theta_0)$ and a semi-implicit timestep scheme

Step 5: if $\|\varphi^{(n+1)} - \varphi^{(n)}\|_{L^2} < \varepsilon_2$ then **STOP**
else $n := n+1$
go to **Step 1**

$\varepsilon_1 > 0$ in Step 2 and $\varepsilon_2 > 0$ in Step 5 are prescribed convergence parameters.

For the implementation we consider $\Omega = (0,1) \times (0,1)$ such that the sides are parallel to the axes Ox_1 and Ox_2 (the horizontal and vertical axes). We introduce equidistant discretization nodes for both axes corresponding to Ω. Thus, the domain Ω is approximated by a grid of $(N+1) \times (N+1)$ equidistant nodes, namely

$$\{(x_1^i, x_2^j) : x_1^i = (i-1)\Delta x_1, x_2^j = (j-1)\Delta x_2, i, j \in \{1,\ldots,N+1\}, \Delta x_1 = \Delta x_2 = 1/N\}.$$

The interval $[0, T]$ is also discretized by $M + 1$ equidistant nodes, $t^k = (k - 1)\Delta t, k = 1, 2, \dots M, M + 1, \Delta t = T/M$. We take M and N to be even. We denote by $\varphi_{i,j} = \varphi(x_1^i, x_2^j),\ i, j \in \{1, 2, \dots, N + 1\}$.

In order to approximate the solution of the parabolic system from Step 1 we use a finite difference method, an implicit one, descending with respect to time levels.

We denote by $h = \Delta x_1 = \Delta x_2$, $p_{i,j}^k = p(x_1^i, x_2^j, t^k)$, $a_{i,j} = a(x_1^i, x_2^j)$, $G_{i,j}^k = \Delta t L H_\varepsilon(\varphi_{i,j})(1 + p_{i,j}^{k+1}) H_\varepsilon(1 + p_{i,j}^{k+1})$, $k \in \{1, \dots, M + 1\}$. The numerical scheme is

$$\begin{cases} \frac{p_{i,j}^{k+1} - p_{i,j}^k}{\Delta t} + d\frac{p_{i-1,j}^k - 2p_{i,j}^k + p_{i+1,j}^k}{h^2} + d\frac{p_{i,j-1}^k - 2p_{i,j}^k + p_{i,j+1}^k}{h^2} & \\ \qquad + a_{i,j} p_{i,j}^k - G_{i,j}^{k+1} = 0, & i, j \in \{2, \dots, N\}, \\ & k \in \{M, \dots, 1\}, \\ p_{i,1}^k = p_{i,2}^k,\ p_{i,N+1}^k = p_{i,N}^k,\ p_{1,j}^k = p_{2,j}^k,\ p_{N+1,j}^k = p_{N,j}^k, & i, j \in \{1, \dots, N + 1\}, \\ & k \in \{M, \dots, 1\}, \\ p_{i,j}^{M+1} = 0, & i, j \in \{1, \dots, N + 1\}. \end{cases}$$

We take the diffusion coefficient $d = 1$ for the implementation, and denote by $\lambda = \Delta t / h^2$. For the interior nodes we get

$$(1 + 4\lambda - \Delta t\, a_{i,j}) p_{i,j}^k - \lambda p_{i-1,j}^k - \lambda p_{i+1,j}^k - \lambda p_{i,j-1}^k - \lambda p_{i,j+1}^k = p_{i,j}^{k+1} - \Delta t G_{i,j}^{k+1}$$

for $i, j \in \{2, \dots, N\}, k \in \{M, \dots, 1\}$. By using the Neumann conditions on the boundary, the numerical scheme becomes

$$\begin{cases} (1 + 2\lambda - \Delta t\, a_{2,2}) p_{2,2}^k - \lambda p_{2,3}^k - \lambda p_{3,2}^k = p_{2,2}^{k+1} - \Delta t G_{2,2}^{k+1}, \qquad i = 2,\ j = 2 \\ (1 + 3\lambda - \Delta t\, a_{2,j}) p_{2,j}^k - \lambda p_{3,j}^k - \lambda p_{2,j-1}^k - \lambda p_{2,j+1}^k = p_{2,j}^{k+1} - \Delta t G_{2,j}^{k+1}, \\ \qquad\qquad i = 2,\ 2 < j < N \\ (1 + 2\lambda - \Delta t\, a_{2,N}) p_{2,N}^k - \lambda p_{3,N}^k - \lambda p_{2,N-1}^k = p_{2,N}^{k+1} - \Delta t G_{2,N}^{k+1},\ i = 2,\ j = N \\ (1 + 3\lambda - \Delta t\, a_{i,2}) p_{i,2}^k - \lambda p_{i,3}^k - \lambda p_{i-1,2}^k - \lambda p_{i+1,2}^k = p_{i,2}^{k+1} - \Delta t G_{i,2}^{k+1}, \\ \qquad\qquad 2 < i < N,\ j = 2 \\ (1 + 4\lambda - \Delta t\, a_{i,j}) p_{i,j}^k - \lambda p_{i-1,j}^k - \lambda p_{i+1,j}^k - \lambda p_{i,j-1}^k - \lambda p_{i,j+1}^k = p_{i,j}^{k+1} - \Delta t G_{i,j}^{k+1}, \\ \qquad\qquad 2 < i < N,\ 2 < j < N \\ (1 + 3\lambda - \Delta t\, a_{i,N}) p_{i,N}^k - \lambda p_{i,N-1}^k - \lambda p_{i-1,N}^k - \lambda p_{i+1,N}^k = p_{i,N}^{k+1} - \Delta t G_{i,N}^{k+1}, \\ \qquad\qquad 2 < i < N,\ j = N \\ (1 + 2\lambda - \Delta t\, a_{N,2}) p_{N,2}^k - \lambda p_{N,3}^k - \lambda p_{N-1,2}^k = p_{N,2}^{k+1} - \Delta t G_{N,2}^{k+1},\ i = N,\ j = 2 \\ (1 + 3\lambda - \Delta t\, a_{N,j}) p_{N,j}^k - \lambda p_{N-1,j}^k - \lambda p_{N,j-1}^k - \lambda p_{N,j+1}^k = p_{N,j}^{k+1} - \Delta t G_{N,j}^{k+1}, \\ \qquad\qquad i = N,\ 2 < j < N \\ (1 + 2\lambda - \Delta t\, a_{N,N}) p_{N,N}^k - \lambda p_{N-1,N}^k - \lambda p_{N,N-1}^{(k)} = p_{N,N}^{k+1} - \Delta t G_{N,N}^{k+1}, \\ \qquad\qquad i = N,\ j = N. \end{cases} \tag{11}$$

We denote by

$$x^k = (p^k_{2,2}, p^k_{2,3}, \ldots, p^k_{2,N}, p^k_{3,2}, p^k_{3,3}, \ldots, p^k_{3,N}, \ldots, p^k_{N,2}, p^k_{N,3}, \ldots, p^k_{N,N})^T$$

the vector formed by the values of p at time level k for the interior nodes. This is a vector of dimension $(N-1)^2$. We also use the following notations $P = p^{k+1}_{i,j}$, $G = \Delta t G^{k+1}_{i,j}$, $E_1 = \Delta t(-a_{i,j})$, $E_2 = 1+2\lambda+E_1$, $E_3 = 1+3\lambda+E_1$, $E_4 = 1+4\lambda+E_1$. This quantities must be evaluated at each time step $k = M, M-1, \ldots, 1$ and for all $i, j \in \{2, \ldots, N\}$. The algebraic linear system to solve at each time step $k = M, M-1, \ldots, 1$ is of the form $Ax^k = B$, with the system matrix A of dimension $(N-1)^2 \times (N-1)^2$ and the vector of constant terms B of dimension $(N-1)^2$. Based on (11) and using also the final condition, for each time level $k = M, M-1, \ldots, 1$ we generate the matrix A and the vector B with the following algorithm: we denote by q the row index of matrix A; at the beginning of each time iteration we make the initializations: $q = 0$, $A = 0_{(N-1)^2\times(N-1)^2}$, and $B = 0_{(N-1)^2\times 1}$. Then, for i from 2 to N and for j from 2 to N, after the evaluation of E_1, E_2, E_3, E_4, G, P, we start the construction of A and B. The index q is incremented for each i and j. Therefore,

- if i = 2 and j = 2 then q = q + 1; A(q,1) = E_2; A(q,2) = -λ; A(q,N) = -λ; B(q) = P - G;
- if i = 2 and 2 < j < N then q = q + 1; A(q,j-2) = -λ; A(q,j-1) = E_3; A(q,N+j-2) = -λ; A(q,j) = -λ; B(q) = P - G;
- if i = 2 and j = N then q = q + 1; A(q,N-2) = -λ; A(q,N-1) = E_2; A(q,2*N-2) = -λ; B(q) = P - G;
- if 2 < i < N and j = 2 then q = q + 1; A(q,(i-3)*(N-1)+1) = -λ; A(q,(i-2)*(N-1)+1) = E_3; A(q,(i-1)*(N-1)+1) = -λ; A(q,(i-2)*(N-1)+2)= -λ; B(q) = P - G;
- if 2 < i < N and 2 < j< N then q = q + 1; A(q,(i-3)*(N-1)+j-1) = -λ; A(q,(i-1)*(N-1)+j-1) = -λ; A(q,(i-2)*(N-1)+j-1) = E_4; A(q,(i-2)*(N-1)+j-2) = -λ; A(q,(i-2)*(N-1)+j) = -λ; B(q) = P - G;
- if 2 < i < N and j = N then q = q + 1; A(q,(i-2)*(N-1)) = -λ; A(q,(i-2)*(N-1)+N-2) = -λ; A(q,(i-1)*(N-1)) = E_3; A(q,i*(N-1)) = -λ; B(q) =P - G;
- if i = N and j = 2 then q = q + 1; A(q,(N-3)*(N-1)+1) = -λ; A(q,(N-2)*(N-1)+1) = E_2; A(q,(N-2)*(N-1)+2)= -λ; B(q) = P - G;
- if i = N and 2 < j < N then q = q + 1; A(q,(N-3)*(N-1)+j-1) = -λ; A(q,(N-2)*(N-1)+j-2) = -λ; A(q,(N-2)*(N-1)+j-1) = E_3; A(q,(N-2)* (N-1)+j) = -λ; B(q) = P - G;

- `if i = N and j = N then q = q + 1; A(q,N*(N-2)) = -λ; A(q,(N-2)*(N-1)) = -λ; A(q,(N-1)*(N-1)) =` E_2`; B(q) = P - G;`

Then, the resulting algebraic linear system is solved by Gaussian elimination. The solution obtained is a vector D of dimension $(N-1)^2$. Therefore, we get the corresponding solution $p_{i,j}$ at time step k.

By using the boundary condition, the solution is completed for $i = 1, i = N+1$ and $j \in \{1, \dots, N+1\}$ and for $j = 1, j = N+1$ and $i \in \{1, \dots, N+1\}$. Now we have the complete solution $p^k_{i,j}$ and we can proceed with the time step $k-1$.

The integrals from Step 1 are numerical computed using Simpson's method corresponding to the discrete grid. For each iteration $n = 1, 2, 3, \dots,$ we have to evaluate the first integral

$$F^{(n)} = \int_{\Omega} f^{(n)}(x)dx,$$

where

$$f^{(n)}(x) = y_0(x)p^{(n)}(x, 0), x \in \Omega.$$

In order to approximate this integral we first calculate, for all $i \in \{1, \dots, N+1\}$,

$$r(i) = \frac{h}{3}\left[f^{(n)}(x_1^i, x_2^1) + f^{(n)}(x_1^i, x_2^{N+1}) + 4\sum_{j=1}^{[N/2]} f^{(n)}(x_1^i, x_2^{2j}) + 2\sum_{j=1}^{[(N-2)/2]} f^{(n)}(x_1^i, x_2^{2j+1}) \right],$$

and then

$$F^{(n)} \approx \frac{h}{3}\left[r(1) + r(N+1) + 4\sum_{i=1}^{[N/2]} r(2i) + 2\sum_{i=1}^{[(N-2)/2]} r(2i+1) \right].$$

To numerically evaluate of the second integral we must approximate $|\nabla\varphi(x_1^i, x_2^j)|$. In order to do this, we use central difference both in x_1 and in x_2 direction.

$$|\nabla\varphi(x_1^i, x_2^j)| = \sqrt{(\partial_{x_1}\varphi(x_1^i, x_2^j))^2 + (\partial_{x_2}\varphi(x_1^i, x_2^j))^2}$$

$$= \sqrt{\frac{(\varphi_{i+1,j} - \varphi_{i-1,j})^2 + (\varphi_{i,j+1} - \varphi_{i,j-1})^2}{4h^2}}, \; i, j \in \{2, \dots, N\}$$

$$|\nabla\varphi(x_1^1, x_2^j)| = \sqrt{\frac{(\varphi_{2,j} - \varphi_{1,j})^2 + (\varphi_{1,j+1} - \varphi_{1,j})^2}{h^2}}, \; j \in \{2, \dots, N\}$$

$$|\nabla\varphi(x_1^{N+1}, x_2^j)| = \sqrt{\frac{(\varphi_{N+1,j} - \varphi_{N,j})^2 + (\varphi_{N+1,j+1} - \varphi_{N+1,j})^2}{h^2}}, \; j \in \{2, \ldots, N\}$$

$$|\nabla\varphi(x_1^i, x_2^1)| = \sqrt{\frac{(\varphi_{i+1,1} - \varphi_{i,1})^2 + (\varphi_{i,2} - \varphi_{i,1})^2}{h^2}}, \; i \in \{2, \ldots, N\}$$

$$|\nabla\varphi(x_1^i, x_2^{N+1})| = \sqrt{\frac{(\varphi_{i+1,N+1} - \varphi_{i,N+1})^2 + (\varphi_{i,N+1} - \varphi_{i,N})^2}{h^2}}, \; i \in \{2, \ldots, N\}$$

$$|\nabla\varphi(x_1^1, x_2^1)| = |\nabla\varphi(x_1^2, x_2^1)|, |\nabla\varphi(x_1^1, x_2^{N+1})| = |\nabla\varphi(x_1^1, x_2^N)|,$$

$$|\nabla\varphi(x_1^{N+1}, x_2^1)| = |\nabla\varphi(x_1^{N+1}, x_2^2)|, |\nabla\varphi(x_1^{N+1}, x_2^{N+1})| = |\nabla\varphi(x_1^{N+1}, x_2^N)|.$$

The parabolic system from Step 3 is approximated also using a finite difference method, but now ascending with respect to time levels. For each iteration $n = 1, 2, \ldots$ and for each time level $k = 1, 2, 3, \ldots, M$, the matrix of the resulting algebraic system is the same as matrix A previously determinated, with $B(q) = p_{i,j}^k, q \in \{1, \ldots, (N-1)^2\}$, and $E_1 = \Delta t(-a_{i,j} + G_{i,j})$, $G_{i,j} = LH_\varepsilon(\varphi_{i,j})H_\varepsilon(1 + p_{i,j}^{k+1}) + LH_\varepsilon(\varphi_{i,j})(1 + p_{i,j}^{k+1})\delta_\varepsilon(1 + p_{i,j}^{k+1})$, which are evaluated for each $i, j \in \{2, \ldots N\}$. The resulting algebraic linear system is solved by Gaussian elimination. By using the boundary conditions we complete the solution of the parabolic system for each time level.

Numerical Examples

We consider a normal initial population density $y_0(x_1, x_2) = \frac{1}{2\pi}e^{-\frac{x_1^2+x_2^2}{2}}$, where $(x_1, x_2) \in \Omega$. Let the diffusion coefficient be $d = 1$, the final time $T = 1$, $L = 1$, and the regularization parameter $\varepsilon = 1$. We take the space discretization step and the time discretization step to be equal $\Delta x_1 = \Delta x_2 = \Delta t = 0.05$. For the convergence tests we consider $\varepsilon_1 = \varepsilon_2 = 0.001$.
In the following figure, the white area represents the subregion ω that provides a small value for J.

Test 1 We take the natural growth rate of the population to be a constant, e.g. $a(x_1, x_2) = 3$, $(x_1, x_2) \in \Omega$. The initialization of φ is made by $\varphi^{(0)}(x_1, x_2) = 0.25 - \sqrt{(x_1 - 0.5)^2 + (x_2 - 0.5)^2}$, $(x_1, x_2) \in \Omega$. We penalize the length of $\partial\omega$ by $\alpha = 0.4$ and the area of ω by $\beta = 0.6$. The corresponding results are shown in Fig. 1.

Test 2 We use the same input data from Test 1 and the initialization of φ with $\varphi^{(0)}(x_1, x_2) = sin(3\pi x_1)sin(3\pi x_2)$, $(x_1, x_2) \in \Omega$, a function that produce a initial checkerboard shape. We penalize the length of $\partial\omega$ by $\alpha = 0.5$ and the area of ω by $\beta = 0.5$. The results are shown in Fig. 2.

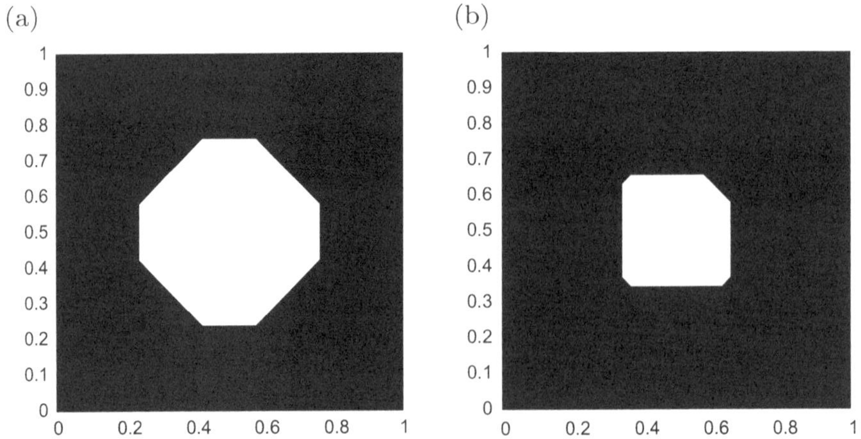

Fig. 1 The representation of initial and final iterations of ω for $\alpha = 0.4$ and $\beta = 0.6$. (**a**) Initial ω. (**b**) Final ω

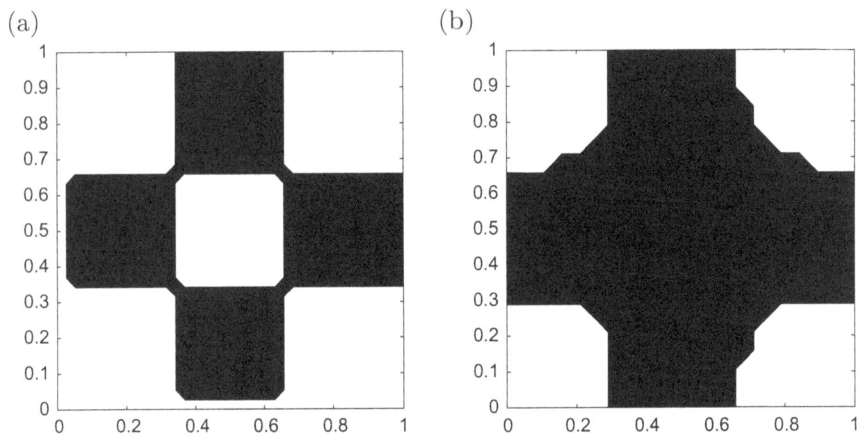

Fig. 2 The representation of initial and final iterations of ω for $\alpha = 0.5$ and $\beta = 0.5$. (**a**) Initial ω. (**b**) Final ω

In both examples, the algorithm ends when the first condition in Step 2 is fulfilled. The total number of iterations in Test 1 was 26, and in Test 2 was 233.

In Fig. 3a and b it can be seen that the algorithm provides a smaller value for J at each iteration.

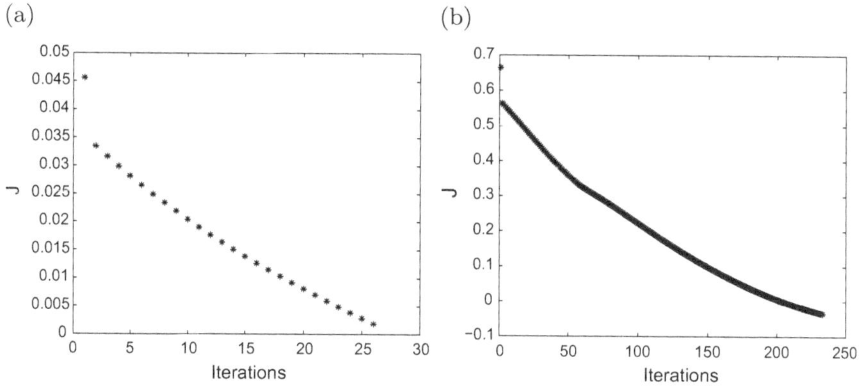

Fig. 3 The representation of J as function of iterations. (**a**) Test 1. (**b**) Test 2

3 Eradicating an Age-Structured Pest Population with Diffusion

Consider here an age-structured population dynamics with diffusion and logistic term:

$$\begin{cases} \partial_t y(x,a,t) + \partial_a y(x,a,t) + \mu(a)y(x,a,t) - d\Delta y(x,a,t) = \\ \quad -\mathcal{M}\left(\int_0^A y(x,a,t)da\right) y(x,a,t) \\ \quad -\chi_\omega(x)u(x,t)y(x,a,t), \qquad x \in \Omega,\ a \in (0,A),\ t \in (0,+\infty) \\ \partial_\nu y(x,a,t) = 0, \qquad x \in \partial\Omega,\ a \in (0,A),\ t \in (0,+\infty) \\ y(x,0,t) = \int_0^A \beta(a)y(x,a,t)da, \qquad x \in \Omega,\ t \in (0,+\infty) \\ y(x,a,0) = y_0(x,a), \qquad x \in \Omega,\ a \in (0,A). \end{cases} \tag{12}$$

Here $A \in (0,+\infty)$ is the maximal age for the population species and $y(x,a,t)$ is the population density at position x, age a and time t; $d \in (0,+\infty)$ is the diffusion coefficient, $\mu(a)$ is the mortality rate and $\beta(a)$ is the fertility rate for individuals of age a; $y_0(x,a)$ is the initial density of population at position x and age a. $u(x,t)$ is a harvesting effort (the control) and is localized in the subregion ω; u does not depend on age.

Assume that Ω, ω satisfy the same assumptions as in the introduction, and that the following hypotheses are satisfied as well

(H1') $\beta \in C([0,A]), \beta(a) \geq 0, \forall a \in [0,A]$;
(H2') $\mu \in C([0,A)), \mu(a) \geq 0, \forall a \in [0,A], \int_0^A \mu(a)da = +\infty$;
(H3') $y_0 \in L^\infty(\Omega \times (0,A)), y_0(x,a) \geq 0$ a.e. in $\Omega \times (0,A)$;
(H4') $\mathcal{M} : [0,+\infty) \longrightarrow [0,+\infty)$ is continuously differentiable, $\mathcal{M}'(r) > 0, \forall r > 0, \mathcal{M}(0) = 0, \lim_{r\to+\infty} \mathcal{M}(r) = +\infty$.

For any $u \in L^{\infty}_{loc}(\overline{\omega}\times[0,+\infty))$, such that $L \geq u(x,t) \geq 0$ a.e., there exists a unique solution y^u to (12) (here $L \in (0,+\infty)$ is a constant; L is the maximal affordable effort). This solution is nonnegative (see Aniţa et al. 2016).

Our goal is to eradicate this population which is considered to be a pest population.

Definition 1 We say that the population is eradicable (zero-stabilizable) if for any y_0 satisfying the hypothesis (H3') there exists $u \in L^{\infty}_{loc}(\overline{\omega}\times[0,+\infty))$, satisfying $L \geq u(x,t) \geq 0$ a.e., such that

$$\lim_{t\to+\infty} y^u(\cdot,\cdot,t) = 0 \text{ in } L^{\infty}(\Omega\times(0,A)).$$

(and $y^u(x,a,t) \geq 0$ a.e. in $\Omega\times(0,A)\times(0,+\infty)$).

Note that this is a problem of zero-stabilization with control and state constraints. Denote by $r^* \in \mathbb{R}$ the solution to the equation

$$\int_0^A \beta(a)e^{-\int_0^a \mu(\tau)d\tau - ra}da = 1$$

(for the existence and uniqueness of r^* we refer to Aniţa 2000) and let λ_1^{ω} be the principal eigenvalue for

$$\begin{cases} -d\Delta\phi = -\chi_{\omega}L\phi + \lambda\phi, & x\in\Omega \\ \partial_{\nu}\phi = 0, & x\in\partial\Omega. \end{cases}$$

Theorem 4

(i) If the population is eradicable then

$$\lambda_1^{\omega} \geq r^*.$$

(ii) If $\lambda_1^{\omega} > r^*$ then the population is eradicable and the harvesting effort $u \equiv L$ diminishes exponentially the population.

Proof (i) Assume that the population is eradicable and let

$$y_0(x,a) = h_0(a)g_0(x),$$

with $h_0 \in C([0,A]), h_0(a) > 0, \forall a \in [0,A)$, to be specified later, $g_0 \in L^{\infty}(\Omega)$, $g_0(x) > 0$ a.e. in Ω.

Let $u \in L^{\infty}_{loc}(\overline{\omega}\times[0,+\infty))$, $L \geq u(x,t) \geq 0$ a.e., such that

$$\lim_{t\to+\infty} y^u(\cdot,\cdot,t) = 0 \text{ in } L^{\infty}(\Omega\times(0,A)).$$

The unique solution y^u to (12) may be written as

$$y^u(x,a,t) = h(a,t)g(x,t),$$

where h is the solution to

$$\begin{cases} \partial_t h(a,t) + \partial_a h(a,t) + \mu(a)h(a,t) = -r^* h(a,t), & a \in (0,A),\ t \in (0,+\infty) \\ h(0,t) = \int_0^A \beta(a)h(a,t)da, & t \in (0,+\infty) \\ h(a,0) = h_0(a), & a \in (0,A), \end{cases} \tag{13}$$

and g is the solution to

$$\begin{cases} \partial_t g(x,t) - d\Delta g(x,t) = r^* g(x,t) \\ \quad -\mathcal{M}\left(\int_0^A h(a,t)g(x,t)da\right) g(x,t) \\ \quad -\chi_\omega(x)u(x,t)g(x,t), & x \in \Omega,\ t \in (0,+\infty) \\ \partial_\nu g(x,t) = 0, & x \in \partial\Omega,\ t \in (0,+\infty) \\ g(x,0) = g_0(x), & x \in \Omega. \end{cases} \tag{14}$$

It is known that the set of all time-independent solutions for the first two equations in (13) is a real vector space of dimension 1 and that there exists a time independent solution $\tilde{h}$ satisfying $\tilde{h}(a) > 0$, for all $a \in [0, A)$ (Aniţa 2000).

If we consider $h_0 = \tilde{h}$, then

$$y^u(x,a,t) = \tilde{h}(a)g(x,t),\ x \in \Omega,\ a \in (0,A),\ t \in (0,+\infty),$$

where g is the solution to

$$\begin{cases} \partial_t g(x,t) - d\Delta g(x,t) = r^* g(x,t) \\ \quad -\mathcal{M}\left(Hg(x,t)\right) g(x,t) \\ \quad -\chi_\omega(x)u(x,t)g(x,t), & x \in \Omega,\ t \in (0,+\infty) \\ \partial_\nu g(x,t) = 0, & x \in \partial\Omega,\ t \in (0,+\infty) \\ g(x,0) = g_0(x), & x \in \Omega. \end{cases} \tag{15}$$

Here $H = \int_0^A \tilde{h}(a)da$.

The eradicability for (12) implies the nonnegative zero-stabilizability for (15). However, the nonnegative zero-stabilizability for (15) implies that

$$\lambda_1^\omega \geq r^*.$$

This follows as in Aniţa et al. (2013) by using of the comparison results for the solutions to parabolic equations.

(ii) If $\lambda_1^\omega > r^*$, then we consider $u(x,t) = L$ a.e. in $\omega \times (0, +\infty)$. Using the comparison result for linear age-structured population dynamics (see Aniţa 2000) we get that

$$y(x,a,t) \leq \tilde{y}(x,a,t) \text{ a.e.,} \tag{16}$$

where $\tilde{y}$ is the solution to

$$\begin{cases} \partial_t y(x,a,t) + \partial_a y(x,a,t) + \mu(a)y(x,a,t) & \\ \quad -d\Delta y(x,a,t) = -\chi_\omega(x)Ly(x,a,t), & x \in \Omega,\ a \in (0,A),\ t \in (0,+\infty) \\ \partial_\nu y(x,a,t) = 0, & x \in \partial\Omega,\ a \in (0,A),\ t \in (0,+\infty) \\ y(x,0,t) = \int_0^A \beta(a)y(x,a,t)da, & x \in \Omega,\ t \in (0,+\infty) \\ y(x,a,0) = y_0(x,a), & x \in \Omega,\ a \in (0,A). \end{cases}$$

Let $h_0(a) = 1, \forall a \in [0,A]$, $g_0(x) = ||y_0||_\infty$ a.e. $x \in \Omega$.

Using again the comparison result for linear age-structured population dynamics we get that

$$\tilde{y}(x,a,t) \leq h(a,t)g(x,t) \text{ a.e.,} \tag{17}$$

where h is the solution to (13) and g is the solution to (14) corresponding to $\mathcal{M} \equiv 0$ and $u \equiv L$. Since $h(\cdot,t) \to \bar{h}$ in $L^\infty(0,A)$ as $t \to +\infty$ (Aniţa 2000), and $g(\cdot,t) \to 0$ in $L^\infty(\Omega)$ as $t \to +\infty$ (because $\lambda_1^\omega > r^*$), we get by (16) and (17) that

$$\lim_{t\to+\infty} y^L(\cdot,\cdot,t) = 0 \text{ in } L^\infty(\Omega \times (0,A)),$$

and the conclusion. □

Since our goal was actually to eradicate a pest population corresponding to a initial density y_0 with a harvesting effort less or equal than L (tacking into account the above theorem) and since we have however to pay a certain cost to harvest in a subdomain ω, we can consider the following related optimal control problem

$$\underset{\omega}{Minimize} \int_0^A \int_\Omega y(x,a,T)dx\, da + \alpha\, length(\partial\omega) + \beta\, area(\omega),$$

where $T > 0$ is a certain moment and y is the solution to (12) corresponding to $u \equiv L$. The cost to be paid for the control is included in $\beta\, area(\omega)$.

This problem may be investigated by using the level set method described in Sect. 2 and rewriting it in the following form

$$\underset{\varphi}{Minimize} \int_0^A \int_\Omega y_\varphi(x,a,T)dx\, da + \alpha \int_\Omega \delta(\varphi(x))|\nabla\varphi(x)|dx + \beta \int_\Omega H(\varphi(x))dx,$$

where y_φ is the solution to

$$\begin{cases} \partial_t y(x,a,t) + \partial_a y(x,a,t) + \mu(a)y(x,a,t) - d\Delta y(x,a,t) = \\ \quad -\mathcal{M}\left(\int_0^A y(x,a,t)da\right) y(x,a,t) \\ \quad -H(\varphi(x))Ly(x,a,t), & x \in \Omega,\ a \in (0,A),\ t \in (0,+\infty) \\ \partial_\nu y(x,a,t) = 0, & x \in \partial\Omega,\ a \in (0,A),\ t \in (0,+\infty) \\ y(x,0,t) = \int_0^A \beta(a)y(x,a,t)da, & x \in \Omega,\ t \in (0,+\infty) \\ y(x,a,0) = y_0(x,a), & x \in \Omega,\ a \in (0,A), \end{cases}$$

with φ the implicit function of ω. The approach is similar to the one in Sect. 2. We will approximate this problem using the mollified version of the Heaviside function, H_ε, and its derivative by the mollified function δ_ε.

Actually, if we denote by

$$\Psi(\varphi) = \int_0^A \int_\Omega y_\varphi(x,a,T)dx\, da + \alpha \int_\Omega \delta_\varepsilon(\varphi(x))|\nabla\varphi(x)|dx + \beta \int_\Omega H_\varepsilon(\varphi(x))dx,$$

for a small but fixed $\varepsilon > 0$, the harvesting problem to be investigated is

$$\underset{\varphi}{Minimize}\ \Psi(\varphi),$$

where $\varphi : \overline{\Omega} \longrightarrow \mathbb{R}$ is a smooth function and y_φ is the solution to

$$\begin{cases} \partial_t y(x,a,t) + \partial_a y(x,a,t) + \mu(a)y(x,a,t) - d\Delta y(x,a,t) = \\ \quad -\mathcal{M}\left(\int_0^A y(x,a,t)da\right) y(x,a,t) \\ \quad -H_\varepsilon(\varphi(x))Ly(x,a,t), & x \in \Omega,\ a \in (0,A),\ t \in (0,+\infty) \\ \partial_\nu y(x,a,t) = 0, & x \in \partial\Omega,\ a \in (0,A),\ t \in (0,+\infty) \\ y(x,0,t) = \int_0^A \beta(a)y(x,a,t)da, & x \in \Omega,\ t \in (0,+\infty) \\ y(x,a,0) = y_0(x,a), & x \in \Omega,\ a \in (0,A). \end{cases}$$

By following the same lines as in Sect. 2 we can get the directional derivative of Ψ. We reach a similar conclusion as in Sect. 2 concerning the gradient descent with respect to φ:

$$\begin{cases} \partial_\theta \varphi(x,\theta) = \delta_\varepsilon(\varphi(x,\theta))[-\alpha \operatorname{div}\left(\frac{\nabla\varphi(x,\theta)}{|\nabla\varphi(x,\theta)|}\right) + \beta \\ \qquad -L\int_0^A \int_0^T r(a,x,t)y_\varphi(a,x,t)da\, dt], & x \in \Omega,\ \theta > 0 \\ \frac{\delta_\varepsilon(\varphi(x,\theta))}{|\nabla\varphi(x,\theta)|} \partial_\nu \varphi(x,\theta) = 0, & x \in \partial\Omega,\ \theta > 0. \end{cases}$$

where θ is an artificial time and r is the solution to

$$\begin{cases} \partial_t r(x,a,t) + \partial_a r(x,a,t) - \mu(a) r(x,a,t) + d\Delta r(x,a,t) = \\ \qquad \mathcal{M}'\left(\int_0^A y(x,a,t)da\right) \int_0^A r(x,a,t) y(x,a,t)da \\ \qquad + \mathcal{M}\left(\int_0^A y(x,a,t)da\right) r(x,a,t) + LH_\varepsilon(\varphi(x)) r(x,a,t) \\ \qquad -\beta(a) r(x,0,t), \qquad\qquad x \in \Omega,\ a \in (0,A),\ t \in (0,+\infty) \\ \partial_\nu r(x,a,t) = 0, \qquad\qquad x \in \partial\Omega,\ a \in (0,A),\ t \in (0,+\infty) \\ r(x,A,t) = 0, \qquad\qquad x \in \Omega,\ t \in (0,+\infty) \\ r(x,a,T) = 1, \qquad\qquad x \in \Omega,\ a \in (0,A). \end{cases}$$

Final Comments

In a typical optimal harvesting problem, as the one considered in Sect. 1, it is obvious the fact that harvesting of a species as a commercial resource has to be optimized by acting on a subdomain ω of the entire "world" Ω. On the contrary, the relevance of the "harvesting" problem considered in this section is due to the fact that, while the concern for pest control would be the eradication of a pest in the entire domain of interest Ω, on the other hand, very often, such domain is practically unknown, or difficult to reach for the implementation of suitable environmental programmes. This is the main reason that has led the authors to suggest that implementation of such programmes might be concretely carried out only in a chosen subregion $\omega \subset \Omega$, where an effective optimal eradication is both practically and financially affordable. From a mathematical point of view this problem has been investigated by using the level set method.

We have been inspired by a series of papers by the same authors regarding the control of epidemics due to interaction of the human population with a polluted environment (see e.g. Aniţa and Capasso (2009) and related references).

Our hope is that additional investigations may lead us to offer solutions for a larger class of pest control as in vector borne diseases, which are usually described in terms of reaction-diffusion systems.

Acknowledgements This work was supported by the CNCS-UEFISCDI (Romanian National Authority for Scientific Research) grant 68/2.09.2013, PN-II-ID-PCE-2012-4-0270: "Optimal Control and Stabilization of Nonlinear Parabolic Systems with State Constraints. Applications in Life Sciences and Economics". The authors are indebted to the referee for the valuable comments and suggestions to improve the paper.

References

S. Aniţa, *Analysis and Control of Age-Dependent Population Dynamics* (Kluwer Academic Publishers, Dordrecht, 2000)

S. Aniţa, V. Capasso, A stabilization strategy for a reaction-diffusion system modelling a class of spatially structured epidemic systems (think globally, act locally). Nonlinear Anal. Real World Appl. **10**, 2026–2035 (2009)

S. Aniţa, V. Arnăutu, V. Capasso, *An Introduction to Optimal Control Problems in Life Sciences and Economics: From Mathematical Models to Numerical Simulation with MATLAB* (Birkhäuser, Basel, 2011)

L.-I. Aniţa, S. Aniţa, V. Arnăutu, Internal null stabilization for some diffusive models of population dynamics. Appl. Math. Comput. **219**, 10231–10244 (2013)

S. Aniţa, V. Capasso, A.-M. Moşneagu, Regional control in optimal harvesting of population dynamics. Nonlinear Anal. **147**, 191–212 (2016)

V. Arnăutu, A.-M. Moşneagu, Optimal control and stabilization for some Fisher-like models. Numer. Funct. Anal. Optim. **36**(5), 567–589 (2015)

V. Barbu, *Mathematical Methods in Optimization of Differential Systems* (Kluwer Academic Publishers, Dordrecht, 1994)

A.O. Belyakov, V.M. Veliov, On optimal harvesting in age-structured populations, Research Report 2015–08, ORCOS, TU Wien, 2015

A. Bressan, G.M. Coclite, W. Shen, A multidimensional optimal-harvesting problem with measure-valued solutions. SIAM J. Control Optim. **51**, 1186–1202 (2013)

D. Bucur, G. Buttazzo, *Variational Methods in Some Shape Optimization Problems, Notes of Courses Given by the Teachers at the School* (Scuola Normale Superiore, Pisa, 2002)

T.F. Chan, L.A. Vese, Active contours without edges. IEEE Trans. Image Process. **10**, 266–277 (2001)

M.C. Delfour, J.-P. Zolesio, *Shapes and Geometries*. Metrics, Analysis, Differential Calculus and Optimization, 2nd edn. (SIAM, Philadelphia, 2011)

K.R. Fister, S. Lenhart, Optimal harvesting in an age-structured predator-prey model. Appl. Math. Optim. **54**, 1–15 (2006)

P. Getreuer, T.F. Chan, L.A. Vese, Segmentation. IPOL J. Image Process. Online **2**, 214–224 (2012)

M.E. Gurtin, L.F. Murphy, On the optimal harvesting of age-structured populations: some simple models. Math. Biosci. **55**, 115–136 (1981)

M.E. Gurtin, L.F. Murphy, On the optimal harvesting of persistent age-structured populations. J. Math. Biol. **13**, 131–148 (1981)

Z.R. He, Optimal harvesting of two competing species with age dependence. Nonlinear Anal. Real World Appl. **7**, 769–788 (2006)

A. Henrot, M. Pierre, *Variation et Optimisation de Formes*. Mathématiques et Applications (Springer, Berlin, 2005)

N. Hritonenko, Y. Yatsenko, Optimization of harvesting age in integral age-dependent model of population dynamics. Math. Biosci. **195**, 154–167 (2005)

S. Lenhart, Using optimal control of parabolic PDEs to investigate population questions, NIMBioS, 9–11 Apr 2014. https://www.fields.utoronto.ca/programs/scientific/13-14/BIOMAT/presentations/lenhartToronto3.pdf

S. Lenhart, J.T. Workman, Optimal Control Applied to Biological Models (Chapman and Hall, Boca Raton, Fl, 2007)

Z. Luo, Optimal harvesting problem for an age-dependent n-dimensional food chain diffusion model. Appl. Math. Comput. **186**, 1742–1752 (2007)

Z. Luo, W.T. Li, M. Wang, Optimal harvesting control problem for linear periodic age-dependent population dynamics. Appl. Math. Comput. **151**, 789–800 (2004)

L.F. Murphy, S.J. Smith, Optimal harvesting of an age-structured population. J. Math. Biol. **29**, 77–90 (1990)

S. Osher, R. Fedkiw, *Level Set Methods and Dynamic Implicit Surfaces* (Springer, New York, 2003)

J.A. Sethian, *Level Set Methods and Fast Marching Methods: Evolving Interfaces in Computational Geometry, Fluid Mechanics, Computer Vision, and Materials Science* (Cambridge University Press, Cambridge, 1999)

J. Sokolowski, J.-P. Zolesio, *Introduction to Shape Optimization* (Springer, Berlin, 1992)

C. Zhao, M. Wang, P. Zhao, Optimal harvesting problems for age-dependent interacting species with diffusion. Appl. Math. Comput. **163**, 117–129 (2005)

C. Zhao, P. Zhao, M. Wang, Optimal harvesting for nonlinear age-dependent population dynamics. Math. Comput. Model. **43**, 310–319 (2006)

Kolibri 360 GmbH
Ferdinand-Jühlke-Straße 7
99095 Erfurt, Deutschland
produktsicherheit@kolibri360.de